Other Reference Books of Interest by McGraw-Hill

Handbooks

BAKER • *C Mathematical Function Handbook*
BENSON • *Audio Engineering Handbook*
BENSON • *Television Engineering Handbook*
CHEN • *Computer Engineering Handbook*
COOMBS • *Printed Circuits Handbook*
DI GIACOMO • *Digital Bus Handbook*
DI GIACOMO • *VLSI Handbook*
FINK AND CHRISTIANSEN • *Electronics Engineers' Handbook*
HARPER • *Electronic Packaging and Interconnection Handbook*
HICKS • *Standard Handbook of Engineering Calculations*
INGLIS • *Electronic Communications Handbook*
JURAN AND GRYNA • *Juran's Quality Control Handbook*
KAUFMAN AND SEIDMAN • *Handbook of Electronics Calculations*
RORABAUGH • *Digital Filter Designer's Handbook*
TUMA • *Engineering Mathematics Handbook*
WAYNANT • *Electro-Optics Handbook*
WILLIAMS AND TAYLOR • *Electronic Filter Design Handbook*

Other

ANTOGNETTI • *Power Integrated Circuits*
BEST • *Phase-Locked Loops*
BUCHANAN • *CMOS/TTL Digital Systems Design*
BUCHANAN • *BiCMOS/CMOS Systems Design*
BYERS • *Printed Circuit Board Design with Microcomputers*
ELLIOTT • *Integrated Circuits Fabrication Technology*
HECHT • *The Laser Guidebook*
MUN • *GaAs Integrated Circuits*
SILICONIX • *Designing with Field-Effect Transistors*
SZE • *VLSI Technology*
TSUI • *LSI/VLSI Testability Design*
WATERS • *Active Filter Design*
WOBSCHALL • *Circuit Design for Electronic Instrumentation*
WYATT • *Electro-Optical System Design*

To order or receive additional information on these or any other McGraw-Hill titles, please call 1-800-822-8158 in the United States. In other countries, contact your local McGraw-Hill representative.

MH93

Semiconductor Device Modeling With SPICE

Giuseppe Massobrio
Department of Electronics (DIBE)
University of Genova
Genova, Italy

Paolo Antognetti
Department of Electronics (DIBE)
University of Genova
Genova, Italy

Second Edition

McGraw-Hill, Inc.
New York San Francisco Washington, D.C. Auckland Bogotá
Caracas Lisbon London Madrid Mexico City Milan
Montreal New Delhi San Juan Singapore
Sydney Tokyo Toronto

Library of Congress Cataloging-in-Publication Data

Semiconductor device modeling with SPICE / Giuseppe Massobrio, Paolo Antognetti — 2nd. ed.
      p.   cm.
      Rev. ed. of: Semiconductor device modeling with SPICE / Paolo Antognetti, editor, Giuseppe Massobrio, coeditor. c1988.
      Includes bibliographical references (p.   ) and index.
      ISBN 0-07-002469-3
      1. SPICE (Computer file)   2. Semiconductors—Design and construction—Data processing.   3. Semiconductors—Computer simulation.   4. Computer-aided design.   I. Massobrio, Giuseppe. II. Antognetti, Paolo.
TK7871.85.S4454   1993
621.3815′2′028553—dc20   93-9726
                                                                       CIP

Copyright © 1993 by McGraw-Hill, Inc. All rights reserved. Printed in the United States of America. Except as permitted under the United States Copyright Act of 1976, no part of this publication may be reproduced or distributed in any form or by any means, or stored in a database or retrieval system, without the prior written permission of the publisher.

3 4 5 6 7 8 9 0   DOC/DOC   9 8 7 6

The sponsoring editor for this book was Stephen S. Chapman, the editing supervisor was Joseph Bertuna, and the production supervisor was Pamela A. Pelton. It was set in Century Schoolbook by Datapage International Ltd.

Printed and bound by R. R. Donnelley & Sons Company.

Information contained in this work has been obtained by McGraw-Hill, Inc., from sources believed to be reliable. However, neither McGraw-Hill nor its authors guarantees the accuracy or completeness of any information published herein, and neither McGraw-Hill nor its authors shall be responsible for any errors, omissions, or damages arising out of use of the information. This work is published with the understanding that McGraw-Hill and its authors are supplying information, but are not attempting to render engineering or other professional services. If such services are required, the assistance of an appropriate professional should be sought.

Contents

List of Physical Parameters — ix
Foreword *by Robert W. Dutton, Stanford University* — xi
Preface — xiii

Chapter 1. *PN*-Junction Diode and Schottky Diode — 1

1.1 DC Current-Voltage Characteristics — 1
1.2 Static Model — 11
1.3 Large-Signal Model — 19
1.4 Small-Signal Model — 23
1.5 Schottky Diode and Its Implementation in SPICE2 — 26
1.6 Temperature and Area Effects on the Diode Model Parameters — 28
1.7 SPICE3 Models — 31
1.8 HSPICE Models — 31
1.9 PSPICE Models — 40
References — 42

Chapter 2. Bipolar Junction Transistor (BJT) — 45

2.1 Transistor Conventions and Symbols — 46
2.2 Ebers-Moll Static Model — 48
2.3 Ebers-Moll Large-Signal Model — 62
2.4 Ebers-Moll Small-Signal Model — 70
2.5 Gummel-Poon Static Model — 74
2.6 Gummel-Poon Large-Signal Model — 96
2.7 Gummel-Poon Small-Signal Model — 101
2.8 Temperature and Area Effects on the BJT Model Parameters — 105
2.9 Power BJT Model — 109
2.10 SPICE3 Models — 118
2.11 HSPICE Models — 118
2.12 PSPICE Models — 128
References — 129

Chapter 3. Junction Field-Effect Transistor (JFET) — 131

- 3.1 Static Model — 132
- 3.2 Large-Signal Model and Its Implementation in SPICE2 — 146
- 3.3 Small-Signal Model and Its Implementation in SPICE2 — 148
- 3.4 Temperature and Area Effects on the JFET Model Parameters — 150
- 3.5 SPICE3 Models — 152
- 3.6 HSPICE Models — 152
- 3.7 PSPICE Models — 156
- References — 158

Chapter 4. Metal-Oxide-Semiconductor Transistor (MOST) — 161

- 4.1 Structure and Operating Regions of the MOST — 161
- 4.2 LEVEL1 Static Model — 168
- 4.3 LEVEL2 Static Model — 180
- 4.4 LEVEL1 and LEVEL2 Large-Signal Model — 195
- 4.5 LEVEL3 Static Model — 202
- 4.6 LEVEL3 Large-Signal Model — 210
- 4.7 Comments on the Three Models — 211
- 4.8 The Effect of Series Resistances — 212
- 4.9 Small-Signal Models — 213
- 4.10 The Effect of Temperature on the MOST Model Parameters — 215
- 4.11 BSIM1 Model — 216
- 4.12 BSIM2 Model — 234
- 4.13 SPICE3 Models — 240
- 4.14 HSPICE Models — 242
- 4.15 PSPICE Models — 246
- References — 247

Chapter 5. BJT Parameter Measurements — 251

- 5.1 Input and Model Parameters — 251
- 5.2 Parameter Measurements — 253
- References — 265

Chapter 6. MOST Parameter Measurements — 267

- 6.1 LEVEL1 Model Parameters — 267
- 6.2 LEVEL2 Model (Long-Channel) Parameters — 273
- 6.3 LEVEL2 Model (Short-Channel) Parameters — 279
- 6.4 LEVEL3 Model Parameters — 289
- 6.5 Measurements of Capacitance — 295
- 6.6 BSIM Model Parameter Extraction — 298
- References — 298

Chapter 7. Noise and Distortion — 299

- 7.1 Noise — 299
- 7.2 Distortion — 309
- References — 324

Chapter 8. The SPICE Program — 325

- 8.1 SPICE2 Capabilities — 325
- 8.2 SPICE2 Structure — 333
- 8.3 Convergence Problems — 354
- 8.4 SPICE2 Operation — 357
- 8.5 Linked-List Specification — 361
- 8.6 SPICE3 Capabilities Versus SPICE2 — 371
- 8.7 HSPICE Capabilities Versus SPICE2 — 371
- 8.8 PSPICE Capabilities Versus SPICE2 — 372
- References — 373

Chapter 9. Metal-Semiconductor Field-Effect Transistor, Ion-Sensitive Field-Effect Transistor, and Semiconductor-Controlled Rectifier — 375

- 9.1 MESFET — 376
- 9.2 ISFET — 390
- 9.3 THYRISTOR — 400
- References — 408

Appendix A. *PN* Junction — 411

- A.1 Elements of Semiconductor Physics — 411
- A.2 Physical Operation of the *PN* Junction — 421
- References — 444

Appendix B. MOS Junction — 445

- B.1 MOS Junction — 445
- References — 469

Appendix C. MS Junction — 471

- C.1 MS Junction — 471
- References — 473

Index 475

List of Physical Parameters

Parameter	Keyname	Value	Units
Boltzmann's constant	k	1.38×10^{-23}	J/K
		8.62×10^{-5}	eV/K
Electronic charge	q	1.60×10^{-19}	C
Thermal voltage at 300 K	kT/q	25.9×10^{-3}	V
Avogadro's number	N	6.02×10^{23}	atoms/mole
Energy gap at 300 K	E_g	Si = 1.12	eV
		Ge = 0.67	eV
		GaAs = 1.43	eV
		$SiO_2 = \sim 9$	eV
		$Si_3N_4 = 4.7$	eV
Electron mobility at 300 K	μ_n	Si = 1500	cm^2/(V·s)
		Ge = 3900	cm^2/(V·s)
		GaAs = 8500	cm^2/(V·s)
		$SiO_2 = 20$	cm^2/(V·s)
Hole mobility at 300 K	μ_p	Si = 600	cm^2/(V·s)
		Ge = 1900	cm^2/(V·s)
		GaAs = 400	cm^2/(V·s)
		$SiO_2 = \sim 10^{-8}$	cm^2/(V·s)
Electron diffusion constant	$D_n = kT\mu_n/q$	Computed from above parameters	cm^2/s
Hole diffusion constant	$D_p = kT\mu_p/q$	Computed from above parameters	cm^2/s
Intrinsic carrier concentration	n_i	Si = 1.45×10^{10}	cm^{-3}
		Ge = 2.4×10^{13}	cm^{-3}
		GaAs = 9×10^6	cm^{-3}
Permittivity of free space	ε_0	8.85×10^{-14}	F/cm
Relative permittivity	ε_r	Si = 11.8	
		Ge = 15.8	
		GaAs = 13.1	
		$SiO_2 = 3.9$	
		$Si_3N_4 = 7.5$	
		$Al_2O_3 = 7.8$	

Foreword

The SPICE program is used worldwide as an essential computer-aid for circuit design. One can simulate circuits prior to their fabrication and predict detailed performance. However, an essential part of using SPICE is the accurate extraction and understanding of the device model parameters. Namely, the variety of devices used in an integrated circuit includes diodes, FETs, and bipolar transistors. Each device has special features of its operation, and more importantly, the models each have special features and nuances. Moreover, to model integrated circuits over their entire range of operation, effects such as temperature and bias dependence must be considered.

This book can be an invaluable aid in effectively using SPICE and understanding the many details with the device models. Because the emphasis of the book is SPICE-based, the language and examples are well-focused toward that end. The evolving state of device models for scaled-down technology leaves a gap between classical device modeling texts and engineering needs for circuit design. This book helps to bridge this gap by addressing directly the model constraints and assumptions used in SPICE. In many cases there are discrepancies between the physics, models, and measurements. This book does a good job of pointing out many of these pitfalls. There are also examples of both device model limitations and their occurrence in circuit applications where the problems become apparent. Finally, the discussion of the internal structure of SPICE and its execution provide unique documentation to help the SPICE user understand how the models are implemented.

In total this book is a most helpful reference for SPICE models as well as a bridge between device physics and the models needed for circuit design. The material is well-focused toward application and provides quick reference to a host of useful facts. This book will be of great value to both practicing engineers as well as students.

Robert W. Dutton
Stanford University

Preface

Everybody involved in CAD (computer-aided design) for electronic circuits knows how critical circuit simulation is to the design and optimization of a circuit. Today, SPICE (Simulation Program with Integrated Circuit Emphasis) is the circuit simulator most designers are familiar with, after more than two decades of running on all sorts of computers around the world, after solving critical problems in extremely complex circuits, SPICE can be regarded as the de facto standard in circuit simulation. Designers and engineers consider SPICE the "Berkeley connection," from the university where it originated in the early seventies. Professor D. O. Pederson can be considered the "father" (or "godfather") of SPICE. Engineers and designers are all indebted to Pederson and his students for giving us (free, by the way) such a powerful design aid. All SPICE lovers also know how critical the input model of the transistor is in order to obtain an experiment-matching output of SPICE; this is another way of saying "garbage in, garbage out."

With the explosion of VLSI, designers of integrated circuits are blossoming everywhere, in university courses, research labs, and system houses, as well as, obviously, semiconductor houses. Many new VLSI designers will not need to use circuit simulation, since design methodologies, like gate arrays and cell libraries, do not present the need to simulate the VLSI system at the circuit level. However, for more advanced or full-custom designs, or for improving and optimizing cells, certainly some circuit optimization is needed. It can be obtained from simulators like SPICE.

Today the emphasis is more on logic and system aspects; thus many designers lack the insight into the real transistor, into the physical mechanisms taking place across the junctions, and therefore have some difficulty keeping track of all the parameters which represent the complete model of the transistor (bipolar or MOS). Furthermore, even if one knows the physical meaning of the different parameters, it is often hard to answer the question, "For my transistor, how do I know the right value for this particular model parameter?" In fact, there are so many physical interactions inside the transistor that a correct parameter measurement setup requires a careful examination of the model and of the experimental procedure.

This book has been written as a reference source for the electronic circuit designer (VLSI or PCB); it covers all the aspects of transistor models that a SPICE user, new or old, may want or need to know—from physics to models to parameter measurements and SPICE use. Certainly there are many books (college-level electronic circuits handbooks or textbooks) that provide details on transistor action and different modeling aspects; some other books may have an appendix describing a circuit simulation program and may mention SPICE, showing an input or output listing. However, for a designer, it is certainly not easy to put things together from different sources, with different notations, in order to efficiently use SPICE. (Here SPICE is referred to with a general meaning, since today there are several, widely used versions of SPICE: SPICE2, SPICE3, HSPICE, and PSPICE. All models used in these versions are referenced in the book.)

The first four chapters of this book deal with the physical devices (*pn* junction and Schottky diodes, bipolar, junction field-effect, and metal-oxide-semiconductor transistors), starting from the basic aspects and taking the reader step by step through the SPICE model; in fact, the reader does not need to know much about the transistor, since the text is written in a simple and understandable way. If some physical background is needed for either bipolar or MOS transistors, it can be found in the appendixes.

Chapters 5 and 6 show how to obtain the numerical values of the parameters of the transistor (bipolar and MOS) model. The parameters are obviously those of the SPICE model, so that SPICE users do not have to do any complicated conversion of parameters. Chapter 7 deals with noise and distortion effects in SPICE.

Chapter 8 deals specifically with problems related to the actual use of SPICE; they can also be of great help to those wanting to improve SPICE to adapt it to the particular needs of the designer. Chapter 9 deals with models for MESFET, ISFET, and THYRISTOR.

Acknowledgments

Special thanks are due to the professors and researchers of the University of California, Berkeley, who gave us the possibility to draw useful information from their valuable *ERL-Memoranda* on SPICE. The authors are also grateful to Meta-Software, Inc. (Campbell, Calif.) and MicroSim Corp. (Irvine, Calif.) for the material on HSPICE and PSPICE, respectively.

P. Antognetti
Genova, Italy

Chapter 1

PN-Junction Diode and Schottky Diode

This chapter emphasizes the basic aspects of a real *pn*-junction (*diode*) behavior, and establishes *static, large-signal, small-signal,* and *thermal* physical-mathematical models and how they are implemented in *SPICE2* (*S*imulation *P*rogram with *I*ntegrated *C*ircuit *E*mphasis) [1] and in other SPICE2-based programs. The *noise* and *distortion* models are described in Chap. 7.

SPICE2 diode models are directly regulated by the subroutines *READIN* (it reads and processes the .MODEL statement), *MODCHK* (it performs one-time processing of the model parameters and assigns the parameter default values), *DIODE* (it computes all the model equations), and *TMPUPD* (it updates the temperature-dependent parameters of the models), in addition to the subroutines concerning each type of analysis. The noise and distortion models are directly regulated by the subroutines *NOISE* and *DISTO*, respectively (see Chap. 7).

1.1 DC Current-Voltage Characteristics

1.1.1 DC characteristics of the ideal diode

Consider the ideal monodimensional *pn* structure (*diode*) shown in Fig. 1-1, whose behavior is analyzed in Appendix A. Here the simplifying assumptions on which the ideal diode equations have been obtained are summarized for convenience.

1. The uniform doping and electrical neutrality in the *p*- and *n*-type regions as shown in Fig. A-5, where N_A is the acceptor concentra-

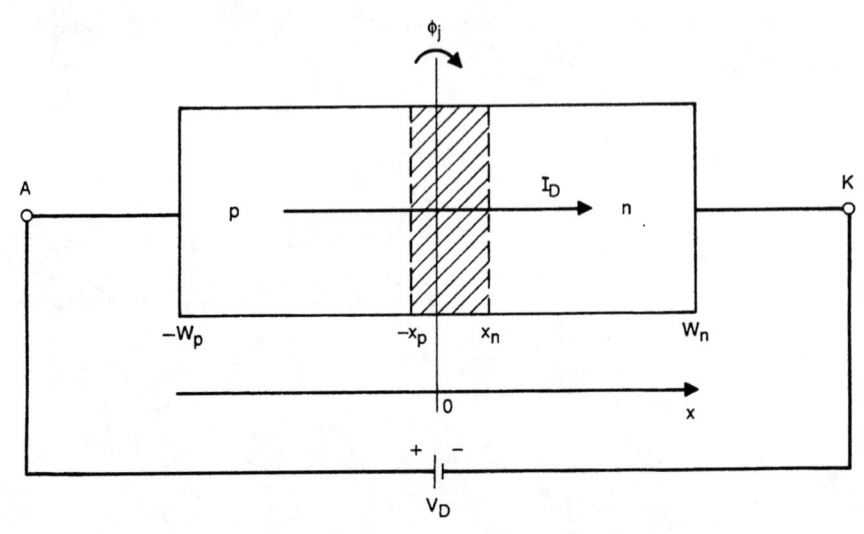

Figure 1-1 PN junction under nonequilibrium conditions: forward bias.

tion in the p-type region and N_D is the donor concentration in the n-type region.

2. The abrupt depletion-region approximation: The built-in potential and applied voltages are supported by a dipole layer with abrupt boundaries, and outside the boundaries the semiconductor is assumed to be neutral.
3. The Boltzmann approximation: Throughout the depletion region, the Boltzmann relations [see Eqs. (A-31)] are valid.
4. The low-injection assumption: The injected minority-carrier densities are small compared with the majority-carrier densities.
5. No generation current exists in the depletion region, and the electron and hole currents are constant through this region.

From Eqs. (A-54) and (A-59), it follows that the ideal diode current is

$$I_D = I_S(e^{qV_D/kT} - 1) \tag{1-1}$$

where

$$I_S = qA_J n_i^2 \left(\frac{D_p}{N_D L_p} + \frac{D_n}{N_A L_n} \right)$$

$$I_S = qA_J n_i^2 \left(\frac{D_n}{N_A W'_p} + \frac{D_p}{N_D W'_n} \right) \tag{1-2}$$

have been defined as the diode *reverse saturation current* for the long-base and short-base diode, respectively.

In the above equations, kT/q is the thermal voltage; q is the electronic charge; k is the Boltzmann constant; A_J is the area of the transverse section of the diode; n_i is the intrinsic concentration; D_n and D_p are the electron and hole diffusion coefficients; L_n and L_p are the electron and hole diffusion lengths; $W'_n = (W_n - x_n)$ and $W'_p = (W_p - x_p)$ are the neutral n- and p-type regions; N_A and N_D are the acceptor and donor concentrations; V_D is the applied voltage (*bias*). The ideal current-voltage relation is shown in Fig. 1-2 in a linear plot.

In the reverse direction the current saturates at $-I_S$. The term *saturation* indicates that I_D approaches an asymptote and becomes independent of the voltage V_D.

Last, the electrical symbol for the ideal diode is shown in Fig. 1-3.

Equations (1-2) allow much simpler relations in the case of the one-sided abrupt junction. For example, for a p^+n diode, Eqs. (1-2)

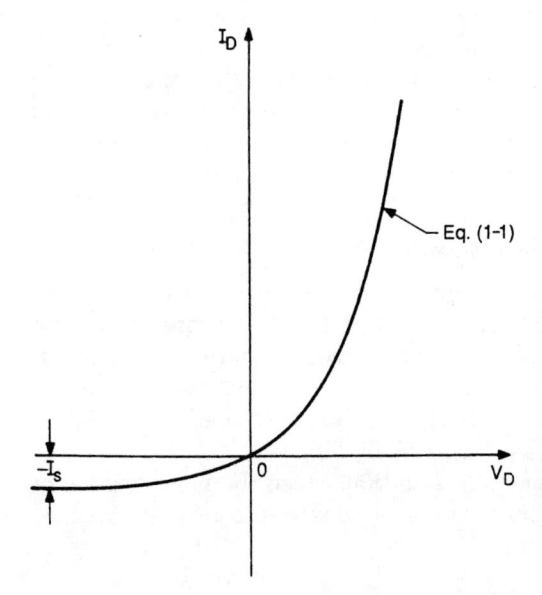

Figure 1-2 Ideal diode characteristic.

$$I_D = I_S(e^{qV_D/kT} - 1)$$

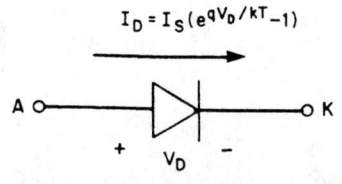

Figure 1-3 Circuit representation of the ideal diode.

become simply

$$I_S \simeq \frac{qA_J n_i^2 D_p}{L_p N_D}$$

$$I_S \simeq \frac{qA_J n_i^2 D_p}{W'_n N_D}$$
(1-3)

being, in a p^+n junction, $N_A \gg N_D$.

The physical justification for this simplification has considerable importance, as it is the basic physical characteristic of the bipolar transistor. In fact if $N_A \gg N_D$, it follows that $|dp_n/dx| \gg |dn_p/dx|$, so that Eqs. (A-54) and (A-59) become

$$I_D \simeq I_{p,\text{diff}}(x_n)$$
(1-4)

Equation (1-4) means that, in the case of greatly asymmetrical doping, the region which is more heavily doped (p region) emits by diffusion a large number of majority carriers (holes), so by itself it will guarantee the supply of carriers to the current I_D of the diode. In that case the n-type region, being less doped, is poor in majority carriers; therefore, its emission to the adjacent p^+ region is negligible, and it has no relevant influence on the total current I_D.

1.1.2 Limitations of the ideal diode model

The experimental results for most junction diodes show a range of the forward and reverse static characteristics for which there is good agreement with the idealized pn-junction diode. There are, however, differences over a significant range of biases for which the ideal diode equation becomes inaccurate, whether for forward bias ($V_D > \phi_0$) or for reverse bias ($V_D < \phi_0$), as is shown in Fig. 1-4.

In this section the main mechanisms that cause the performance of diodes to differ from the predictions of idealized analysis are presented, including the following:

1. Carrier generation-recombination in the depletion region.
2. High-level injection.
3. Voltage drop associated with the electric field in the neutral regions.
4. Internal breakdown associated with high reverse voltage.

a. Carrier generation-recombination in the depletion region. One assumption that defines the idealized pn-junction diode model is that the

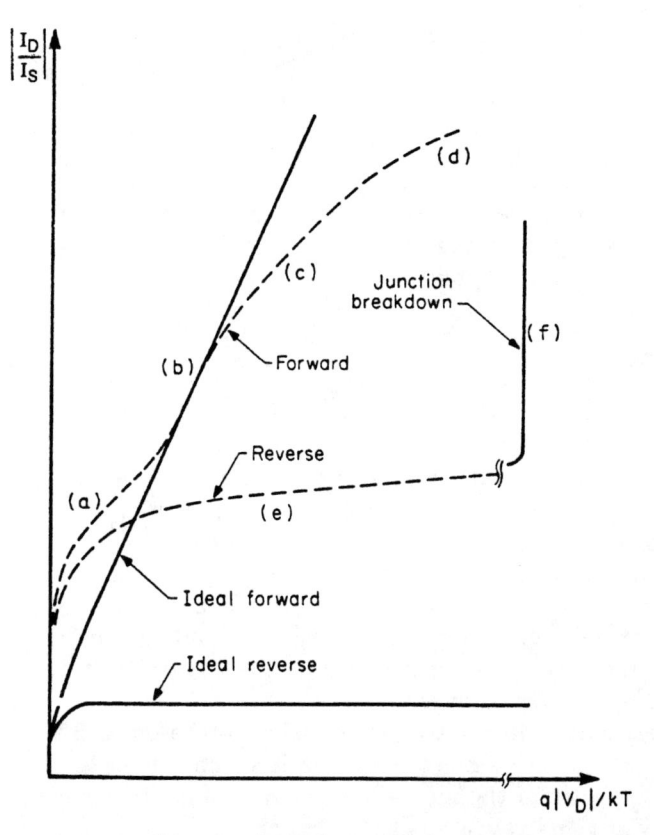

Figure 1-4 Current-voltage characteristic of a real silicon diode: (a) generation-recombination current region, (b) diffusion (ideal) current region, (c) high-level injection region, (d) series resistance effect, (e) reverse leakage current due to generation-recombination and surface effects, and (f) breakdown region. (*From S. M. Sze, Physics of Semiconductor Devices. Copyright © 1969 by John Wiley & Sons, Inc. Used by permission.*)

hole and electron current components are constant through the depletion region.

Like the neutral regions of the diode, the depletion region contains generation-recombination centers. Unlike the neutral regions, it is a region of steep impurity gradients and very rapidly changing populations of carriers. Since injected carriers, under forward bias, must cross through this region, some carriers may be lost by recombination. Under reverse bias, generation of carriers in this region leads to excess current above the saturation value predicted by the ideal diode equations [2].

To find expressions for generation and recombination in the depletion region, we must use the *Shockley-Hall-Read* (*SHR*) theory in conjunction with Eq. (A-74). The resulting recombination-generation function $U = G - R$ can be positive or negative; i.e.,

$$V_D > 0 \rightarrow pn > n_i^2 \rightarrow U > 0 \rightarrow \text{recombination}$$

$$V_D < 0 \rightarrow pn < n_i^2 \rightarrow U < 0 \rightarrow \text{generation}$$

The current arising from recombination and generation in the depletion region is given by the integral of the recombination-generation rate across the depletion region; i.e.,

$$I_{r,g} = qA_J \int_{-x_p}^{x_n} U\,dx = \begin{cases} I_r = f(e^{qV_D/2kT}) & \text{for } V_D > 0 \\ I_g = f\left(\sqrt{\dfrac{qV_D}{kT}}\right) & \text{for } V_D < 0 \end{cases} \quad (1\text{-}5)$$

Thus, unlike current arising from carrier diffusion and recombination in the neutral regions, the current resulting from recombination in the depletion region varies with applied voltage as $e^{qV_D/2kT}$; this different exponential behavior can be observed in real diodes, especially at low currents [see Fig. 1-4, region (a)]. The generation current in the depletion region, on the contrary, is only a slight function of the reverse bias, varying roughly as the square root of applied voltage [see Fig. 1-4, region (e)].

It is possible to estimate the relative importance of the contributions to the diode current from the recombination and generation currents by taking the ratio of the ideal-diode current [Eq. (1-1)] and Eqs. (1-5) under forward and reverse bias [3].

Now, if the total current of the *real* diode (when the generation-recombination processes are taken into account) is written as $I_D = [I_D(\text{ideal}) + I_{r,g}]$, then the current on the reverse-biased diode is generated predominantly in the depletion region, while the current in the forward-biased diode is dominated by the ideal diode diffusion current, especially as bias increases.

In general, an expression that well describes the effects due to the generation-recombination processes is given by the empirical relation

$$I_D = I_S(e^{qV_D/nkT} - 1) \quad (1\text{-}6)$$

where n, the *emission coefficient*, is usually about 2 if the recombination current dominates, and equal to 1 for the ideal diode behavior (see Fig. 1-4, region b).

b. High-level injection. It is known that for *low-level injection* the change in the majority-carrier concentration due to the injected minority carriers is so small in the regions outside the depletion region that its effect can be neglected. Thus, for a uniformly doped p^+n diode there is no electric field in the n region if current flows, and injected carriers move only by diffusion.

At *high-level injection* (see Sec. A.1.3a), on the contrary, there is a significant alteration of the majority-carrier concentration outside the depletion region, which gives rise to an electric field. Thus carrier motion is influenced by both diffusion and drift in this region. The presence of the electric field results in a voltage drop across this region. Hence only a fraction of the applied voltage appears across the junction.

Consider an abrupt p^+n junction with a wide, uniformly doped n region.

For charge neutrality, it is assumed that, under high-level condition

$$n_n = p_n + N_D \simeq p_n \tag{1-7}$$

$$\frac{dn_n}{dx} \simeq \frac{dp_n}{dx} \tag{1-8}$$

Furthermore, in the p^+n junction, the electron current I_n is negligible with respect to the hole current I_p. Therefore, from Eq. (A-19b),

$$I_n = qA_J\left(\mu_n n_n E + D_n \frac{dn_n}{dx}\right) \simeq 0 \tag{1-9}$$

The electric field is then obtained from Eq. (1-9) as

$$E = -\frac{D_n}{\mu_n}\frac{1}{n_n}\frac{dn_n}{dx} = -\frac{D_p}{\mu_p}\frac{1}{p_n}\frac{dp_n}{dx} \tag{1-10}$$

when taking into account Eqs. (1-7) and (1-8) and Einstein's relations [see Eqs. (A-20)].

In reality, at high-level injection, the condition of neutrality is not rigorously observed; there exists a space charge that creates the electric field given in Eq. (1-10). However, since n_n is very high, a very small electric field is sufficient to balance the diffusion current and satisfy Eq. (1-9). The departure from the neutrality condition is therefore negligible, and the assumption of neutrality is acceptable. Knowing the expression for the electric field allows us to determine the hole current

$$I_p = qA_J\left(\mu_p p_n E - D_p \frac{dp_n}{dx}\right) = -2qA_J D_p \frac{dp_n}{dx} \tag{1-11}$$

Regarding the transport of carriers under high-level injection conditions, it can be concluded that the contribution to the current of the electric field is equal to that of diffusion; therefore, it is as if the current were due only to diffusion, with doubled diffusion constant.

Under the assumption of linear distribution of the carriers (short-base diode), it follows that

$$\frac{dp_n}{dx} = \text{const.} = -\frac{p_n(x_n)}{W'_n} \tag{1-12}$$

and the profile of the mobile carriers (minority and majority) is linear, the doping profile being constant, provided that the condition of high-level injection is verified.

Last, from Eq. (A-74) and charge-neutrality conditions at high-level injection [see Eq. (1-7)], it follows that

$$p_n(x_n) = n_n(x_n) = n_i \, e^{qV_D/2kT} \tag{1-13}$$

Taking into account Eqs. (1-11) and (1-12), the expression for the p^+n diode current can be obtained from Eq. (1-13). Then

$$I_D = I_p = \frac{2qA_J D_p n_i}{W'_n} e^{qV_D/2kT} \tag{1-14}$$

In Eq. (1-14) the current depends again exponentially on the applied voltage V_D, but with the coefficient $q/2kT$ instead of q/kT [see Fig. 1-4, region (c)].

c. Voltage drop associated with the electric field in the neutral regions.* The injection or extraction of excess minority carriers across the depletion region is invariably accompanied by an electric field in the neutral regions. Consequently, the voltage drops in the neutral regions which are associated with this electric field are linearly dependent on the total current. These voltage drops can be evaluated by integrating the electric field E over the neutral regions. Then, if V'_D is the voltage applied to the diode and V_D is the drop across the pn junction, it can be written that

$$V'_D = V_D + \int_{\substack{\text{neutral}\\\text{regions}}} (-E) \, dx \tag{1-15}$$

---

*The material in this section is taken from Semiconductor Electronics Education Committee [3]. Copyright © 1964, Education Development Center, Inc., Newton, Mass. Used by permission.

Because V_D is related to the total current I_D by Eq. (1-1), which is the result of the idealized model, Eq. (1-15) may be written as

$$V'_D = \underbrace{\frac{kT}{q} \ln\left(1 + \frac{I_D}{I_S}\right)}_{V_D} + \underbrace{r_S I_D}_{V_{r_S}} \qquad (1\text{-}16)$$

where r_S is the series resistance given by

$$r_S = \frac{1}{I_D} \int_{\substack{\text{neutral} \\ \text{regions}}} (-E)\,dx \qquad (1\text{-}17)$$

Equation (1-16) suggests that, for small forward currents, the total diode voltage V'_D should vary logarithmically with I_D, in accordance with the idealized model. For large forward currents, on the contrary, the voltage should increase linearly with I_D because $I_D r_S$ increases faster than $(kT/q)\ln(1 + I_D/I_S)$, and thus it dominates V'_D, as shown in Fig. 1-4, region (d).

d. Junction breakdown. At large reverse voltages, the I_D-V_D curve for real diodes departs from the theoretically predicted relationship [Eq. (1-1)] or shape (Fig. 1-2). This deviation is due to carrier multiplication effects occurring in the depletion region when a critical value of electric field (breakdown field) is exceeded [see Fig. 1-4, region (f)].

Two mechanisms of diode breakdown can be recognized: the *avalanche breakdown* and *Zener breakdown*.

1. *Avalanche breakdown.* A thermally generated carrier (e.g., an electron) is accelerated by the electric field in the depletion region and acquires kinetic energy. If this kinetic energy is at least equal to the energy band gap E_g [4, 5], when the electron collides with a lattice atom in an inelastic collision, the kinetic energy it loses can be sufficient to disrupt a covalent bond. The result is that in addition to the original electron, a new electron-hole pair has now been generated. These carriers can now be accelerated by the field and, on picking up sufficient kinetic energy ($\geq E_g$) can generate another electron-hole pair on collisions with lattice atoms. Thus, each new carrier may, in turn, produce additional carriers through collision and action of breaking bonds: this mechanism is referred to as *avalanche multiplication*. This process is also usually referred to as *avalanche breakdown*; however, this does not mean that the process is necessarily destructive. The resulting high reverse currents (Fig. 1-5, dotted curve) generate heat through Joulean heating (I^2R losses), and it is the dissipation of this heat that is critical to the life of the

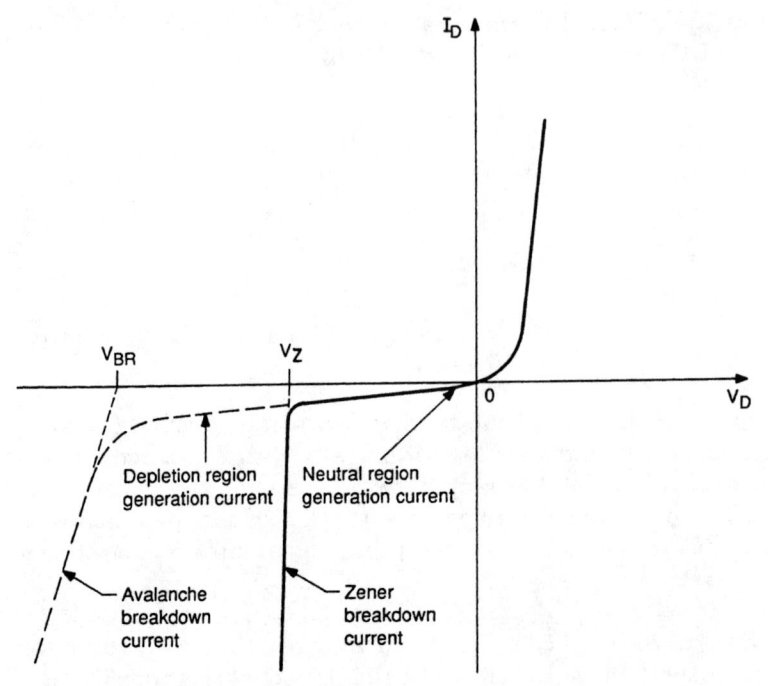

Figure 1-5 Diode characteristic, showing the effects of avalanche breakdown (dotted line) and Zener breakdown (solid line).

device. The lack of a well-designed dissipation mechanism leads to a destructive breakdown, usually in the form of the vaporization of the metallic contacts [6]. However some devices, such as IMPATT diodes and avalanche photodiodes, do normally operate in the avalanche breakdown mode.

The values of the avalanche breakdown field E_{BR} for silicon and gallium arsenide are about 3×10^5 and 4×10^5 V/cm, respectively. To evaluate the dependence of the breakdown voltage V_{BR} on the diode properties, we can substitute E_{BR} for the maximum field E_{max} [Eq. (A-41)] in the depletion region. Also assuming that the applied voltage V_D is much greater than the built-in potential ϕ_0 (assumption that is true at breakdown), we find the following expression for the breakdown voltage:

$$|V_{BR}| = E_{BR}^2 \frac{\varepsilon_s}{2q}\left(\frac{1}{N_A} + \frac{1}{N_D}\right) \qquad (1\text{-}18)$$

From Eq. (1-18) it follows that to increase the breakdown voltage, one has to reduce the lesser of the two doping concentrations in the diode [6]. The low doping concentrations ensure that the depletion region width is large.

2. *Zener breakdown.* The I_D-V_D characteristic of a diode when Zener breakdown occurs is shown in Fig. 1-5 (solid line). The abrupt rise in current is due to the onset of *tunneling* [5, 6]. This condition can arise in a reverse-biased *pn* junction with high doping concentrations on both regions of the junction. The high doping concentrations ensure that the depletion region width is small. The voltage V_Z at which Zener breakdown occurs is usually in the neighborhood of 3–6 V and it is generally much lower than that at which avalanche breakdown occurs. It must also be noted that if the doping concentrations are such as to make the zero-bias depletion region width large, it may not be possible to increase the reverse bias to V_Z and so to obtain the necessary physical properties (separation of the energy levels <5 nm [5, 6]) for tunneling, because avalanche breakdown will occur first.

With respect to avalanche breakdown, Zener breakdown does not involve collisions of carriers with lattice atoms as does avalanche breakdown; moreover, the conductance dI_D/dV_D in Zener breakdown is high and makes these devices (Zener diodes) suitable as voltage regulator circuits.

1.2 Static Model

1.2.1 Static model of the ideal diode and its implementation in SPICE2

a. DC forward characteristic. The *dc forward characteristic* of the ideal diode is modeled by the nonlinear current source I_D, whose value is determined by Eq. (1-1), rewritten for convenience in SPICE2 implemented form [1]

$$I_D = I_S(e^{qV_D/kT} - 1) + V_D \, \text{GMIN} \quad \text{for} \quad V_D \geq -5\frac{kT}{q} \qquad (1\text{-}19)$$

where V_D is that portion of the applied voltage that appears across the junction. The model derived from Eq. (1-19) is outlined in Fig. 1-6.

The dc forward characteristic is then determined by only one model parameter to be specified in the .MODEL statement [7] of SPICE2 the *saturation current* I_S. It should be noted that I_S, in SPICE2 is a constant and so does not properly model the voltage dependence of the depletion region current, which is due to the varying width of the depletion region [8].

It is important to note that to aid convergence a conductance GMIN is added by default in SPICE2 in parallel with every *pn*

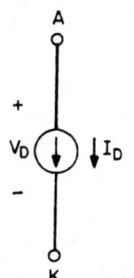

Figure 1-6 SPICE2 ideal diode static model.

junction. The value of this conductance is a program parameter that can be set by the user with the .OPTIONS statement. The default value for GMIN is 10^{-12} mho; moreover, the user *cannot* set GMIN to zero.

b. DC reverse characteristic. The *dc reverse characteristic* of the ideal diode is modeled in SPICE2 by a nonlinear current source I_D, as shown in the following equations:

$$I_D = f(V_D) = \begin{cases} I_S(e^{qV_D/kT} - 1) + V_D \text{GMIN} & \text{for } -5\dfrac{kT}{q} \leq V_D \leq 0 \\ -I_S + V_D \text{GMIN} & \text{for } V_D < -5\dfrac{kT}{q} \end{cases}$$

(1-20)

Figure 1-7 shows the plot of the dc forward and reverse characteristics of an ideal diode using SPICE2 default parameter values (see Table 1-1).

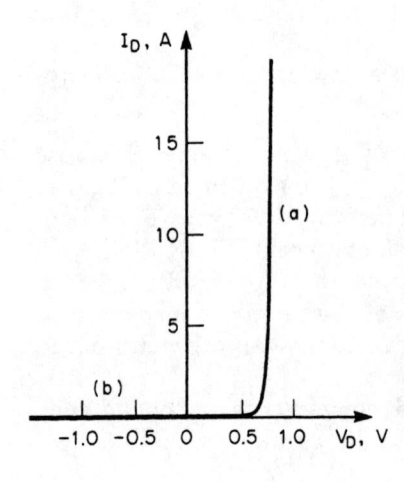

Figure 1-7 SPICE2 output plot by using default parameter values for (a) ideal diode dc forward characteristic and (b) ideal diode dc reverse characteristic.

TABLE 1-1 SPICE2 Diode Model Parameters

Symbol	SPICE2G keyword	Parameter name	Default value	Unit
I_S	IS	Saturation current	10^{-14}	A
r_S	RS	Ohmic resistance	0	Ω
n	N	Emission coefficient	1	
τ_D	TT	Transit time	0	s
$C_D(0)$	CJO	Zero-bias junction capacitance	0	F
ϕ_0	VJ	Junction potential	1	V
m	M	Grading coefficient	0.5	
E_g	EG	Energy gap: 1.11 for Si 0.69 for SBD 0.67 for Ge	1.11	eV
p_t	XTI	Saturation current temperature exponent: 3.0 for pn-junction diode 2.0 for SBD	3.0	
FC	FC	Coefficient for forward-bias depletion capacitance formula	0.5	
BV	BV	Reverse breakdown voltage (positive number)	∞	V
IBV	IBV	Reverse breakdown current (positive number)	10^{-3}	A
k_f	KF	Flicker-noise coefficient	0	
a_f	AF	Flicker-noise exponent	1	
T	*	Nominal temperature for simulation and at which all input data are assumed to have been measured	27	°C

*Remember, the temperature T is regarded as an operating condition and not as a model parameter.

1.2.2 Static model of the real diode and its implementation in SPICE2

In this section the secondary effects present in the real diode and their implementation in the SPICE2 diode model are considered.

a. Small bias: Effect of the carrier recombination-generation. The recombination-generation processes in the depletion region (see Sec. 1.1.2a) are described in SPICE2 through a relation like Eq. (1-6). In particular, the diode dc characteristic becomes

$$I_D = f(V_D) = \begin{cases} I_S(e^{qV_D/nkT} - 1) + V_D \, \text{GMIN} & \text{for } V_D \geq -5\dfrac{nkT}{q} \\ -I_S + V_D \, \text{GMIN} & \text{for } V_D < -5\dfrac{nkT}{q} \end{cases} \quad (1\text{-}21)$$

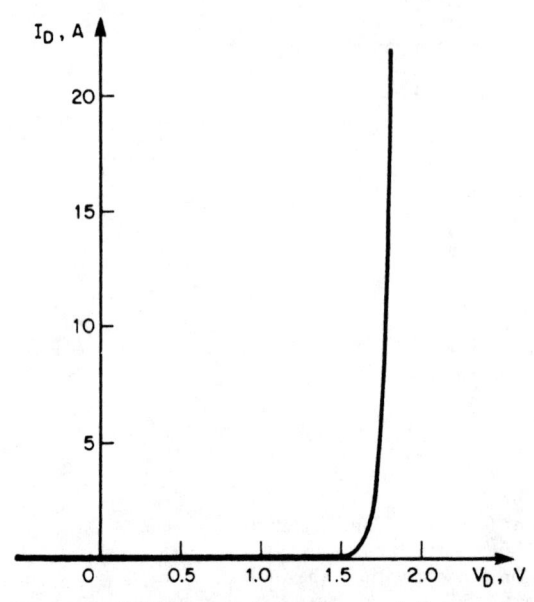

Figure 1-8 SPICE2 output plot for a real diode under recombination condition ($n = 2$). The other parameters have SPICE2 default values.

where n, the emission coefficient, is the model parameter that the user of SPICE2 has to specify in the .MODEL statement to simulate the recombination-generation processes. It must be noted that n, in SPICE2, is a constant for the whole static characteristic computation.

The dc characteristic of the diode is also modeled by the nonlinear current source I_D as shown in Fig. 1-6, whose value is now given by Eq. (1-21).

Figure 1-8 shows the SPICE2 output plot of the dc characteristic of a real diode under recombination condition ($n = 2$). All the parameters used for the simulation are the same as those used in the ideal diode case.

b. Large bias: Effect of the ohmic resistance and high-level injection. At higher levels of bias, the diode current deviates from the ideal characteristic given by Eq. (1-1). This deviation is due to the presence of a resistance associated with the neutral regions and contacts (see Sec. 1.1.2c), as well as any equivalent resistance associated with high-level injection (see Sec. 1.1.2b). It should be noted that high-level injection is not explicitly implemented in the SPICE2 diode model. Therefore, the effects of both ohmic resistance and high-level

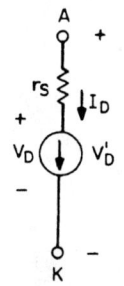

Figure 1-9 SPICE2 real diode static model at high-level injection.

injection are modeled by the ohmic resistance r_S. In this case, the dc characteristic of the diode is modeled by the nonlinear current source I_D and the linear resistor r_S, so that

$$V'_D = r_S I_D + V_D \qquad (1\text{-}22)$$

where I_D is given by Eq. (1-21). This model is outlined in Fig. 1-9.

Figure 1-10 shows the plot of the dc forward characteristic of a real diode under high-level injection conditions (simulated by the r_S series ohmic resistance). All the parameters used for the simulation are the same as those used for the ideal diode case except for the r_S value, which is different from zero.

c. Large reverse bias: Effect of the junction breakdown. A more significant deviation of the real diode reverse characteristic from the ideal one [Eq. (1-1)] occurs when the breakdown processes are considered.

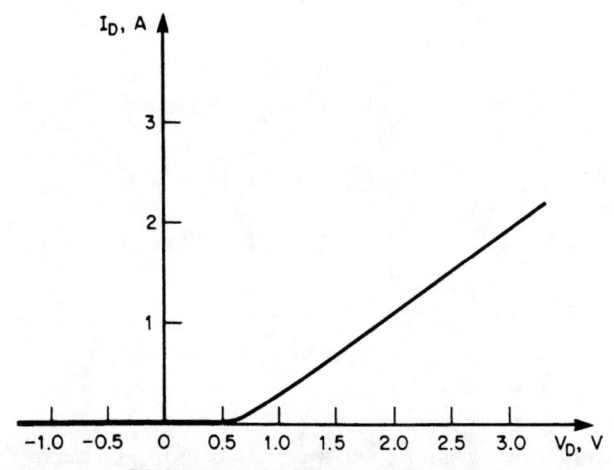

Figure 1-10 SPICE2 output plot for a real diode dc forward characteristic under high-level injection conditions with $r_S = 1\,\Omega$. The other parameters have SPICE2 default values.

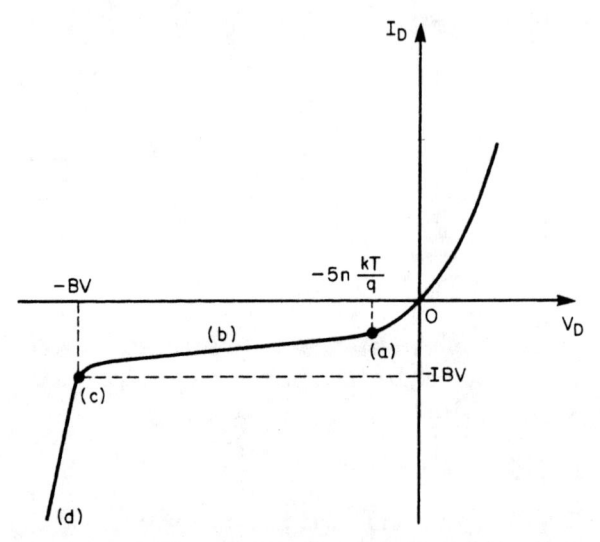

Figure 1-11 SPICE2 reverse characteristic of the real diode.

These processes are taken into account (implemented) in SPICE2, by dividing the reverse characteristic into four distinct regions [from (a) to (d) in Fig. 1-11] according to the value of the applied voltage V_D.

Also in this case, the diode is modeled by a nonlinear current source I_D according to the following equations:

$$I_D = f(V_D)$$

$$= \begin{cases} I_S(e^{qV_D/nkT} - 1) + V_D \text{GMIN} & \text{for } -5\dfrac{nkT}{q} \leq V_D \leq 0 \\[6pt] -I_S + V_D \text{GMIN} & \text{for } -\text{BV} < V_D < -5\dfrac{nkT}{q} \\[6pt] -\text{IBV} & \text{for } V_D = -\text{BV} \\[6pt] -I_S\left(e^{-q(\text{BV} + V_D)/kT} - 1 + \dfrac{q\text{BV}}{kT}\right) & \text{for } V_D < -\text{BV} \end{cases}$$

(1-23)

It must be noted that since regions (c) and (d) of Fig. 1-11 are defined by different analytical expressions, problems of convergence for voltage values V_D close to both regions can arise. In fact, since from Eq. (1-23) it follows that

$$I_D(-\text{BV}) = -\text{IBV} \tag{1-24}$$

Eq. (1-23), too, must converge to

$$\lim_{V_D \to -\text{BV}} I_D = -I_S \frac{q\text{BV}}{kT} = -\text{IBV} \tag{1-25}$$

The numerical value for IBV and BV is therefore of extreme importance, as it characterizes point (c) in Fig. 1-11.

The diode forward and reverse characteristics (*diode static model*) are modeled in SPICE2 by the following parameters, which the user can specify in the .MODEL statement [7]:

IS   saturation current (I_S)
RS   ohmic resistance (r_S)
N   emission coefficient (n)
BV   breakdown voltage (BV) (it is a positive number)
IBV breakdown current (IBV) (it is a positive number)

Figure 1-12 shows the dc reverse characteristic of the ideal diode using SPICE2 default parameter values.

Figure 1-13 shows the dc reverse characteristic of the real diode with specified BV and IBV parameter values, while the other parameter values are the same as those used for the ideal diode case.

In order to simulate a *Zener* diode, the macromodel shown in Fig. 1-14 can be used. Diode D_1 and the threshold voltage source V_0

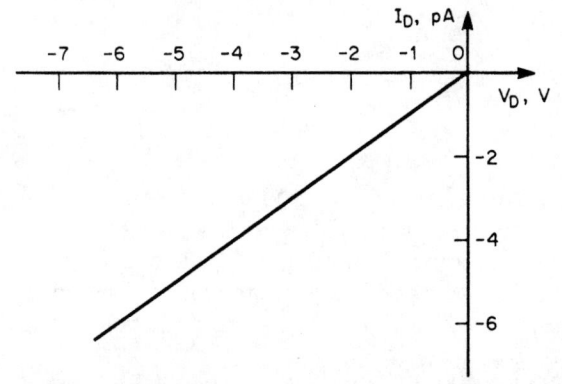

Figure 1-12 SPICE2 output plot using default parameter values for an ideal diode dc reverse characteristic.

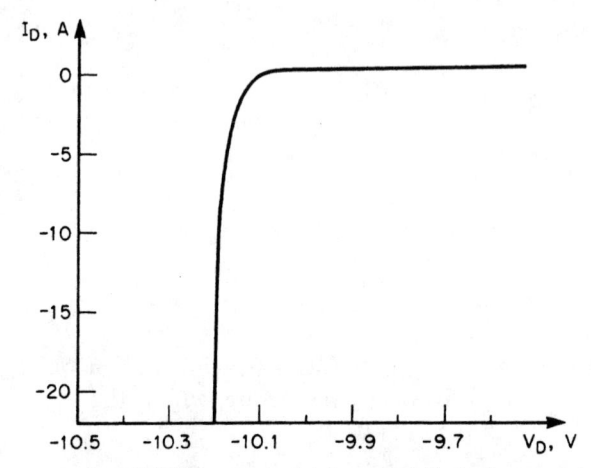

Figure 1-13 SPICE2 output plot for a real diode dc reverse characteristic with BV = 10 V and IBV = 10 mA. The other parameter values are SPICE2 default values.

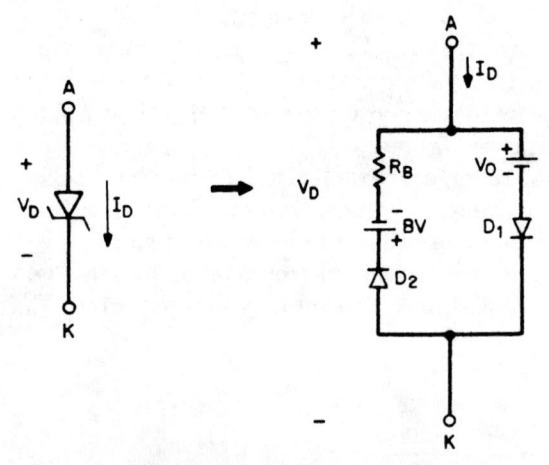

Figure 1-14 Macromodel of Zener diode.

represent the normal forward and reverse behavior [see region (b) in Fig. 1-11]. Diode D_2, the voltage source BV, and the resistance R_B define the breakdown region [region (d) in Fig. 1-11]. Diode D_2 does not conduct until $V_D = -\text{BV}$; if the reverse voltage is also increased, this diode becomes forward-biased and the reverse current crosses the resistor R_B.

A Zener diode physical model has been proposed and implemented in SPICE2 by Laha and Smart [9]. In this model, the I_D-V_D characteristic is described by [9]

$$I_D = f(V_D)$$

$$= \begin{cases} I_S(e^{qV_D/nkT} - 1) & \text{for } V_D > 0 \\ \dfrac{qI_S}{nkT} V_D & \text{for } -n\dfrac{kT}{q} \leq V_D \leq 0 \\ -I_S & \text{for } -\text{BV} < V_D < -n\dfrac{kT}{q} \\ -e^{-K_1(V_1+V_D)} - e^{-K_2(V_2+V_D)} \\ \quad + e^{-K_1 V_1} + e^{-K_2 V_2} & \text{for } V_D \leq -\text{BV} \end{cases} \quad (1\text{-}26)$$

The first relation in Eq. (1-26) is the familiar Ebers-Moll relation; the second and third relations are a piecewise linear approximation to the first one for $V_D \leq 0$ in order to avoid the computation of exponentials in regions (a) and (b) of Fig. 1-11; the fourth relation describes reverse breakdown as a sum of two exponentials. Also for this model, in order to maintain continuity between regions (b) and (d), an appropriate value of BV must be selected by equating the two last relations in Eq. (1-26). If the parameters K_1, V_1, K_2, and V_2 are such that $\text{BV} > -nkT/q$, regions (a) and (b) are not taken into account and region (d) extends to $V_D = 0$.

If breakdown is neglected, the model reduces to an ordinary diode model. In addition, the accuracy of the breakdown model can be chosen by acting on the double exponential in the last relation of Eq. (1-26)—e.g., by setting $K_2 = 0$. A new limiting algorithm is also presented to ensure numerical stability during the iterative solution of the circuit equations.

1.3 Large-Signal Model

1.3.1 Large-signal model of the diode

Having defined the dc relationships for the *pn*-junction diode, let us consider the charge-storage effects of the device. If there were no charge storage, the device would be infinitely fast; that is, there would be no charge inertia, and currents could be changed in zero time.

There are two forms of charge storage: charge storage in the depletion region due to the dopant concentrations, i.e.,

$$Q'_j = \sqrt{\dfrac{2q\varepsilon_s(\phi_0 - V_D)}{\dfrac{1}{N_A} + \dfrac{1}{N_D}}} \quad (1\text{-}27)$$

and charge storage due to the minority-carrier charges injected into the neutral regions, i.e.,

$$Q'_d = (Q'_p + Q'_n) = \tau_D I_D \tag{1-28}$$

In the above equations, ϕ_0 is the built-in potential; Q'_p [Eq. (A-65)] and Q'_n are the charges per unit area resulting from holes stored in the n-type region and electrons stored in the p-type region, respectively; τ_D is the transit time of the diode. Both of these charges give rise to two small-signal capacitances per unit area, usually called the *junction capacitance* C'_j and the *diffusion capacitance* C'_d, respectively.

As explained in Sec. A.2.4, to which the reader can refer to for a wide description, the *junction capacitance per unit area* C'_j [Eq. (A-63)] is

$$C'_j = \frac{dQ'_j}{dV_D} = \sqrt{\frac{q\varepsilon_s}{2(\phi_0 - V_D)\left(\frac{1}{N_A} + \frac{1}{N_D}\right)}} = \frac{C'_j(0)}{\sqrt{1 - \frac{V_D}{\phi_0}}} \tag{1-29}$$

where $C'_j(0)$ is the capacitance per unit area at equilibrium (i.e., at $V_D = 0$), and the *diffusion capacitance per unit area* C'_d is

$$C'_d = \frac{dQ'_d}{dV_D} = \frac{q}{nkT}\tau_D I_S e^{qV_D/nkT} \tag{1-30}$$

Equation (1-27) can be suitably written in terms of the zero-bias capacitance $C'_j(0)$, i.e.,

$$Q'_j = 2\phi_0 C'_j(0) \sqrt{1 - \frac{V_D}{\phi_0}} \tag{1-31}$$

We can now define a *total charge* Q'_D,

$$Q'_D = Q'_j + Q'_d \tag{1-32}$$

or equivalently its capacitance C'_D,

$$C'_D = \frac{dQ'_D}{dV_D} = \frac{d(Q'_j + Q'_d)}{dV_D} \tag{1-33}$$

as the charge-storage element of the diode. Thus the diode *large-signal* model can be represented by the circuit of Fig. 1-15.

Two important observations can be made. For reverse bias and small forward bias, Q'_j is the dominant charge storage. For moderate

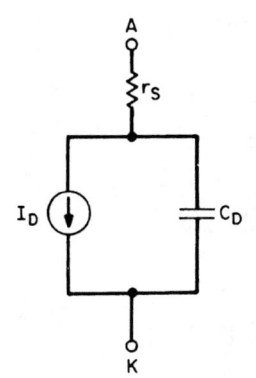

Figure 1-15 SPICE2 diode large-signal model.

forward bias and beyond, the injected charge Q'_d dominates; hence the effects of Q'_j become negligible.

This second observation is extremely important, since as V_D approaches ϕ_0, the junction capacitance [see Eq. (1-29)] becomes infinite whereas charge Q'_j becomes zero [see Eq. (1-27)]. Of course, in this case, Eq. (1-29) is no longer valid.

A more exact analysis [10] of the behavior of C'_j as a function of V_D gives the result shown in Fig. 1-16.

For forward-bias voltages up to about $\phi_0/2$, the values of C'_j predicted by Eq. (1-29) are very close to the more accurate value. Some computer programs approximate C'_j for $V_D > \phi_0/2$ by a linear extrapolation of Eq. (1-29) [11].

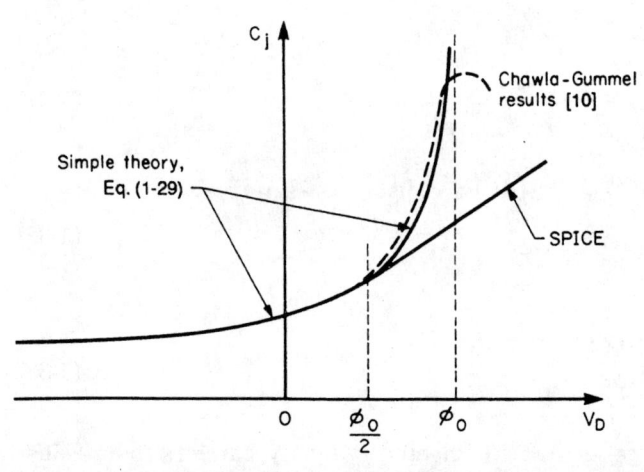

Figure 1-16 Behavior of diode junction capacitance C_j as a function of the bias voltage V_D. (*From P. R. Gray and R. G. Meyer, Analysis and Design of Analog Integrated Circuits,* copyright © 1977 by John Wiley & Sons, Inc. Used by permission.)

1.3.2 Large-signal model of the diode and its implementation in SPICE2

Having established the basic relationships of charge-storage effects, we can now discuss in a straightforward way the details of the large-signal model implemented in SPICE2.

Using the total charges and capacitances, instead of their definitions per unit area, the charge-storage element $Q_D = Q_j + Q_d$ in SPICE2 is determined by the following relations:

$$Q_D = \begin{cases} \underbrace{\tau_D I_D}_{Q_d} + \underbrace{C_j(0) \int_0^{V_D} \left(1 - \frac{V}{\phi_0}\right)^{-m} dV}_{Q_j} & \text{for } V_D < \text{FC} \times \phi_0 \\ \underbrace{\tau_D I_D}_{Q_d} + \underbrace{C_j(0) F_1 + \frac{C_j(0)}{F_2} \int_{\text{FC} \times \phi_0}^{V_D} \left(F_3 + \frac{mV}{\phi_0}\right) dV}_{Q_j} & \text{for } V_D \geq \text{FC} \times \phi_0 \end{cases} \quad (1\text{-}34)$$

These elements can be defined equivalently by the capacitance relations

$$C_D = \frac{dQ_D}{dV_D} = \begin{cases} \underbrace{\tau_D \frac{dI_D}{dV_D}}_{C_d} + \underbrace{C_j(0) \left(1 - \frac{V_D}{\phi_0}\right)^{-m}}_{C_j} & \text{for } V_D < \text{FC} \times \phi_0 \\ \underbrace{\tau_D \frac{dI_D}{dV_D}}_{C_d} + \underbrace{\frac{C_j(0)}{F_2} \left(F_3 + \frac{mV_D}{\phi_0}\right)}_{C_j} & \text{for } V_D \geq \text{FC} \times \phi_0 \end{cases} \quad (1\text{-}35)$$

where F_1, F_2, and F_3 are SPICE2 constants whose values are

$$F_1 = \frac{\phi_0}{1-m}[1-(1-\text{FC})^{1-m}]$$

$$F_2 = (1-\text{FC})^{1+m} \quad (1\text{-}36)$$

$$F_3 = 1 - \text{FC}(1+m)$$

The stored charge Q_d due to injected minority carriers is modeled by I_S, the saturation current; n, the emission coefficient ($n = 1$ for the ideal diode); and τ_D, the transit time, as follows from Eqs. (1-34) and (1-21).

It is important to note that Q_d, is proportional to I_D so that it is modeled by a highly nonlinear capacitor.

The stored charge Q_j is modeled by the parameter $C_j(0)$, the diode junction capacitance at zero bias ($V_D = 0$); m, the junction grading coefficient; ϕ_0, the junction potential (built-in voltage); and FC, the forward-bias junction capacitance coefficient.

Typically, ϕ_0 may range from 0.2 to 1 V, and m is 0.33 for a linearly graded junction or 0.5 for an abrupt junction.

FC is a factor between 0 and 1, which determines how the junction capacitance is calculated when the junction is forward-biased (by default, FC = 0.5); thus, when $V_D \geq FC \times \phi_0$, SPICE2 approximates C_j [see Eq. (1-29)] by a linear extrapolation (see Fig. 1-16).

The SPICE2 *large-signal* model for the diode is also shown in Fig. 1-15, when Eqs. (1-35) are taken into account.

To summarize, the new model parameters needed to describe the capacitance C_D (or the *large-signal model*) and that the user of SPICE2 has to specify in the .MODEL statement [7] are

TT	Transit time (τ_D)
CJO	Zero-bias junction capacitance [$C_j(0)$]
M	Grading coefficient (m)
VJ	Junction potential (ϕ_0)
FC	Coefficient for forward-bias depletion capacitance formula (FC)

1.4 Small-Signal Model

The models of the diode discussed so far maintain the strictly nonlinear nature of the physics of the device, and therefore are able to represent, both statically and dynamically, the behavior of the diode for both types of bias V_D. In some circuit situations, however, the characteristics of the device must be represented only in a restricted range of currents and voltages.

In particular, for small variations around the operating point (op) which is fixed by a constant source, the nonlinear characteristic of the diode can be *linearized* so that the incremental current of the diode becomes proportional to the incremental voltage, if the variations are sufficiently small.

The development of an incremental linear model is not necessarily linked to a graphic interpretation of the relationships between the variables; the mathematical development can be based on the linearization of a nonlinear functional relationship, by means of the expansion in *Taylor series* truncated after the first-order term.

1.4.1 Small-signal model of the diode and its implementation in SPICE2

The *small-signal* model can be found by considering small-voltage variations of V_D around a given dc voltage V_D^* and writing $V_D = V_D^* + \Delta V_D$, as indicated in Fig. 1-17.

The interest is in the corresponding small change of current, which is proportional to the change in voltage if the voltage change is small enough. This linear relationship can be used to define a *small-signal conductance* $g_D = dI_D/dV_D$.

By using Eq. (1-21), which may be expanded in a Taylor series about V_D^*, we have

$$I_D = I_S(e^{qV_D^*/nkT} - 1) + I_S e^{qV_D^*/nkT}\left[\frac{q\,\Delta V_D}{nkT} + \frac{1}{2!}\left(\frac{q\,\Delta V_D}{nkT}\right)^2 + \cdots\right] \quad (1\text{-}37)$$

The first term on the right-hand side of Eq. (1-37) is the operating-point current I_D^*; the second term is the small-signal component of the current, which is approximately linearly related to V_D if the second term in the brackets is much less than the first one. Conse-

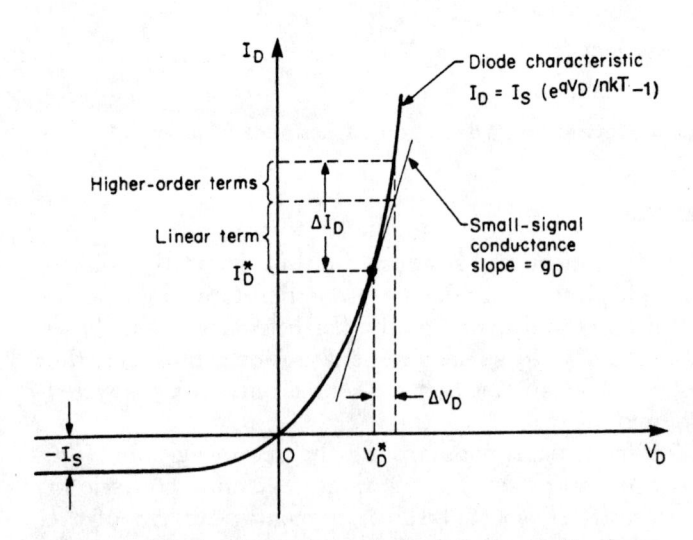

Figure 1-17 Relationship between diode static I_D vs. V_D characteristic and the small-signal conductance. (*From Semiconductor Electronics Education Committee [3]. Copyright © 1964, Education Development Center, Inc., Newton, Mass. Used by permission.*)

quently, the *small-signal conductance* is

$$g_D = \left.\frac{dI_D}{dV_D}\right|_{op} = \frac{qI_S}{nkT}e^{qV_D^*/nkT} \qquad (1\text{-}38)$$

Figure 1-17 shows that g_D is equal to the slope of the static characteristic at the operating point. With moderate reverse bias, the current I_D is negative and approaches the saturation current, so that the conductance g_D is zero. On the other hand, for moderate forward bias, the current I_D is much larger than the saturation current, and the conductance is approximately proportional to I_D.

Further, the charge Q_D stored in the diode changes an amount dQ_D for a small voltage dV_D; so if $Q_D = Q_j + Q_d$, it can be written

$$\begin{aligned} C_j &= \left.\frac{dQ_j}{dV_D}\right|_{op} \\ C_d &= \left.\frac{dQ_d}{dV_D}\right|_{op} \end{aligned} \qquad (1\text{-}39)$$

From Eqs. (1-34) and (1-38), it follows that

$$C_D = \left.\frac{dQ_D}{dV_D}\right|_{op} = \begin{cases} \underbrace{\tau_D g_D}_{C_d} + \underbrace{C_j(0)\left(1 - \frac{V_D}{\phi_0}\right)^{-m}}_{C_j} & \text{for } V_D < FC \times \phi_0 \\ \underbrace{\tau_D g_D}_{C_d} + \underbrace{\frac{C_j(0)}{F_2}\left(F_3 + \frac{mV_D}{\phi_0}\right)}_{C_j} & \text{for } V_D \geq FC \times \phi_0 \end{cases} \qquad (1\text{-}40)$$

where F_2 and F_3 are expressed by Eqs. (1-36).

The SPICE2 *small-signal* linearized model for the diode is shown in Fig. 1-18.

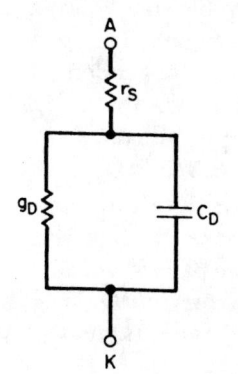

Figure 1-18 SPICE2 diode small-signal model.

Both g_D and C_D are evaluated at the dc operating point which is computed in SPICE2 before the ac analysis is initiated, by performing a dc analysis with all capacitors open-circuited and all inductors short-circuited [7].

It can be seen that no new model parameters are needed to perform an ac analysis (*small-signal model*) using SPICE2. In fact, it is sufficient that the user of SPICE2 specifies the static and large-signal model parameters; then, their linearization (small-signal model) is automatically performed inside SPICE2. This consideration is valid also for all the other active devices implemented in SPICE2.

1.5 Schottky Diode and Its Implementation in SPICE2

A rectifying barrier is usually formed when a metal is in contact with a semiconductor (e.g., n-type substrate). These structures, described in Appendix C, are electrically similar to abrupt one-sided p^+n junctions, but several important differences make these diodes, called *Schottky Barrier Diodes (SBDs)*, attractive and useful.

First, the SBD operates as a majority-carrier device under low-level injection conditions. As a consequence, storage time (due to storage of minority carriers) is eliminated and a fast response is obtained.

Second, for a fixed voltage bias, the current through the device is typically more than two orders of magnitude greater than it is for the junction device, as shown in Fig. 1-19.

A detailed understanding of the Schottky diode can be found in [6]. In this case, it is sufficient to say that the SBD current I_D (*static model*) is given by

$$I_D = I_S(e^{qV_D/nkT} - 1) \qquad (1\text{-}41)$$

where

$$I_S = A^* T^2 e^{-q\phi_B/kT} \qquad (1\text{-}42)$$

in which A^* is a constant and ϕ_B is the Schottky barrier voltage, depending only on the material (metal–semiconductor). The constant A^*, called the *effective Richardson constant*, is independent of temperature but it depends on the particular semiconductor under consideration and on the crystal orientation. For n-type silicon, A^* is approximately $250\,\text{A}\cdot\text{K}^{-2}$, while for n-type GaAs is $8.1\,\text{A}\cdot\text{K}^{-2}$.

Equation (1-41) is similar to Shockley's equation for the *pn* junction [see Eq. (1-6)]. However, Eq. (1-42), representing the saturation current, is quite different from Eqs. (1-2) and is a function of the barrier voltage.

Reverse-bias breakdown in Schottky diodes formed on *lightly doped* substrates occurs through avalanching in high-field regions, just as

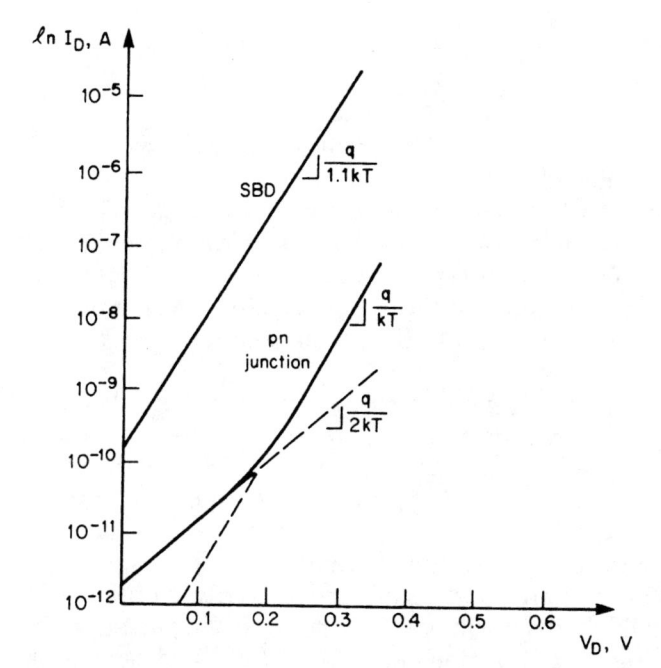

Figure 1-19 Current-voltage characteristic for Schottky barrier diode (SBD) and junction diode (pn).

in a pn-junction diode (see Sec. 1.1.2d). However, Schottky diodes tend to present lower breakdown voltage than pn-junction diodes [6]. For Schottky diodes formed on more *heavily doped* substrates, breakdown may occur through tunneling. Again, this situation is quite comparable to that of Zener breakdown in a pn-junction diode (see Sec. 1.1.2d).

As far as the *large-signal model* is concerned, since there is very little injection, and then storage, of minority carriers into the semiconductor of a SBD, the minority-carrier charge-storage effects are dominated almost entirely by the junction capacitance given by Eq. (A-64) and here rewritten for convenience

$$C'_j = \frac{\varepsilon_s}{W} \qquad (1\text{-}43)$$

where W is the depletion region width, given by

$$W = \sqrt{\frac{2\varepsilon_s}{qN_D}(\phi_{s0} - V_D)} \qquad (1\text{-}44)$$

In Eq. (1-44), the potential ϕ_{s0} expresses the equilibrium difference in energy for electrons in the metal compared to electrons in the semiconductor, i.e.,

$$\phi_{s0} = \phi_B - \phi_n \tag{1-45}$$

where ϕ_n is given by Eq. (A-29a). The potential ϕ_{s0} is analogous to ϕ_0 for the pn-junction diode.

Thus, SBDs are usually referred to as majority-carrier devices, that is, the forward-bias current in these devices is characterized by electron emission from the n-type semiconductor into the metal, rather than by hole injection into the semiconductor. Then, minority-carrier charge storage effects in SBDs can generally be ignored [$\tau_D = 0$ in Eq. (1-28)].

Moreover, also the *small-signal model* of the Schottky diode includes only the junction capacitance C_j in parallel with a resistor defining the differential conductance g_D.

A model for the Schottky diode is not implemented explicitly in SPICE2; however, it can be simulated using the pn-junction diode model present in SPICE2 by setting the minority-carrier storage time TT to zero, specifying a value of the saturation current IS obtained by Eq. (1-42), setting the built-in voltage VJ to ϕ_{s0}, choosing an appropriate value for the zero-bias junction capacitance CJO, and setting the emission coefficient N to 1 and the grading coefficient M to 0.5.

The previous pn-junction diode circuit models of Figs. 1-9, 1-15, and 1-18 (static, large-signal, and small-signal, respectively) apply just as well to the Schottky diode provided that the above assumptions are taken into account.

1.6 Temperature and Area Effects on the Diode Model Parameters

This section analyzes the effects that temperature and area factor have on the diode model implemented in SPICE2. However, in order to more easily understand the relations introduced, there follows a brief description of the physical quantities that have marked dependence on temperature.

1.6.1 Temperature dependence of the diode characteristic

In many applications of junction diodes, the temperature dependence of the diode characteristic is important. Although the diode equation contains the absolute temperature explicitly in the exponent qV_D/kT, the principal temperature dependence of the characteristic results from the extremely strong implicit temperature dependence of the saturation currents I_S [Eqs. (1-2)].

The factor in parentheses of Eqs. (1-2) is not strongly temperature-dependent. Therefore, the principal temperature dependence of the saturation current is that of n_i^2, which is

$$n_i^2(T) = BT^3 \, e^{-E_g(T)/kT} \qquad (1\text{-}46)$$

where B is a constant and E_g is also a function of temperature, according to the general relation [5]

$$E_g(T) = E_g(0) - \frac{\alpha T^2}{\beta + T} \qquad (1\text{-}47)$$

For Si experimental results give $\alpha = 7.02 \times 10^{-4}$, $\beta = 1108$, and $E_g(0) = 1.16$ eV.

Last, it is necessary to emphasize that even the built-in voltage ϕ_0 [see Eqs. (A-29)] is dependent on temperature both explicitly and through n_i [see Eq. (1-46)].

1.6.2 Temperature dependence of SPICE2 diode model parameters

All input data for SPICE2 .MODEL statement are assumed to have been measured at 27°C (300 K). The simulation also assumes a *nominal temperature* (*TNOM*) of 27°C that can be changed with the TNOM option of the .OPTIONS control statement.

The circuits can be simulated at temperatures other than TNOM by using a .TEMP control statement. Temperature appears explicitly in the exponential term of the diode model equation.

In addition, four diode parameters are modified in SPICE2 to reflect changes in the temperature of pn and SBD diodes. They are I_S, the saturation current; ϕ_0, the junction potential; $C_D(0)$, the zero-bias junction capacitance; and FC, the coefficient for forward-bias depletion capacitance formula.

The temperature dependence of I_S in the SPICE2 diode model is determined by

$$I_S(T_2) = I_S(T_1) \left(\frac{T_2}{T_1}\right)^{\text{XTI}/n} \exp\left[-\frac{qE_g(300)}{nkT_2}\left(1 - \frac{T_2}{T_1}\right)\right] \qquad (1\text{-}48)$$

where two new model parameters are introduced: XTI, the saturation current temperature exponent, and E_g, the energy gap.

It is important to note that these parameters assume different values in the simulation of *pn* diodes or SBDs; that is, XTI = 3 for a *pn* diode and XTI = 2 for an SBD; $E_g(300\text{ K}) = 1.11$ eV for a *pn* diode (Si) and $E_g(300\text{ K}) = 0.69$ for an SBD.

The effect of temperature on ϕ_0 is modeled by

$$\phi_0(T_2) = \frac{T_2}{T_1}\phi_0(T_1) - 2\frac{kT_2}{q}\ln\left(\frac{T_2}{T_1}\right)^{1.5} - \left[\frac{T_2}{T_1}E_g(T_1) - E_g(T_2)\right] \quad (1\text{-}49)$$

where $E_g(T_1)$ and $E_g(T_2)$ are described by the general Eq. (1-47).

The temperature dependence of $C_j(0)$ is determined by

$$C_j(T_2) = C_j(T_1)\left\{1 + m\left[400 \times 10^{-6}(T_2 - T_1) - \frac{\phi_0(T_2) - \phi_0(T_1)}{\phi_0(T_1)}\right]\right\} \quad (1\text{-}50)$$

The temperature dependence of the coefficient for forward-bias depletion capacitance formula is determined by

$$\text{FCPB}(T_2) = \text{FC} \times \phi_0 = \text{FCPB}(T_1)\frac{\phi_0(T_2)}{\phi_0(T_1)} \quad (1\text{-}51)$$

$$F_1(T_2) = F_1(T_1)\frac{\phi_0(T_2)}{\phi_0(T_1)}$$

In the preceding equations, T_1 and T_2 must be considered as follows:

- If TNOM in the .OPTIONS statement is not specified, then T_1 = TNOM = TREF = 27°C (300 K); this is the default value for TNOM.
- If a TNOM$_{\text{new}}$ value is specified, then T_1 = TNOM = 27°C; T_2 = TNOM$_{\text{new}}$.

These two cases are valid only if one temperature is requested.

If more than one temperature is requested (using the .TEMP statement), then T_1 = TNOM = 27°C and T_2 is the first temperature specified in the .TEMP statement. Afterward, T_1 is the last working temperature (T_2) and T_2 is the next temperature in the .TEMP statement to be calculated.

It is very important to note that the temperature T is regarded as an operating condition and *not* as a model parameter, so that it is not possible to simulate the behavior of devices in the same circuit at different temperatures at the same time. In other words, the *same* temperature T is used in the *whole* circuit to be simulated.

1.6.3 Area dependence of the SPICE2 diode model parameters

The AREA factor used in SPICE2 for the diode model determines the number of equivalent parallel devices of a specified model. The diode model parameters affected by the AREA factor are I_S, r_S, and $C_D(0)$; that is,

$$I_S = I_S \times \text{AREA}$$

$$C_j(0) = C_j(0) \times \text{AREA} \qquad (1\text{-}52)$$

$$r_S = r_S / \text{AREA}$$

The default value for the AREA parameter is 1.

1.7 SPICE3 Models

SPICE3 [12, 13] diode models are based directly on SPICE2 diode models. Thus the *static model* is determined by the parameters IS, N, and RS. Reverse breakdown is modeled by an exponential increase in the reverse diode current and is determined by the parameters BV and IBV (both of which are positive numbers). Charge storage effects (*large-signal model*) are modeled by the transit time TT, and the nonlinear depletion region capacitance which is determined by the parameters CJO, VJ, FC, and M. The *small-signal model* is automatically computed inside the program by the linearization of the static and large-signal model equations (see Sec. 1.4). The *temperature dependence* of the saturation current IS is defined by the parameters EG and XTI, and the temperature updating relations for the other parameters follow the ones described in Sec. 1.6.2. The values specified for the parameters are assumed to have been measured at the temperature TNOM, which can be specified in the .OPTIONS statement. This value of temperature can be overridden by the specification of the new (with respect to SPICE2) .MODEL statement parameter TNOM (in units of °C) or of the device element statement option TEMP. The *noise* and *distortion models* are described in Secs. 7.1.3a and 7.2.2c, respectively. The parameters affected by the AREA factor follow the same relations as indicated in Sec. 1.6.3. The equivalent circuits for the static, large-signal, small-signal, noise, and distortion models are shown in Figs. 1-9, 1-15, 1-19, 7-3, and 7-12, respectively.

1.8 HSPICE Models

HSPICE [14] diode models include the *junction* diode model and the *Fowler-Nordheim* model.

The *junction diode* model has two variations: the *geometric* and *nongeometric*. The *geometric* junction diode is used to model IC-based silicon diffused diodes, Schottky barrier diodes, and Zener diodes. The geometric model allows for *pn* junction poly and metal capacitance dimensions to be specified for a particular IC process and technology. The *nongeometric* junction diode is used to model discrete diode devices such as standard and Zener diodes. The nongeometric model allows for the scaling of currents, resistances, and capacitances by the use of dimensionless area parameters [14].

The *Fowler-Nordheim diode* defines tunneling current flow through insulators (see Sec. 1.8.6). It is used for the modeling of diode effects in nonvolatile EEPROM memory [14].

The LEVEL parameter in the .MODEL statement selects the type of diode model used. In particular the user of HSPICE has to set LEVEL = 1 for the nongeometric junction diode model, LEVEL = 2 for the Fowler-Nordheim diode model, and LEVEL = 3 for the geometric junction diode model. If the LEVEL parameter is omitted, the model defaults to LEVEL = 1. Element parameters [14], if repeated in the .MODEL statement as model parameters, take precedence over the model parameters. Parameters common to both element (D) and .MODEL statements are: AREA, PJ, M, LM, LP, WM, WP, W, and L (see Sec. 1.8.5).

In the following, HSPICE diode static, large-signal, small-signal, and thermal models are taken into account. The noise model is considered in Chap. 7.

1.8.1 Junction diode static model (LEVEL = 1,3)

As in SPICE2, the HSPICE diode *static model* equations are split in three regions (see Fig. 1-11): forward-bias, reverse-bias, and breakdown regions.

The ideal exponential equation obtained in the analysis of the diode model is adapted to define the HSPICE static model in both the forward and reverse regions. Thus, the equations for the forward and reverse bias are, respectively [14]

$$I_D = \begin{cases} \dfrac{I_D^*}{1+\left(\dfrac{I_D^*}{I_{KF}}\right)^{1/2}} & \text{for } V_D \geq -10\dfrac{nkT}{q} \\[2ex] \dfrac{I_D^*}{1+\left(\dfrac{I_D^*}{I_{KR}}\right)^{1/2}} & \text{for } V_D < -10\dfrac{nkT}{q} \end{cases} \quad (1\text{-}53)$$

where I_D^* is defined as [14]

$$I_D^* = \begin{cases} I_S(e^{qV_D/nkT} - 1) & \text{for } V_D \geq -10\dfrac{nkT}{q} \\ I_S(e^{qV_D/nkT} - 1) - I_S[e^{-q(\text{BV} + V_D)/nkT} - 1] & \text{for } V_D < -10\dfrac{nkT}{q} \end{cases} \tag{1-54}$$

The second of the Eqs. (1-54), can be better understood if the two regions of the reverse bias characteristic [regions (b) and (d) in Fig. 1-11] are considered. Thus

$$I_D^* = \begin{cases} -I_S & \text{for } -\text{BV} < V_D < -10\dfrac{nkT}{q} \\ -I_S e^{-q(\text{BV} + V_D)/nkT} & \text{for } V_D < -\text{BV} \end{cases} \tag{1-55}$$

The reverse breakdown voltage BV specified in the .MODEL statement is adjusted as follows [14].
If we consider

$$I_{\text{break}} = -I_S(e^{-q\text{BV}/nkT} - 1) \tag{1-56}$$

and if the .MODEL parameter IBV, which defines the current at breakdown voltage, is greater than I_{break}, then

$$\text{BV}_{\text{adjusted}} = \text{BV} - \frac{nkT}{q}\ln\left(\frac{\text{IBV}}{I_{\text{break}}}\right) \tag{1-57}$$

otherwise

$$\text{IBV} = I_{\text{break}} \tag{1-58}$$

From the above relations, it can be seen that HSPICE introduces (in comparison with SPICE2) two new parameters for the .MODEL statement: IKF (I_{KF}), the forward knee current and IKR (I_{KR}), the reverse knee current, which model the forward and reverse high-level injection effects (see Secs. 1.1.2b and 1.2.2b), respectively. The other parameters [IS (I_S), N (n), BV, IBV, RS (r_S)] maintain their meaning and keywords encountered in SPICE2 (see Table 1-1). The keyname used in the text is indicated in parentheses. If IKF and IKR are not specified (default values), then $I_D = I_D^*$ and its expression defaults to SPICE2 relations [see Eqs. (1-21)]. The parameter RS should be also set to simulate the resistance of the neutral regions (see Secs. 1.1.2c and 1.2.2b). The choice of the thermal voltage value that determines the limit between the forward- and reverse-bias regions is different in

HSPICE and SPICE2: in HSPICE this limit is set to $-10nkT/q$, while in SPICE2 it was $-5nkT/q$.

The equivalent circuit of the HSPICE diode static model is shown in Fig. 1-9, when the above equations are considered for the nonlinear current source I_D.

1.8.2 Junction diode large-signal model (LEVEL = 1,3)

The HSPICE diode *large-signal model* is defined as in SPICE2 by the capacitor C_D as shown in Fig. 1-15. However, the HSPICE capacitance expression for C_D is different from the analogous one of SPICE2 [see Eq. (1-35)]. In HSPICE the capacitance C_D is a combination of the diffusion capacitance C_d, junction capacitance C_j, metal capacitance C_m, and poly capacitance C_p, i.e., [14]

$$C_D = C_d + C_j + C_m + C_p \tag{1-59}$$

The *diffusion capacitance* C_d is modeled by the same relations as in SPICE2 [see Eq. (1-30), and the first term in each of the Eqs. (1-35)]. Thus, C_d is modeled by the transit time parameter TT (see Table 1-1) when it is specified in the .MODEL statement.

The *junction capacitance* C_j is split into two components: the junction bottom area capacitance C_{ja} and the junction periphery capacitance C_{jp}. The formula for both the capacitances is similar, except that each has its own model parameters. The *junction bottom area capacitance* C_{ja} is equivalent to the C_j capacitance in SPICE2 [see Eq. (1-29) and the second term in each of the Eqs. (1-35)]. Thus, C_{ja} is described by the parameters CJO (CJ), VJ (PB), M (MJ), and FC that the user of HSPICE can specify in the .MODEL statement. For the meaning of each parameter, refer to Table 1-1, while in parentheses are shown other equivalent keywords which can be used in HSPICE. The *junction periphery capacitance* C_{jp} is given by [14]

$$C_{jp} = \begin{cases} C_{jp}(0)\left(1 - \dfrac{V_D}{\phi_{0p}}\right)^{-m_{jsw}} & \text{for } V_D < \text{FCS} \times \phi_{0p} \\ \dfrac{C_{jp}(0)}{F_2}\left(F_3 + \dfrac{m_{jsw} V_D}{\phi_{0p}}\right) & \text{for } V_D \geq \text{FCS} \times \phi_{0p} \end{cases} \tag{1-60}$$

Equation (1-60) contains four new parameters to be specified in the .MODEL statement of HSPICE: CJP (C_{jp}), the zero-bias periphery capacitance in [F]; PHP (ϕ_{0p}), the periphery junction contact potential in [V]; MJSW (m_{jsw}), the periphery junction grading coefficient; FCS, the coefficient for the forward-bias depletion periphery capacitance formula. The constants F_2 and F_3 follow Eqs. (1-36), when

m and FC are substituted by m_{jsw} and FCS, respectively. For the above HSPICE parameters, the keyname used in the text is indicated in parentheses. The total junction capacitance is then

$$C_j = C_{ja} + C_{jp} \qquad (1\text{-}61)$$

The *poly capacitance* C_p and the *metal capacitance* C_m are determined by the following equations [14].

$$C_p = \frac{\varepsilon_{ox}}{X_{P0}}(W_P + X_P)(L_P + X_P)M \qquad (1\text{-}62)$$

and

$$C_m = \frac{\varepsilon_{ox}}{X_{M0}}(W_M + X_M)(L_M + X_M)M \qquad (1\text{-}63)$$

where ε_{ox} is the oxide dielectric constant. Equations (1-62) and (1-63), valid only for LEVEL = 3, contain parameters (not present in SPICE2) that the user of HSPICE has to specify in the .MODEL statement. These parameters are: XOI (X_{P0}), the thickness of the polysilicon to bulk oxide in [Å]; WP (W_P), the width of polysilicon in [m]; XP (X_P), which accounts for masking and etching effects in [m]; LP (L_P), the length of polysilicon in [m]; XOM (X_{M0}), the thickness of the metal to bulk oxide in [Å]; WM (W_M), the width of metal in [m], XM (X_M), which accounts for masking and etching effects, in [m]; LM (L_M), the length of metal in [m]; M, the multiplier factor to simulate multiple diodes. In parentheses the keyname used in the text is indicated.

1.8.3 Junction diode small-signal model (LEVEL = 1,3)

As in SPICE2, the HSPICE small-signal model is automatically computed inside the program by the linearization of the static and large-signal model equations (see Sec. 1.4.1); thus no new model parameters are needed to be specified in the .MODEL statement by the user of HSPICE. The small-signal model equivalent circuit is shown in Fig. 1-18.

1.8.4 Temperature effects on the diode model parameters (LEVEL = 1,3)

Both LEVEL1 and LEVEL3 diode models contain the parameters TLEV and TLEVC to select different temperature equations for the calculation of temperature effects on energy gap, saturation current, breakdown voltage, transit time, contact potential, junction capacitance, grading coefficient, and ohmic series resistance. In the following, we consider only the equations selected with TLEV and TLEVC set to their default values (zero). Alternative equations for some of the temperature dependent parameters can be found in [14].

As in SPICE2, the temperature dependence of the *energy gap* E_g (EG) in HSPICE follows Eq. (1-47), while the temperature dependence of the *saturation current* I_S (IS) is determined by Eq. (1-48).

The *breakdown voltage* BV depends on temperature through the following relation [14].

$$\text{BV}(T_2) = \text{BV}(T_1) - \text{TCV}(T_2 - T_1) \qquad (1\text{-}64)$$

where TCV, the breakdown voltage temperature coefficient, is a parameter to be specified in the .MODEL statement of HSPICE.

The temperature dependence of the transit time τ_D (TT) is determined by [14]

$$\text{TT}(T_2) = \text{TT}(T_1)[1 + \text{TTT1}(T_2 - T_1) + \text{TTT2}(T_2 - T_1)^2] \qquad (1\text{-}65)$$

where TTT1 and TTT2 are the first- and second-order temperature coefficients, respectively, which the user of HSPICE has to specify in the .MODEL statement.

As in SPICE2, the *area junction contact potential* ϕ_0 (PB) in HSPICE depends on temperature through Eq. (1-49). An analogous relation holds for the HSPICE parameter ϕ_{0p} (PHP), *the periphery junction contact potential*.

The temperature dependence of the *zero-bias junction capacitance per unit junction bottomwall area* $C_j(0)$ (CJCO) follows Eq. (1-50), while the *zero-bias junction capacitance per unit junction periphery* C_{jsw} (CJSW) is determined by [14]

$$C_{jsw}(T_2) = C_{jsw}(T_1)\left\{1 + m_{jsw}\left[400 \times 10^{-6}(T_2 - T_1) - \frac{\phi_{0p}(T_2) - \phi_{0p}(T_1)}{\phi_{0p}(T_1)}\right]\right\}$$

$$(1\text{-}66)$$

The temperature dependence of the *grading coefficient m* (MJ) is determined by [14]

$$m(T_2) = m(T_1)[1 + \text{TM1}(T_2 - T_1) + \text{TM2}(T_2 - T_1)^2] \qquad (1\text{-}67)$$

where TM1 and TM2 are the first- and second-order temperature coefficient, respectively, which the user of HSPICE has to specify in the .MODEL statement.

The ohmic *series resistance* r_S (RS) depends on temperature through the following expression [14]

$$RS(T_2) = RS(T_1)[1 + TRS(T_2 - T_1)] \quad (1\text{-}68)$$

where TRS is the resistance temperature coefficient measured in [°C^{-1}], a parameter that has to be specified in the .MODEL statement of HSPICE. The HSPICE keyword is indicated in parentheses. The meaning of T_1 and T_2 is explained in Sec. 1.6.2, provided that the user of HSPICE keeps in mind what it follows. The temperature of a circuit for an HSPICE run can be specified with either the .TEMP statement or the TEMP parameter in either the .DC, .AC, or .TRAN statements [14]. The circuit simulation temperature set by either of these is compared against the reference temperature set by the TNOM option. TNOM defaults to 25°C, unless the option SPICE2 is used, where it defaults to 27°C [7]. The derating of component values and model parameters is calculated by using the difference of the circuit simulation temperature and the reference temperature, TNOM.

Since elements and models within a circuit can often be operating at different temperatures, an element temperature parameter, DTEMP, and a model reference parameter, TREF, can be used. Specifying DTEMP in an element statement causes the element temperature for the simulation to be

Element temperature = Circuit temperature + DTEMP

The DTEMP value can be specified in the diode element statement; the DTEMP value defaults to zero.

By specifying TREF in the model statement, the model reference temperature is changed (TREF overrides TNOM). The derating of the model parameters is based on the difference of the circuit simulators, temperature, and TREF, instead of TNOM.

The possibility of using a parameter such as DTEMP is a remarkable improvement introduced in the HPSPICE program.

1.8.5 Scaling, area, and periphery dependence of the diode model parameters (LEVEL = 1,3)

Scaling, which allows selected model and element parameters to be scaled, is controlled through the SCALE and SCALM options [14] in the .OPTIONS statement. SCALE = x is used to scale element (keyword D, see Ref. [14]) parameters. Default is $x = 1$; element parameters are then entered with units of meters. If compatibility with SPICE2 is needed, set $x = 10^{-2}$ (centimeter scale). SCALM = x is used to scale .MODEL parameters; the value x follows the same "rule" as in SCALE. In addition, the scaling factor is determined by the units of the parameter: e.g., if the units of a .MODEL parameter are meters square, then the parameter is multiplied by SCALM; if the units are 1/meter, the parameter is divided by SCALM. The area and periphery dependence of the parameters is, on the other hand, affected by AREA, PJ, and M factors.

a. LEVEL = 1 Model. Scaling for LEVEL = 1 involves the use of the junction area (AREA), junction periphery (PJ), and the dimensionless multiplier factor (M) to simulate multiple diodes. For AREA and M, the default is 1; for PJ, the default is 0. Both the AREA and PJ parameters are measured in dimensionless values. As these parameters can be specified in both element and .MODEL statements, the user of HSPICE must keep in mind that the parameter in the element statement overrides the .MODEL one. The periphery junction parameter, PJ, is multiplied by M to obtain its effective scaled value PJ_{eff}; i.e.

$$PJ_{eff} = PJ \cdot M \qquad (1\text{-}69)$$

As a consequence, the zero-bias junction capacitance CJP and the sidewall saturation current ISW, are then multiplied by PJ_{eff} to obtain their effective scaled values.

The parameters AREA and M are used to obtain the effective area value, $AREA_{eff}$, through the relation

$$AREA_{eff} = AREA \cdot M \qquad (1\text{-}70)$$

As a consequence, the parameters CJO, IKF, IKR, IBV, and IS are then multiplied by $AREA_{eff}$ to obtain their effective scaled values; RS, on the other hand, is divided by $AREA_{eff}$.

b. LEVEL = 3 Model. LEVEL = 3 model scaling is affected by the options and parameters SCALM, SCALE, SHRINK, and M. The LEVEL = 3 element (keyword D, see Ref. [14]) parameters affected by SCALE include: AREA, PJ, LM, LP, WM, WP, W, and L. The .MODEL parameters affected by SCALM include: AREA, PJ, LM, LP, WM, WP, W, L, XP, XW, XM, IS, ISW, IBV, IKF, IKR, RS, CJO, and CJP. For the parameters common to both element and .MODEL statements, it must be noted that the element parameters, if repeated in the .MODEL statement, take precedence over the .MODEL parameters.

If the AREA and PJ parameters are specified in the element statement, then the scaled effective area and periphery junction element parameters are determined by

$$AREA_{eff} = AREA \cdot M \cdot SCALE^2 \cdot SHRINK^2$$
$$PJ_{eff} = PJ \cdot M \cdot SCALE \cdot SHRINK \qquad (1\text{-}71)$$

or, if the W (width of the diode in meters) and L (length of the diode in meters) are specified, by

$$AREA_{eff} = W_{eff} \cdot L_{eff} \cdot M$$
$$PJ_{eff} = 2(W_{eff} + L_{eff}) \cdot M \qquad (1\text{-}72)$$

where

$$W_{eff} = W \cdot SCALE \cdot SHRINK + XW_{eff}$$
$$L_{eff} = L \cdot SCALE \cdot SHRINK + XW_{eff} \qquad (1\text{-}73)$$

The .MODEL parameter SHRINK is a shrink dimensionless factor which defaults to unity; on the other hand, $AREA_{eff}$ and PJ_{eff} are not dimensionless anymore.

If the AREA and PJ .MODEL parameters are specified, and the element ones are not, SCALM as the scaling factor instead of SCALE must be used in the above equations. Obviously, SCALM must be used for all the parameters not included in the element statement [14].

The general rule that all currents and capacitances are multiplied by $AREA_{eff}$ or PJ_{eff} (for periphery parameters only), and that all resistances are divided by $AREA_{eff}$, is always valid, provided that the original units of the parameters are kept. This implies to "merge" SCALE, SCALM, $AREA_{eff}$, and PJ_{eff} on the basis of the relations expressed by Eqs. (1-71) to (1-73), to obtain the correct units for the scaled parameters. Explicit relations for all the parameters affected by scaling can be found in Ref. [14]. Finally, the polysilicon and metal capacitor dimensions are determined by multiplying each by SCALE or SCALM if specified as .MODEL parameters. The SHRINK factor is also used as multiplier except for the parameters (XP, XM) which account for masking and etching effects.

1.8.6 Fowler-Nordheim diode static model

Diode model parameter LEVEL = 2 selects the Fowler-Nordheim (FN) model. Fowler-Nordheim diodes are formed as a metal-insulator-semiconductor or as a semiconductor-insulator-semiconductor layer device. The insulator is sufficiently thin (100 Å) to permit tunneling of carriers. This element models electrically alterable memory cells, air-gap switches, and other insulation breakdown devices [14].

The FN diode *static model* is defined by the following equations in the forward and reverse operation, respectively

$$I_D = \begin{cases} \text{AREA}_{\text{eff}} J_F \left(\dfrac{V_D}{t_{\text{ox}}}\right)^2 e^{-t_{\text{ox}} E_F / V_D} & \text{for } V_D \geq 0 \\ -\text{AREA}_{\text{eff}} J_R \left(\dfrac{V_D}{t_{\text{ox}}}\right)^2 e^{-t_{\text{ox}} E_R / V_D} & \text{for } V_D < 0 \end{cases} \quad (1\text{-}74)$$

where J_F (JF) is the forward FN current coefficient in [A/m^2], J_R (JR) is the reverse FN current coefficient in [A/m^2], E_F (EF) is the forward critical electric field in [V/cm], E_R (ER) is the reverse critical electric field in [V/cm], t_{ox} (TOX) is the thickness of oxide layer in [Å]. The HSPICE keyword for the .MODEL statement is indicated in parentheses. The effective area is determined as the product of the scaled effective length and width [see Eq. (1-72)].

1.8.7 Fowler-Nordheim diode large-signal model

The FN *large-signal model* is defined by a constant capacitance, whose value is given by

$$C_D = \text{AREA}_{\text{eff}} \frac{\varepsilon_{\text{ox}}}{t_{\text{ox}}} \quad (1\text{-}75)$$

1.9 PSPICE Models

PSPICE [15–17] diode models include the standard junction, Schottky, and Zener diodes. PSPICE models are based on SPICE2 diode models, with new features for modeling the recombination and high-level effects. In the following, PSPICE diode *static*, *large-signal*, *small-signal*, and *thermal models* are taken into account. The noise model is considered in Chap. 7.

1.9.1 Static model

It is a set of equations, like the Shockley equation [Eq. (1-1)], that defines the *static model* of the diode or equivalently, the value of the nonlinear current source I_D in the equivalent circuit of Fig. 1-9. Thus, if we define I_{FWD} as the forward current and I_{REV} as the reverse current, we can write [15]

$$I_D = I_{\text{FWD}} - I_{\text{REV}} = (K_{\text{hli}} I_F + K_{\text{gen}} I_R) - I_{\text{REV}} \quad (1\text{-}76)$$

In Eq. (1-76), I_F is the normal current given by

$$I_F = I_S (e^{q V_D / nkT} - 1) \quad (1\text{-}77)$$

K_{hli} is the high-level injection factor given by

$$K_{hli} = \begin{cases} \sqrt{\dfrac{I_{KF}}{I_{KF}+I_F}} & \text{for } I_{KF} > 0 \\ 1 & \text{otherwise} \end{cases} \qquad (1\text{-}78)$$

I_R is the recombination current given by

$$I_R = I_{SR}(e^{qV_D/n_R kT} - 1) \qquad (1\text{-}79)$$

K_{gen} is the generation factor given by

$$K_{gen} = \sqrt{\left[\left(1-\frac{V_D}{\phi_0}\right)^2 + 0.005\right]^m} \qquad (1\text{-}80)$$

and I_{REV} is the reverse current given by

$$I_{REV} = \text{IBV}\, e^{-q(BV+V_D)/kT} \qquad (1\text{-}81)$$

From the above equations, it can be seen that PSPICE introduces, in comparison with SPICE2, three new parameters for the .MODEL statement: IKF (I_{KF}), the high-injection knee current measured in [A]; ISR (I_{SR}), the recombination current parameter measured in [A], and NR (n_R), the emission coefficient for ISR. These parameters model the high-level injection and recombination effects. The other parameters [IS (I_S), N (n), BV, IBV, RS (r_S), VJ (ϕ_0), and M (m)] maintain their meaning and keywords encountered in SPICE2 (see Table 1-1). The keyname used in the text is indicated in parentheses. If IKF and ISR are not specified (default values), then $I_{FWD} = I_F$ and its expression defaults to the SPICE2 relation [see Eqs. (1-21)]. The parameter RS (see Fig. 1-9) should be also set to simulate the resistance of the neutral regions.

1.9.2 Large-signal model

The PSPICE diode *large-signal model* is based directly on the SPICE2 diode large-signal model, and it is defined by the capacitor C_D as shown in Fig. 1-15. Thus, to take into account the charge-storage effects, the user of PSPICE has to specify, in the .MODEL statement, the parameters as described in Sec. 1.3.2.

1.9.3 Small-signal model

As in SPICE2, the PSPICE *small-signal model* is automatically computed inside the program by the linearization of the static and large-signal model equations (see Sec. 1.4). The small-signal model equivalent circuit is shown in Fig. 1-18.

1.9.4 Temperature effects on the diode model parameters

The PSPICE parameters common to SPICE2 maintain their dependence of temperature as explained in Sec. 1.6.2. In addition three new relations have been introduced in PSPICE to take into account the temperature dependence of ISR, IKF, and RS. The recombination current ISR follows directly Eq. (1-48), while for the high-level injection knee current IKF, we have

$$I_{KF}(T_2) = I_{KF}(T_1)[1 + \text{TIKF}(T_2 - T_1)] \qquad (1\text{-}82)$$

and, for the resistance RS of the neutral region, we have

$$r_S(T_2) = r_S(T_1)[1 + \text{TRS1}(T_2 - T_1) + \text{TRS2}(T_2 - T_1)^2] \qquad (1\text{-}83)$$

In Eqs. (1-82) and (1-83), TIKF is the I_{KF} temperature coefficient measured in [°C^{-1}]; TRS1 and TRS2 are the r_S temperature coefficients, measured in [°C^{-1}] and [°C^{-2}], respectively. All the above coefficients are model parameters that the user of PSPICE can specify in the .MODEL statement.

1.9.5 Area dependence of the diode model parameters

The PSPICE diode model parameters affected by the AREA factor are IS, RS, CJO, ISR, IKF, and IBV. The first three parameters are common to SPICE2 and thus have already been modeled; the last three parameters follow the same AREA dependence of IS in Eq. (1-52).

References

1. L. W. Nagel, SPICE2: A Computer Program to Simulate Semiconductor Circuits, Electronics Research Laboratory, Rep. No. ERL-M520, University of California, Berkeley, 1975.
2. R. S. Muller and T. J. Kamins, *Device Electronics for Integrated Circuits*, Wiley, New York, 1977.
3. P. E. Gray, D. DeWitt, A. R. Boothroyd, and J. F. Gibbons, *Physical Electronics and Circuits Models of Transistors*, Vol. 2, Semiconductor Electronics Education Committee, Wiley, New York, 1964.
4. A. S. Grove, *Physics and Technology of Semiconductor Devices*, Wiley, New York, 1967.
5. S. M. Sze, *Physics of Semiconductor Devices*, Wiley, New York, 1969.
6. D. L. Pulfrey and N. G. Tarr, *Introduction to Microelectronic Devices*, Prentice-Hall, Englewood Cliffs, N.J., 1989.

7. A. Vladimirescu, K. Zhang, A. R. Newton, D. O. Pederson, and A. Sangiovanni-Vincentelli, *SPICE Version 2 G User's Guide*, Dept. EECS, University of California, Berkeley, 1981.
8. G. Massobrio, Modulazione della Corrente Inversa di Saturazione del Diodo e del Modello del Reach-Through in SPICE, *Pixel*, **6**, 1985.
9. A. Laha and D. Smart, A Zener Diode Model with Application to SPICE2, *IEEE J. Solid-State Circuits*, **16**, 1981.
10. B. R. Chawla and H. K. Gummel, Transition Region Capacitance of Diffused pn Junctions, *IEEE Trans. Electron Devices*, ED-18, 1971.
11. P. R. Gray and R. G. Meyer, *Analysis and Design of Analog Integrated Circuits*, Wiley, New York, 1977.
12. T. I. Quarles, The SPICE3 Implementation Guide, Electronics Research Laboratory, Rep. No. ERL-M44, University of California, Berkeley, 1989.
13. T. I. Quarles, *SPICE3 Version 3C1 Users Guide*, Electronics Research Laboratory, Rep. No. ERL-M46, University of California, Berkeley, 1989.
14. Meta-Software, Inc., *HSPICE User's Manual*, Campbell, Calif., 1991.
15. MicroSim Corporation, *PSPICE User's Manual*, Irvine, Calif., 1989.
16. P. W. Tuinenga, *SPICE; A Guide to Circuit Simulation and Analysis Using PSPICE*, Prentice-Hall, Englewood Cliffs, N.J., 1988.
17. M. H. Rashid, *SPICE for Circuits and Electronics Using PSPICE*, Prentice-Hall, Englewood Cliffs, N.J., 1990.

Chapter

2

Bipolar Junction Transistor (BJT)

In Chap. 1, it was shown that a *pn* junction under forward bias conducts current because holes are injected from the *p* region and electrons are injected from the *n* region. These are majority carriers in each region, and their supply is not limited near the junction; consequently, current rises rapidly as voltage is increased. Current is much smaller under the reverse-bias condition because it is carried only by minority carriers that are generated either in the depletion region or in other nearby regions. The current flowing by a reverse-biased junction will, however, be increased if the number of minority carriers in the vicinity of the junction is increased. One means of enhancing the minority-carrier concentration in the vicinity of a reverse-biased *pn* junction is simply to locate a forward-biased *pn* junction very close to it. This method is especially advantageous because it brings the minority-carrier concentration under electrical control, that is, under the control of the bias applied to the nearby forward-biased junction. Modulating the current-flow properties of one *pn* junction by changing the bias on a nearby junction is called *bipolar transistor action* [1].

Transistors may be divided into two classes: *bipolar transistors* and *unipolar transistors* (see Chap. 3). In bipolar transistors, both positive and negative free carriers take part in the device operation; hence

---

Note: The material contained in this chapter is based in part upon the book *Modeling the Bipolar Transistor*, by Ian E. Getreu, copyright © Tektronix, Inc., 1976. All rights reserved. Reproduced with permission.

the term *bipolar*. In unipolar transistors, the current is carried only by the free majority carriers in the conducting channel; hence the term *unipolar*.

The Bipolar Junction Transistor (BJT) has the physical configuration of two *pn* junctions back to back with a thin *p*-type or *n*-type region between them. In accordance with the order of the layers of the two junctions, we have two typical structures: *npn* or *pnp* configuration. The middle layer, through which the minority carriers pass, is called the *base* of the BJT; one outer layer in which the minority carriers originate is called the *emitter*; and the other outer layer is called the *collector*.

In our analysis of BJT operation, we will use the one-dimensional model of an *npn* transistor. The operation of a *pnp* device is similar in every respect if the roles of holes and electrons are interchanged and the polarities of the terminal currents and voltages are reversed.

This chapter establishes *static, large-signal, small-signal,* and *thermal* physical-mathematical models of a BJT, and how the models are implemented in SPICE2 [2] and other SPICE2-based programs. The *noise* and *distortion* models are described in Chap. 7.

SPICE2 BJT models are directly regulated by the subroutines *READIN* (it reads and processes the .MODEL statement), *MODCHK* (it performs one-time processing of the model parameters and assigns the parameter default values), *BJT* (it computes all the model equations), and *TMPUPD* (it updates the temperature-dependent parameters of the models), in addition to the subroutines concerning each type of analysis. The noise and distortion models are directly regulated by the subroutines *NOISE* and *DISTO*, respectively (see Chap. 7).

The BJT model analysis follows the SPICE2 classification, i.e., (1) the *Ebers-Moll* model [3] and (2) the *Gummel-Poon* model [4], even if item (1) contains something more than the original Ebers-Moll model and item (2) contains something different from the integral charge-control on which the Gummel-Poon model is based.

2.1 Transistor Conventions and Symbols

The circuit symbols for *npn* and *pnp* BJTs are shown in Fig. 2-1. The emitter shows an arrow indicating the actual current direction under normal operation in which the emitter region injects minority carriers into the base region.

To define the circuit currents and voltages of a BJT, the emitter, base, and collector currents, as well as the emitter-base, emitter-collector, and base-collector voltages, must be specified. Independent of the actual direction of current flow, all the terminal currents are

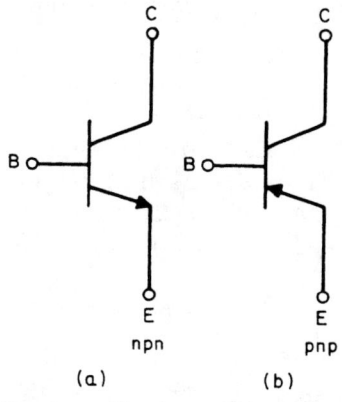

Figure 2-1 Circuit symbols for the BJT: (a) *npn* type and (b) *pnp* type.

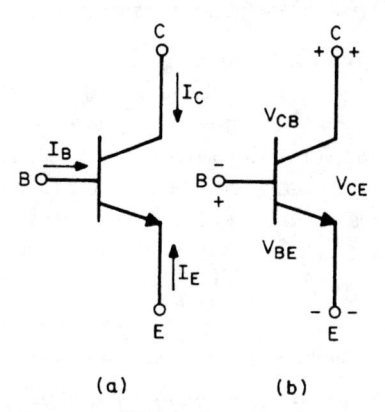

Figure 2-2 Terminal variables for the BJT: (a) terminal currents and (b) interterminal voltages.

conventionally defined as *positive* when *directed inward*, as illustrated in Fig. 2-2a, while voltages between the terminals are defined according to the notation of Fig. 2-2b [5].

Currents and voltages are related through the basic circuit laws, valid for both *npn* and *pnp* BJTs. From *Kirchhoff's current law* (*KCL*)

$$I_E + I_B + I_C = 0 \qquad (2\text{-}1a)$$

and from *Kirchhoff's voltage law* (*KVL*)

$$V_{EB} + V_{BC} + V_{CE} = 0 \qquad (2\text{-}1b)$$

Figure 2-3 is a useful aid for understanding transistor behavior, because it defines the regions of device operation in terms of the applied junction voltages.

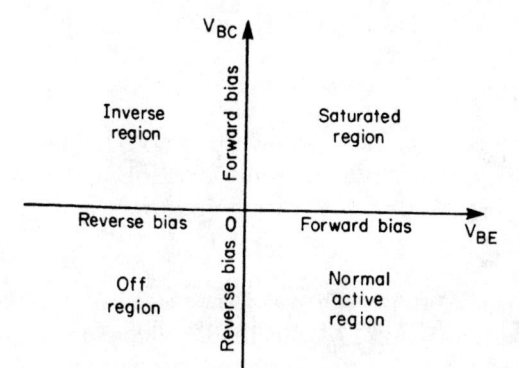

Figure 2-3 Regions of operation for a BJT as defined by base-emitter and base-collector bias polarities.

2.2 Ebers-Moll Static Model

The Bipolar Junction Transistor (BJT) can be considered to be an *interacting* pair of *pn* junctions, and the approach to the problem is basically the same as that used for the diode. Two diode solutions are needed, and the boundary conditions for one diode are the same as the boundary conditions for the other. The Ebers-Moll equations express this formally. They provide the emitter and collector currents I_E and I_C of the transistor in terms of the emitter and collector diode voltages V_{BE} and V_{BC}.

2.2.1 Static model and its implementation in SPICE2

a. Equation formulation. Currently there are two versions of the Ebers-Moll model: the *injection version* (Fig. 2-4) and the *transport version* (Fig. 2-5). These versions are mathematically the same; however, since the second version is preferred for computer simulation, it will be the base for the analysis of the BJT behavior.

In the injection version (Fig. 2-4), the reference currents are I_F and I_R, the currents through the diodes. Thus it can be written

$$I_F = I_{ES}(e^{qV_{BE}/kT} - 1) \tag{2-2a}$$

$$I_R = I_{CS}(e^{qV_{BC}/kT} - 1) \tag{2-2b}$$

where I_{ES} and I_{CS} are the base-emitter and base-collector saturation currents, respectively.

The terminal currents can now be expressed in terms of I_F and I_R as follows:

$$\begin{aligned} I_C &= \alpha_F I_F - I_R \\ I_E &= -I_F + \alpha_R I_R \\ I_B &= (1 - \alpha_F)I_F + (1 - \alpha_R)I_R \end{aligned} \tag{2-3}$$

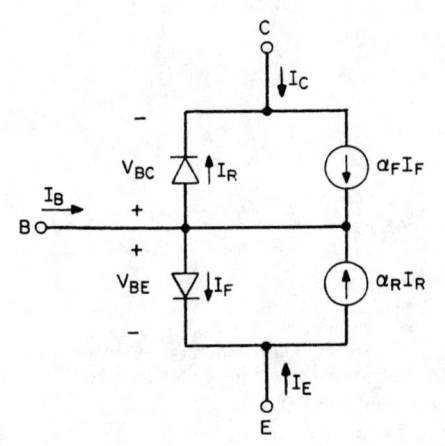

Figure 2-4 Ebers-Moll static model for an *npn* ideal transistor: injection version.

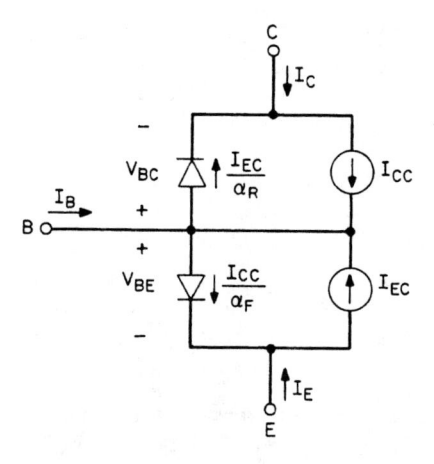

Figure 2-5 Ebers-Moll static model for an *npn* ideal transistor: transport version.

where α_F and α_R are, respectively, the large-signal forward and reverse current gains of a common-base BJT.

A simple intuitive method that justifies this model can be obtained by inspection. The diodes represent the BJT's base-emitter and base-collector junctions. I_F (I_R) is the current that would flow across the base-emitter (base-collector) junction for a given V_{BE} (V_{BC}) if the collector (emitter) region were replaced by an ohmic contact without disturbing the base. The two current-dependent current sources model the coupling between the two junctions, physically represented by the base region.

Consider now the BJT to be biased in the forward active region. In this situation the base-collector diode can be approximated by an open circuit, and thus the model reduces to the $\alpha_F I_F$ current source and the base-emitter diode. I_F represents the total current through the base-emitter junction, while α_F is the fraction of that current that is collected at the base-collector junction.

A dual consideration is valid for a BJT operating in the reverse active region; it is sufficient to consider I_R and α_R instead of I_F and α_F and to change the role of the two junctions. From Eqs. (2-1) to (2-3), it follows that four parameters are needed to describe the model of Fig. 2-4, which, however, can be reduced by one when the reciprocity property is applied. This property gives [3]

$$\alpha_F I_{ES} = \alpha_R I_{CS} \equiv I_S \qquad (2\text{-}4)$$

where I_S is the BJT saturation current.

The physical meaning of I_S is the following. As explained in Appendix A, the saturation current in a *pn* junction is due to two components. As far as the BJT is concerned, the $\alpha_F I_{ES}$ term represents the portion of the emitter-base saturation current (I_{ES}) coming from the analysis of the base region. Similarly, $\alpha_R I_{CS}$ is the portion of the collector-base saturation current (I_{CS}) coming from the analysis of the base region, too. Equation (2-4) means that this analysis of the base region is the same for both I_{ES} and I_{CS} and that I_S is *the common portion of both saturation currents* [6].

Finally, the following familiar relations must be considered.

$$\beta_F = \frac{\alpha_F}{1-\alpha_F}$$

$$\beta_R = \frac{\alpha_R}{1-\alpha_R}$$
(2-5)

where β_F and β_R are, respectively, the large-signal forward and reverse current gain of a common-emitter BJT.

In the transport version (Fig. 2-5), the reference currents I_{CC} and I_{EC} are those flowing through the model's current sources and are given by

$$I_{CC} = I_S(e^{qV_{BE}/kT} - 1) \tag{2-6}$$

$$I_{EC} = I_S(e^{qV_{BC}/kT} - 1) \tag{2-7}$$

The BJT's terminal currents are then given by

$$I_C = I_{CC} - \frac{I_{EC}}{\alpha_R}$$

$$I_E = -\frac{I_{CC}}{\alpha_F} + I_{EC} \tag{2-8}$$

$$I_B = \left(\frac{1}{\alpha_F} - 1\right)I_{CC} + \left(\frac{1}{\alpha_R} - 1\right)I_{EC}$$

A change in the model topology of Fig. 2-5 can be made. The change, as shown in Fig. 2-6, consists of replacing the two reference current sources with a single current source (I_{CT}) between the emitter and the collector, defined by the expression

$$I_{CT} = I_{CC} - I_{EC} = I_S(e^{qV_{BE}/kT} - e^{qV_{BC}/kT}) \tag{2-9}$$

Equation (2-9) causes a change in the equations for the diode saturation currents. Thus the diode currents become

$$\frac{I_{CC}}{\beta_F} = \frac{I_S}{\beta_F}(e^{qV_{BE}/kT} - 1) \tag{2-10}$$

$$\frac{I_{EC}}{\beta_R} = \frac{I_S}{\beta_R}(e^{qV_{BC}/kT} - 1) \tag{2-11}$$

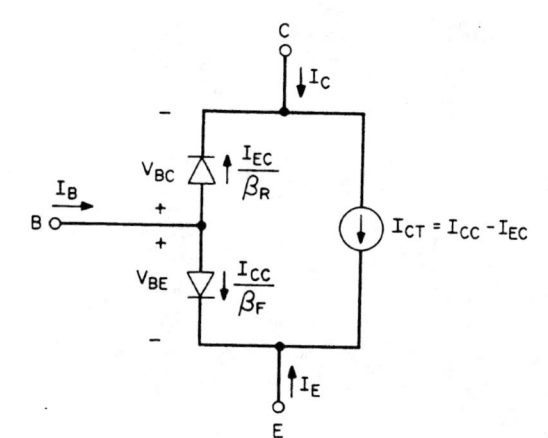

Figure 2-6 SPICE2 BJT Ebers-Moll static model.

The BJT's terminal currents are then given by

$$I_C = I_{CT} - \frac{I_{EC}}{\beta_R}$$

$$I_E = -\frac{I_{CC}}{\beta_F} - I_{CT} \quad (2\text{-}12)$$

$$I_B = \frac{I_{CC}}{\beta_F} + \frac{I_{EC}}{\beta_R}$$

b. Implementation in SPICE2. The model's terminal currents for the four regions of Fig. 2-3 can now be written as they are implemented in SPICE2.

As outlined in Sec. 1.2.1a, to aid convergence, a small conductance GMIN (its default value is 10^{-12} mho) is added automatically by SPICE2 [2] and put in parallel with every pn junction. This yields the following equations.

Normal active region

For $V_{BE} > -5\frac{kT}{q}$ and $V_{BC} \leq -5\frac{kT}{q}$:

$$I_C = I_S \left(e^{qV_{BE}/kT} + \frac{1}{\beta_R} \right) + \left[V_{BE} - \left(1 + \frac{1}{\beta_R}\right) V_{BC} \right] \text{GMIN}$$

$$I_B = I_S \left[\frac{1}{\beta_F}(e^{qV_{BE}/kT} - 1) - \frac{1}{\beta_R} \right] + \left(\frac{V_{BE}}{\beta_F} + \frac{V_{BC}}{\beta_R} \right) \text{GMIN}$$

(2-13)

Inverse region

For $V_{BE} \leq -5\frac{kT}{q}$ and $V_{BC} > -5\frac{kT}{q}$:

$$I_C = -I_S\left[e^{qV_{BC}/kT} + \frac{1}{\beta_R}(e^{qV_{BC}/kT} - 1)\right]$$
$$+ \left[V_{BE} - \left(1 + \frac{1}{\beta_R}\right)V_{BC}\right]\text{GMIN} \quad (2\text{-}14)$$

$$I_B = -I_S\left[\frac{1}{\beta_F} - \frac{1}{\beta_R}(e^{qV_{BC}/kT} - 1)\right] + \left(\frac{V_{BE}}{\beta_F} + \frac{V_{BC}}{\beta_R}\right)\text{GMIN}$$

Saturated region

For $V_{BE} > -5\frac{kT}{q}$ and $V_{BC} > -5\frac{kT}{q}$:

$$I_C = I_S\left[(e^{qV_{BE}/kT} - e^{qV_{BC}/kT}) - \frac{1}{\beta_R}(e^{qV_{BC}/kT} - 1)\right]$$
$$+ \left[V_{BE} - \left(1 + \frac{1}{\beta_R}\right)V_{BC}\right]\text{GMIN}$$
$$\quad (2\text{-}15)$$
$$I_B = I_S\left[\frac{1}{\beta_F}(e^{qV_{BE}/kT} - 1) + \frac{1}{\beta_R}(e^{qV_{BC}/kT} - 1)\right]$$
$$+ \left(\frac{V_{BE}}{\beta_F} + \frac{V_{BC}}{\beta_R}\right)\text{GMIN}$$

Off region

For $V_{BE} \leq -5\frac{kT}{q}$ and $V_{BC} \leq -5\frac{kT}{q}$:

$$I_C = \frac{I_S}{\beta_R} + \left[V_{BE} - \left(1 + \frac{1}{\beta_R}\right)V_{BC}\right]\text{GMIN}$$
$$\quad (2\text{-}16)$$
$$I_B = -I_S\left(\frac{\beta_F + \beta_R}{\beta_F\beta_R}\right) + \left(\frac{V_{BE}}{\beta_F} + \frac{V_{BC}}{\beta_R}\right)\text{GMIN}$$

For a *pnp* transistor, the voltage and current polarities must be changed appropriately. In SPICE2, the parameter values of the model are always considered to be positive, and the appropriate sign changes are implemented internally by the program; the user need only specify the transistor type *npn* or *pnp* in the .MODEL statement [7].

Although it is very simple in form and requires, at most, three parameters—β_F, β_R, and I_S—the Ebers-Moll model just derived is accurate and is recommended when the BJT is working as a dc switch or in a specific and narrow bias range. In fact, this principle must be taken into account: Use the simplest model that will do the job. This saves modeling effort and computer time, and the results are easier to understand.

Figure 2-7 shows the I_C vs. V_{CE} characteristics of the Ebers-Moll model for the ideal BJT as SPICE2 outputs when the values of the default parameters (see Table 2-1) are used for $I_B = 0.1$ A and $I_B = 0.5$ A.

The three parameters I_S, β_F, and β_R are user-specified parameters. The saturation current I_S (which, by its definition, takes into account base doping density and neutral basewidth) is the extrapolated intercept current of $\ln I_C$ vs. V_{BE} in the forward region and $\ln I_E$ vs. V_{BC} in the reverse region, as shown in Fig. 2-8. The parameters β_F and β_R are the *maximum* forward and reverse common-emitter current gains, respectively, which are assumed not to vary with the operating point.

To summarize, the ideal model just described is characterized in SPICE2 by the following three parameters the user of SPICE2 can specify in the .MODEL statement [7]:

IS   Saturation current (I_S)
BF   Ideal maximum forward current gain (β_F)
BR   Ideal maximum reverse current gain (β_R)

The keyname used in the text is indicated in parentheses.

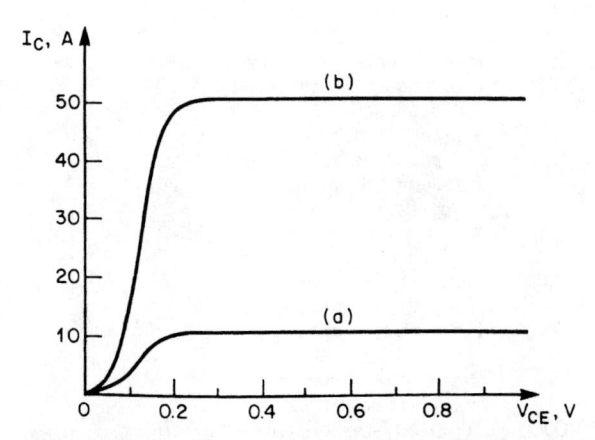

Figure 2-7 SPICE2 I_C vs. V_{CE} characteristics of the Ebers-Moll model of the ideal BJT when default parameter values are used for $I_B = 0.1$ A and $I_B = 0.5$ A [curve (b)].

TABLE 2-1 SPICE2 BJT Model Parameters

Symbol	SPICE2G keyword	SPICE2E/F keyword	Parameter name	Default value	Unit
I_S	IS	IS/JS	Saturation current	10^{-16}	A
β_F	BF	BF/BF	Ideal maximum forward current gain	100	
β_R	BR	BR/BR	Ideal maximum reverse current gain	1	
n_F	NF	.../NF	Forward current emission coefficient	1	
n_R	NR	.../NR	Reverse current emission coefficient	1	
C_2	$\text{ISE} = C_2 I_S$	C2/JLE	Base-emitter leakage saturation current	0	A
C_4	$\text{ISC} = C_4 I_S$	C4/JLC	Base-collector leakage saturation current	0	A
I_{KF}	IKF	IK/JBF	Corner for forward β high-current roll-off	∞	A
I_{KR}	IKR	IKR/JBR	Corner for reverse β high-current roll-off	∞	A
n_{EL}	NE	NE/NLE	Base-emitter leakage emission coefficient	1.5	
n_{CL}	NC	NC/NLC	Base-collector leakage emission coefficient	2	
V_A	VAF	VA/VBF	Forward Early voltage	∞	V
V_B	VAR	VB/VBR	Reverse Early voltage	∞	V
r_C	RC	RC/RC	Collector resistance	0	Ω
r_E	RE	RE/RE	Emitter resistance	0	Ω
r_B	RB	RB/RB	Zero-bias base resistance	0	Ω
r_{BM}	RBM	.../RBM	Minimum base resistance at high currents	RB	Ω
I_{rB}	IRB	.../JRB	Current where base resistance falls halfway to its minimum value	∞	A
τ_F	TF	TF/TF	Ideal forward transit time	0	s
τ_R	TR	TR/TR	Ideal reverse transit time	0	s
$X_{\tau F}$	XTF	.../XTF	Coefficient for bias dependence of TF	0	
$V_{\tau F}$	VTF	.../VTF	Voltage describing V_{BC} dependence of TF	∞	V
$I_{\tau F}$	ITF	.../JTF	High-current parameter for effect on TF	0	A
$P_{\tau F}$	PTF	.../PTF	Excess phase at $f = 1/2\pi\tau_F$	0	°
C_{JE}	CJE	CJE/CJE	Zero-bias base-emitter depletion capacitance	0	F
ϕ_E	VJE	PE/VJE	Base-emitter built-in potential	0.75	V
m_E	MJE	ME/MJE	Base-emitter junction grading coefficient	0.33	
C_{JC}	CJC	CJC/CJC	Zero-bias base-collector depletion capacitance	0	F
ϕ_C	VJC	PC/VJC	Base-collector built-in potential	0.75	V

TABLE 2-1 SPICE2 BJT Model Parameters *(Continued)*

Symbol	SPICE2G keyword	SPICE2E/F keyword	Parameter name	Default value	Unit
m_C	MJC	MC/MJC	Base-collector junction grading coefficient	0.33	
C_{JS}	CJS	CCS/CJS	Zero-bias collector-substrate capacitance	0	F
ϕ_S	VJS	.../VJS	Substrate-junction built-in potential	0.75	V
m_S	MJS	.../MJS	Substrate-junction exponential factor	0	
X_{CJC}	XCJC	.../CDIS	Fraction of base-collector depletion capacitance connected to internal base node	1	
FC	FC	FC	Coefficient for forward-bias depletion capacitance formula	0.5	
$X_{T\beta}$	XTB	.../TB	Forward and reverse β temperature coefficient	0	
X_{TI}	XTI	PT/PT	Saturation current temperature exponent	3	
E_g	EG	EG/EG	Energy gap for temperature effect on I_S	1.11	eV
k_f	KF	KF/KF	Flicker-noise coefficient	0	
a_f	AF	AF/AF	Flicker-noise exponent	1	
T	T*	T/T	Nominal temperature for simulation and at which all input data are assumed to have been measured	27	°C

*Remember, the temperature T is regarded as an operating condition and not as a model parameter.

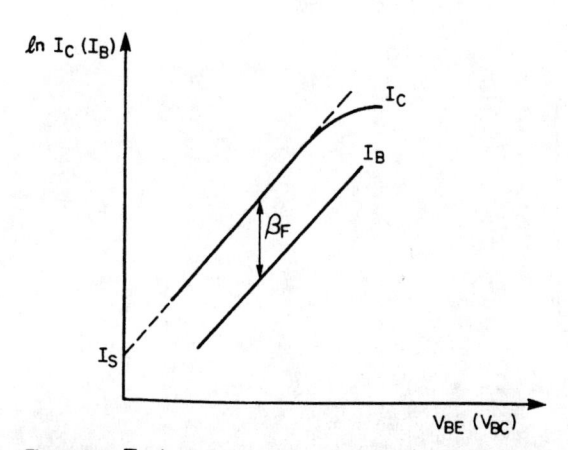

Figure 2-8 Typical current vs. voltage plots to show the meaning of I_S and β_F.

2.2.2 Static model: Second-order effects and their implementation in SPICE2

The ideal BJT model described in the preceding section is very simple because of the following reasons:

1. It neglects parasitic base, collector, and emitter resistances.
2. It neglects the dependence of I_S on V_{BC} (Early effect, base-width modulation).
3. It assumes *ideal* base current components (Boltzmann-type) or β_F and β_R current-independent.
4. It neglects high-level effects in the base and collector regions.

In this section, an improved and more accurate Ebers-Moll dc model of the BJT is developed. We introduce four new parameters, r_E, r_B, r_C, and V_A, which take into account the base, collector, and emitter resistances and the dependence of I_S on V_{BC}.

a. Ohmic resistances. The inclusion of three constant resistors (r_C, r_E, and r_B) improves the dc characterization of the model. They represent the transistor's ohmic resistances from its active region to its collector, emitter, and base terminals, respectively. These resistors are included in the model as shown in Fig. 2-9. The internal nodes of these resistances are denoted by the letters C, E, and B in the model circuit. The voltages used in describing the two ideal diodes

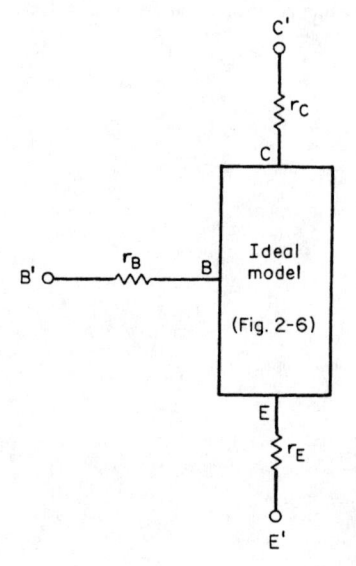

Figure 2-9 BJT Ebers-Moll static model: Effect of the ohmic resistances [6].

are the *internal voltages*. In the following, the effects of ohmic resistances on the static model are described [6].

Collector resistance r_C. The effect of the *collector resistance r_C* is shown in Fig. 2-10, in which collector characteristics of the actual model (solid lines) are compared to collector characteristics of the ideal model (dashed lines). Resistance r_C decreases the slope of the curves in the saturated region for low collector-emitter voltages.

The collector resistance can limit the current-handling capability of a BJT, and it also affects the maximum operating frequency at high currents.

Emitter resistance r_E. The emitter is the most heavily doped region in most present-day transistors in order to produce a high emitter injection efficiency and therefore a high β_F. For this reason, the dominant component of *emitter resistance r_E* is normally the contact resistance (usually on the order of 1 Ω). r_E, which is often neglected, can normally be assumed to have a small, constant value.

Its main effect is a reduction in the voltage seen by the emitter-base junction by a factor of $r_E I_E$. In this effect on V_{BE}, r_E is equivalent to a base resistance of $(1+\beta_F)r_E$. Therefore r_E affects the collector current as well as the base current, as shown in Fig. 2-11. This effect can be significant, and r_E can cause substantial errors in the determination of r_B. Resistance r_E can also affect the collector characteristics in the saturation region if the transistor has a low r_C value.

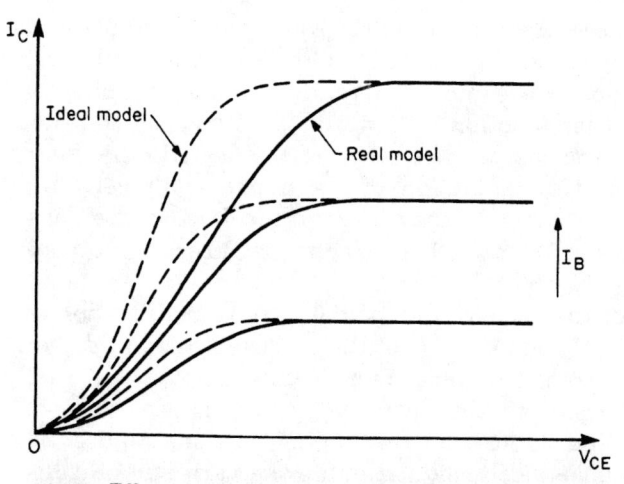

Figure 2-10 Effect of the collector resistance on the I_C vs. V_{CE} characteristics. Dashed lines represent $r_C = 0$. Solid lines represent $r_C \neq 0$.

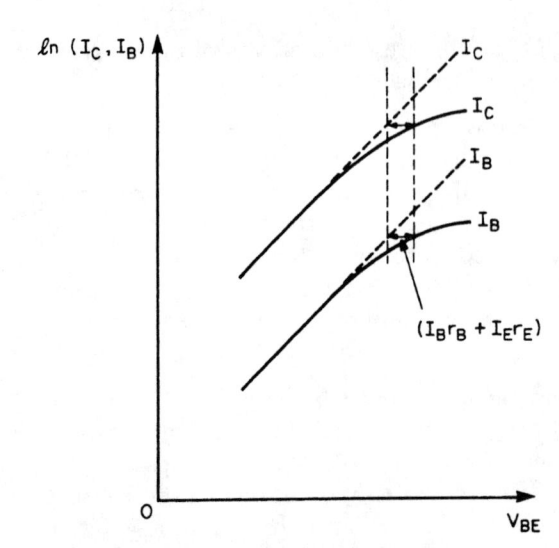

Figure 2-11 The effect of r_B and r_E on the $\ln I_C$ and I_B vs. V_{BE} characteristics of the real model [6].

Base resistance r_B. *Base resistance r_B* is an important model parameter. Its greatest impact is its effect on the small-signal and transient responses. It is also one of the most difficult parameters to measure accurately, partly because of its strong dependence on operating point (due to crowding) and partly because of the error introduced by the small but finite value of r_E.

In the Ebers-Moll model, r_B is assumed to be constant. The dc effect of r_B is seen on the $\ln I_C$ and $\ln I_B$ vs. V_{BE} curve, as illustrated in Fig. 2-11.

b. Base-width modulation (Early effect). Base-width modulation (the so-called *Early effect*) is the change in base width, W_B, that results from a change in the collector-base junction voltage. In the normal active region, the emitter-base junction is forward-biased and the collector-base junction is reverse-biased. The width of the depletion region of a *pn* junction is a strong function of the applied voltage. Large variations in V_{BC}, for example, may cause the collector-base depletion region to vary significantly. This, in turn, changes the normally thin base width, as shown in Fig. 2-12.

The total effect of the base-width modulation is a modification (as a function of V_{BC}) of I_S and thereby of the collector current, β_F, and τ_F through the base transit time τ_{BF} (see Sec. 2.3.1).

Only one extra parameter, the *Early voltage V_A*, is used to model base-width modulation in the forward active region. Figure 2-13 shows the effect of base-width modulation on the variation of collector current I_C with collector-emitter voltage (I_C vs. V_{CE})—a nonzero slope in the normal active region. (The dashed lines illustrate the

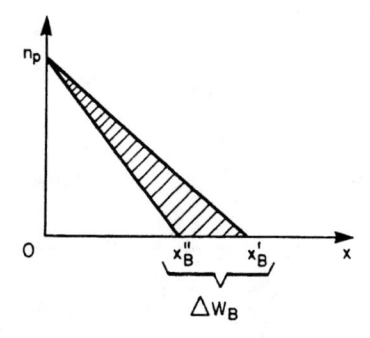

Figure 2-12 Effect of increases in V_{BC} on the collector-base depletion region and base width of an npn BJT.

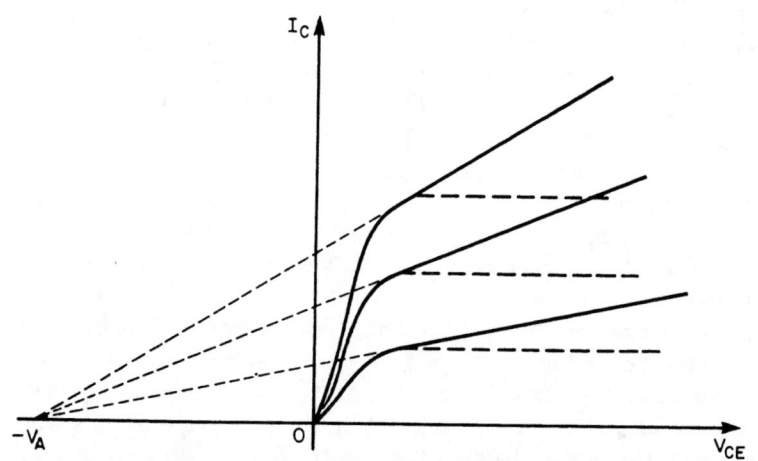

Figure 2-13 The effect of base-width modulation on the I_C vs. V_{CE} characteristics. The dashed lines illustrate the zero slope obtained with the ideal model, where base-width modulation is not included.

zero slope obtained with the ideal model, where base-width modulation is not included.)

The analysis, which assumes that the transistor is operated in the linear region, first determines the effect of base-width modulation on the base width and then on the three base-width-related parameters. The results of the analysis are as follows [6]:

$$W_B(V_{BC}) = W_B(0)\left(1 + \frac{V_{BC}}{V_A}\right) \tag{2-17}$$

$$I_S(V_{BC}) = \frac{I_S(0)}{1 + V_{BC}/V_A} \simeq I_S(0)\left(1 - \frac{V_{BC}}{V_A}\right) \tag{2-18}$$

$$\beta_F(V_{BC}) = \frac{\beta_F(0)}{1 + V_{BC}/V_A} \simeq \beta_F(0)\left(1 - \frac{V_{BC}}{V_A}\right) \tag{2-19}$$

$$\tau_{BF}(V_{BC}) = \tau_{BF}(0)\left(1 + \frac{V_{BC}}{V_A}\right)^2 \tag{2-20}$$

where V_A is defined as

$$V_A = \begin{cases} \left[\dfrac{1}{W_B(0)} \dfrac{dW_B}{dV_{BC}}\bigg|_{V_{BC}=0}\right]^{-1} & \text{for } npn \text{ transistor} \quad (2\text{-}21a) \\[2ex] \left[-\dfrac{1}{W_B(0)} \dfrac{dW_B}{dV_{BC}}\bigg|_{V_{BC}=0}\right]^{-1} & \text{for } pnp \text{ transistor} \quad (2\text{-}21b) \end{cases}$$

and $I_S(0)$, $\beta_F(0)$, $\tau_{BF}(0)$ are the values of the parameters at $V_{BC}=0$.

The difference between the definitions for the *npn* and *pnp* transistors lies only in the sign of V_{BC}. For an *npn* transistor in the normal active region, V_{BC} is negative (i.e., reverse bias). An increase in V_{BC} (a decrease in the reverse bias) results in an increase in the base width, and the derivative in Eq. (2-21a) is positive. The minus sign in Eq. (2-21b) preserves the positive nature of V_A for a *pnp* transistor.

Equation (2-17) describes the (assumed linear) variation of the base width with V_{BC}. Equations (2-18) to (2-20), which give the variation of the three model parameters with V_{BC}, follow directly from Eq. (2-17) and the assumptions that the constant-base doping relationships $I_S \propto 1/W_B$, $\beta_F \propto 1/W_B$, and $\tau_{BF} \propto W_B^2$ are also approximately valid in general.

The second forms of Eqs. (2-18) and (2-19) (obtained using the binomial expansion assuming $|V_{BC}| \ll V_A$) are preferred computationally, since the first forms become infinite at $V_{BC} = V_A$ ($V_{BC} = V_A$ may not necessarily be the correct value but could be a temporary value while the computer is iterating to the solution).

V_A has no physical counterpart in the circuit model, only a mathematical effect whereby existing equations are modified. (This process of altering equations or parameters without altering the *form* of the equivalent circuit will be observed for other effects in the section on the Gummel-Poon model.)

With the exception of τ_{BF}, which is described later, the total effect of base-width modulation is accounted for if I_S and β_F are modified as in Eqs. (2-18) and (2-19). The expressions for I_{CT} and I_B [Eqs. (2-9) and (2-12), respectively] then become

$$I_{CT} = \frac{I_S(0)}{1 + V_{BC}/V_A}(e^{qV_{BE}/kT} - e^{qV_{BC}/kT}) \quad (2\text{-}22)$$

$$I_B = \frac{I_S(0)}{\beta_F(0)}(e^{qV_{BE}/kT} - 1) + \frac{I_S(0)}{\beta_R}(e^{qV_{BC}/kT} - 1) \quad (2\text{-}23)$$

In the first term of Eq. (2-23), the similar dependence of I_S and β_F on V_{BC} (since both are assumed to be proportional to $1/W_B$) results in a cancellation with respect to the V_A parameter. Therefore, in the normal active region where the second term is negligible, I_B is independent of V_{BC} and thus of Early effect. A similar analysis for base-width modulation when the device is operated in the inverse mode is included in the discussion of the Gummel-Poon model (see Sec. 2.5.4).

The Early voltage V_A can be obtained directly from the I_C vs. V_{CE} characteristics. The slope of these characteristics in the normal active region, g_o, is obtained from Eqs. (2-12) and (2-22) by first dropping the (negligible) second term and then differentiating with respect to V_{BC} (V_{BE} is assumed constant). The result is

$$g_o = \frac{dI_C}{dV_{CE}}\bigg|_{V_{BE}=\text{const.}} \simeq \frac{I_C(0)}{V_A} \qquad (2\text{-}24)$$

The geometrical interpretation of Eq. (2-24) shows that V_A is obtained from the intercept of the extrapolated slope on the V_{CE} axis (as shown in the curve in Fig. 2-13). For example, a slope of $(50\text{ k}\Omega)^{-1}$ at $I_C(0) = 1\text{ mA}$ gives, from Eq. (2-24), $V_A = 50\text{ V}$. A more detailed description of the geometrical interpretation is given in Ref. [6].

c. Implementation in SPICE2. The physics-based static Ebers-Moll model for the *real* BJT requires, besides the parameters affecting the *ideal* model, four new model parameters that the user can specify in the .MODEL statement. These parameters are

RC   Collector resistance (r_C)
RE   Emitter resistance (r_E)
RB   Zero-bias base resistance (r_B); it can be dependent on high current (see Sec. 2.5.5)
VAF   Forward Early voltage (V_A)

The keyname used in the text is indicated in parentheses.

The Early voltage V_A affects the basic model equations for I_C and I_B [see Eqs. (2-13) to (2-16)].

To show how these equations are modified in SPICE2, the more general relationships of Eqs. (2-15) are taken into account; the other equations follow immediately.

$$\begin{aligned} I_C &= I_S \left[(e^{qV_{BE}/kT} - e^{qV_{BC}/kT}) \left(1 - \frac{V_{BC}}{V_A}\right) - \frac{1}{\beta_R}(e^{qV_{BC}/kT} - 1) \right] \\ &\quad + \left[(V_{BE} - V_{BC})\left(1 - \frac{V_{BC}}{V_A}\right) - \frac{V_{BC}}{\beta_R} \right] \text{GMIN} \\ I_B &= I_S \left[\frac{1}{\beta_F}(e^{qV_{BE}/kT} - 1) + \frac{1}{\beta_R}(e^{qV_{BC}/kT} - 1) \right] \\ &\quad + \left(\frac{V_{BE}}{\beta_F} + \frac{V_{BC}}{\beta_R}\right) \text{GMIN} \end{aligned} \qquad (2\text{-}25)$$

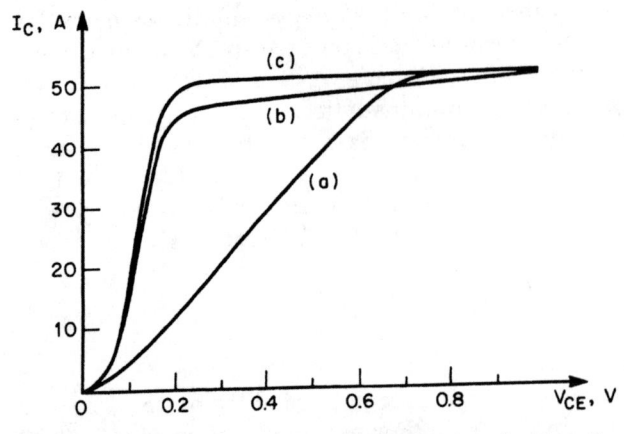

Figure 2-14 SPICE2 I_C vs. V_{CE} characteristics showing the second-order effects on the Ebers-Moll model: (a) $r_C = 10$ mΩ, (b) $V_A = 10$ V, and (c) default values, with $I_B = 0.5$ A. The other parameters have default values.

Obviously the models of Figs. 2-6 and 2-9 are still valid if the terminal currents are changed appropriately in order to model the above-described effects.

Figure 2-14 shows the SPICE2-generated I_C vs. V_{CE} characteristics that plot the effects of the parasitic collector resistance and the Early effect on the Ebers-Moll model.

2.3 Ebers-Moll Large-Signal Model

Having defined the complete set of dc transport relationships for the BJT (or better, the Ebers-Moll static model), we will now consider the charge-storage effects of the device. The importance of these effects has been outlined in Sec. 1.3.1, which you may refer to in order to understand the physics presiding over the large-signal behavior of the device (keep in mind that the BJT can be considered to be an interacting pair of two *pn* junctions).

2.3.1 Large-signal model

Charge storage in the BJT is modeled by the introduction of three types of capacitors: two nonlinear junction capacitors, two nonlinear diffusion capacitors, and a constant substrate capacitor.

a. Storage charges Q_{DE} and Q_{DC}. The charge associated with the *mobile carriers* in a BJT is modeled by the *diffusion capacitances*. This charge is divided into two components: one associated with the reference collector current source, I_{CC}, and the other with the reference emitter current source, I_{EC}. Each component is represented by a capacitor.

To evaluate the diffusion capacitance associated with I_{CC}, the total mobile charge associated with this current must be considered. Therefore, the base-emitter junction is assumed to be forward-biased and $V_{BC} = 0$.

For the simplified one-dimensional case of constant base doping, negligible base recombination, and low-level injection of the BJT of Fig. 2-15, the total mobile charge associated with I_{CC}, Q_{DE}, can be written as the sum of the individual minority charges:

$$Q_{DE} = Q_E + Q_{JE} + Q_{BF} + Q_{JC} \qquad (2\text{-}26)$$

where Q_E is the mobile minority charge stored in the neutral emitter region, Q_{JE} is the mobile minority charge in the emitter-base depletion region associated with I_{CC} (normally considered to be zero), Q_{BF} is the minority mobile charge stored in the neutral base region, and Q_{JC} is the mobile minority charge in the collector-base depletion region associated with I_{CC}. Because of charge neutrality, there will be identical majority charges stored in the neutral regions. However,

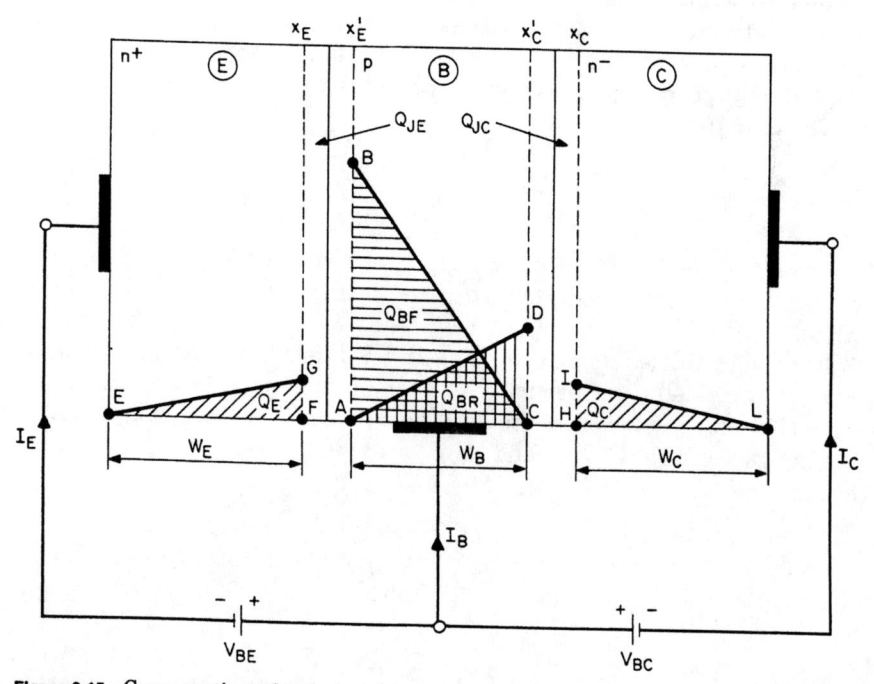

Figure 2-15 Cross section of an n^+pn^- transistor indicating the location of the charge components.

to determine diffusion capacitance, only one (minority or majority) needs to be considered.

From Eq. (2-26) the total mobile charge associated with I_{CC} can also be expressed as [6]

$$Q_{DE} = (\tau_E + \cancelto{0}{\tau_{EB}} + \tau_{BF} + \tau_{CB})I_{CC} \equiv \tau_F I_{CC} \qquad (2\text{-}27)$$

where τ_E is the emitter delay; τ_{EB} is the emitter-base depletion region transit time; τ_{BF} is the base transit time; τ_{CB} is the base-collector depletion region transit time; and τ_F is the total forward transit time (assumed here to be a constant), which represents the mean time for the minority carriers to cross (diffuse) the neutral base region from the emitter to the collector.

A similar analysis of the total mobile charge associated with I_{EC} results in

$$Q_{DC} = Q_C + Q_{JC} + Q_{BR} + Q_{JE} \qquad (2\text{-}28)$$

where Q_C is the mobile minority charge stored in the neutral collector region, Q_{JC} is the mobile minority charge in the collector-base depletion region associated with I_{EC}, Q_{BR} is the minority mobile charge stored in the neutral base region, and Q_{JE} is the mobile minority charge in the emitter-base depletion region associated with I_{EC}. If charge Q_{JC} is assumed to be zero, then it follows from Eq. (2-28) that [6]

$$Q_{DC} = (\tau_C + \cancelto{0}{\tau_{CB}} + \tau_{BR} + \tau_{EB})I_{EC} \equiv \tau_R I_{EC} \qquad (2\text{-}29)$$

where τ_C is the collector delay, τ_{BR} is the reverse base transit time, and τ_R is the total reverse transit time (assumed to be a constant here also).

For the saturated mode (i.e., V_{BE} and V_{BC} are both forward-biased), both Q_{DE} and Q_{DC} are assumed to occur independently and the total minority charge stored in the BJT is the sum of all the components.

The two charges Q_{DE} and Q_{DC} are modeled by two *nonlinear* capacitors

$$\begin{aligned} C_{DE} &\equiv \frac{dQ_{DE}}{dV_{BE}} = \frac{d(\tau_F I_{CC})}{dV_{BE}} \\ C_{DC} &\equiv \frac{dQ_{DC}}{dV_{BC}} = \frac{d(\tau_R I_{EC})}{dV_{BC}} \end{aligned} \qquad (2\text{-}30)$$

as shown in Fig. 2-16.

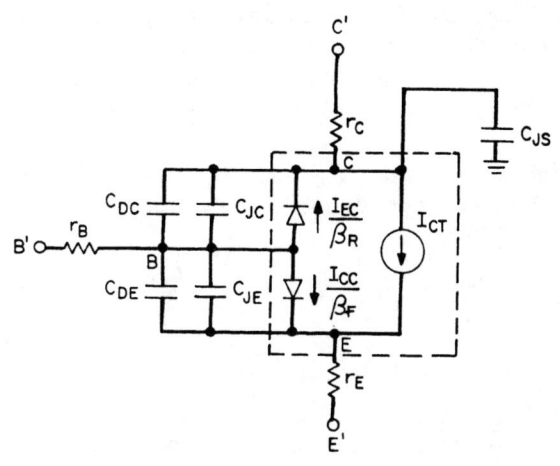

Figure 2-16 SPICE2 BJT Ebers-Moll large-signal model.

b. Fixed charges Q_{JE} and Q_{JC}. The incremental *fixed charges* Q_{JE} and Q_{JC} stored in the BJT's depletion regions for incremental changes in the associated junction voltages can be modeled by two *junction capacitances* (sometimes called *depletion capacitances*). These capacitances, denoted by C_{JE} for the base-emitter junction and C_{JC} for the base-collector junction, are included in the model as shown in Fig. 2-16. Each junction capacitance is a nonlinear function of the voltage across the junction with which it is associated.

It can be seen (see Sec. A.2.4) that both these capacitances have the following form:

$$C_J = \frac{C_J(0)}{(1 - V/\phi)^m} \qquad (2\text{-}31)$$

Since the actual junctions are usually somehow between these two cases, the capacitance model assumes the parameter m as an exponent, which should usually be between 0.5 and 0.33. The doping profile at emitter-base junctions is usually *Gaussian* and closer to an abrupt junction, and m should be close to 0.5. The collector-base junctions, on the other hand, are more like linearly graded junctions, and m is close to 0.33. Then, for a step (or abrupt) junction and for a linear (or graded) junction, the variation of the emitter and collector junction capacitances with base-emitter and base-collector junction voltages can be written for an *npn* transistor as follows:

$$C_{JE}(V_{BE}) = \frac{C_{JE}(0)}{(1 - V_{BE}/\phi_E)^{m_E}}$$

$$C_{JC}(V_{BC}) = \frac{C_{JC}(0)}{(1 - V_{BC}/\phi_C)^{m_C}} \qquad (2\text{-}32)$$

where $C_{JE}(0)$ and $C_{JC}(0)$ are the values of the emitter-base and collector-base junction capacitances at $V_{BE} = 0$ or $V_{BC} = 0$, ϕ_E and ϕ_C are the emitter-base and collector-base barrier potentials, and m_E and m_C are the emitter-base and collector-base capacitance gradient factors.

To obtain the fixed charges Q_{JE} and Q_{JC}, it is necessary to integrate the junction capacitances with respect to their voltages, i.e.,

$$Q_{JE} = \int_0^{V_{BE}} C_{JE}\, dV = \frac{C_{JE}(0)}{1 - m_E}\left(1 - \frac{V_{BE}}{\phi_E}\right)^{1-m_E}$$
$$Q_{JC} = \int_0^{V_{BC}} C_{JC}\, dV = \frac{C_{JC}(0)}{1 - m_C}\left(1 - \frac{V_{BC}}{\phi_C}\right)^{1-m_C}$$
(2-33)

It is important to note that the physical considerations that have led to the junction capacitance in the diode analysis are still valid for obtaining C_{JE} and C_{JC} for the BJT.

Figure 2-17 shows three plots of the variation of the junction capacitance as a function of voltage. The solid line represents Eqs. (2-32). Under forward bias these equations predict infinite capacitance when the internal junction voltage equals the built-in voltage. It has been shown by Chawla and Gummel [8] that under forward bias, the depletion approximation is no longer valid and that Eqs. (2-32) no longer apply. The dashed line in Fig. 2-17 shows the (noninfinite) variation of the junction capacitance obtained by Chawla and Gummel. An expression has been fitted to this curve by Gummel and

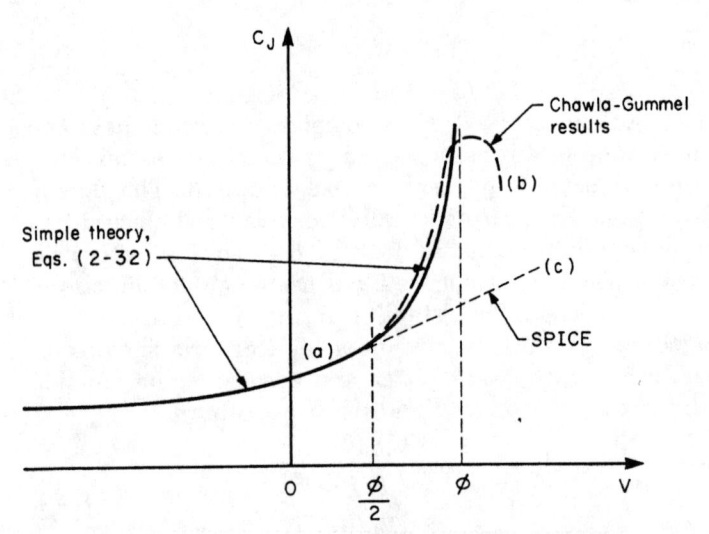

Figure 2-17 Plot of the variation of junction capacitances with voltage [6].

Poon [9]. The third curve in Fig. 2-17 represents the straight-line approximation made by usual computer programs for $V > \phi/2$. The equation for this straight line, obtained by matching slopes at $\phi/2$, is given by

$$C_J = 2^m C_J(0) \left[2m \frac{V}{\phi} + (1-m) \right] \quad \text{for } V \geq \frac{\phi}{2} \quad (2\text{-}34)$$

where the junction subscripts have been omitted. This approximation, while avoiding the infinite capacitance, is not as accurate as the Chawla-Gummel curve [8]. However, it is acceptable because under forward bias, the diffusion capacitances, described next, are dominant and inherently include the effect of the mobile charges in the depletion regions [6].

SPICE2 uses a straight-line approximation for C_J similar to the line (c) of Fig. 2-17; Eq. (2-34) is replaced by the following general relationships:

$$C_J = C_J(0)\left(1 + m\frac{V}{\phi}\right) \quad \text{for } V \geq 0 \quad (2\text{-}35)$$

Besides C_{JE} and C_{JC}, another capacitance must be taken into account in the design of integrated circuits: the *substrate capacitance* C_{JS}. Although it is actually a junction capacitance in the way it varies with the epitaxial-layer–substrate potential, it is modeled here as a constant-value capacitor. This representation is adequate for most cases, since the epitaxial-layer–substrate junction is reverse-biased for isolation purposes.

2.3.2 Large-signal model and its implementation in SPICE2

Having established the basic relationships of the charge-storage effects, we can discuss the details of the large-signal model implemented in SPICE2 in a straightforward way.

The charge-storage elements $Q_{BE} = Q_{DE} + Q_{JE}$ and $Q_{BC} = Q_{DC} + Q_{JC}$, or, better, the capacitors C_{BE} and C_{BC}, shown in Fig. 2-18 define the SPICE2, charge-storage model of the BJT.

The charges Q_{BE} and Q_{BC} are determined by the following relations:

$$Q_{BE} = \begin{cases} \underbrace{\tau_F I_{CC}}_{Q_{DE}} + \underbrace{C_{JE}(0) \int_0^{V_{BE}} \left(1 - \frac{V}{\phi_E}\right)^{-m_E} dV}_{Q_{JE}} & \text{for } V_{BE} < \text{FC} \times \phi_E \\[2em] \underbrace{\tau_F I_{CC}}_{Q_{DE}} + C_{JE}(0)F_1 + \underbrace{\frac{C_{JE}(0)}{F_2} \int_{\text{FC} \times \phi_E}^{V_{BE}} \left(F_3 + \frac{m_E V}{\phi_E}\right) dV}_{Q_{JE}} & \text{for } V_{BE} \geq \text{FC} \times \phi_E \end{cases} \quad (2\text{-}36)$$

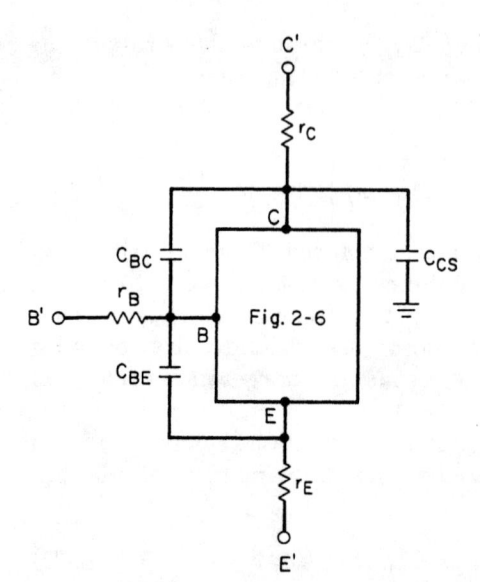

Figure 2-18 SPICE2 BJT Ebers-Moll large-signal model.

$$Q_{BC} = \begin{cases} \underbrace{\tau_R I_{EC}}_{Q_{DC}} + \underbrace{C_{JC}(0) \int_0^{V_{BC}} \left(1 - \frac{V}{\phi_C}\right)^{-m_C} dV}_{Q_{JC}} & \text{for } V_{BC} < \text{FC} \times \phi_C \\ \underbrace{\tau_R I_{EC}}_{Q_{DC}} + \underbrace{C_{JC}(0) F_1 + \frac{C_{JC}(0)}{F_2} \int_{\text{FC} \times \phi_C}^{V_{BC}} \left(F_3 + \frac{m_C V}{\phi_C}\right) dV}_{Q_{JC}} & \text{for } V_{BC} \geq \text{FC} \times \phi_C \end{cases} \quad (2\text{-}37)$$

$$Q_{CS} = \begin{cases} C_{JS}(0) \int_0^{V_{CS}} \left(1 - \frac{V}{\phi_S}\right)^{-m_S} dV & \text{for } V_{CS} < 0 \\ C_{JS}(0) \int_0^{V_{CS}} \left(1 + \frac{m_S V}{\phi_S}\right) dV & \text{for } V_{CS} \geq 0 \end{cases} \quad (2\text{-}38)$$

These charge-storage elements can be equivalently represented by the following SPICE2 voltage-dependent capacitance equations.

$$C_{BE} = \frac{dQ_{BE}}{dV_{BE}} = \begin{cases} \underbrace{\tau_F \frac{dI_{CC}}{dV_{BE}}}_{C_{DE}} + \underbrace{C_{JE}(0) \left(1 - \frac{V_{BE}}{\phi_E}\right)^{-m_E}}_{C_{JE}} & \text{for } V_{BE} < \text{FC} \times \phi_E \\ \underbrace{\tau_F \frac{dI_{CC}}{dV_{BE}}}_{C_{DE}} + \underbrace{\frac{C_{JE}(0)}{F_2} \left(F_3 + \frac{m_E V_{BE}}{\phi_E}\right)}_{C_{JE}} & \text{for } V_{BE} \geq \text{FC} \times \phi_E \end{cases} \quad (2\text{-}39)$$

$$C_{BC} = \frac{dQ_{BC}}{dV_{BC}} = \begin{cases} \underbrace{\tau_R \dfrac{dI_{EC}}{dV_{BC}}}_{C_{DC}} + \underbrace{C_{JC}(0)\left(1 - \dfrac{V_{BC}}{\phi_C}\right)^{-m_C}}_{C_{JC}} & \text{for } V_{BC} < \text{FC} \times \phi_C \\ \underbrace{\tau_R \dfrac{dI_{EC}}{dV_{BC}}}_{C_{DC}} + \underbrace{\dfrac{C_{JC}(0)}{F_2}\left(F_3 + \dfrac{m_C V_{BC}}{\phi_C}\right)}_{C_{JC}} & \text{for } V_{BC} \geq \text{FC} \times \phi_C \end{cases} \quad (2\text{-}40)$$

$$C_{CS} = \begin{cases} C_{JS}(0)\left(1 - \dfrac{V_{CS}}{\phi_S}\right)^{-m_S} & \text{for } V_{CS} < 0 \\ C_{JS}(0)\left(1 + \dfrac{m_S V_{CS}}{\phi_S}\right) & \text{for } V_{CS} > 0 \end{cases} \quad (2\text{-}41)$$

where, for the base-emitter junction,

$$F_1 = \frac{\phi_E}{1 - m_E}[1 - (1 - \text{FC})^{1 - m_E}]$$
$$F_2 = (1 - \text{FC})^{1 + m_E} \quad (2\text{-}42)$$
$$F_3 = 1 - \text{FC}(1 + m_E)$$

and for the base-collector junction

$$F_1 = \frac{\phi_C}{1 - m_C}[1 - (1 - \text{FC})^{1 - m_C}]$$
$$F_2 = (1 - \text{FC})^{1 + m_C} \quad (2\text{-}43)$$
$$F_3 = 1 - \text{FC}(1 + m_C)$$

FC is a factor between 0 and 1, which is used to compute the voltage (FC $\times \phi_E$ and FC $\times \phi_C$) in the forward-bias region, beyond which the capacitance is modeled by a linear extrapolation. This has been done to prevent infinite capacitances at $V = \phi_E$ and at $V = \phi_C$ and also to guarantee a continuous function for the capacitances and derivatives. The default value for FC is set by SPICE2 to 0.5.

The physics-based Ebers-Moll model for the real BJT requires new model parameters that the user of SPICE2 can specify in the .MODEL statement. These parameters, which take into account the large-signal effects, are

CJE   Base-emitter zero-bias depletion capacitance (C_{JE})
CJC   Base-collector zero-bias depletion capacitance (C_{JC})
CJS   Collector-substrate zero-bias capacitance (C_{JS}); it is assumed constant

VJE   Base-emitter built-in potential (ϕ_E)
VJC   Base-collector built-in potential (ϕ_C)
VJS   Substrate junction built-in potential (ϕ_S)
TF   Ideal total forward transit time (τ_F); it is assumed constant
TR   Ideal total reverse transit time (τ_R); it is assumed constant
FC   Coefficient for forward-bias depletion capacitance (FC)
MJE   Base-emitter junction grading factor (m_E)
MJC   Base-collector junction grading factor (m_C)
MJS   Collector-substrate junction grading factor (m_S)

The keyname used in the text is indicated in parentheses.

The junction gradient factors, even if included in Eqs. (2-36) to (2-41), are not considered in the actual Ebers-Moll model, since they affect only the Gummel-Poon model (see Sec. 2.6.1). For the Ebers-Moll model these parameters are fixed at their default values.

2.4 Ebers-Moll Small-Signal Model

The models of the BJT discussed so far maintain the strictly nonlinear nature of the physics of the device and therefore are able to represent, both statically and dynamically, the behavior of the BJT for both types of bias.

In some circuit situations, however, the characteristics of the device must be represented only in a restricted range of currents and voltages. In particular, for small variations around the operating point (op), which is fixed by a constant source, the nonlinear characteristic of the BJT can be *linearized*, so that the incremental current becomes proportional to the incremental voltage if the variations are sufficiently small.

The development of an incremental linear model is not necessarily linked to a graphic interpretation of the relationships between the variables; the mathematical development can be based on the linearization of a nonlinear functional relationship by means of the expansion in a Taylor series truncated after the first-order terms, as obtained for the diode small-signal model.

2.4.1 Small-signal model and its implementation in SPICE2

a. Parameter definition. When transistors are biased in the active region and used for amplification, it is often worthwhile to approximate their behavior under conditions of small voltage variations at the base-emitter junction. If these variations are smaller than the thermal voltage kT/q, it is possible to represent the transistor by a linear equivalent circuit. This representation can be of great aid in the design of amplifying circuits. It is called the *small-signal transistor model*.

When a transistor is biased in the active mode, collector current is related to base-emitter voltage by Eq. (2-12), which is repeated here for convenient reference (assuming $V_{BE} \gg kT/q$ and $V_{BC} \leq 0$):

$$I_C = I_S \, e^{qV_{BE}/kT} \qquad (2\text{-}44)$$

Hence, if V_{BE} varies incrementally, I_C also varies according to

$$\left.\frac{dI_C}{dV_{BE}}\right|_{op} = \frac{qI_S}{kT} e^{qV_{BE}/kT} = \left.\frac{qI_C}{kT}\right|_{op} \equiv g_{mF} \qquad (2\text{-}45)$$

This derivative is recognized as the *transconductance* and is given the usual symbol g_{mF}.

The variation of base current with base-emitter voltage can be found directly from Eq. (2-44), i.e.,

$$\left.\frac{dI_B}{dV_{BE}}\right|_{op} = \left.\frac{d(I_C/\beta_F)}{dV_{BE}}\right|_{op} = \frac{1}{\beta_F}\frac{I_S q}{kT} e^{qV_{BE}/kT} = \frac{g_{mF}}{\beta_F} \equiv g_\pi \qquad (2\text{-}46)$$

It must be noted that β_F is here a constant (variations of β_F are described in Sec. 2.5.1).

From the analysis of Sec. 2.2.2b, it is known that the voltage across the collector-base junction influences collector current chiefly as a result of the Early effect. The variation of I_C with V_{BC} is shown in Eq. (2-24) to be the ratio of the collector current I_C to the Early voltage V_A. In terms of the small-signal parameters,

$$\left.\frac{dI_C}{dV_{BC}}\right|_{op} = \left.\frac{I_C}{|V_A|}\right|_{op} = \frac{g_{mF}kT}{q|V_A|} \equiv g_o \qquad (2\text{-}47)$$

Any change in base minority charge results in a change in base current as well as in collector current. Thus, the variation in V_{BC} results in a change in I_B; this effect can be modeled by inclusion of a resistor, r_μ, from collector to base. If V_{BE} is assumed constant, we can write

$$\left.\frac{dI_B}{dV_{BC}}\right|_{op} = \left.\frac{d(I_C/\beta_R)}{dV_{BC}}\right|_{op} = \frac{g_{mR}}{\beta_R} \equiv g_\mu \qquad (2\text{-}48)$$

An additional effect has to be considered: the base resistance. Current-crowding effects cause an overall dc base resistance that is a function of collector current. In the Ebers-Moll model, however, this variation is not taken into account; thus

$$g_x = \frac{1}{r_B} \tag{2-49}$$

Expressing the charge-storage elements as voltage-dependent small-signal capacitances (linearized capacitances) yields

$$\left.\frac{dQ_{BE}}{dV_{BE}}\right|_{op} = \underbrace{\left.\frac{d(Q_{DE} + Q_{JE})}{dV_{BE}}\right|_{op}}_{C_\pi}$$

$$= \underbrace{\tau_F \frac{qI_S}{kT} e^{qV_{BE}/kT}}_{\tau_F g_{mF}} + \underbrace{C_{JE}(0) \left(1 - \frac{V_{BE}}{\phi_E}\right)^{-m_E}}_{C_{JE}(V_{BE})} \tag{2-50}$$

$$\left.\frac{dQ_{BC}}{dV_{BC}}\right|_{op} = \underbrace{\left.\frac{d(Q_{DC} + Q_{JC})}{dV_{BC}}\right|_{op}}_{C_\mu} = \underbrace{\tau_R \frac{qI_S}{kT} e^{qV_{BC}/kT}}_{\tau_R g_{mR}} + \underbrace{C_{JC}(0) \left(1 - \frac{V_{BC}}{\phi_C}\right)^{-m_C}}_{C_{JC}(V_{BC})} \tag{2-51}$$

This linearization is performed automatically inside SPICE2.

Combination of the above small-signal circuit elements yields the small-signal model of the bipolar transistor shown in Fig. 2-19. This is valid for both *npn* and *pnp* devices and is called the *hybrid-π model*.

The elements for this linear hybrid-π model, shown in Fig. 2-19, are summarized as follows:

$$g_{mF} = \frac{q}{kT} I_S e^{qV_{BE}/kT}$$

$$r_\pi = \frac{\beta_F}{g_{mF}}$$

$$r_o = \frac{q}{kT g_{mF}} V_A$$

$$g_{mR} = \frac{q}{kT} I_S e^{qV_{BC}/kT} \tag{2-52}$$

$$r_\mu = \frac{\beta_R}{g_{mR}}$$

$$r_x = r_B$$

$$C_\pi = g_{mF} \tau_F + C_{JE}(V_{BE})$$

$$C_\mu = g_{mR} \tau_R + C_{JC}(V_{BC})$$

$$C_{CS} = C_{JS}(V_{CS})$$

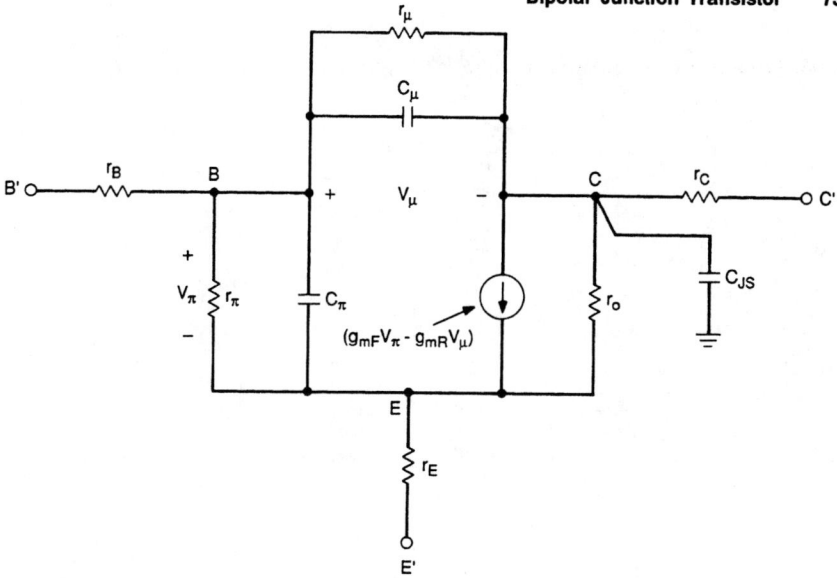

Figure 2-19 SPICE2 BJT Ebers-Moll small-signal model.

In the normal region of operation, the reverse transconductance g_{mR} is essentially zero, so that resistance r_μ can be regarded as infinite and capacitance $C_\mu = C_{JC}(V_{BC})$.

b. Implementation in SPICE2. Equations (2-52) are expressed in SPICE2 by the following relations, which define the BJT SPICE2 Ebers-Moll small-signal model:

$$g_m = \begin{cases} \dfrac{qI_S}{kT} e^{qV_{BE}/kT} + \text{GMIN} - g_o & \text{for } V_{BE} > -5\dfrac{kT}{q} \\ -\dfrac{I_S}{V_{BE}} + \text{GMIN} - g_o & \text{for } V_{BE} \leq -5\dfrac{kT}{q} \end{cases} \quad (2\text{-}53)$$

$$g_\pi \equiv \dfrac{1}{r_\pi} = \begin{cases} \dfrac{I_S}{\beta_F} \dfrac{q}{kT} e^{qV_{BE}/kT} + \dfrac{\text{GMIN}}{\beta_F} & \text{for } V_{BE} > -5\dfrac{kT}{q} \\ -\dfrac{I_S}{\beta_F V_{BE}} + \dfrac{\text{GMIN}}{\beta_F} & \text{for } V_{BE} \leq -5\dfrac{kT}{q} \end{cases} \quad (2\text{-}54)$$

$$g_o \equiv \dfrac{1}{r_o} = \begin{cases} \dfrac{qI_S}{kT} e^{qV_{BC}/kT} + \text{GMIN} & \text{for } V_{BC} > -5\dfrac{kT}{q} \\ -\dfrac{I_S}{V_{BC}} + \text{GMIN} & \text{for } V_{BC} \leq -5\dfrac{kT}{q} \end{cases} \quad (2\text{-}55)$$

$$g_\mu \equiv \frac{1}{r_\mu} = \begin{cases} \dfrac{I_S}{\beta_R}\dfrac{q}{kT}e^{qV_{BC}/kT} + \dfrac{\text{GMIN}}{\beta_R} & \text{for } V_{BC} > -5\dfrac{kT}{q} \\[6pt] -\dfrac{I_S}{\beta_R V_{BC}} + \dfrac{\text{GMIN}}{\beta_R} & \text{for } V_{BC} \leq -5\dfrac{kT}{q} \end{cases} \quad (2\text{-}56)$$

$$g_x = \frac{1}{r_B} \quad (2\text{-}57)$$

$$C_\pi \equiv C_{BE} = \begin{cases} \tau_F \dfrac{qI_S}{kT}e^{qV_{BE}/kT} + C_{JE}(0)\left(1 - \dfrac{V_{BE}}{\phi_E}\right)^{-m_E} \\ \hspace{6em} \text{for } V_{BE} < \text{FC} \times \phi_E \\[6pt] \tau_F \dfrac{qI_S}{kT}e^{qV_{BE}/kT} + \dfrac{C_{JE}(0)}{F_2}\left(F_3 + \dfrac{m_E V_{BE}}{\phi_E}\right) \\ \hspace{6em} \text{for } V_{BE} \geq \text{FC} \times \phi_E \end{cases}$$
$$(2\text{-}58)$$

$$C_\mu \equiv C_{BC} = \begin{cases} \tau_R \dfrac{qI_S}{kT}e^{qV_{BC}/kT} + C_{JC}(0)\left(1 - \dfrac{V_{BC}}{\phi_C}\right)^{-m_C} \\ \hspace{6em} \text{for } V_{BC} < \text{FC} \times \phi_C \\[6pt] \tau_R \dfrac{qI_S}{kT}e^{qV_{BC}/kT} + \dfrac{C_{JC}(0)}{F_2}\left(F_3 + \dfrac{m_C V_{BC}}{\phi_C}\right) \\ \hspace{6em} \text{for } V_{BC} \geq \text{FC} \times \phi_C \end{cases}$$
$$(2\text{-}59)$$

$$C_{CS} = \begin{cases} C_{JS}(0)\left(1 - \dfrac{V_{CS}}{\phi_S}\right)^{-m_S} & \text{for } V_{CS} < 0 \\[6pt] C_{JS}(0)\left(1 + \dfrac{m_S V_{CS}}{\phi_S}\right) & \text{for } V_{CS} > 0 \end{cases} \quad (2\text{-}60)$$

where F_2 and F_3 are derived from Eqs. (2-42) and (2-43).

In Eqs. (2-53) and (2-55), the Early effect is not considered.

2.5 Gummel-Poon Static Model

The Ebers-Moll model lacks a representation of many important second-order effects present in actual devices; the two most important effects are those of *low current* and *high-level injection*. The low-current effects results from additional base current due to recombination that degrades current gain. The effects of high-level injection also reduce current gain and in addition cause an increase in τ_F and τ_R. To account for these second-order effects, the *Gummel-Poon model* has been implemented in SPICE2.

The following derivation of the Gummel-Poon static model proceeds basically in three major steps according to the modeling of the following second-order effects:

1. Low-current effects (low-current drop in β)
2. Complete description of base-width modulation
3. High-level injection

It must be noted that item 1 affects modifications in the expression for I_B, while items 2 and 3 are incorporated by modifying existing equations (for I_{CT} and C_{DE} only).

Involved in the derivation of these two last effects is a new definition of I_S in terms of the internal physics of the BJT.

It should be emphasized that the Gummel-Poon model derivation (like the previous Ebers-Moll model) assumes a one-dimensional transistor structure.

2.5.1 β variation with current

As described in Sec. 1.1.2, recombination in the depletion region leads to modified diode relationships for the junction currents. These can be modeled by adding four parameters for the Ebers-Moll model in order to define base current in terms of a superposition of ideal diode and nonideal diode components. These extra components of base current affect the low-current drop in the current gain β (discussed next), allowing introduction of the four new parameters.

In general, there are three regions of interest in the variation of β with current. Figure 2-20 shows a typical variation of β_F with I_C. Region 1 is the low-current region in which β_F increases with I_C. Region 2 is the midcurrent region in which β_F is constant (β_{FM}).

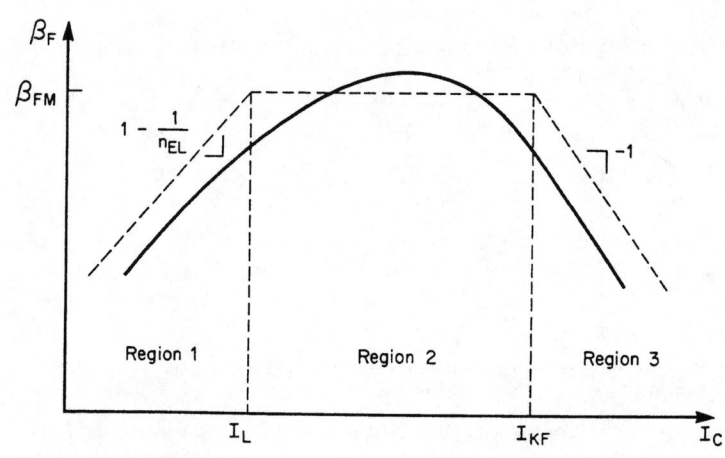

Figure 2-20 Graph of β_F vs. I_C.

Region 3 is the high-current region in which β_F drops as the current is increased. Before we analyze these regions, several points should be noted about Fig. 2-20.

The curve in Fig. 2-20 is drawn for constant V_{BC}, in this case, $V_{BC} = 0$.

In the following, it will be assumed that all data correspond to $V_{BC} = 0$. Nonzero V_{BC} data can be reduced appropriately by the application of Eq. (2-19). For simplicity, most of the following analysis considers only the variation of β_F with I_C. A similar analysis can be performed for the variation of β_R with I_E at constant V_{BE}.

It will be shown that region 1 is governed by additional components of I_B, while region 3 results from a change in I_C. This information is not evident from Fig. 2-20. Therefore, an alternative form of presenting the above information is used. This alternative form is a plot of $\ln I_C$ and $\ln I_B$ as a function of V_{BE}, as shown in Fig. 2-21. Because of the logarithmic nature of the vertical axis, β_F is obtained directly from the plot as the distance between the I_C and I_B curves. Not only is it evident from this plot what causes the variation of β_F with I_C, but all the model parameters needed to characterize this variation can be obtained directly from it. It is still assumed that $V_{BC} = 0$ for all points.

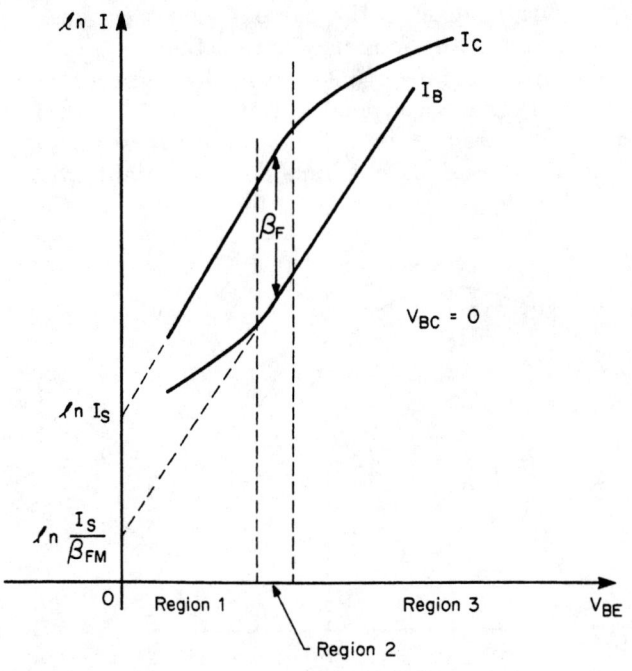

Figure 2-21 Graph of $\ln I_C$ and $\ln I_B$ vs. V_{BE}, with $V_{BC} = 0$ [6].

For some transistors, there may not appear to be a region in which β_F is constant. For these transistors, regions 1 and 3 have simply overlapped. The analysis and subdivision into the three regions is still valid, since the model parameters can still be obtained from Fig. 2-21 even when region 2 in the β_F vs. I_C curve does not exist.

An analysis of the regions is now performed, after Getreu [6].

a. Region 2: Midcurrents. In this region, the Ebers-Moll model for the ideal BJT holds; the β_F used in this model applies only to region 2 and is now called β_{FM}. The two currents in this region (for $V_{BC} = 0$) are given by

$$I_C = I_S(0)(e^{qV_{BE}/kT} - 1)$$

$$I_B = \frac{I_S(0)}{\beta_{FM}(0)}(e^{qV_{BE}/kT} - 1)$$ (2-61)

Values of $\beta_{FM}(0)$ and $I_S(0)$ can be obtained directly from Fig. 2-21.

b. Region 1: Low currents. The drop in β_F at low currents is caused by extra components of I_B that until now have been ignored. For the normal active region with $V_{BC} = 0$, there are three extra components, which are caused by the following:

1. The recombination of carriers at the surface
2. The recombination of carriers in the emitter-base depletion region
3. The formation of emitter-base surface channels

All three components have a similar variation with base-emitter voltage V_{BE}. These three components should be added to the base current of Eq. (2-61). Fortunately, this can be simplified. A composite current can be made of all three extra components. It has the form [6]

$$I_B = I_{S,\text{composite}}(e^{qV_{BE}/n_{EL}kT} - 1)$$ (2-62)

where n_{EL} is called the *nonideal low-current base-emitter emission coefficient* and lies between 1 and 4.

For most cases, a fit to Eq. (2-62) can be made with reasonable accuracy. Since channeling and surface recombination can both be made small with careful processing, the dominant component is normally the recombination in the emitter-base depletion region, and n_{EL} is normally close to 2. Therefore, at $V_{BC} = 0$, the base current is approximated by

$$I_B = \frac{I_S(0)}{\beta_{FM}(0)}(e^{qV_{BE}/kT} - 1) + C_2 I_S(0)(e^{qV_{BE}/n_{EL}kT} - 1) \quad (2\text{-}63)$$

where the term $I_{S,\text{composite}}$ in Eq. (2-62) has been replaced by $C_2 I_S(0)$; that is, it has simply been normalized to $I_S(0)$. The two additional model parameters are C_2 (the *forward low-current nonideal base current coefficient*) and n_{EL}.

When the base-collector junction is forward-biased, there will generally be three similar additional components of I_B at low-current levels: surface recombination, collector-base depletion region recombination, and collector-base channeling. In a similar way, they can be lumped together into a composite component that depends on V_{BC}. The expression for I_B then becomes

$$I_B = \frac{I_S(0)}{\beta_{FM}(0)}(e^{qV_{BE}/kT}-1) + C_2 I_S(0)(e^{qV_{BE}/n_{EL}kT}-1)$$

$$+ \frac{I_S(0)}{\beta_{RM}}(e^{qV_{BC}/kT}-1) + C_4 I_S(0)(e^{qV_{BC}/n_{CL}kT}-1) \quad (2\text{-}64a)$$

where the extra model parameters n_{CL} (the *nonideal low-current base-collector emission coefficient*), and C_4 (the *reverse low-current nonideal base current coefficient*) have been introduced.

The additional components of base current I_B are included in the circuit model by means of two nonideal diodes, as shown in Fig. 2-22.

The expression for I_C then becomes

$$I_C = I_S(0)(e^{qV_{BE}/kT} - e^{qV_{BC}/kT}) - \frac{I_S(0)}{\beta_{RM}}(e^{qV_{BC}/kT}-1)$$

$$- C_4 I_S(0)(e^{qV_{BC}/n_{CL}kT}-1) \quad (2\text{-}64b)$$

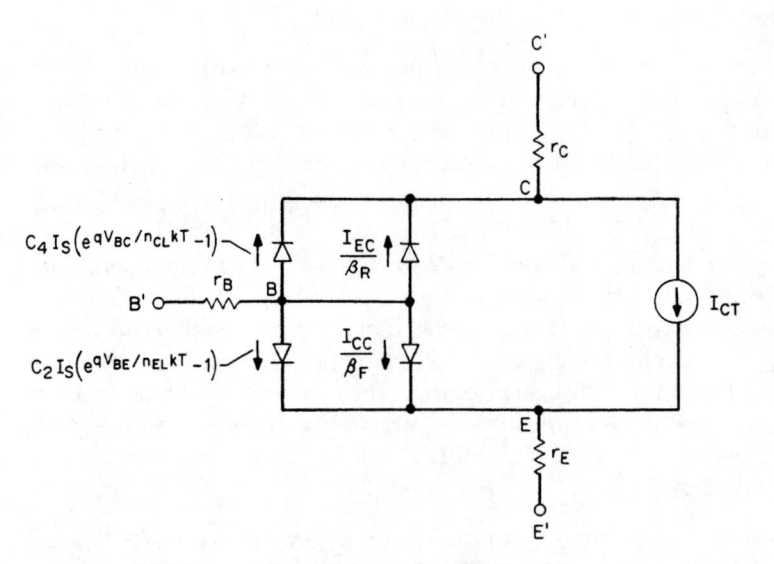

Figure 2-22 BJT Gummel-Poon static model: Effects of the extra components of I_B.

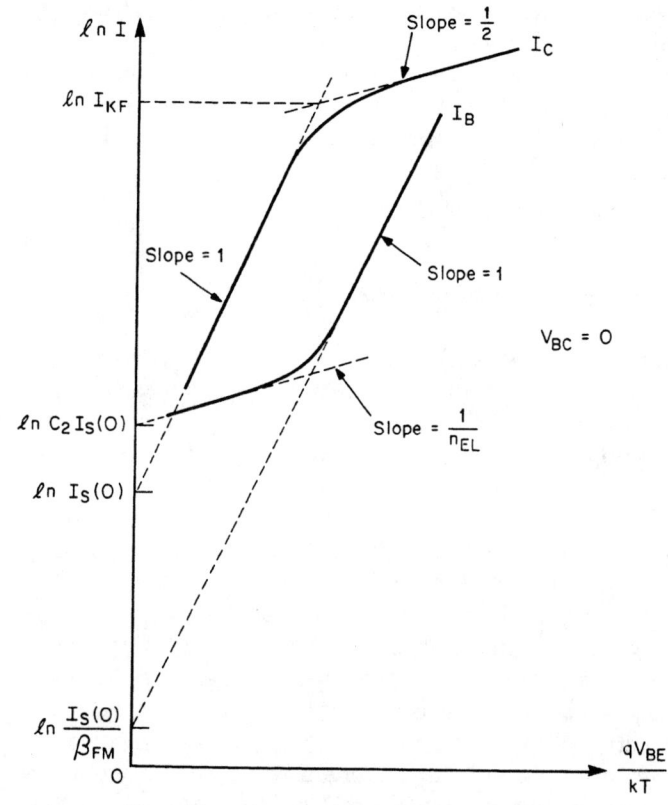

Figure 2-23 Plot of $\ln I_C$ and $\ln I_B$ vs. qV_{BE}/kT for $V_{BC} = 0$ [6].

The plot of $\ln I_B$ vs. qV_{BE}/kT for $V_{BC} = 0$, shown in Fig. 2-23, illustrates the two components of base current I_B: the ideal component with the slope of 1 and the nonideal component with the slope of $1/n_{EL}$. The extrapolation of these straight-line components to the line defined by $V_{BE} = 0$ gives the values of $C_2 I_S(0)$ and $I_S(0)/\beta_{FM}(0)$.

A similar plot of $\ln I_B$ as a function of V_{BC} for inverse operation yields values for the model parameters C_4 and n_{CL}. A typical value for C_2 (and C_4) is 10^3, and a typical value for n_{EL} (and n_{CL}) is 2.

c. Region 3: High currents. The drop in β at high currents, caused by the effects of high-level injection in the base, will be discussed later.

2.5.2 Physical definition of I_S

In this section, as pointed out previously, a new expression for I_S is obtained from an examination of the current-density equations in the

BJT. This new expression inherently incorporates the effects of base-width modulation and high-level injection.

The derivation starts with the one-dimensional, dc equations for the electron current density J_n [see Eq. (A-19b)] and the hole current density J_p [see Eq. (A-19a)] in an npn transistor:

$$J_n = q\mu_n n(x)E(x) + qD_n \frac{dn(x)}{dx}$$

$$J_p = q\mu_p p(x)E(x) - qD_p \frac{dp(x)}{dx}$$

(2-65)

where $E(x)$ is the electric field, $n(x)$ is the free electron concentration, and $p(x)$ is the hole concentration. No restriction is placed on the variation of the carrier concentrations, so the following analysis applies for any doping profile. Equations (2-65) apply for both high- and low-level injection. At this point it is assumed that the hole current is zero. This is not exactly the case, but it is normally a reasonable approximation. The approximation is justified by showing that there is no place where a large hole current could go.

The *base-emitter junction* is normally designed for high emitter-injection efficiency (high emitter doping with respect to the base doping). This means that the current injected into the emitter from the base is small even when the emitter-base junction is forward-biased [6].

In the normal active region, the *collector-base junction* is reverse-biased, and therefore no significant current flows across it from the base to the collector. In the inverse and saturated regions of operation, however, the collector-base junction is forward-biased. It is assumed that for most cases of interest, the hole current that flows from the base to the collector is still small.

Obviously, then, the following analysis will be reasonably valid only for the normal active region and when the device is weakly saturated. For cases of strong saturation and inverse operation, it may not hold. However, if the J_p term is retained, it will be multiplied by $n(x)/p(x)$, which reduces its effect even further, since $n(x) \ll p(x)$, except at high-level injection.

Then if the hole current is assumed to be zero, the second expression of Eqs. (2-65) becomes [6]

$$J_p = q\mu_p p(x)E(x) - qD_p \frac{dp(x)}{dx} = 0 \qquad (2\text{-}66)$$

If Eq. (2-66) is solved for the field E, it yields

$$E(x) = \frac{D_p}{\mu_p} \frac{1}{p(x)} \frac{dp(x)}{dx} \qquad (2\text{-}67)$$

Substituting Eq. (2-67) in the first expression of Eqs. (2-65) results in

$$J_n = q\mu_n \left[\frac{D_p}{\mu_p} \frac{1}{p(x)} \frac{dp(x)}{dx} \right] n(x) + qD_n \frac{dn(x)}{dx} \quad (2\text{-}68)$$

or

$$p(x)J_n = qD_n n(x) \frac{dp(x)}{dx} + qD_n p(x) \frac{dn(x)}{dx} \quad (2\text{-}69)$$

where the Einstein relationship [see Eqs. (A-20)] has been used. The multiplication of Eq. (2-68) by $p(x)$ is the mathematical trick that, with the following use of the differential product rule, will result in the importance of the majority-carrier concentration, $p(x)$, in the base. The use of the product rule on Eq. (2-69) results in

$$p(x)J_n = qD_n \frac{d}{dx}[n(x)p(x)] \quad (2\text{-}70)$$

Both sides of this equation are now integrated from x_E to x_C, where x_E is the position of the emitter side of the emitter-base depletion region and x_C is the position of the collector side of the collector-base depletion region, as illustrated in Fig. 2-24. Figure 2-24 also defines x'_E and x'_C, the positions of the base sides of these depletion regions [6].

Since current density J_n is constant for dc and is independent of x, assuming negligible recombination in the base region (recombination

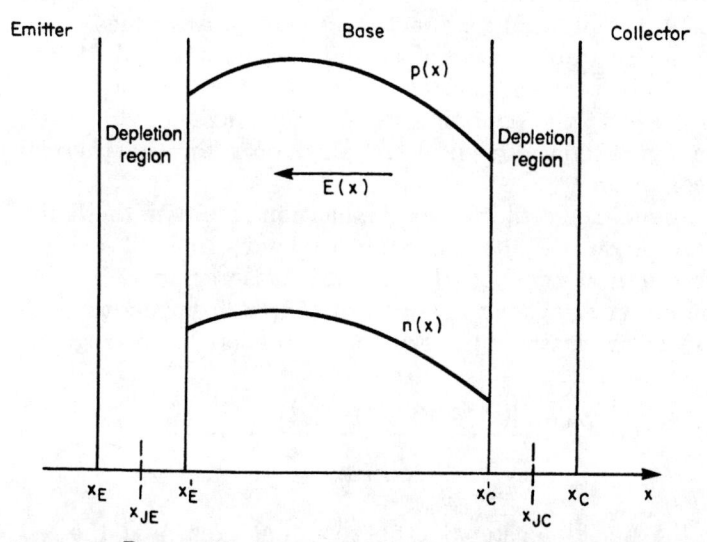

Figure 2-24 Base profile for an *npn* transistor defining the integral limits x_E, x_C, x'_E, x'_C [6].

in the depletion regions is independently modeled by the two non-ideal diodes in Fig. 2-22, so it is assumed to be zero here), J_n is taken out of the integral. Then

$$J_n \int_{x_E}^{x_C} p(x)\,dx = qD_n \int_{x_E}^{x_C} \frac{d}{dx}[n(x)p(x)]\,dx$$

$$= qD_n[n(x_C)p(x_C) - n(x_E)p(x_E)]$$

$$J_n = \frac{qD_n[n(x_C)p(x_C) - n(x_E)p(x_E)]}{\int_{x_E}^{x_C} p(x)\,dx} \tag{2-71}$$

Equation (2-71) can be simplified further by applying Boltzmann statistics to the *pn* products [see Eqs. (A-74)].

$$n(x_C)p(x_C) = n_i^2\, e^{qV_{BC}/kT}$$
$$n(x_E)p(x_E) = n_i^2\, e^{qV_{BE}/kT} \tag{2-72}$$

Therefore

$$J_n = -\frac{qD_n n_i^2 (e^{qV_{BE}/kT} - e^{qV_{BC}/kT})}{\int_{x_E}^{x_C} p(x)\,dx} \tag{2-73}$$

At this point, the *depletion approximation* is made [6]. This assumes that there are no (or negligible) mobile carriers in a depletion region (see Sec. A.2.1*b*). That is, the field experienced by the carriers and the thickness of the depletion region are such that the carriers are transported instantaneously across it. This approximation, as pointed out in Sec. 2.3.1, is not valid for junctions under forward bias.

The analysis that follows therefore first assumes the depletion approximation to be true and then later fixes up the solution to take the mobile charges in the depletion regions into account. In effect, the depletion approximation will be applied only for the thermal equilibrium condition.

The application of the depletion approximation results in the limits of the integral in Eq. (2-73) being replaced by x'_C and x'_E and the integration being performed in the neutral base region only. The change in the integration limits results, of course, from the assumption that $p(x)$ is approximately zero inside the depletion regions. Equation (2-73) can therefore be rewritten as

$$I_n \simeq -\frac{qD_n A_J n_i^2 [(e^{qV_{BE}/kT} - 1) - (e^{qV_{BC}/kT} - 1)]}{\int_{x'_E}^{x'_C} p(x)\,dx} \tag{2-74}$$

where A_J is the one-dimensional cross-sectional area, and the -1 term has been introduced for convenience.

I_n represents the total dc minority current in the positive x direction that results from minority carriers injected into the base at the emitter and/or the collector. It is represented in the model of Fig. 2-22 by the current generator I_{CT}. (The other components of the collector current in Fig. 2-22 are components of base current resulting from the injection of holes, for an npn transistor, from the base toward the collector.)

The equation for I_{CT}, previously obtained in the Ebers-Moll model [see Eq. (2-9)], can be written as

$$I_{CT} = I_S[(e^{qV_{BE}/kT} - 1) - (e^{qV_{BC}/kT} - 1)] \qquad (2\text{-}75)$$

The sign of I_{CT} is opposite to that of I_n [see Eq. (2-74)] since I_n has the opposite direction (out of the collector terminal) to that assumed for I_C (into the collector region from the terminal).

A direct comparison of Eqs. (2-74) and (2-75) yields a physical definition of I_S. In the Ebers-Moll model, I_S has been considered a fundamental constant of the device. Yet the integral in Eq. (2-74) is not constant in that, under high-level injection, $p(x)$, the majority-carrier concentration, is a function of the applied bias. To reconcile this difference, a new symbol, I_{SS}, is used in the Gummel-Poon model and is defined from Eq. (2-74) under low-level injection conditions only [6].

At low-level injection, the combination of Eqs. (2-74) and (2-75) gives

$$I_{CT(\text{low-level})} = \frac{qD_n n_i^2 A_J}{\displaystyle\int_{x_E'}^{x_C'} N_A(x)\,dx}[(e^{qV_{BE}/kT} - 1) - (e^{qV_{BC}/kT} - 1)] \qquad (2\text{-}76)$$

where $p(x)$ has been replaced by $N_A(x)$, since, at low-current levels in the neutral base region, $p(x)$ approximates $N_A(x)$.

Before the definition of I_{SS} is made, though, more attention must be paid to the limits of the integration, x_E' and x_C'. Because of the variation of the depletion region widths with applied voltage, x_E' and x_C' are functions of the appropriate bias voltage (and, in fact, will be seen later to incorporate the effects of base-width modulation).

The fundamental constant I_{SS} is therefore defined at *zero* V_{BE} and V_{BC} as

$$I_{SS} = \frac{qD_n n_i^2 A_J}{\displaystyle\int_{x_{E0}'}^{x_{C0}'} N_A(x)\,dx} \qquad (2\text{-}77)$$

where x_{E0}' and x_{C0}' are the values of x_E' and x_C' when the applied juntion voltages are zero. The fundamental nature of I_{SS} is seen immediately from Eq. (2-77), since it is uniquely determined once the base-doping profile is fixed. It must be noted that in SPICE2, I_S and I_{SS} coincide, as they are specified in the .MODEL statement [7] by the same parameter IS.

In the derivation of Eq. (2-77), the diffusion constant D_n has been assumed to be constant and independent of x. In practice, this assumption is not valid. The diffusion constant should be included in the denominator integral. Instead, D_n is interpreted in Eq. (2-77) as an effective diffusion constant in the base.

At this point in the analysis, the depletion approximation can be removed. The application of the depletion approximation allowed the above derivation to concentrate solely on the majority carriers in the neutral base region. As a result, the mobile carriers in the depletion regions have been ignored. If these carriers are now taken into account, the definition of I_{SS} has to be updated as [6]

$$I_{SS} = \frac{qD_n n_i^2 A_J}{\int_{x_{E0}}^{x_{C0}} p_0(x)\, dx} \tag{2-78}$$

where $p_0(x)$ is the equilibrium hole concentration in the neutral base and depletion regions.

2.5.3 Q_B concept

The general expression for I_S (which is now a function of bias voltages) can be obtained in terms of the zero-bias constant I_{SS}. However, before this is done, it is worthwhile to make a few definitions and to introduce some new concepts.

When multiplied by q and A_J, the integral in Eq. (2-74) represents the total *majority* charge in the neutral base region and is given the symbol Q_B:

$$Q_B = \int_{x_E}^{x_C'} qA_J p(x)\, dx \tag{2-79}$$

The *zero-bias* majority base charge, Q_{B0}, is defined by

$$Q_{B0} = \int_{x_{E0}}^{x_{C0}'} qA_J N_A(x)\, dx \tag{2-80}$$

Finally, the *normalized* majority base charge, q_b, is defined as

$$q_b = \frac{Q_B}{Q_{B0}} \tag{2-81}$$

In the following analysis, all charge normalizations are with respect to Q_{B0} and are represented by q with an appropriate subscript.

The above definitions [Eqs. (2-79) to (2-81)] can be used to find the new definition of I_S. If Eq. (2-74) is multiplied and divided by Q_{B0} and I_n is replaced by $-I_{CT}$, then [6]

$$I_{CT} = \frac{qD_n n_i^2 A_J}{\int_{x_E}^{x_C} p(x)\, dx\, qA_J \int_{x_{E0}}^{x_{C0}} N_A(x)\, dx} \cdot qA_J \int_{x_{E0}}^{x_{C0}} N_A(x)\, dx \; [(e^{qV_{BE}/kT} - 1) - (e^{qV_{BC}/kT} - 1)]$$

(2-82)

Combining the first term in the numerator with the second term in the denominator, and using Eqs. (2-77) to (2-81), gives

$$I_{CT} = \frac{I_{SS} Q_{B0}}{Q_B}[(e^{qV_{BE}/kT} - 1) - (e^{qV_{BC}/kT} - 1)]$$

$$= \frac{I_{SS}}{q_b}[(e^{qV_{BE}/kT} - 1) - (e^{qV_{BC}/kT} - 1)] \qquad (2\text{-}83)$$

Equation (2-83) is the new equation introduced by the Gummel-Poon model, and the new concept introduced is the fundamental importance of the (normalized) majority charge in the base, q_b. The old saturation current I_S (which was assumed constant in the Ebers-Moll model) has been replaced by the new term I_{SS}/q_b, where I_{SS} is the fundamental constant (defined at zero-bias condition) and q_b is a variable that still needs to be determined. The rest of the Gummel-Poon derivation involves the determination of q_b as a function of the bias conditions.

If the depletion approximation is removed, an updated expression for Q_B and Q_{B0}, including the depletion regions (as seen for I_{SS}), follows:

$$Q_B = \int_{x_E}^{x_C} qA_J p(x)\, dx \qquad (2\text{-}84)$$

$$Q_{B0} = \int_{x_{E0}}^{x_{C0}} qA_J p_0(x)\, dx \simeq \int_{x_{E0}}^{x_{C0}} qA_J N_A(x)\, dx \qquad (2\text{-}85)$$

where $p_0(x)$ is the equilibrium hole concentration in the neutral base and depletion regions. The second form of Eq. (2-85), which assumes that the depletion approximation is valid for the equilibrium case, naturally agrees with the previous definition given in Eq. (2-80).

2.5.4 Solution for q_b

When Q_B is normalized with respect to Q_{B0} and is split into its five components, it can be written in the following form:

$$q_b = \frac{Q_B}{Q_{B0}} = 1 + \frac{C_{JE}}{Q_{B0}} V_{BE} + \frac{C_{JC}}{Q_{B0}} V_{BC} + \frac{\tau_{BF}}{Q_{B0}} I_{CC} + \frac{\tau_{BR}}{Q_{B0}} I_{EC} \quad (2\text{-}86)$$

(See Ref. [6] for a complete treatment of the components of Q_B; here only the SPICE2 user-oriented results are described.)

Equation (2-86) can be put into a more expanded form as

$$q_b = 1 + \underbrace{\frac{V_{BE}}{V_B}}_{q_e} + \underbrace{\frac{V_{BC}}{V_A}}_{q_c} + \underbrace{\frac{\tau_{BF}}{Q_{B0}} I_{SS} \frac{e^{qV_{BE}/kT} - 1}{q_b}}_{q_{bF}} + \underbrace{\frac{\tau_{BR}}{Q_{B0}} I_{SS} \frac{e^{qV_{BC}/kT} - 1}{q_b}}_{q_{bR}} \quad (2\text{-}87)$$

where

$$\frac{Q_{B0}}{C_{JE}} = V_B \quad \text{(reverse Early voltage)}$$

$$\frac{Q_{B0}}{C_{JC}} = V_A \quad \text{(forward Early voltage)}$$

$$I_{CC} = \frac{I_{SS}}{q_b}(e^{qV_{BE}/kT} - 1) \quad \text{[see Eqs. (2-6)]}$$

$$I_{EC} = \frac{I_{SS}}{q_b}(e^{qV_{BC}/kT} - 1) \quad \text{[see Eqs. (2-7)]}$$

Note that q_e and q_c both model the effects of base-width modulation, while q_{bF} and q_{bR} both model the effects of high-level injection.

To simplify the algebra, the following definitions can be made:

$$q_b = q_1 + \frac{q_2}{q_b} \quad (2\text{-}88)$$

where

$$q_1 = 1 + q_e + q_c = 1 + \frac{V_{BE}}{V_B} + \frac{V_{BC}}{V_A}$$

$$q_2 = \frac{\tau_{BF} I_{SS}(e^{qV_{BE}/kT} - 1) + \tau_{BR} I_{SS}(e^{qV_{BC}/kT} - 1)}{Q_{B0}} \quad (2\text{-}89)$$

By definition, q_1 is unity at zero bias; this reflects the fact that the physical base width and the electrical base width are equal to zero bias. If both junctions are forward-biased, the electrical base width is larger than the physical base width, and q_1 is larger than unity. Conversely, if both junctions are reverse-biased, then q_1 is less than unity, since the electrical base width is less than the physical base width. Thus q_1 models the effects of base-width modulation.

The q_2 component of base charge accounts for the excess majority-carrier base charge that results from injected minority carriers. Excess majority-carrier base charge is insignificant until the injected minority carrier is comparable to the zero-bias majority-carrier charge, that is, until high-level injection is reached. Thus q_2 models the effects of high-level injection.

Equation (2-88) results in a quadratic expression for q_b,

$$q_b^2 - q_b q_1 - q_2 = 0 \tag{2-90}$$

which gives

$$q_b = \frac{q_1}{2} + \sqrt{\left(\frac{q_1}{2}\right)^2 + q_2} \tag{2-91}$$

where the negative solution has been ignored because q_b is greater than zero.

From Eq. (2-91) it follows that

$$q_b \simeq q_1 \quad \text{if } q_2 \ll \frac{q_1^2}{4} \quad \text{(low-level injection)}$$

$$q_b \simeq \sqrt{q_2} \quad \text{if } q_2 \gg \frac{q_1^2}{4} \quad \text{(high-level injection)} \tag{2-92}$$

a. High-level injection solution. In the normal active region at high-injection levels, $q_b \simeq \sqrt{q_2}$. To simplify the situation, consider the case of $V_{BC} = 0$ (that is, $q_{bR} = 0$). The extension of the result to nonzero q_{bR} will be made by analogy.

For zero q_{bR}

$$q_b = \sqrt{q_2} \simeq \sqrt{\frac{\tau_{BF} I_{SS}}{Q_{B0}} e^{qV_{BE}/kT}} = \sqrt{\frac{\tau_{BF} I_{SS}}{Q_{B0}}} e^{qV_{BE}/2kT} \tag{2-93}$$

Therefore, solving for I_C ($I_C = I_{CC}$ in this situation),

$$I_C = \frac{I_{SS}(e^{qV_{BE}/kT} - 1)}{\sqrt{\tau_{BF} I_{SS}/Q_{B0}} \, e^{qV_{BE}/2kT}} \simeq \sqrt{\frac{Q_{B0} I_{SS}}{\tau_{BF}}} e^{qV_{BE}/2kT} \tag{2-94}$$

Therefore

$$I_C \propto e^{qV_{BE}/2kT} \tag{2-95}$$

b. Simplification of q_2. The simplification of the coefficient $\sqrt{Q_{B0} I_{SS}/\tau_{BF}}$ arises from a consideration of the ln I_C vs. V_{BE} characteristics at the two extremes: high- and low-level injection.

Figure 2-23 shows the variation of $\ln I_C$ as a function of qV_{BE}/kT. The low-current asymptote is given approximately by the following (for $q_e \simeq q_c \simeq 0$):

$$I_C \simeq I_{SS} \, e^{qV_{BE}/kT} \qquad (2\text{-}96)$$

The high-current asymptote is given by Eq. (2-94). The intersection of those two asymptotes defines the knee current I_{KF} and the *knee voltage* V_{KF}.

From Eq. (2-94), it follows that

$$I_{KF} = \sqrt{\frac{Q_{B0} I_{SS}}{\tau_{BF}}} \, e^{qV_{KF}/2kT} \quad \text{(high-level)} \qquad (2\text{-}97)$$

while from Eq. (2-96), it follows that

$$I_{KF} = I_{SS} \, e^{qV_{KF}/kT} \quad \text{(low-level)} \qquad (2\text{-}98)$$

The solution of Eqs. (2-97) and (2-98) yields

$$I_{KF} = \frac{Q_{B0}}{\tau_{BF}} \qquad (2\text{-}99)$$

A similar analysis for the inverse region defines I_{KR}, the knee current for $\ln I_E$ vs. V_{BC} in the inverse region, as

$$I_{KR} = \frac{Q_{B0}}{\tau_{BR}} \qquad (2\text{-}100)$$

Additional understanding of the Gummel-Poon parameters is obtained by inspection of the *asymptotic* behavior of the short-circuit current gain as shown in Fig. 2-20. The current gain is essentially constant at a value of β_{FM} for collector currents greater than I_L, falls off with a slope of $1 - 1/n_{EL}$ for collector currents less than I_L, and falls off with a slope of -1 for collector currents greater than I_{KF}. I_L, the low-current breakpoint, is given by the equation

$$I_L = I_S (C_2 \beta_{FM})^{n_{EL}/(n_{EL}-1)} \qquad (2\text{-}101)$$

This, of course, is only an asymptotic relation.

c. Final solution. As a result of Eqs. (2-99) and (2-100), q_b can be finally written as

$$q_b = \frac{q_1}{2} + \sqrt{\left(\frac{q_1}{2}\right)^2 + q_2} \qquad (2\text{-}102)$$

where

$$q_1 = 1 + \frac{V_{BE}}{V_B} + \frac{V_{BC}}{V_A}$$

$$q_2 = \frac{I_{SS}}{I_{KF}}(e^{qV_{BE}/kT} - 1) + \frac{I_{SS}}{I_{KR}}(e^{qV_{BC}/kT} - 1)$$

(2-103)

It must be noticed that all parameters in the above expressions (V_B, V_A, I_{SS}, I_{KF}, and I_{KR}) are measurable from plots of $\ln I_C$ vs. V_{BE} in the normal active region, from plots of $\ln I_E$ vs. V_{BC} in the inverse region, and from I_C vs. V_{CE} and I_E vs. V_{CE} characteristics. Once these parameters are known, q_1, q_2, and, therefore, q_b can be easily determined by the above expressions.

The preceding derivation of the Gummel-Poon model differs from that of Ref. [4] in the integration limits used. Whereas the preceding derivation integrated from the outside of the depletion regions, Gummel and Poon integrated over virtually the entire transistor.

The reason for this difference is widely explained in Ref. [6]. Two basic differences result. They are different definitions of I_{SS} and the knee currents I_{KF} and I_{KR}.

In the definition of I_{SS} [see Eq. (2-78)], the Gummel-Poon approach results in the denominator integration being performed over virtually the entire transistor. Since the base majority carriers are minority carriers in the neutral emitter and collector regions, this difference should have a negligible effect. This difference in definition is not important for the device characterization, since I_{SS} is determined experimentally [6].

In the equations for I_{KF} and I_{KR} [Eqs. (2-99) and (2-100)], the terms τ_{BF} and τ_{BR} occur. If the integration is performed over the entire transistor, the definitions of τ_{BF} and τ_{BR} would include the emitter delay τ_E and the collector delay τ_C, respectively (since Q_E would be added to Q_{BF}, and Q_C would be added to Q_{BR}). Again, since I_{KF} and I_{KR} are both experimentally determined, this difference in τ_{BF} and τ_{BR} is not important for the device characterization.

2.5.5 Current dependence of the base resistance

An improvement to the Gummel-Poon model [4] has been introduced in SPICE2. It takes into account the dependence of the base resistance on the current (*current crowding*).

The base resistance between the external and the internal base nodes comes from two separate resistances [1]. The external constant resistance r_B (*extrinsic base resistance*) consists of the contact resistance and the sheet resistance of the external base region. The resistance r_{BM} of the internal base (active base) region (*intrinsic base resistance*), which is the part of the base lying directly under the emitter, is a function of the base current. The dependence of this resistance on the device current arises from nonzero base-region resistivity, which in turn precipitates nonuniform biasing of the base-emitter junction.

It is possible to analyze this effect exactly, but the mathematics become relatively tedious and might obscure the relevant physical mechanisms; therefore, we will consider only the results of this analysis.

Thus, it can be shown that the total base resistance can be expressed as [10]

$$r_{BB'} = r_{BM} + 3(r_B - r_{BM})\left(\frac{\tan z - z}{z \tan^2 z}\right) \quad (2\text{-}104)$$

where r_{BM} is the minimum base resistance that occurs at high currents; r_B is the base resistance at zero bias (small base currents); and z is a variable of base resistivity, thermal voltage, and intrinsic base length.

In order to reduce the computational complexity in calculating z, an approximation method is used to represent $\cos z$ by the first two terms of its McLaurin series. The value of z from this approximation is

$$z = \frac{-1 + \sqrt{1 + 144 I_B/\pi^2 I_{rB}}}{(24/\pi^2)\sqrt{I_B/I_{rB}}} \quad (2\text{-}105)$$

where I_{rB} is the current where the base resistance falls halfway to its minimum value.

In Fig. 2-25 the collector current is shown as a function of the base-emitter voltage, giving the deviation from ideal behavior at high currents due to the base resistance.

2.5.6 Static model summary

The improvements to the Ebers-Moll static model provided by the Gummel-Poon static model influence the following effects.

1. *Low-current drop in β*. This drop is caused by extra components of I_B [see Eqs. (2-64)], which can be described by four model parameters, C_2, n_{EL} (for β_F) and C_4, n_{CL} (for β_R). Two nonideal diodes were added to the Ebers-Moll static circuit model (see Fig. 2-22).

2. *Base-width modulation*. Base-width modulation is due to a change in the collector-base and emitter-base junction voltages, which causes a variation of the collector-base and emitter-base depletion

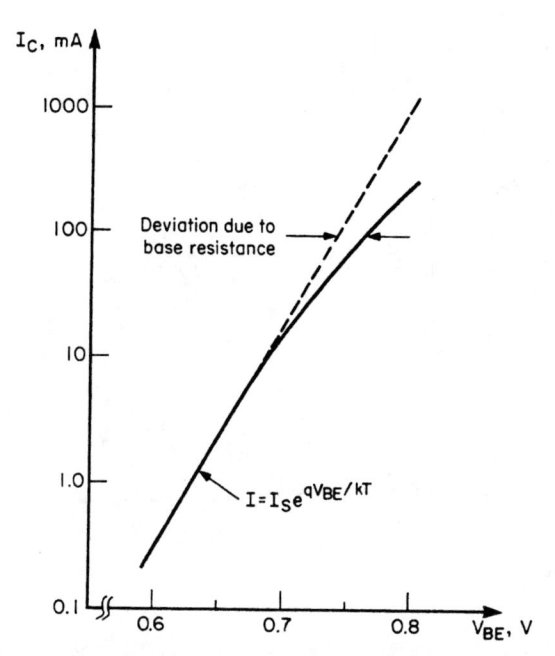

Figure 2-25 Collector current as a function of base-emitter voltage showing the deviation from ideal behavior at high currents due to the base resistance.

regions. This effect is described by two parameters, V_A (forward Early voltage) and V_B (reverse Early voltage), which are identified in the q_c and q_e components of q_b [see Eq. (2-87)], respectively.

3. *High-level injection.* At high-level injection, the injection of minority carriers into the base region is significant with respect to the majority-carrier concentration. Since space-charge neutrality is maintained in the base, the total majority-carrier concentration is increased by the same amount as the total minority-carrier concentration. The effect of the excess majority carriers is represented by a changing in the slope of the I_C vs. V_{BE} characteristic (see Fig. 2-23) above the so-called knee current I_{KF} (for the normal active region) [see Eq. (2-97)]. Another parameter, I_{KR}, is defined for the reverse active region.

4. *Base resistance.* The base resistance is demonstrated to be current-dependent and is modeled by a combination of r_B, r_{BM}, and I_{rB} [see Eq. (2-104)].

It must be noted that involved in the derivation of the Gummel-Poon static model is a new definition of $I_S = I_{SS}/q_b$. Another feature provided by the Gummel-Poon model is the different approach to the

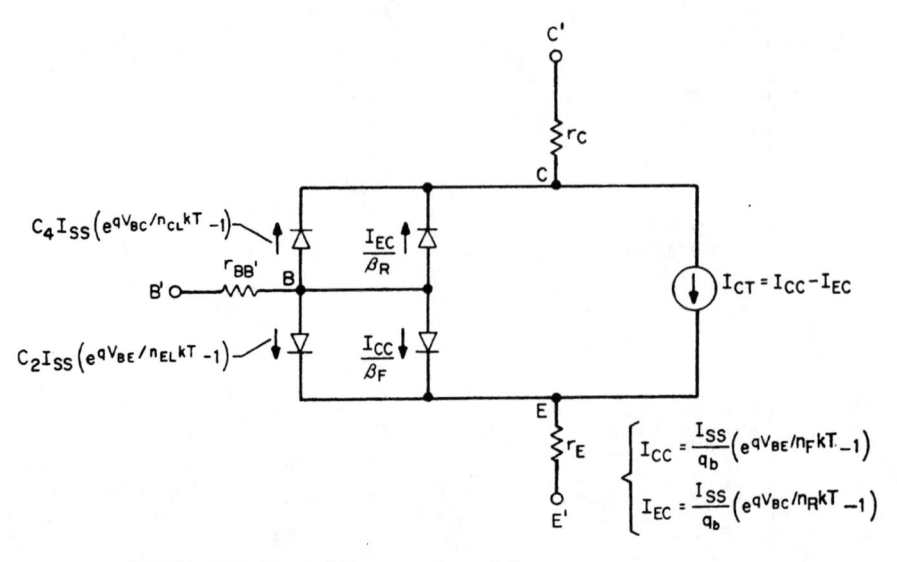

Figure 2-26 SPICE2 BJT Gummel-Poon static model.

device that is used, namely, how the analysis of the majority carriers in the base can be used to generate the preceding equations.

The complete Gummel-Poon model that accounts for the above-described effects is shown in Fig. 2-26.

2.5.7 Static model and its implementation in SPICE2

This section shows how the Gummel-Poon static model is implemented in SPICE2 [2].

The dc characteristics of the BJT are defined by the following equations when the GMIN term is also considered (see Sec. 1.2.1a).

Normal active region

For $V_{BE} > -5\dfrac{n_F kT}{q}$ and $V_{BC} \le -5\dfrac{n_R kT}{q}$:

$$I_C = \frac{I_S}{q_b}\left(e^{qV_{BE}/n_F kT} + \frac{q_b}{\beta_R}\right) + C_4 I_S + \left[\frac{V_{BE}}{q_b} - \left(\frac{1}{q_b} + \frac{1}{\beta_R}\right)V_{BC}\right]\text{GMIN} \tag{2-106}$$

$$I_B = I_S\left[\frac{1}{\beta_F}(e^{qV_{BE}/n_F kT} - 1) - \frac{1}{\beta_R}\right] + C_2 I_S(e^{qV_{BE}/n_{EL} kT} - 1)$$

$$- C_4 I_S + \left(\frac{V_{BE}}{\beta_F} + \frac{V_{BC}}{\beta_R}\right)\text{GMIN}$$

Inverse region

For $V_{BE} \leq -5\dfrac{n_F kT}{q}$ and $V_{BC} > -5\dfrac{n_R kT}{q}$:

$$I_C = -\frac{I_S}{q_b}\left[e^{qV_{BC}/n_R kT} + \frac{q_b}{\beta_R}(e^{qV_{BC}/n_R kT} - 1)\right] - C_4 I_S(e^{qV_{BC}/n_{CL} kT} - 1)$$

$$+ \left[\frac{V_{BE}}{q_b} - \left(\frac{1}{q_b} + \frac{1}{\beta_R}\right)V_{BC}\right]\text{GMIN}$$

$$I_B = -I_S\left[\frac{1}{\beta_F} - \frac{1}{\beta_R}(e^{qV_{BC}/n_R kT} - 1)\right] - C_2 I_S$$

$$+ C_4 I_S(e^{qV_{BC}/n_{CL} kT} - 1) + \left(\frac{V_{BE}}{\beta_F} + \frac{V_{BC}}{\beta_R}\right)\text{GMIN}$$

(2-107)

Saturated region

For $V_{BE} > -5\dfrac{n_F kT}{q}$ and $V_{BC} > -5\dfrac{n_R kT}{q}$:

$$I_C = \frac{I_S}{q_b}\left[(e^{qV_{BE}/n_F kT} - e^{qV_{BC}/n_R kT}) - \frac{q_b}{\beta_R}(e^{qV_{BC}/n_R kT} - 1)\right]$$

$$- C_4 I_S(e^{qV_{BC}/n_{CL} kT} - 1) + \left[\frac{V_{BE}}{q_b} - \left(\frac{1}{q_b} + \frac{1}{\beta_R}\right)V_{BC}\right]\text{GMIN}$$

$$I_B = I_S\left[\frac{1}{\beta_F}(e^{qV_{BE}/n_F kT} - 1) + \frac{1}{\beta_R}(e^{qV_{BC}/n_R kT} - 1)\right]$$

$$+ C_2 I_S(e^{qV_{BE}/n_{EL} kT} - 1) + C_4 I_S(e^{qV_{BC}/n_{CL} kT} - 1)$$

$$+ \left(\frac{V_{BE}}{\beta_F} + \frac{V_{BC}}{\beta_R}\right)\text{GMIN}$$

(2-108)

Off region

For $V_{BE} \leq -5\dfrac{n_F kT}{q}$ and $V_{BC} \leq -5\dfrac{n_R kT}{q}$:

$$I_C = \frac{I_S}{\beta_R} + C_4 I_S + \left[\frac{V_{BE}}{q_b} - \left(\frac{1}{q_b} + \frac{1}{\beta_R}\right)V_{BC}\right]\text{GMIN}$$

$$I_B = -I_S\left(\frac{\beta_F + \beta_R}{\beta_F \beta_R}\right) - C_2 I_S - C_4 I_S + \left(\frac{V_{BE}}{\beta_F} + \frac{V_{BC}}{\beta_R}\right)\text{GMIN}$$

(2-109)

The SPICE2 BJT Gummel-Poon model is the same as in Fig. 2-26 when the new expressions for I_C and I_B [see Eqs. (2-106) to (2-109)] are used.

The extra parameters required for the Gummel-Poon model, as well as those for the Ebers-Moll model, can be specified in the .MODEL statement as:

C2   Forward low-current nonideal base current coefficient (C_2)
C4   Reverse low-current nonideal base current coefficient (C_4)
NE   Nonideal low-current base-emitter emission coefficient (n_{EL})
NC   Nonideal low-current base-collector emission coefficient (n_{CL})
VAF   Forward Early voltage (V_A)
VAR   Reverse Early voltage (V_B)
IKF   Corner for forward β high-current roll-off (I_{KF})
IKR   Corner for reverse β high-current roll-off (I_{KR})
RB   Zero-bias resistance (r_B)
RBM   Minimum base resistance at high currents (r_{BM})
IRB   Current where base resistance falls halfway to its minimum value (I_{rB})
ISE   Nonideal base-emitter saturation current ($I_{SE} = C_2 I_S$)
ISC   Nonideal base-collector saturation current ($I_{SC} = C_4 I_S$)
NF   Forward current emission coefficient (n_F)
NR   Reverse current emission coefficient (n_R)

The keyname used in the text is indicated in parentheses. The ISE and ISC parameters are used in version G of SPICE2 instead of the C_2 and C_4 parameters. The parameter n_F models the forward collector current at low currents; it is the collector current exponential factor that specifies the slope of the I_C vs. V_{BE} characteristic on the logarithmic scale; n_R has exactly the same effect in the reverse mode.

The base charge q_b, as explained in the previous sections, is the total majority-carrier charge in the base region divided by the zero-bias majority-carrier charge and is defined by Eqs. (2-102) and (2-103).

SPICE2 makes the following modification in Eq. (2-102) to separate the effect of q_1 and q_2. The modification has a net effect of slightly more β roll-off at high collector current.

Thus

$$q_b = \frac{q_1}{2}[1 + \sqrt{1 + 4q_2}] \qquad (2\text{-}110)$$

In SPICE2, q_1 is approximated by the equation

$$q_1 = \left(1 - \frac{V_{BC}}{V_A} - \frac{V_{BE}}{V_B}\right)^{-1} \qquad (2\text{-}111)$$

while q_2 has different expressions according to the regions of operation, as follows.

Normal active region

For $V_{BE} > -5\dfrac{n_F kT}{q}$ and $V_{BC} \leq -5\dfrac{n_R kT}{q}$:

$$q_2 = \dfrac{I_S}{I_{KF}}(e^{qV_{BE}/n_F kT} - 1) - \dfrac{I_S}{I_{KR}} + \left(\dfrac{V_{BE}}{I_{KF}} + \dfrac{V_{BC}}{I_{KR}}\right)\text{GMIN} \quad (2\text{-}112)$$

Inverse region

For $V_{BE} \leq -5\dfrac{n_F kT}{q}$ and $V_{BC} > -5\dfrac{n_R kT}{q}$:

$$q_2 = \dfrac{I_S}{I_{KR}}(e^{qV_{BC}/n_R kT} - 1) - \dfrac{I_S}{I_{KF}} + \left(\dfrac{V_{BE}}{I_{KF}} + \dfrac{V_{BC}}{I_{KR}}\right)\text{GMIN} \quad (2\text{-}113)$$

Saturated region

For $V_{BE} > -5\dfrac{n_F kT}{q}$ and $V_{BC} > -5\dfrac{n_R kT}{q}$:

$$q_2 = \dfrac{I_S}{I_{KF}}(e^{qV_{BE}/n_F kT} - 1) + \dfrac{I_S}{I_{KR}}(e^{qV_{BC}/n_R kT} - 1) + \left(\dfrac{V_{BE}}{I_{KF}} + \dfrac{V_{BC}}{I_{KR}}\right)\text{GMIN}$$
$$(2\text{-}114)$$

Off region

For $V_{BE} \leq -5\dfrac{n_F kT}{q}$ and $V_{BC} \leq -5\dfrac{n_R kT}{q}$:

$$q_2 = -I_S\left(\dfrac{I_{KF} + I_{KR}}{I_{KF} I_{KR}}\right) + \left(\dfrac{V_{BE}}{I_{KF}} + \dfrac{V_{BC}}{I_{KR}}\right)\text{GMIN} \quad (2\text{-}115)$$

It must be noted that since V_A, V_B, I_{KF}, and I_{KR} cannot be zero-valued, SPICE2 interprets a zero value for these parameters as an infinite value.

The current dependence of the base resistance is modeled in SPICE2 as follows:

$$r_{BB'} = \begin{cases} r_{BM} + \dfrac{r_B - r_{BM}}{q_b} & \text{for } I_{rB} = 0 \\ r_{BM} + 3(r_B - r_{BM})\left(\dfrac{\tan z - z}{z \tan^2 z}\right) & \text{for } I_{rB} \neq 0 \end{cases} \quad (2\text{-}116)$$

where z is given by Eq. (2-105).

Figure 2-27 shows SPICE2 output plots of I_C vs. V_{CE} characteristics of a BJT with (a) $C_2 = 10^3$, (b) $I_{KF} = 5$ A, and (c) $n_F = 1.3$. The simulation has been made with $I_B = 0.5$ A, while the other parameter values (for each case) have been kept to their SPICE2 default values.

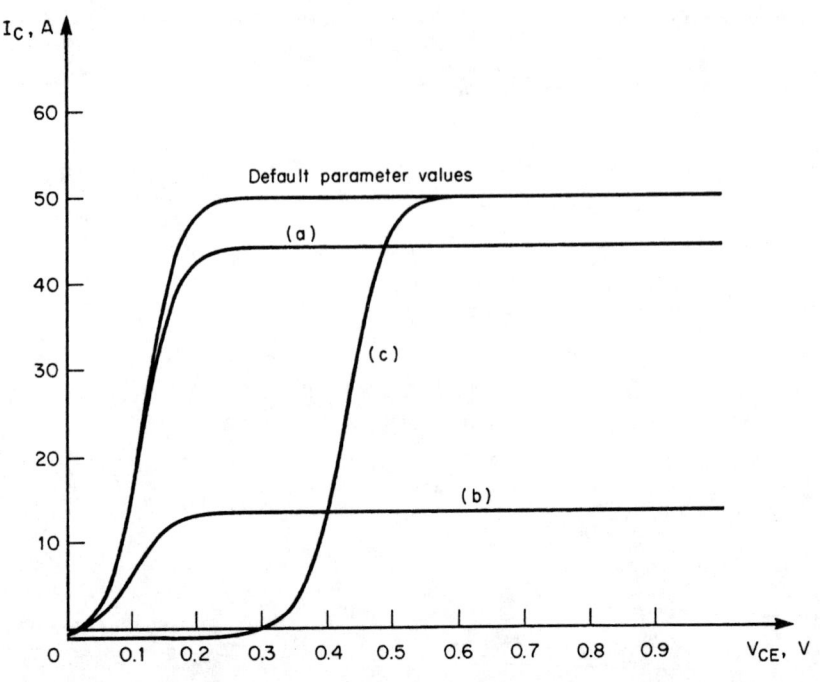

Figure 2-27 SPICE2 output plots of I_C vs. V_{CE} characteristics of a BJT with (a) $C_2 = 10^3$, (b) $I_{KF} = 5$ A, and (c) $n_F = 1.3$. The simulation has been made with $I_B = 0.5$ A, while the other parameter values (for each case) have been kept to their SPICE2 default values.

2.6 Gummel-Poon Large-Signal Model

2.6.1 Large-signal model and its implementation in SPICE2

The SPICE2 Gummel-Poon large-signal circuit model is topologically the same as that of Fig. 2-18.

The nonlinear charge elements, or equivalently the voltage-dependent capacitances, are determined by Eqs. (2-36) to (2-43) (used also for the Ebers-Moll model) when the grading coefficients m_E, m_C, and m_S are taken into account (usually they vary between 0.33 and 0.5), and I_{EC} and I_{CC} are considered as functions of I_{SS} and q_b (see Sec. 2.5.4). Besides, in the Gummel-Poon large-signal model, three added effects have to be accounted for with respect to the Ebers-Moll model: distributed base-collector capacitance, τ_F modulation (transit charge), and distributed phenomena in the base region (excess phase).

a. Distributed base-collector capacitance. In order to find a better approximation of the distributed resistance and capacitance network at the base-collector junction, the junction capacitance is divided into two sections. Parameter X_{CJC}, which varies between 0 and 1, specifies the ratio of this partitioning. The capacitance $X_{CJC}C_{JC}$ is placed between the internal base node and the collector. $(1 - X_{CJC})C_{JC}$ is the capacitance from external base to collector, while C_{JC} is the total base-collector capacitance. This parameter is usually important only at very high frequencies [10].

The implementation in SPICE2 is determined by the following relations:

$$Q_{BX} = \begin{cases} C_{JC}(0)(1-X_{CJC}) \int_0^{V_{BX}} \left(1 - \frac{V}{\phi_C}\right)^{-m_C} dV & \text{for } V_{BX} < FC \times \phi_C \\ C_{JC}(0)(1-X_{CJC})F_1 \\ + \frac{C_{JC}(0)(1-X_{CJC})}{F_2} \int_{FC \times \phi_C}^{V_{BX}} \left(F_3 + \frac{m_C V}{\phi_C}\right) dV \\ \phantom{C_{JC}(0)(1-X_{CJC})F_1} \text{for } V_{BX} \geq FC \times \phi_C \end{cases} \quad (2\text{-}117)$$

These charge-storage elements can be equivalently represented in SPICE2 by the following voltage-dependent capacitance equations:

$$C_{JX} = \begin{cases} C_{JC}(0)(1-X_{CJC})\left(1 - \frac{V_{BX}}{\phi_C}\right)^{-m_C} & \text{for } V_{BX} < FC \times \phi_C \\ \frac{C_{JC}(0)(1-X_{CJC})}{F_2}\left(F_3 + \frac{m_C V_{BX}}{\phi_C}\right) & \text{for } V_{BX} \geq FC \times \phi_C \end{cases} \quad (2\text{-}118)$$

where F_1, F_2, and F_3 are given by Eq. (2-43).

Figure 2-28 shows the complete BJT large-signal model with the addition of the effect of the distributed base-collector capacitance.

b. τ_F modulation (transit charge). The behavior of τ_F vs. I_C is shown in Fig. 2-29. The variation of τ_F at high collector currents is usually determined by an empirical equation derived from the gain-bandwidth product, f_T, vs. the collector current, I_C, at various collector-emitter voltages V_{CE}. Three distinct regions are observed in the f_T characteristics.

At low currents, f_T is dominated by the junction capacitances and the forward transconductance g_{mF}; since g_{mF} increases with current, f_T goes up with an increase in I_C. This variation has already been taken into account.

At the midrange, f_T is at its peak value and is almost constant; here the transit time is the time required by the carriers to transverse the base region and collector depletion region. The base-emitter diffusion

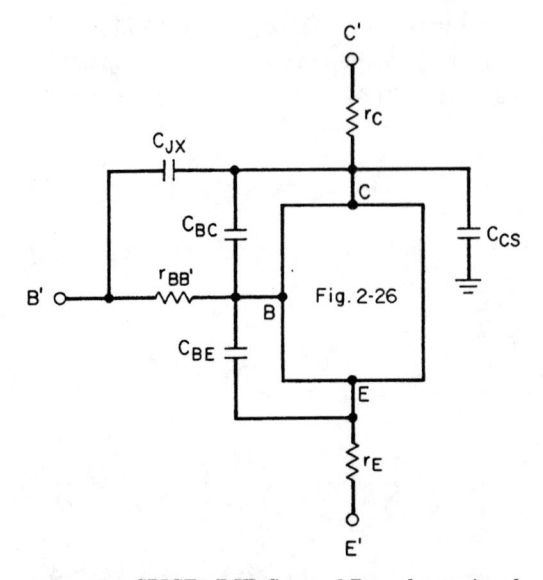

Figure 2-28 SPICE2 BJT Gummel-Poon large-signal model showing the distributed base-collector capacitance C_{JX}.

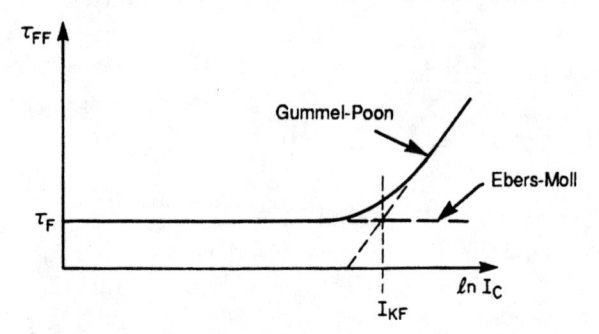

Figure 2-29 τ_{FF} vs. $\ln I_C$ to show high-current effects on charge storage.

capacitance increases with current, canceling the increase in forward g_{mF}, resulting in a definite limit for f_T. Thus, the ideal maximum τ_F is

$$\tau_F = \frac{1}{2\pi f_T} \qquad (2\text{-}119)$$

At high currents, f_T and then τ_F become a function of I_C and V_{CE} and do not remain constant (see Fig. 2-29). Physically, anomalies such as base pushout, lateral spreading, space-charge-limited current flow, and quasi-saturation increase the transit time and decrease f_T.

This effect is modeled by the following empirical function [10]:

$$\text{ATF} \equiv 1 + X_{\tau F}\, e^{V_{BC}/1.44 V_{\tau F}} \left(\frac{I_{CC}}{I_{CC} + I_{\tau F}}\right)^2 \qquad (2\text{-}120)$$

This function multiplies τ_F in the charge equations. The constant 1.44 simply gives the interpretation to $V_{\tau F}$ as the value of V_{BC} where the exponential equals $\frac{1}{2}$. $X_{\tau F}$ (a SPICE2 parameter) controls the total fall-off of f_T; $V_{\tau F}$ (a SPICE2 parameter) dominates the change in f_T with respect to V_{CE}; $I_{\tau F}$ (a SPICE2 parameter) dominates the change in f_T with respect to current.

It can be demonstrated [10] that

$$f_T = \frac{1}{2\pi \tau_F \left[\text{ATF} + \dfrac{2(\text{ATF}-1)I_{\tau F}}{I_{CC} + I_{\tau F}} + \dfrac{kT}{q} n_F \dfrac{\text{ATF}-1}{1.44 V_{\tau F}} \right]} \qquad (2\text{-}121)$$

At low currents or high V_{CE} (ATF = 1), Eq. (2-121) reduces to

$$f_T = \frac{1}{2\pi \tau_F} \qquad (2\text{-}122)$$

which is the familiar Ebers-Moll expression given in Eq. (2-119).

At large I_C, such that $I_{CC}/(I_{CC} + I_{\tau F}) \simeq 1$ and V_{CE} is moderate, Eq. (2-121) reduces to

$$f_T \big|_{I_C \to \infty} = \frac{1}{2\pi \tau_F [1 + X_{\tau F}(e^{V_{BC}/1.44 V_{\tau F}} + e^{V_{BC}/1.44 V_{\tau F}} n_F kT/q\, 1.44 V_{\tau F})]}. \qquad (2\text{-}123)$$

Thus, the high-current asymptote for a given V_{BC} is determined by parameters $X_{\tau F}$ and $V_{\tau F}$. Likewise, the asymptotic dependence of Eq. (2-121) in the extremes of V_{BC} are

$$f_T = \frac{1}{2\pi \tau_F [1 + X_{\tau F}(1 + n_F kT/q\, 1.44 V_{\tau F})]} \quad \text{for } I_C \to \infty,\, V_{BC} \to 0 \qquad (2\text{-}124)$$

where $\text{ATF} \simeq 1 + X_{\tau F}$ for $I_C \to \infty,\, V_{BC} \to 0$ $\qquad (2\text{-}125)$

Furthermore,

$$f_T = \frac{1}{2\pi \tau_F} \quad \text{for } V_{BC} \to \infty \qquad (2\text{-}126)$$

Thus, the maximum possible fall-off in f_T is controlled by parameter $X_{\tau F}$.

In SPICE2 this effect is expressed by the following expression for charge or, equivalently, for its capacitance:

$$Q_{\tau FF} = \tau_{FF} \frac{I_{CC}}{q_b}$$

$$C_{\tau FF} = \frac{dQ_{\tau FF}}{dV_{BC}} \qquad (2\text{-}127)$$

where τ_{FF} is the modulated transit time given by

$$\tau_{FF} = \tau_F \left[1 + X_{\tau F} \left(\frac{I_{CC}}{I_{CC} + I_{\tau F}} \right)^2 e^{V_{BC}/1.44 V_{\tau F}} \right] \qquad (2\text{-}128)$$

and τ_F is the ideal forward transit time.

Thus the forward and reverse transit charge is modeled by τ_F, $X_{\tau F}$, $V_{\tau F}$, $I_{\tau F}$, and τ_R.

c. Distributed phenomena in the base region (excess phase). The measured phase shift in the forward transverse current from actual devices often exceeds the phase shift predicted by the finite lumped set of poles and zeros within the model. This is due to the distributed phenomena in the base region. Therefore, an extra phase delay is inserted in the model:

$$I_{FX} = I_{CC} \text{ (with excess phase)} = I_{CC} \phi(s) \qquad (2\text{-}129)$$

Unnecessary small time steps during transient analysis can be avoided by modeling the excess phase with a second-order *Bessel function*. This function is an all-pass filter in the frequency domain with no appreciable effect on magnitude. In the time domain, this polynomial resembles a time-domain delay for a Gaussian curve, which is similar to the physical phenomenon exhibited by the transistor action [10]:

$$\phi(s) = \frac{3\omega_0^2}{s^2 + 2\omega_0 s + 3\omega_0^2} \qquad (2\text{-}130)$$

It is important to realize that the equation for $\phi(s)$ is given in the frequency domain; s is the complex frequency variable in the transfer domain (s plane) and ω_0 is a parameter. The phase shift of this function is

$$\theta = \arctan \frac{3\omega_0 \omega}{3\omega_0^2 - \omega^2} \qquad (2\text{-}131)$$

For $\omega < \omega_0$, the phase may be written from the binomial expansion of the arctangent as $\theta = \omega/\omega_0$. Assuming $P_{\tau F}$ is the phase delay at idealized maximum bandwidth ($\omega = 1/\tau_F$), it can now be written

$$\omega_0 = \frac{1}{P_{\tau F}\tau_F} \qquad (2\text{-}132)$$

In SPICE2 ac analysis, this phase shift is implemented simply by adding the appropriate linear phase to the forward transverse term of the collector current I_{CC}. The transient analysis, however, is more complicated. Considering the inverse Laplace transform of $\phi(s)$, it can be written

$$\frac{d^2 I_{FX}}{dt^2} + 3\omega_0 \frac{dI_{FX}}{dt} + 3\omega_0^2 I_{FX} = 3\omega_0^2 I_{CC} \qquad (2\text{-}133)$$

This equation is then implemented by performing a numerical integration algorithm (*backward Euler integration*).

2.7 Gummel-Poon Small-Signal Model

2.7.1 Small-signal model and its implementation in SPICE2

The Gummel-Poon small-signal model is drawn for an *npn* transistor in Fig. 2-30. It is identical in form to that used for the Ebers-Moll model (Fig. 2-19), but the expressions for Eqs. (2-53) to (2-60) are different here.

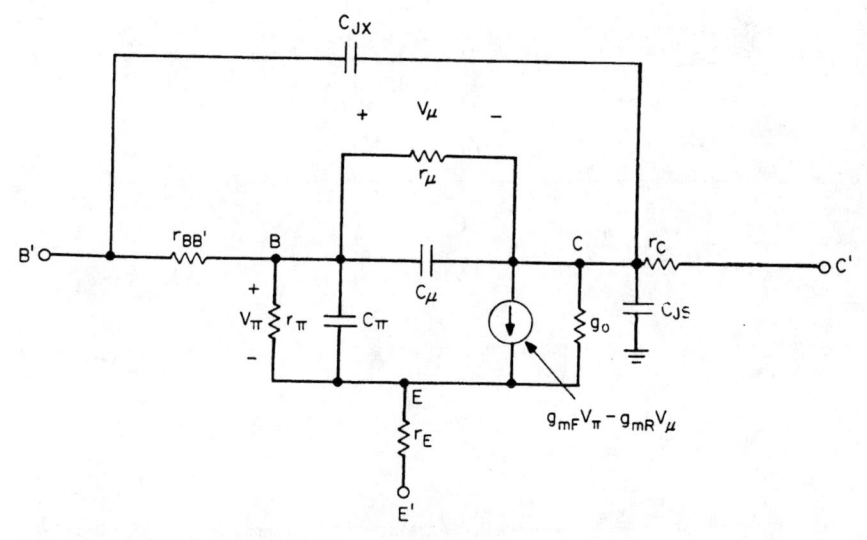

Figure 2-30 SPICE2 BJT Gummel-Poon small-signal model.

Taking into account Eqs. (2-44) to (2-51), the relations for the small-signal parameters defining the SPICE2 Gummel-Poon small-signal model can be written as follows.

For $V_{BE} > -5\dfrac{n_F kT}{q}$ and $V_{BC} \leq -5\dfrac{n_R kT}{q}$:

$$g_m = \dfrac{\dfrac{qI_S}{n_F kT}e^{qV_{BE}/n_F kT}+\text{GMIN}-\dfrac{1}{q_b}[I_S e^{qV_{BE}/n_F kT}+(V_{BE}-V_{BC})\text{GMIN}]\dfrac{dq_b}{dV_{BE}}}{q_b} - g_o \quad (2\text{-}134a)$$

For $V_{BE} \leq -5\dfrac{n_F kT}{q}$ and $V_{BC} > -5\dfrac{n_R kT}{q}$:

$$g_m = \dfrac{\dfrac{-I_S}{V_{BE}}+\text{GMIN}+\dfrac{1}{q_b}[I_S e^{qV_{BC}/n_R kT}-(V_{BE}-V_{BC})\text{GMIN}]\dfrac{dq_b}{dV_{BE}}}{q_b} - g_o \quad (2\text{-}134b)$$

For $V_{BE} > -5\dfrac{n_F kT}{q}$ and $V_{BC} > -5\dfrac{n_R kT}{q}$:

$$g_m = \dfrac{\dfrac{qI_S}{n_F kT}e^{qV_{BE}/n_F kT}+\text{GMIN}-\dfrac{1}{q_b}[I_S(e^{qV_{BE}/n_F kT}-e^{qV_{BC}/n_R kT})+(V_{BE}-V_{BC})\text{GMIN}]\dfrac{dq_b}{dV_{BE}}}{q_b} - g_o \quad (2\text{-}134c)$$

For $V_{BE} \leq -5\dfrac{n_F kT}{q}$ and $V_{BC} \leq -5\dfrac{n_R kT}{q}$:

$$g_m = \dfrac{\dfrac{-I_S}{V_{BE}}+\text{GMIN}-\dfrac{1}{q_b}(V_{BE}-V_{BC})\text{GMIN}\dfrac{dq_b}{dV_{BE}}}{q_b} - g_o \quad (2\text{-}134d)$$

For $V_{BE} > -5\dfrac{n_F kT}{q}$:

$$g_\pi = \dfrac{I_S}{\beta_F}\dfrac{q}{n_F kT}e^{qV_{BE}/n_F kT}+\dfrac{qC_2 I_S}{n_{EL} kT}e^{qV_{BE}/n_{EL} kT}+\dfrac{\text{GMIN}}{\beta_F} \quad (2\text{-}135a)$$

For $V_{BE} \leq -5\dfrac{n_F kT}{q}$:

$$g_\pi = -\dfrac{I_S}{\beta_F V_{BE}}-\dfrac{C_2 I_S}{V_{BE}}+\dfrac{\text{GMIN}}{\beta_F} \quad (2\text{-}135b)$$

For $V_{BE} > -5\dfrac{n_F kT}{q}$ and $V_{BC} \leq -5\dfrac{n_R kT}{q}$:

$$g_o = \dfrac{\dfrac{-I_S}{V_{BC}} + \text{GMIN} + \dfrac{1}{q_b}[I_S\, e^{qV_{BE}/n_F kT} + (V_{BE} - V_{BC})\text{GMIN}]\dfrac{dq_b}{dV_{BC}}}{q_b}$$

(2-136a)

For $V_{BE} \leq -5\dfrac{n_F kT}{q}$ and $V_{BC} > -5\dfrac{n_R kT}{q}$:

$$g_o = \dfrac{\dfrac{qI_S}{n_R kT} e^{qV_{BC}/n_R kT} + \text{GMIN} - \dfrac{1}{q_b}[I_S\, e^{qV_{BC}/n_R kT} - (V_{BE} - V_{BC})\text{GMIN}]\dfrac{dq_b}{dV_{BC}}}{q_b}$$

(2-136b)

For $V_{BE} > -5\dfrac{n_F kT}{q}$ and $V_{BC} > -5\dfrac{n_R kT}{q}$:

$$g_o = \dfrac{\dfrac{qI_S}{n_R kT} e^{qV_{BC}/n_R kT} + \text{GMIN} + \dfrac{1}{q_b}[I_S(e^{qV_{BE}/n_F kT} - e^{qV_{BC}/n_R kT}) + (V_{BE} - V_{BC})\text{GMIN}]\dfrac{dq_b}{dV_{BC}}}{q_b}$$

(2-136c)

For $V_{BE} \leq -5\dfrac{n_F kT}{q}$ and $V_{BC} \leq -5\dfrac{n_R kT}{q}$:

$$g_o = \dfrac{\dfrac{-I_S}{V_{BC}} + \text{GMIN} + \dfrac{1}{q_b}(V_{BE} - V_{BC})\text{GMIN}\dfrac{dq_b}{dV_{BC}}}{q_b}$$

(2-136d)

For $V_{BC} > -5\dfrac{n_R kT}{q}$:

$$g_\mu = \dfrac{I_S}{\beta_R}\dfrac{q}{n_R kT} e^{qV_{BC}/n_R kT} + \dfrac{qC_4 I_S}{n_{CL} kT} e^{qV_{BC}/n_{CL} kT} + \dfrac{\text{GMIN}}{\beta_R}$$

(2-137a)

For $V_{BC} \leq -5\dfrac{n_R kT}{q}$:

$$g_\mu = -\dfrac{I_S}{\beta_R V_{BC}} - \dfrac{C_4 I_S}{V_{BC}} + \dfrac{\text{GMIN}}{\beta_R}$$

(2-137b)

For $I_{rB} = 0$:

$$g_x = \frac{1}{r_{BM} + (r_B - r_{BM})/q_b} \qquad (2\text{-}138a)$$

For $I_{rB} \neq 0$:

$$g_x = \frac{1}{r_{BM} + 3(r_B - r_{BM})[(\tan z - z)/(z \tan^2 z)]} \qquad (2\text{-}138b)$$

For $V_{BE} < \text{FC} \times \phi_E$:

$$C_\pi \equiv C_{BE} = \tau_F \frac{qI_S}{kT} e^{qV_{BE}/kT} + C_{JE}(0)\left(1 - \frac{V_{BE}}{\phi_E}\right)^{-m_E} \qquad (2\text{-}139a)$$

For $V_{BE} \geq \text{FC} \times \phi_E$:

$$C_\pi \equiv C_{BE} = \tau_F \frac{qI_S}{kT} e^{qV_{BE}/kT} + \frac{C_{JE}(0)}{F_2}\left(F_3 + \frac{m_E V_{BE}}{\phi_E}\right) \qquad (2\text{-}139b)$$

For $V_{BC} < \text{FC} \times \phi_C$:

$$C_\mu \equiv C_{BC} = \tau_R \frac{qI_S}{kT} e^{qV_{BC}/kT} + C_{JC}(0)\left(1 - \frac{V_{BC}}{\phi_C}\right)^{-m_C} \qquad (2\text{-}140a)$$

For $V_{BC} \geq \text{FC} \times \phi_C$:

$$C_\mu \equiv C_{BC} = \tau_R \frac{qI_S}{kT} e^{qV_{BC}/kT} + \frac{C_{JC}(0)}{F_2}\left(F_3 + \frac{m_C V_{BC}}{\phi_C}\right) \qquad (2\text{-}140b)$$

For $V_{CS} < 0$:

$$C_{CS} = C_{JS}(0)\left(1 - \frac{V_{CS}}{\phi_S}\right)^{-m_S} \qquad (2\text{-}141a)$$

For $V_{CS} \geq 0$:

$$C_{CS} = C_{JS}(0)\left(1 + \frac{m_S V_{CS}}{\phi_S}\right) \qquad (2\text{-}141b)$$

For $V_{BX} < \text{FC} \times \phi_C$:

$$C_{JX} = C_{JC}(0)(1 - X_{CJC})\left(1 - \frac{V_{BX}}{\phi_C}\right)^{-m_C} \qquad (2\text{-}142a)$$

For $V_{BX} \geq FC \times \phi_C$:

$$C_{JX} = \frac{C_{JC}(0)(1 - X_{CJC})}{F_2}\left(F_3 + \frac{m_C V_{BX}}{\phi_C}\right) \quad (2\text{-}142b)$$

where F_2 and F_3 are given by Eqs. (2-42) and (2-43) and q_b is given by Eqs. (2-110) to (2-115).

The preceding equations are not as simple, and therefore not as intuitive, as those for the Ebers-Moll model; however, they are basically equivalent.

2.8 Temperature and Area Effects on the BJT Model Parameters

This section analyzes the effects that temperature and area factor have on the BJT model implemented in SPICE2. In order to more easily understand the relations introduced, refer to Sec. 1.6.1 (remember that a BJT can be considered an interacting pair of *pn* junctions).

2.8.1 Temperature dependence of SPICE2 BJT model parameters

All input data for SPICE2 is assumed to have been measured at 27°C (300 K). The simulation also assumes a nominal temperature (TNOM) of 27°C that can be changed with the TNOM option of the .OPTIONS control statement. The circuits can be simulated at temperatures other than TNOM by using a .TEMP control statement.

Temperature appears explicitly in the exponential term of the BJT model equations.

In addition, SPICE2 modifies BJT parameters to reflect changes in the temperature. They are I_S, the saturation current; C_2 (or I_{SE}), the base-emitter leakage saturation current; C_4 (or I_{SC}), the base-collector leakage saturation current; ϕ_E, the base-emitter built-in potential; ϕ_C, the base-collector built-in potential; β_F, the forward current gain; β_R, the reverse current gain; C_{JC}, the zero-bias base-collector depletion capacitance; C_{JE}, the zero-bias base-emitter depletion capacitance; and FC, the coefficient for forward-bias depletion capacitance formula.

The temperature dependence of I_S in the SPICE2 BJT model is determined by

$$I_S(T_2) = I_S(T_1)\left(\frac{T_2}{T_1}\right)^{X_{TI}} \exp\left[-\frac{qE_g(300)}{kT_2}\left(1 - \frac{T_2}{T_1}\right)\right] \quad (2\text{-}143)$$

where two new model parameters are introduced: X_{TI}, the saturation current temperature exponent, and E_g, the energy gap.

The effect of temperature on ϕ_J is modeled as follows:

$$\phi_E(T_2) = \frac{T_2}{T_1}\phi_E(T_1) - 2\frac{kT_2}{q}\ln\left(\frac{T_2}{T_1}\right)^{1.5} - \left[\frac{T_2}{T_1}E_g(T_1) - E_g(T_2)\right] \quad (2\text{-}144)$$

$$\phi_C(T_2) = \frac{T_2}{T_1}\phi_C(T_1) - 2\frac{kT_2}{q}\ln\left(\frac{T_2}{T_1}\right)^{1.5} - \left[\frac{T_2}{T_1}E_g(T_1) - E_g(T_2)\right]$$

where $E_g(T_1)$ and $E_g(T_2)$ are described by the following general expression [11]:

$$E_g(T) = E_g(0) - \frac{\alpha T^2}{\beta + T} \quad (2\text{-}145)$$

Experimental results give for Si

$$\alpha = 7.02 \times 10^{-4}$$

$$\beta = 1108$$

$$E_g(0) = 1.16 \text{ eV}$$

The temperature dependence of β_F and β_R is determined by

$$\beta_F(T_2) = \beta_F(T_1)\left(\frac{T_2}{T_1}\right)^{X_{T\beta}}$$

$$\beta_R(T_2) = \beta_R(T_1)\left(\frac{T_2}{T_1}\right)^{X_{T\beta}} \quad (2\text{-}146)$$

Through the parameters I_{SE} and I_{SC}, respectively, the temperature dependence of C_2 and C_4 is determined by

$$I_{SE}(T_2) = I_{SE}(T_1)\left(\frac{T_2}{T_1}\right)^{-X_{T\beta}}\left[\frac{I_S(T_2)}{I_S(T_1)}\right]^{1/n_{EL}}$$

$$I_{SC}(T_2) = I_{SC}(T_1)\left(\frac{T_2}{T_1}\right)^{-X_{T\beta}}\left[\frac{I_S(T_2)}{I_S(T_1)}\right]^{1/n_{CL}} \quad (2\text{-}147)$$

The temperature dependence of C_{JE} and C_{JC} is determined by

$$C_{JE}(T_2) = C_{JE}(T_1)\left\{1 + m_E\left[400 \times 10^{-6}(T_2 - T_1) - \frac{\phi_E(T_2) - \phi_E(T_1)}{\phi_E(T_1)}\right]\right\}$$

(2-148)

$$C_{JC}(T_2) = C_{JC}(T_1)\left\{1 + m_C\left[400 \times 10^{-6}(T_2 - T_1) - \frac{\phi_C(T_2) - \phi_C(T_1)}{\phi_C(T_1)}\right]\right\}$$

The temperature dependence of the coefficient for forward-bias depletion capacitance formula is determined by

$$\text{FCPE}(T_2) = \text{FC} \times \phi_E = \text{FCPE}(T_1)\frac{\phi_E(T_2)}{\phi_E(T_1)}$$

(2-149)

$$F_1(T_2) = F_1(T_1)\frac{\phi_E(T_2)}{\phi_E(T_1)}$$

$$\text{FCPC}(T_2) = \text{FC} \times \phi_C = \text{FCPC}(T_1)\frac{\phi_C(T_2)}{\phi_C(T_1)}$$

(2-150)

$$F_1(T_2) = F_1(T_1)\frac{\phi_C(T_2)}{\phi_C(T_1)}$$

In the previous equations, T_1 must be considered as follows:

- If TNOM in the .OPTIONS statement is not specified, then $T_1 = \text{TNOM} = \text{TREF} = 27°\text{C}$ (300 K); this is the default value for TNOM.

- If a new TNOM_{new} value is specified, then $T_1 = \text{TNOM} = 27°\text{C}$; $T_2 = \text{TNOM}_{\text{new}}$.

These two cases are valid if only one temperature request is made. If more than one temperature is requested (.TEMP statement), then $T_1 = \text{TNOM} = 27°\text{C}$ and T_2 is the first temperature specified in the .TEMP statement. Afterward, T_1 is the last working temperature (T_2) and T_2 is the next temperature in the .TEMP statement to be calculated.

It is very important to note that the temperature T is regarded as an operating condition and *not* as a model parameter so that it is not possible to simulate the behavior of devices in the same circuit at different temperatures at the same time. In other words, the *same* temperature T is used in the *whole* circuit to be simulated.

2.8.2 Area dependence of SPICE2 BJT model parameters

The AREA factor used in SPICE2 for the BJT model determines the number of equivalent parallel devices of a specified model. The BJT model parameters affected by the AREA factor and specified in the device statement are I_S, the saturation current; I_{SE}, the base-emitter leakage saturation current; I_{SC}, the base-collector leakage saturation current; I_{KF}, the forward knee current; I_{KR}, the reverse knee current; I_{rB}, the current where base resistance falls halfway to its minimum value; $I_{\tau F}$, the high-current parameter for effect on τ_F; r_B, the zero-bias resistance; r_{BM}, the minimum base resistance at high currents; r_E, the emitter resistance; r_C, the collector resistance; C_{JC}, the zero-bias base-collector depletion capacitance; C_{JE}, the zero-bias base-emitter depletion capacitance; and C_{JS}, the zero-bias collector-substrate capacitance. In summary,

$$I_S = I_S \times \text{AREA}$$
$$I_{SE} = I_{SE} \times \text{AREA}$$
$$I_{SC} = I_{SC} \times \text{AREA}$$
$$I_{KF} = I_{KF} \times \text{AREA}$$
$$I_{KR} = I_{KR} \times \text{AREA}$$
$$I_{rB} = I_{rB} \times \text{AREA}$$
$$I_{\tau F} = I_{\tau F} \times \text{AREA} \qquad (2\text{-}151)$$
$$C_{JC}(0) = C_{JC}(0) \times \text{AREA}$$
$$C_{JE}(0) = C_{JE}(0) \times \text{AREA}$$
$$C_{JS}(0) = C_{JS}(0) \times \text{AREA}$$
$$r_B = r_B / \text{AREA}$$
$$r_{BM} = r_{BM} / \text{AREA}$$
$$r_E = r_E / \text{AREA}$$
$$r_C = r_C / \text{AREA}$$

The default value for the AREA parameter is 1.

2.9 Power BJT Model

This section presents a model for the power BJT and shows how it can be implemented in SPICE2.

2.9.1 Power BJT overview

Power BJTs must be designed to withstand high voltage, current, and power ratings. Wide operation areas in forward or reverse base-driving condition and a good switching speed are also often required.

The need to support high collector-emitter voltages V_{CE} implies the presence of a high reverse collector-base voltage V_{BC}; a high breakdown voltage base-collector junction is therefore needed.

A significant improvement in the breakdown voltage can be obtained with an $n^+pn^-n^+$ structure, as shown in Fig. 2-31. Here the additional reverse voltage that can be supported because the n^- region is approximately equal to the breakdown voltage of a reverse-biased $p^+n^-n^+$ diode of comparable width. Furthermore, since the movement of the collector depletion region into the diffused base is now inhibited by the grading, failure by punch-through seldom occurs in this device.

In normal operation, the n^- region is fully depleted, and the device behaves like a double diffused n^+pn^+ transistor, with a relatively narrow base width. The frequency response of the structure is thus considerably superior to that of the single diffused transistor. As a result of these many advantages, the $n^+pn^-n^+$ structure is the basic modern high-voltage power BJT.

Power BJT I_C vs. V_{CE} characteristics are divided into three regions, as shown in Fig. 2-32:

1. *Nonsaturation region* (a) to the right of the line marked $R_P + R_{SC}$. R_{SC} is the *collector series resistance* given by

$$R_{SC} = \frac{\rho_C W_C}{A_E} \qquad (2\text{-}152)$$

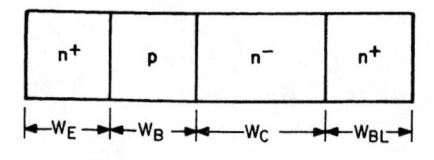

Figure 2-31 Power BJT structure.

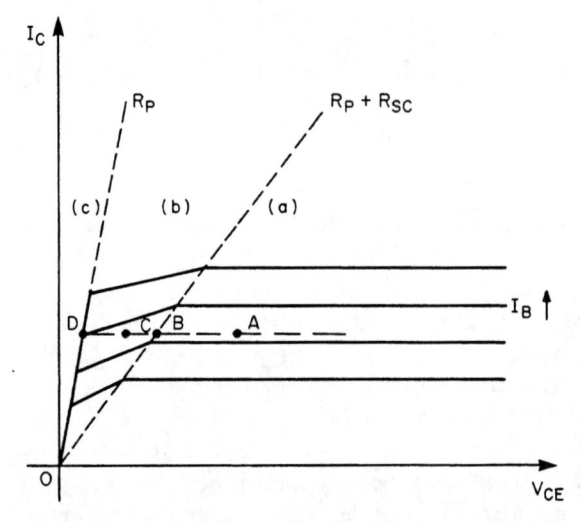

Figure 2-32 Power BJT I_C versus V_{CE} characteristics, showing (a) nonsaturation region, (b) quasi-saturation region, and (c) hard-saturation region.

where ρ_C is the collector resistivity and A_E is the emitter area. R_P is the parasitic series resistance, which consists of internal device resistances, lead and connection resistances, and the voltage difference between V_{BE} and V_{BC}, which are both positive, since emitter-base and collector-base junctions are both forward-biased in saturation.

2. *Quasi-saturation region (b)*, between R_p and $R_P + R_{SC}$.
3. *Saturation region (c)*, formed by the points lying on R_P.

Transistor behavior in the three regions is discussed in the following sections for constant collector current I_C and decreasing V_{CE} [12].

a. Nonsaturation region. At point A of Fig. 2-32, V_{CE} is the sum of the voltage drop in the depletion region and of the ohmic drops in the undepleted collector and across R_P; that is,

$$V_{CE} = \frac{\varepsilon_s (E_0 - E_C)^2}{2qN_D} + E_C W_C + R_P I_C \qquad (2\text{-}153)$$

where

$$E_C = \frac{\rho_C I_C}{A_E} \qquad (2\text{-}154)$$

and E_0 is the electric field in $x = W_B$.

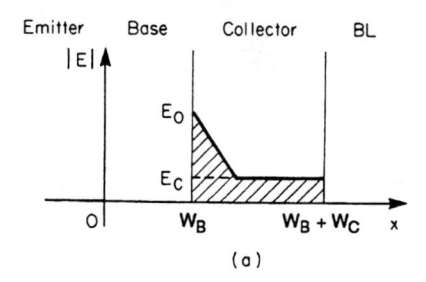

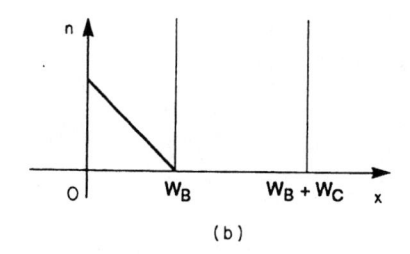

Figure 2-33 Nonsaturation region: (a) electric field and (b) electron stored charge. (*From Antognetti [12]. Used by permission.*)

Figure 2-33 is a representation of the electric field and the electron charge distribution. V_{CE} is given by the dashed area (plus $I_C R_P$ drop) [12].

b. Quasi-saturation region. At point B of Fig. 2-32,

$$V_{CE} = I_C(R_P + R_{SC}) \tag{2-155}$$

The depletion region disappears and V_{CE} is determined only by the ohmic drop on R_{SC}. The electron distribution is the same as in the nonsaturation region, while the electric field varies according to Fig. 2-34.

When V_{CE} is further decreased (point C in Fig. 2-32), the following phenomena occur [12]:

1. The collector-base junction becomes forward-biased.
2. Holes are injected from the base into some part of the collector region (from $x = W_B$ to $x = W_{CIB}$), which behaves as an extended base (see Fig. 2-35). The sum of this extended base and of the metallurgical base is usually called *current-induced base* (*CIB*).
3. In the extended base, the quasi-neutrality condition must be applied.
4. The normally small N_D value is negligible with respect to $n(x)$, and the extended base can be considered as a real base region with $N_A = 0$.

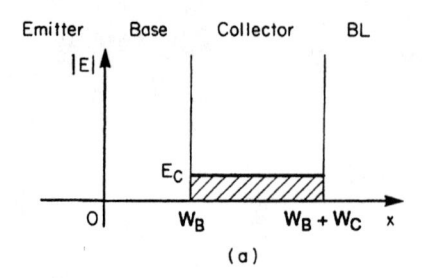

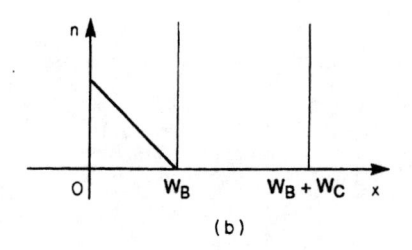

Figure 2-34 Boundary between nonsaturation and quasi-saturation regions: (*a*) electric field and (*b*) electron stored charge. (*From Antognetti [12]. Used by permission.*)

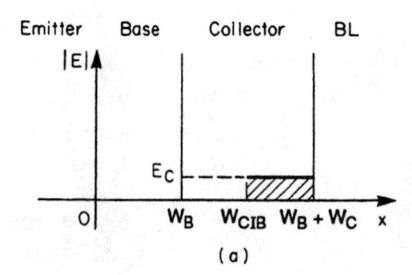

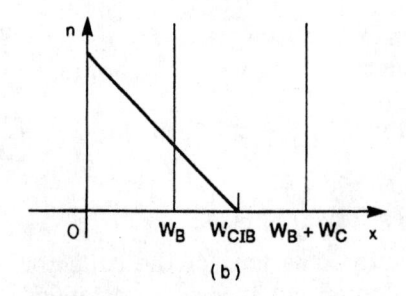

Figure 2-35 Quasi-saturation region: (*a*) electric field and (*b*) electron stored charge. (*From Antognetti [12]. Used by permission.*)

It can be shown [12] that V_{CE} is the sum of ohmic drops on the residual collector and on R_P; i.e.,

$$V_{CE} = R_P I_C + \frac{\rho_C(W_B + W_C - W_{CIB})}{A_E} I_C \qquad (2\text{-}156)$$

c. Saturation region. When the condition $V_{CE} = I_C R_P$ holds, the collector voltage drop is zero and $W_{CIB} = W_B + W_C$. The electron distri-

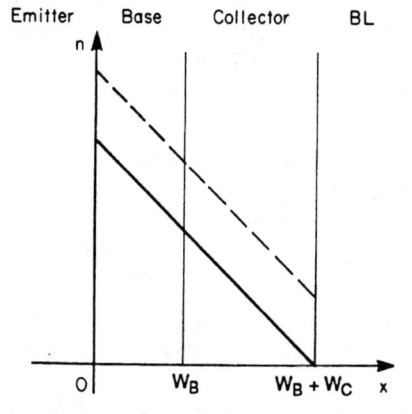

Figure 2-36 Electron stored charge in hard saturation (solid line, $I_B = I_{B,\text{sat}}$) and in oversaturation (dashed line, $I_B > I_{B,\text{sat}}$).

bution, corresponding to point D in Fig. 2-32, is given by the solid line in Fig. 2-36 for $I_B = I_{B,\text{sat}}$, which is the minimum I_B required for the transistor to be in saturation.

Details of power transistor theory are described and developed in Refs. [12,13].

2.9.2 Power BJT quasi-saturation model*

Physical phenomena which determine power BJT performances and which affect the model implementation are briefly discussed before the model itself is described.

Two distinct possibilities are present when the current in an $n^+pn^-n^+$ transistor is increased. First, the increasing voltage drop across the n^- region reduces the available reverse bias at the collector-base junction. Consequently, its depletion region shrinks and eventually collapses. Further increase of collector current drives the device into quasi-saturation, accompanied by an expansion of the base width. This process is commonly encountered for low supply voltages and is known as the *low-field case (LFC)*.

Conversely, increasing collector current reduces the space-charge density in the pn^- depletion region, causing it to expand until it fills the n^- region. Further increase in collector current results in space-charge limited current flow in this region. Still further increase leads to growth in the effective base width, by one or more mechanisms, described by the one- and two-dimensional models.

These processes are commonly encountered with higher supply voltages, which means that the device does not enter quasi-saturation: this process is known as the *high-field case (HFC)*.

---

*The material in this section is taken in part from S. K. Ghandhi, *Semiconductor Power Devices*, copyright © 1977 by John Wiley & Sons, Inc. Reprinted by permission.

Typically, $NW_C = 10^{11}$ cm^{-2} represents the boundary between the low-field case and the high-field case [13]. In both situations, however, it can be shown that the end result is an effective widening of the base width when the collector current density exceeds a specific critical value.

The model proposed here [14–17], which is an extension of the classic Gummel-Poon model used in SPICE2, affects only the low-field case.

Consider an $n^+pn^-n^+$ transistor, fed from a constant collector-base supply voltage [13]. With increasing collector current, the ohmic drop across the undepleted part of the n^- region increases, until all the supply voltage is used up across this region and the depletion region collapses. If I_C^* is the critical current at which this occurs, then from Ohm's law it can be written, if R_P is omitted,

$$I_C^* = \frac{A_E(V_{CB} + \phi_C)}{\rho_C W_C} = \frac{V_{CB} + \phi_C}{R_{SC}} \qquad (2\text{-}157)$$

where ϕ_C is the contact potential of the pn^- junction. This situation is commonly encountered with low supply voltages (LFC).

Transistor action at $I_C = I_C^*$ consists of diffusion through W_B, drift through a low-field n^- region of width W_C, and eventual collection. The pn^- junction becomes forward-biased when the current exceeds I_C^*. This is accompanied by hole injection into the n^- region, together with conductivity modulation of part of it, as in Fig. 2-35.

Thus for $I_C > I_C^*$, the drift region shrinks away from the pn^- junction, reducing its width from W_C to $W_C - W_{CIB}$. For this situation, I_C is obtained by modifying Eq. (2-157), so that

$$I_C = \frac{A_E(V_{CB} + \phi_C)}{\rho_C(W_C - W_{CIB})} \qquad (2\text{-}158)$$

where W_{CIB} is the current-induced base width or, in other words, the modulated part of the n^- region.

Combining Eqs. (2-157) and (2-158) yields:

$$W_{CIB} = W_C \left(1 - \frac{I_C^*}{I_C}\right) \qquad (2\text{-}159)$$

Transistor action at current $I_C > I_C^*$ thus consists of diffusion through a neutral region of width $W_B + W_{CIB}$, drift through a region of width $W_C - W_{CIB}$, and eventual collection.

Gummel introduced the effective base width, $W_{B,\text{eff}}$, as follows:

$$W_{B,\text{eff}} = \begin{cases} W_B & \text{for } I_C < I_C^* \\ W_B + W_{CIB} = W_B + W_C\left(1 - \dfrac{I_C^*}{I_C}\right) & \text{for } I_C > I_C^* \end{cases} \quad (2\text{-}160)$$

Knowing that the base transit time varies as the square of the base width, Gummel then defines an effective base transit time given by the following:

$$\tau_{F,\text{eff}} = B\tau_F \quad (2\text{-}161)$$

where

$$B = \left(\dfrac{W_{B,\text{eff}}}{W_B}\right)^2 \quad (2\text{-}162)$$

B, called the *base pushout factor*, models the effective increase in the base width at high-current levels. Thus, from what was explained before, it characterizes the behavior of the power BJT in the quasi-saturation region.

From Eqs. (2-160) and (2-162), it follows that

$$B = 1 \quad \text{for} \quad I_C < I_C^*$$

$$B = \left[1 + \dfrac{W_C}{W_B}\left(1 - \dfrac{I_C^*}{I_C}\right)\right]^2 \quad (2\text{-}163)$$

$$= \left[1 + \dfrac{W_C}{W_B}\left(1 + \dfrac{V_{CB} + \phi_C}{R_{SC}I_C}\right)\right]^2 \quad \text{for } I_C > I_C^*$$

Various expressions of B can be considered [6], but all of them are very complicated with respect to Eq. (2-163) and thus have never been implemented in any version of SPICE2.

Now B has been defined as a base-widening factor; its implementation in SPICE2 requires only three new parameters:

R_{SC} n^--region resistance
W_C n^--region width
W_B p-base width

2.9.3 A possible Implementation of the B factor in SPICE2

The base-widening effect modifies both static behavior and dynamic behavior of the power BJT.

Furthermore, the analysis and modeling of this phenomenon on the dynamic behavior can become very complicated, since in SPICE2 there is no *explicit* expression for $\tau_{F,\text{eff}}$ (total forward transit time), which is *implicitly* obtained via the following relation:

$$\tau_{F,\text{eff}} = \frac{Q_{DE}(V_{BE})}{I_{CC}(V_{BE})} \quad (2\text{-}164)$$

In Eq. (2-164), Q_{DE} and I_{CC} are separate functions of V_{BE}. By modeling each one independently, an implicit expression for $\tau_{F,\text{eff}}$ can be obtained.

Static and dynamic analyses follow separately, and deviations from the basic Gummel-Poon model in SPICE2 will be pointed out.

a. Dynamic analysis. Referring to Eq. (2-26), and assuming that Q_{JE} and Q_{JC} are negligible (which is often reasonable), then we have the following equation:

$$Q_{DE} \simeq Q_E + Q_{BF} \quad (2\text{-}165)$$

Now Q_E, the mobile minority charge stored in the neutral emitter region, is unaffected by B, but Q_{BF}, the minority mobile charge stored in the neutral base region, is affected by B.

Equation (2-165) can now be rewritten as

$$Q_{DE} \simeq Q_E + Q_{BF}(B) \quad (2\text{-}166)$$

To implement B in Q_{DE}, the preceding expression must be replaced by the following one:

$$Q_{DE} = \underbrace{\tau_E I_{SS}(e^{qV_{BE}/kT} - 1)}_{Q_E} + \underbrace{B\tau_{BF}I_{SS}(e^{qV_{BE}/kT} - 1)}_{Q_{BF}} \quad (2\text{-}167)$$

or

$$Q_{DE} = \underbrace{(\tau_E + B\tau_{BF})}_{\tau_{F,\text{eff}}} I_{SS}(e^{qV_{BE}/kT} - 1) \quad (2\text{-}168)$$

Naturally, if $B = 1$, $\tau_{F,\text{eff}} = \tau_E + \tau_{BF} = \tau_F$, and the original Gummel-Poon model used in SPICE2 can be found again according to

$$Q_{DE} = \tau_F I_{SS}(e^{qV_{BE}/kT} - 1) \quad (2\text{-}169)$$

b. Static analysis. To discuss the influence of B on the static distribution of carriers stored in the base region, the first term of I_{CT} in Eq. (2-83) must be considered. We make

$$I_{CC} = \frac{I_{SS}}{q_b}(e^{qV_{BE}/kT} - 1) \quad (2\text{-}170)$$

Combining Eq. (2-169) with Eq. (2-170) yields

$$Q_{DE} = \tau_F I_{CC} q_b \tag{2-171}$$

and it immediately follows that

$$\frac{Q_{DE}}{I_{CC}} = \tau_F q_b \tag{2-172}$$

Finally, combining Eq. (2-164) with Eq. (2-172) yields

$$\tau_F q_b(B) = \tau_{F,\text{eff}} \tag{2-173}$$

Now, q_b represents an important parameter in modeling the static behavior of the power BJT. In particular, q_{bF} can be regarded as the component of q_b [see Eq. (2-87)], which is affected by the base pushout effect.

By the definition of q_{bF},

$$q_{bF} = \frac{\tau_{BF} I_{CC}}{Q_{B0}} \tag{2-174}$$

and taking into account Eq. (2-170),

$$q_{bF} = \frac{\tau_{BF} I_{SS}(e^{qV_{BE}/kT} - 1)}{Q_{B0} q_b} \tag{2-175}$$

Therefore, from Eq. (2-167),

$$q_{bF} = \frac{B\tau_{BF} I_{SS}(e^{qV_{BE}/kT} - 1)}{Q_{B0} q_b} \tag{2-176}$$

The complete solution for q_b vs. B comes from a combination of Eqs. (2-176) and (2-87).

$$q_b = 1 + \frac{V_{BE}}{V_B} + \frac{V_{BC}}{V_A} + \frac{B\tau_{BF}}{Q_{B0} q_b} I_{SS}(e^{qV_{BE}/kT} - 1) + \frac{\tau_{BR}}{Q_{B0} q_b} I_{SS}(e^{qV_{BC}/kT} - 1) \tag{2-177}$$

If Eqs. (2-99) and (2-100) are taken into account, then

$$q_b = 1 + \frac{V_{BE}}{V_B} + \frac{V_{BC}}{V_A} + \frac{B}{I_{KF} q_b} I_{SS}(e^{qV_{BE}/kT} - 1) + \frac{1}{I_{KR} q_b} I_{SS}(e^{qV_{BC}/kT} - 1) \tag{2-178}$$

Equations (2-163), (2-168), and (2-178) cover the implementation of the base pushout factor B in SPICE2, even if there is no simple method (other than calculation from process and geometric data) for determining τ_E and τ_{BF} parameters. The implementation of Eq. (2-178) in SPICE2 is also subjected to the simplifications pointed out in Eqs. (2-110) and (2-111).

2.10 SPICE3 Models

SPICE3 [18,19] BJT models are based directly on SPICE2 BJT models [Ebers-Moll (E-M) and Gummel-Poon (G-P)]. Thus the *static models* are defined by the parameters IS, BF, NF, ISE, IKF, and NE, which determine the forward current gain characteristics; IS, BR, NR, ISC, IKR, and NC, which determine the reverse current gain characteristics; and VAF and VAR, which determine the output conductance for forward and reverse regions. Three ohmic resistances RB, RC, and RE are included, where RB can be high-current-dependent.

The *large-signal models* are defined by forward and reverse transit times, TF and TR, the forward transit time TF being bias dependent if desired, and nonlinear depletion region capacitances which are determined by CJE, VJE, and MJE for the base-emitter junction, CJC, VJC, and MJC for the base-collector junction, and CJS, VJS, and MJS for the collector-substrate junction.

The *small-signal models* are automatically computed inside the program by the linearization of the static and large-signal model equations. The *temperature dependence* and the updating relations for the parameters follow the ones described in Sec. 2.8.1. The values specified for the parameters are assumed to have been measured at the temperature TNOM, which can be specified in the .OPTIONS statement. This value of temperature can be overridden by the specification of the new (in comparison with SPICE2) .MODEL statement parameter TNOM (in units °C) or by the device element statement option TEMP. The *noise* and *distortion models* are described in Secs. 7.1.3*b* and 7.2.2*d*, respectively.

The parameters affected by the AREA factor follow the same relations as indicated in Sec. 2.8.2. The equivalent circuits for the static, large-signal, small-signal, noise, and distortion models are shown in Figs. 2-9 (E-M), 2-26 (G-P), 2-16 (E-M), 2-28 (G-P), 2-19 (E-M), 2-30 (G-P), 7-4, and 7-13, respectively.

2.11 HSPICE Models

HSPICE [20] BJT models are basically based on SPICE2 BJT models (Ebers-Moll and Gummel-Poon). These models are selected by using the parameter LEVEL = 1 in the .MODEL statement. In addition, HSPICE allows the user to model both vertical and lateral geometries (SUBS parameter), to use a quasi-saturation model for power BJTs (LEVEL = 2 parameter), and to scale the .MODEL parameters as functions of the emitter, base, and collector region area (AREA, AREAB, AREAC parameters). In the following, HSPICE BJT *static, large-signal, small-signal,* and *thermal models* are taken into account. The *noise model* is considered in Chap. 7.

2.11.1 Static model for LEVEL = 1

The equations, the equivalent circuits, and the .MODEL parameters described in the previous sections for the SPICE2 BJT *static models* (Ebers-Moll and Gummel-Poon) are also valid in HSPICE. However, some improvements have been made in Eqs. (2-64): two new parameters, the reverse saturation current for the base-emitter junction I_{BE} (IBE) and the reverse saturation current for the base-collector junction I_{BC} (IBC), both measured in [A], have been introduced in HSPICE. The .MODEL keyword is indicated in parentheses. If both IBE and IBC are specified in the .MODEL statement, HSPICE uses them instead of IS to calculate the appropriate junction components in Eqs. (2-64); otherwise IS is used for both the junctions. It is understood that both IBE and IBC depends on q_b as IS does in the Gummel-Poon model (see Secs. 2.5.2 to 2.5.4).

HSPICE models the substrate junction with a diode connected to either the collector or the base regions depending on whether the BJT has a lateral or vertical geometry. Lateral geometry is implied when the model parameter SUBS = −1, and vertical geometry when SUBS = +1. The lateral transistor substrate diode is connected to the internal base and the vertical transistor substrate diode is connected to the internal collector.

Figure 2-37 shows the HSPICE static and large-signal model, pointing out the influence of the substrate junction for the vertical and lateral geometry.

The substrate current is substrate to collector for vertical BJTs and substrate to base for lateral BJTs. Thus [20],

Vertical BJT

$$I_{SC} = I_{SSUB}(e^{qV_{SC}/n_s kT} - 1) \quad \text{for } V_{SC} > -10\frac{n_s kT}{q}$$

$$I_{SC} = -I_{SSUB} \quad \text{for } V_{SC} \leq -10\frac{n_s kT}{q}$$

(2-179)

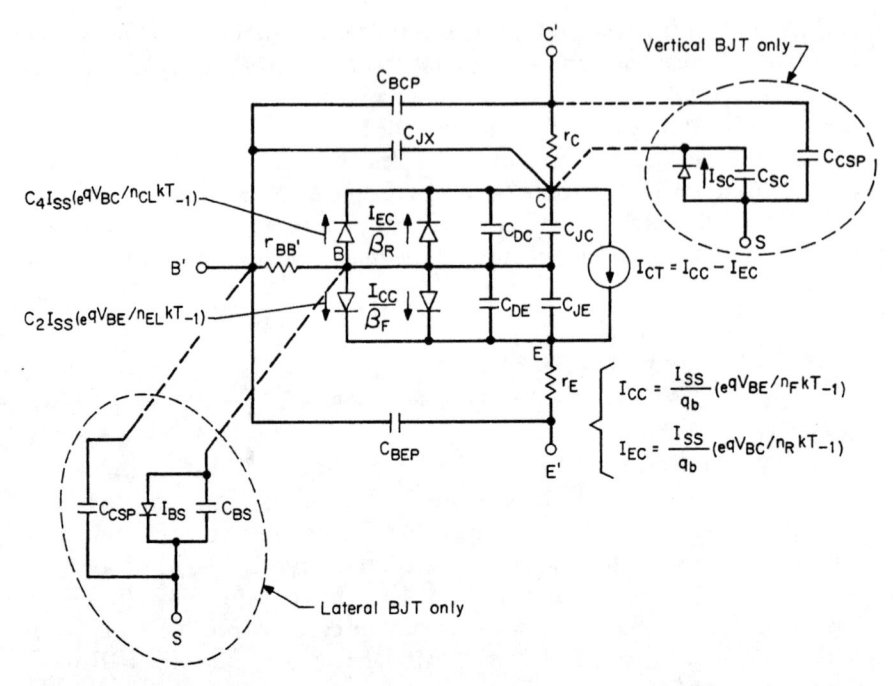

Figure 2-37 HSPICE BJT model showing the influence of the substrate junction for vertical and lateral geometry.

Lateral BJT

$$I_{BS} = I_{SSUB}(e^{qV_{BS}/n_S kT} - 1) \quad \text{for } V_{BS} > -10\frac{n_S kT}{q}$$

$$I_{BS} = -I_{SSUB} \quad \text{for } V_{BS} \leq -10\frac{n_S kT}{q}$$

(2-180)

In the above equations two new parameters for the .MODEL statement have been introduced. They are: I_{SSUB} (ISS), the reverse saturation current bulk-collector or bulk-base, depending on vertical or lateral geometry selection, and measured in [A]; n_S (NS), the substrate current emission coefficient. The HSPICE .MODEL keyword is indicated in parentheses. It also must be noted that the limit between the forward- and reverse-bias regions is set to $-10nkT/q$ while in SPICE2 it was $-5nkT/q$.

As far as the base charge equations are concerned, only Eq. (2-110) has been modified in HSPICE, assuming the form [20]

$$q_b = \frac{q_1}{2}[1 + (1 + 4q_2)^{n_{KF}}] \quad (2-181)$$

where the new parameter n_{KF} (NKF), the base charge exponent (defaults = 0.5), has been introduced for the .MODEL statement of HSPICE.

2.11.2 Large-signal model for LEVEL = 1

The HSPICE BJT *large-signal model* (Fig. 2-37) is characterized by the same equations encountered in the definition of the SPICE2 Gummel-Poon large-signal model. The substrate capacitances C_{BS} and C_{SC} (selected by setting the .MODEL parameter SUBS, SUBS = +1 for vertical geometry and SUBS = −1 for lateral geometry) have expressions defined by Eqs. (2-41) when the appropriate subscripts are considered; however no new model parameter (other than the SPICE2 CJS, VJS, and MJS parameters) is needed to simulate the effects of these two capacitances.

In comparison with SPICE2, HSPICE introduces three new parasitic capacitances (see Fig. 2-37): the external base-collector constant capacitance C_{BCP} (CBCP), the external base-emitter constant capacitance C_{BEP} (CBEP), and the external collector-substrate constant capacitance (for vertical geometry) or base-substrate constant capacitance (for lateral geometry) C_{CSP} (CCSP). All these capacitances are measured in [F], while the .MODEL keyword is indicated in parentheses. It must be taken into account that in HSPICE, the coefficient FC is set to zero when DCAP = 2 option [20] in the .OPTIONS statement is selected (default). This induces changes in the equations used in calculating the depletion capacitance and the limits beyond which the capacitance is modeled by a linear extrapolation.

2.11.3 Small-signal model for LEVEL = 1

As in SPICE2, the HSPICE *small-signal model* is automatically computed inside the program by the linearization of the static and large-signal model equations; thus no new model parameters need be specified in the .MODEL statement by the user of HSPICE. The small-signal model equivalent circuits is shown in Fig. 2-38.

2.11.4 Temperature effects on the BJT model parameters for LEVEL = 1

There are several sets of temperature equations for the HSPICE BJT model parameters which may be selected by setting the parameters TLEV and TLEVC in the .MODEL statement. In the following, we consider only the equations selected with TLEV and TLEVC set to

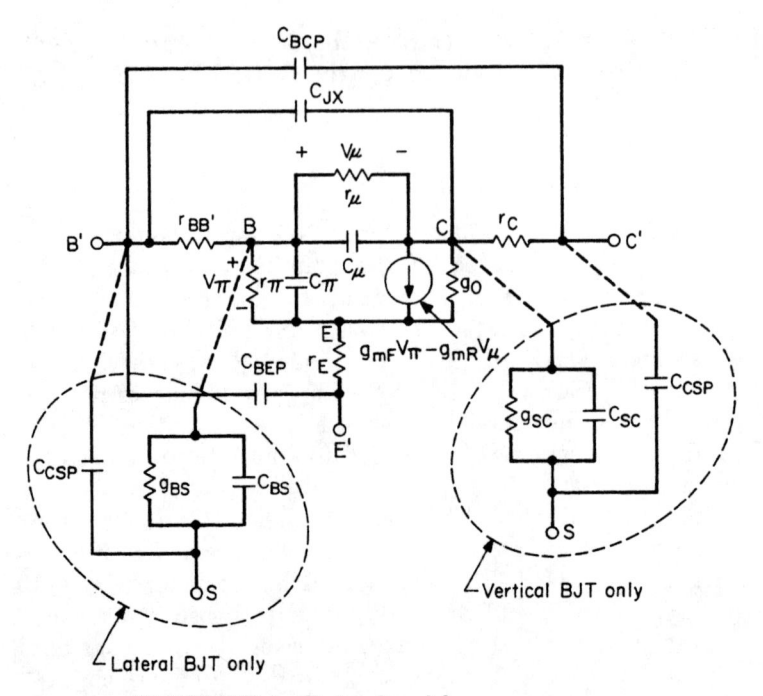

Figure 2-38 HSPICE BJT small-signal model.

their default values (zero). Alternative equations for some of the temperature-dependent parameters can be found in [20].

As in SPICE2, also in HSPICE the temperature dependence of the energy gap E_g (EG) follows Eq. (2-145), while the temperature dependence of β_F (BF) and β_R (BR) is determined by Eqs. (2-146).

The temperature dependence of I_S (IS) is modeled by Eq. (2-143), while I_{BE} (IBE) and I_{BC} (IBC) are determined by

$$I_{BE}(T_2) = I_{BE}(T_1)\left(\frac{T_2}{T_1}\right)^{X_{TI}/n_F} \exp\left[-\frac{qE_g(300)}{n_F k T_2}\left(1 - \frac{T_2}{T_1}\right)\right]$$
$$I_{BC}(T_2) = I_{BC}(T_1)\left(\frac{T_2}{T_1}\right)^{X_{TI}/n_R} \exp\left[-\frac{qE_g(300)}{n_R k T_2}\left(1 - \frac{T_2}{T_1}\right)\right]$$
(2-182)

The parameters I_{SE} (ISE) and I_{SC} (ISC) depend on temperature through Eqs. (2-147), while the temperature dependence of I_{SSUB} (ISS) is determined by

$$I_{SSUB}(T_2) = I_{SSUB}(T_1)\left(\frac{T_2}{T_1}\right)^{-X_{T\beta}}\left[\frac{I_S(T_2)}{I_S(T_1)}\right]^{-n_S} \quad (2\text{-}183)$$

The parameters I_{KF} (IKF), I_{KR} (IKR), and I_{rB} (IRB) are also modified as follows:

$$I_{KF}(T_2) = I_{KF}(T_1)[1 + \text{TIKF1}(T_2 - T_1) + \text{TIKF2}(T_2 - T_1)^2]$$

$$I_{KR}(T_2) = I_{KR}(T_1)[1 + \text{TIKR1}(T_2 - T_1) + \text{TIKR2}(T_2 - T_1)^2] \quad (2\text{-}184)$$

$$I_{rB}(T_2) = I_{rB}(T_1)[1 + \text{TIRB1}(T_2 - T_1) + \text{TIRB2}(T_2 - T_1)^2]$$

where TIKF1, TIKR1, TIRB1 and TIKF2, TIKR2, TIRB2 are the first- and second-order temperature coefficients for the appropriate parameters, respectively.

The following parameters are also modified when corresponding temperature coefficients are specified, regardless of the TLEV value [20]

$$\beta_F(T_2) = \beta_F(T_1)[1 + \text{TBF1}(T_2 - T_1) + \text{TBF2}(T_2 - T_1)^2]$$

$$\beta_R(T_2) = \beta_R(T_1)[1 + \text{TBR1}(T_2 - T_1) + \text{TBR2}(T_2 - T_1)^2]$$

$$V_{AF}(T_2) = V_{AF}(T_1)[1 + \text{TVAF1}(T_2 - T_1) + \text{TVAF2}(T_2 - T_1)^2]$$

$$V_{AR}(T_2) = V_{AR}(T_1)[1 + \text{TVAR1}(T_2 - T_1) + \text{TVAR2}(T_2 - T_1)^2]$$

$$I_{\tau F}(T_2) = I_{\tau F}(T_1)[1 + \text{TITF1}(T_2 - T_1) + \text{TITF2}(T_2 - T_1)^2]$$

$$\tau_F(T_2) = \tau_F(T_1)[1 + \text{TTF1}(T_2 - T_1) + \text{TTF2}(T_2 - T_1)^2]$$

$$\tau_R(T_2) = \tau_R(T_1)[1 + \text{TTR1}(T_2 - T_1) + \text{TTR2}(T_2 - T_1)^2]$$

$$n_F(T_2) = n_F(T_1)[1 + \text{TNF1}(T_2 - T_1) + \text{TNF2}(T_2 - T_1)^2] \quad (2\text{-}185)$$

$$n_R(T_2) = n_R(T_1)[1 + \text{TNR1}(T_2 - T_1) + \text{TNR2}(T_2 - T_1)^2]$$

$$n_E(T_2) = n_E(T_1)[1 + \text{TNE1}(T_2 - T_1) + \text{TNE2}(T_2 - T_1)^2]$$

$$n_C(T_2) = n_C(T_1)[1 + \text{TNC1}(T_2 - T_1) + \text{TNC2}(T_2 - T_1)^2]$$

$$n_S(T_2) = n_S(T_1)[1 + \text{TNS1}(T_2 - T_1) + \text{TNS2}(T_2 - T_1)^2]$$

$$m_{JE}(T_2) = m_{JE}(T_1)[1 + \text{TMJE1}(T_2 - T_1) + \text{TMJE2}(T_2 - T_1)^2]$$

$$m_{JC}(T_2) = m_{JC}(T_1)[1 + \text{TMJC1}(T_2 - T_1) + \text{TMJC2}(T_2 - T_1)^2]$$

$$m_{JS}(T_2) = m_{JS}(T_1)[1 + \text{TMJS1}(T_2 - T_1) + \text{TMJS2}(T_2 - T_1)^2]$$

In the above relations the coefficients ending with 1 are first-order temperature coefficients, and the ones finishing with 2 are second-order temperature coefficients.

The temperature dependence of C_{JE} (CJE) and C_{JC} (CJC) is determined by Eqs. (2-148), while C_{SC} (CJS) follows the same relations when using the subscript S for the corresponding m and ϕ parameters.

The effect of temperature on ϕ_E (VJE) and ϕ_C (VJC) is described by Eqs. (2-144); the temperature dependence of ϕ_S (VJS) also follows Eqs. (2-144) if the subscript S is considered.

Finally, the resistors, as a function of temperature regardless of TLEV value, are determined as follows [20]:

$$r_E(T_2) = r_E(T_1)[1 + \text{TRE1}(T_2 - T_1) + \text{TRE2}(T_2 - T_1)^2]$$
$$r_B(T_2) = r_B(T_1)[\text{TRB1}(T_2 - T_1) + \text{TRB2}(T_2 - T_1)^2]$$
$$r_{BM}(T_2) = r_{BM}(T_1)[1 + \text{TRM1}(T_2 - T_1) + \text{TRM2}(T_2 - T_1)^2]$$
$$r_C(T_2) = r_C(T_1)[1 + \text{TRC1}(T_2 - T_1) + \text{TRC2}(T_2 - T_1)^2]$$

(2-186)

In the above relations, the coefficients ending with 1 are the first-order temperature coefficients, the ones finishing with 2 are the second-order temperature coefficients for the corresponding parameter. For the meaning of T_1 and T_2, refer to Sec. 1.8.4, where the role of the DTEMP parameter which allows the user of HSPICE to simulate some devices (here BJTs) at a different temperature in comparison with the circuit temperature (TNOM) is also explained.

2.11.5 Scaling and area dependence of the BJT model parameters for LEVEL = 1

Scaling is controlled by the element parameters AREA, AREAB, AREAC, and M. The AREA parameter, the normalized emitter area, multiplies all currents and capacitors and divides all resistors. AREAB and AREAC scale the size of the base area and collector area. AREAB or AREAC are used for scaling, depending on whether vertical or lateral geometry is selected (using the SUBS model parameter). For vertical geometry, AREAB is the scaling factor for IBC, ISC, and CJC. For lateral geometry, AREAC is the scaling factor. The scaling factor is AREA for all other parameters. M is the multiplier factor to simulate multiple BJTs.

The scaling of the static model parameters, IBE, IS, ISE, IKF, IKR, and IRB, for both vertical and lateral BJTs is determined by the following formula:

$$I_{\text{eff}} = \text{AREA} \cdot \text{M} \cdot I \qquad (2\text{-}187)$$

where I is either IBE, IS, ISS, ISE, IKF, IKR, ITF, or IRB.

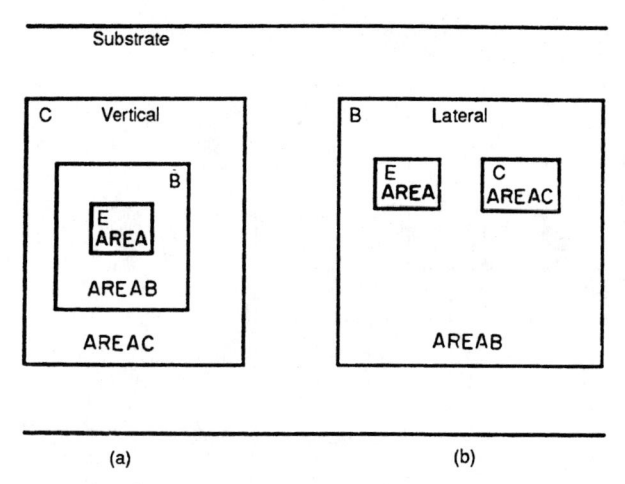

Figure 2-39 Top view of (a) vertical BJT and (b) lateral BJT, showing the role of AREA, AREAB, and AREAC parameters [20].

For both the vertical and lateral geometries, the resistor model parameters RB, RBM, RE, and RC are scaled by the following equation:

$$R_{\text{eff}} = \frac{R}{\text{AREA} \cdot \text{M}} \quad (2\text{-}188)$$

where R is either RB, RBM, RE, or RC.

Figure 2-39 shows the role of the AREA, AREAB, and AREAC parameters on the vertical and lateral BJT.

2.11.6 Quasi-saturation model (LEVEL = 2)

The HSPICE BJT quasi-saturation model is based on Ref. [21]. It is used to model BJTs that exhibit quasi-saturation or base push-out effects (see Sec. 2.9). This model is selected when LEVEL = 2 is set in the .MODEL statement of HSPICE. The equivalent circuit for the quasi-saturation model is shown in Fig. 2-40, where the current source I_{epi} and charge-storage elements C_I and C_X model the quasi-saturation effects.

a. Static model. The epitaxial current value, I_{epi}, is determined by the following equation [20]:

$$I_{\text{epi}} = \frac{F_i - F_x - \ln\left(\frac{1 + F_i}{1 + F_x}\right) + \frac{q(V_{BC} - V_{BCX})}{n_{\text{epi}} kT}}{\frac{q R_{C\text{epi}}}{n_{\text{epi}} kT}\left(1 + \frac{|V_{BC} - V_{BCX}|}{V_o}\right)} \quad (2\text{-}189)$$

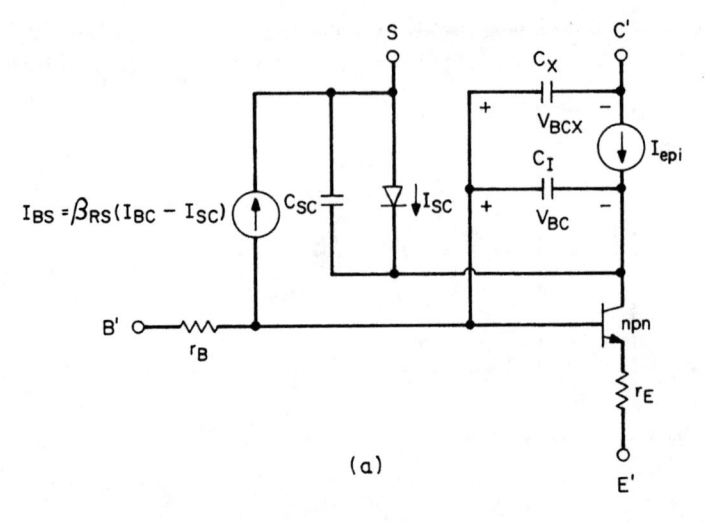

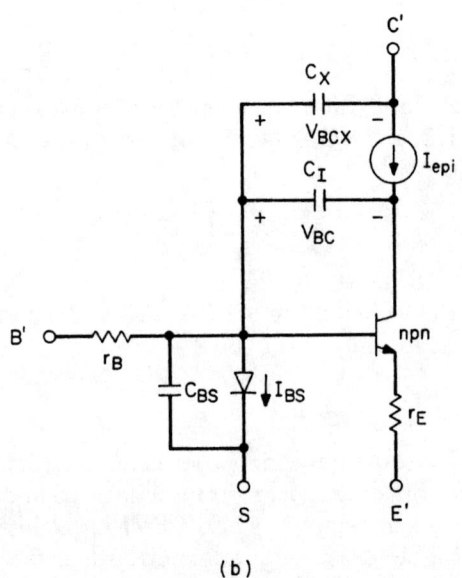

Figure 2-40 HSPICE BJT quasi-saturation model for (a) vertical geometry, and (b) lateral geometry [20].

where
$$F_i = [1 + \gamma_{\text{epi}}\, e^{qV_{BC}/n_{\text{epi}}kT}]^{1/2}$$
$$F_x = [1 + \gamma_{\text{epi}}\, e^{qV_{BCX}/n_{\text{epi}}kT}]^{1/2} \tag{2-190}$$

In Eq. (2-189), n_{epi} (NEPI) is the epitaxial region emission coefficient; $R_{C\text{epi}}$ (RC), which defaults to zero, is the resistance of the epitaxial region under equilibrium condition, measured in $[\Omega]$; V_o

(VO) is the carrier velocity saturation voltage, measured in [V]. A specified zero value (default) for this parameter indicates an infinite value. In Eqs. (2-190), γ_{epi} (GAMMA), which defaults to zero, is the epitaxial doping factor; $\gamma_{epi} = (2n_i/N_{epi})^2$, where N_{epi} is the epitaxial impurity concentration and n_i is the intrinsic concentration. The .MODEL keyword for HSPICE is indicated in parentheses.

b. Large-signal model. The epitaxial charge storage elements are determined by [20]

$$Q_I = Q_{CO}\left(F_i - 1 - \frac{\gamma_{epi}}{2}\right)$$

$$Q_X = Q_{CO}\left(F_x - 1 - \frac{\gamma_{epi}}{2}\right) \tag{2-191}$$

The corresponding capacitances, represented by C_I and C_X in Fig. 2-40, are calculated as follows [20]:

$$C_I = \frac{dQ_I}{dV_{BC}} = \left(\frac{q\gamma_{epi}Q_{CO}}{2n_{epi}kTF_i}\right)e^{qV_{BC}/n_{epi}kT}$$

$$C_X = \frac{dQ_X}{dV_{BCX}} = \left(\frac{q\gamma_{epi}Q_{CO}}{2n_{epi}kTF_x}\right)e^{qV_{BCX}/n_{epi}kT} \tag{2-192}$$

In the above equations a new .MODEL parameter has been introduced: Q_{CO} (QCO), the epitaxial charge factor (defaults = 0 Coul). Setting $Q_{CO} = 0$ inhibits the collector storage portion of the model. Finally, when considering the geometry of Fig. 2-40a, the reverse beta for the substrate, β_{RS} (BRS), must be taken into account. Its default value is the unity. The parameter BRS multiplies the base-collector current IBC: BRS = 0 (default for LEVEL = 1) results in a traditional substrate diode; BRS = 1 (default for LEVEL = 2) results in a parasitic transistor with a gain of 1. Any positive values for BRS creates the vertical parasitic transistor.

c. Temperature and scaling effects on the model parameters. The model parameters of the HSPICE BJT quasi-saturation model are modified for temperature as follows:

$$R_{Cepi}(T_2) = R_{Cepi}(T_1)\left(\frac{T_2}{T_1}\right)^{BEX}$$

$$V_o(T_2) = V_o(T_1)\left(\frac{T_2}{T_1}\right)^{BEXV} \tag{2-193}$$

$$\gamma_{epi}(T_2) = \gamma_{epi}(T_1)\left(\frac{T_2}{T_1}\right)^{X_{TI}} \exp\left[-\frac{qE_g(300)}{kT_2}\left(1 - \frac{T_2}{T_1}\right)\right]$$

where BEX (which defaults to 2.42) and BEXV (which defaults to 1.90) are temperature exponents for R_{Cepi} and V_o, respectively; they are .MODEL parameters. As far as scaling effects are concerned, only Q_{CO} and R_{Cepi} are affected by this dependence. In particular, we have

$$Q_{CO} = Q_{CO} \text{AREAB} \cdot M \quad \text{(vertical geometry)}$$

$$Q_{CO} = Q_{CO} \text{AREAC} \cdot M \quad \text{(lateral geometry)} \quad (2\text{-}194)$$

$$R_{Cepi} = R_{Cepi}/(\text{AREA} \cdot M)$$

2.12 PSPICE Models

PSPICE [22–24] BJT models are based directly on the integral charge-control Gummel-Poon model implemented in SPICE2. In the following, PSPICE BJT *static*, *large-signal*, *small-signal*, and *thermal* models are considered. The *noise* model is described in Chap. 7.

2.12.1 Static model

The PSPICE BJT *static model* is represented by the equivalent circuit of Fig. 2-37 when the capacitances are not considered. The equations describing the static behavior of the BJT are the same as in SPICE2.

2.12.2 Large-signal model

The PSPICE BJT *large-signal model* is represented by the equivalent circuit of Fig. 2-37 when the parasitic capacitances C_{BCP}, C_{BEP}, and C_{CSP} are neglected. The equations describing the charge-storage effects are the same as in SPICE2 and HSPICE.

The substrate node is optional, and if not specified, it defaults to ground. For model types NPN and PNP, the insulation junction capacitance is connected between the intrinsic-collector and substrate nodes. This is the same as in SPICE2 or SPICE3, and works well for vertical BJT geometry.

For lateral BJT geometry, there is a third model [22] (selected by specifying the .MODEL parameter LPNP) in which the insulation junction capacitance is connected between the intrinsic-base and substrate nodes.

2.12.3 Small-signal model

As in SPICE2, the PSPICE BJT *small-signal model* is automatically computed inside the program by the linearization of the static and large-signal model equations; thus no new model parameters need be specified in the .MODEL statement by the user of PSPICE. The small-signal model equivalent circuit is shown in Fig. 2-38 when the parasitic capacitances are neglected.

2.12.4 Temperature and area factor effects on the BJT model parameters

The relations which describe the temperature dependence of the BJT parameters in SPICE2 are also valid in PSPICE. In addition, PSPICE provides formulations for I_{SSUB} (ISS), r_E (RE), r_B (RB), r_C (RC), and r_{BM} (RBM), expressed by Eqs. (2-183) and (2-186).

Scaling is controlled by the element parameter AREA. All the currents and capacitances are multiplied by AREA, and all the resistances are divided by AREA.

References

1. R. S. Muller and T. Kamins, *Device Electronics for Integrated Circuits*, Wiley, New York, 1977.
2. L. W. Nagel, SPICE2: A Computer Program to Simulate Semiconductor Circuits, Electronics Research Laboratory, Rep. No. ERL-M520, University of California, Berkeley, 1975.
3. J. J. Ebers and J. L. Moll, Large-Signal Behavior of Junction Transistors, *Proc. IRE*, 42, 1954.
4. H. K. Gummel and H. C. Poon, An Integral Charge Control Model of Bipolar Transistors, *Bell Syst. Tech. J.*, 49, 1970.
5. W. G. Oldham and S. E. Schwarz, *An Introduction to Electronics*, Holt, Rinehart and Winston, New York, 1972.
6. I. E. Getreu, *Modeling the Bipolar Transistor*, Tektronix, Inc., Beaverton, Oreg., 1976.
7. A. Vladimirescu, K. Zhang, A. R. Newton, D. O. Pederson, and A. Sangiovanni-Vincentelli, *SPICE Version 2G User's Guide*, Dept. EECS, University of California, Berkeley, 1981.
8. B. R. Chawla and H. K. Gummel, Transition Region Capacitance of Diffused *pn* Junctions, *IEEE Trans. Electron Devices*, ED-18, 1971.
9. H. K. Gummel and H. C. Poon, Modeling of Emitter Capacitance, *IEEE Proc. (Lett.)* 57, 1969.
10. E. Khalily, Hewlett-Packard Co., private communication.
11. S. M. Sze, *Physics of Semiconductor Devices*, Wiley, New York, 1969.
12. P. Antognetti (ed.), *Power Integrated Circuits*, McGraw-Hill, New York, 1986.
13. S. K. Ghandhi, *Semiconductor Power Devices*, Wiley, New York, 1977.
14. P. Antognetti, G. Massobrio, G. Sciutto, B. Gabriele, and B. Cotta, Modeling and Simulation of Power Transistors in Electronic Power Converter, *Proc. Eurocon '80 Conf.*, Stuttgart, Germany, 1980.
15. P. Antognetti, G. Massobrio, M. Mazzucchelli and G. Sciutto, Optimum Design of a Power Transistor Inverter Controlled by PWM Technique, *Proc. Motorcon.* Chicago, 1981.
16. G. Massobrio, Modello del Transistore di Potenza per Applicazioni CAD, *Pixel*, 2, 1983.
17. G. Massobrio, Vincoli all'Implementazione in SPICE del Modello del Transistore di Potenza, *Pixel*, 4, 1984.
18. T. I. Quarles, The SPICE3 Implementation Guide, Electronics Research Laboratory, Rep. No. ERL-M44, University of California, Berkeley, 1989.
19. T. I. Quarles, *SPICE3 Version 3C1 Users Guide*, Electronics Research Laboratory, Rep. No. ERL-M46, University of California, Berkeley, 1989.

20. Meta-Software Inc., *HSPICE User's Manual*, Campbell, Calif., 1991.
21. G. M. Kull, L. W. Nagel, Shiu-Wuu Lee, P. Lloyd, E. J. Prendergast, and H. Dirks, A Unified Circuit Model for Bipolar Transistors Including Quasi-Saturation Effects, *IEEE Trans. Electron Devices*, **ED-32**, 1985.
22. MicroSim Corporation, *PSPICE User's Manual*, Irvine, Calif., 1989.
23. P. W. Tuinenga, *SPICE: A Guide to Circuit Simulation and Analysis Using PSPICE*, Prentice-Hall, Englewood Cliffs, N.J., 1988.
24. M. H. Rashid, *SPICE for Circuits and Electronics Using PSPICE*, Prentice-Hall, Englewood Cliffs, N.J., 1990.

Chapter

3

Junction Field-Effect Transistor (JFET)

The *Junction Field-Effect Transistor* (*JFET*) is a semiconductor device that depends for its operation on the control of current by an electric field. The structure of an *n-channel* field-effect transistor is shown in Fig. 3-1a. It consists of a conductive channel that has two ohmic contacts, one acting as the cathode (*source*) and the other as the anode (*drain*), with an appropriate voltage applied between drain and source. The third electrode (*gate*) forms a rectifying junction with the channel. Thus the JFET is basically a voltage-controlled resistor, and its resistance can be varied by changing the width of the depletion region extending into the channel.

Since the conduction process involves predominantly one kind of carrier, the JFET is also called a *unipolar* transistor in order to distinguish it from the bipolar junction transistor (BJT), in which both types of carriers are involved.

The symbols and sign convention for a *p*-channel and an *n*-channel JFET are indicated in Fig. 3-1b. The direction of the arrow at the gate of the JFET indicates the direction in which gate current would flow if the gate junction were forward-biased.

Note that the *n*-channel JFET requires zero or negative gate bias and positive drain voltage. The *p*-channel JFET requires opposite voltage polarities [1–3].

This chapter establishes *static*, *large-signal*, *small-signal*, and *thermal* physical-mathematical models of a JFET, and how the models are implemented in SPICE2 [4] and other SPICE2-based programs. The *noise* model is described in Chap. 7.

SPICE2 JFET models are directly regulated by the subroutines *READIN* (it reads and processes the .MODEL statement), *MODCHK*

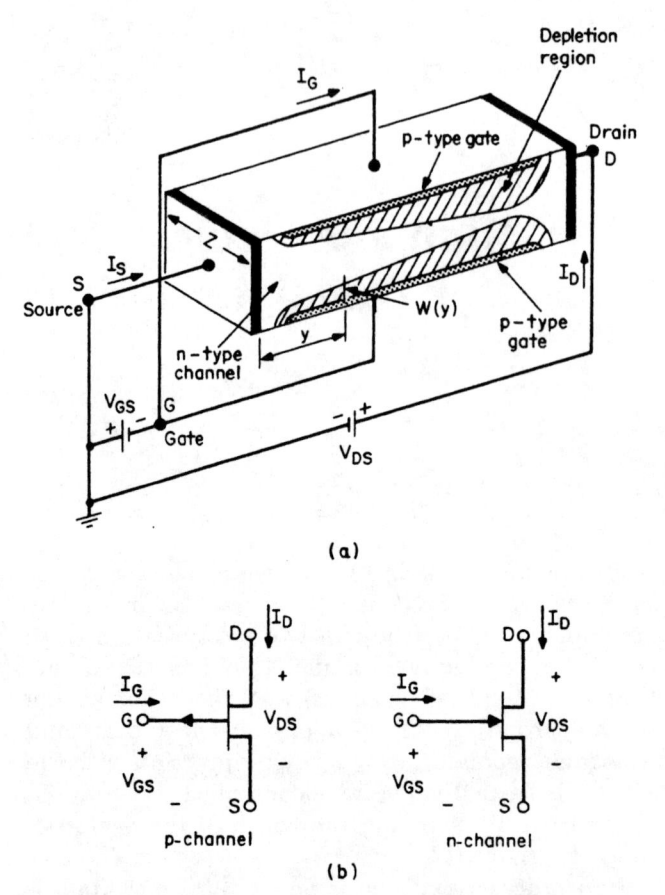

Figure 3-1 (a) Basic structure of an n-channel JFET and (b) JFET symbols and sign convention. (*Adapted from Millman [2]. Used by permission.*)

(it performs one-time processing of the model parameters and assigns the parameter default values), *JFET* (it computes all the model equations), and *TMPUPD* (it updates the temperature-dependent parameters of the models), in addition to the subroutines concerning each type of analysis. The noise model is directly regulated by the subroutine *NOISE* (see Chap. 7).

3.1 Static Model

This section emphasizes the basic aspects of the behavior of a junction field-effect transistor in order to obtain its static model; then, how this model is implemented in SPICE2 is investigated.

3.1.1 DC characteristics*

To analyze the JFET, consider first a very small bias V_{DS} applied to the *drain* electrode (while the *source* is grounded) of the n-channel field-effect transistor of Fig. 3-1a. Under this condition the gate-channel bias (and therefore the width of the gate depletion region) is uniform along the entire channel. The voltage at the gate is V_{GS}. An expanded view of the channel region is shown in Fig. 3-2.

A one-dimensional structure with a gate length L between the source and drain regions and a width Z perpendicular to the plane of the paper is assumed (usually $Z \gg L$). Drain current flows along the dimension y.

A *one-sided step junction* at the gate with the acceptor concentration N_A in the p region much greater than the donor concentration N_D in the channel is also assumed. Therefore, the depletion region extends primarily into the n channel. The distance between the p-type gate and the substrate is d, the thickness of the gate depletion region in the n-type channel is W, and the thickness of the neutral portion of the channel is x_W. In order to focus on the role of the gate, assume that the depletion region at the substrate junction extends primarily into the substrate so that $x_W \simeq d - W$. This is generally true in practice.

The resistance of the channel region can be written as

$$R = \frac{\rho L}{x_W Z} \quad (3\text{-}1)$$

where ρ is the *resistivity* of the channel. Hence, the drain current is

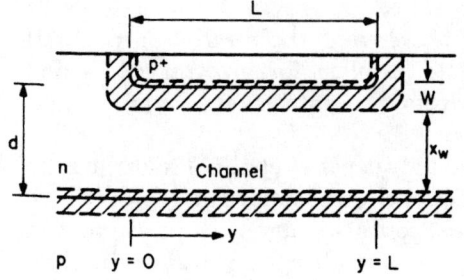

$$I_D = \frac{V_{DS}}{R} = \frac{Z}{L}(q\mu_n N_D x_W V_{DS}) \quad (3\text{-}2)$$

Figure 3-2 Channel region of a JFET showing depletion regions. *(From Muller and Kamins [3]. Copyright © 1977 by John Wiley & Sons, Inc. Used by permission.)*

---

*The material in this section is based in part on R. S. Muller and T. I. Kamins, *Device Electronics for Integrated Circuits*, copyright © 1977 by John Wiley & Sons, Inc. Reprinted by permission.

The dependence on gate voltage is incorporated in Eq. (3-2) by expressing $x_W = d - W$, where W from Eq. (A-39), when $N_A \gg N_D$, is

$$W = \sqrt{\frac{2\varepsilon_s}{qN_D}(\phi_0 - V_{GS})} \qquad (3\text{-}3)$$

and ϕ_0 is the built-in potential. The current can now be written as a function of the gate and drain voltages.

$$I_D = \frac{Z}{L} q\mu_n N_D d \left[1 - \sqrt{\frac{2\varepsilon_s}{qN_D d^2}(\phi_0 - V_{GS})} \right] V_{DS} \qquad (3\text{-}4)$$

The factors in front of the bracketed terms represent the *conductance* G_0 of the n region if it were completely undepleted (the so-called metallurgical channel) so that Eq. (3-4) may be rewritten as

$$I_D = G_0 \left[1 - \sqrt{\frac{2\varepsilon_s}{qN_D d^2}(\phi_0 - V_{GS})} \right] V_{DS} \qquad (3\text{-}5)$$

Hence, at a given gate voltage, a linear relationship between I_D and V_{DS} can be found. This is a consequence of having assumed small applied drain voltages. The square-root dependence on gate voltage in Eq. (3-5) arises from the assumption of an *abrupt gate-channel junction*.

From Eq. (3-5) it can be seen that the current is maximum at zero applied gate voltage and decreases as $|V_{GS}|$ increases. The equation predicts zero current when the gate voltage is large enough to deplete the entire channel region [3].

Now to see the physics behind the device, the restriction of small drain voltages is removed and the problem for arbitrary V_{DS} and V_{GS} values (with the restriction that the gate must always remain reverse-biased) is considered.

With V_{DS} arbitrary, the voltage between the channel and the gate is a function of position y. Consequently, the depletion-region width and therefore the channel cross section also vary with position. The voltage across the depletion region is higher near the drain than near the source in this n-channel device. Therefore, the depletion region is wider near the drain, as shown in Fig. 3-3.

The *gradual channel approximation* will be used. This approximation assumes that the channel and depletion-region widths vary slowly from source to drain so that the depletion region is influenced

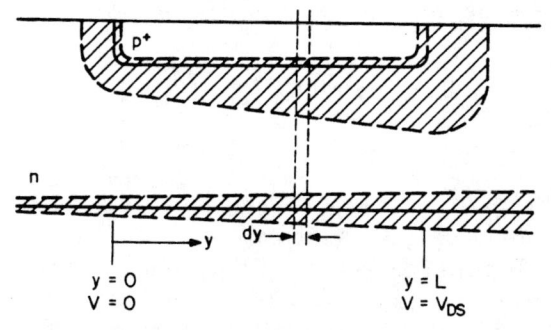

Figure 3-3 Channel region of a JFET showing variation of the width of depletion regions along the channel when the drain voltage is significantly higher than the source voltage. (*From Muller and Kamins [3]. Copyright © 1977 by John Wiley & Sons, Inc. Used by permission.*)

only by fields in the vertical dimension and not by fields extending from drain to source. In other words, the field in the y direction is much less than that in the x direction in the depletion region, and the depletion-region width from a one-dimensional analysis may be found [3].

Within this approximation, an expression for the increment of voltage across a small section of the channel of length dy at y may be written as

$$dV(y) = I_D \, dR(y) = \frac{I_D \, dy}{Zq\mu_n N_D [d - W(y)]} \qquad (3\text{-}6)$$

The width of the depletion region is now controlled by the voltage $[V_{GS} - V(y)]$, where $V(y)$ is the potential in the channel at point y, so that, from Eq. (A-39),

$$W(y) = \sqrt{\frac{2\varepsilon_s}{qN_D} [\phi_0 - V_{GS} + V(y)]} \qquad (3\text{-}7)$$

This expression may be used in Eq. (3-6), which is then integrated from source to drain to obtain the current-voltage relationship for the JFET

$$\frac{I_D \int_0^L dy}{Zq\mu_n N_D} = \int_0^{V_D} \left[d - \sqrt{\frac{2\varepsilon_s}{qN_D}(\phi_0 - V_{GS} + V)} \right] dV \qquad (3\text{-}8)$$

After integrating and rearranging,

$$I_D = G_0 \left\{ V_{DS} - \frac{2}{3} \sqrt{\frac{2\varepsilon_s}{qN_D d^2}} \left[\sqrt{(\phi_0 - V_{GS} + V_{DS})^3} - \sqrt{(\phi_0 - V_{GS})^3} \right] \right\}$$

(3-9)

At low drain voltages, Eq. (3-9) reduces to the simpler expression of Eq. (3-5); at large drain voltages, Eq. (3-9) indicates that the current reaches a maximum and begins decreasing with increasing drain voltage, but this maximum corresponds to the limit of validity of the analysis. Indeed, from Fig. 3-4 it can be seen that as the drain voltage increases, the width of the conducting channel near the drain decreases, until finally the channel is completely depleted in this region (Fig. 3-4b).

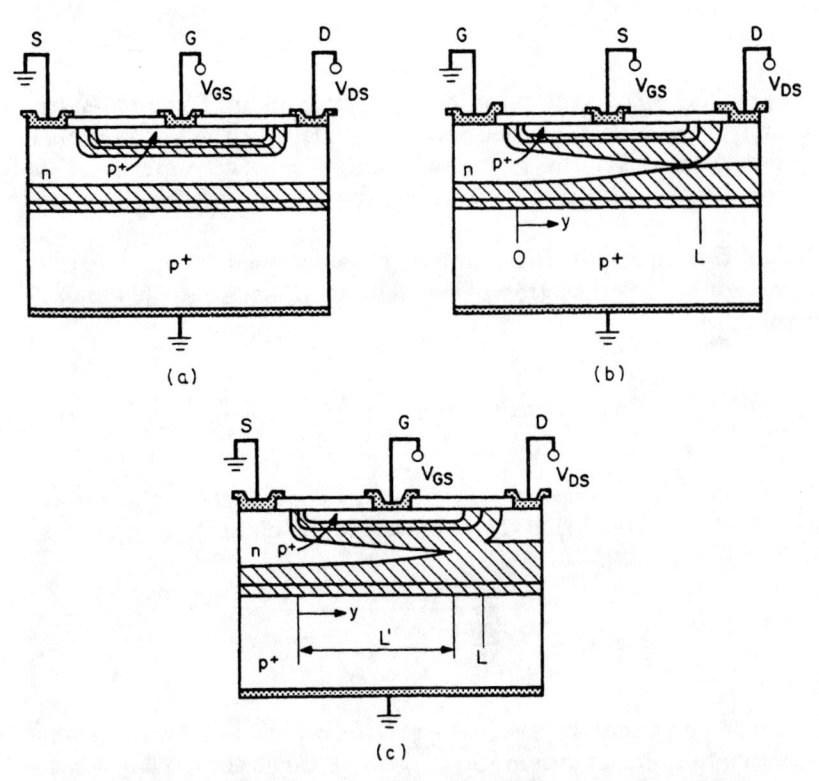

Figure 3-4 Behavior of the depletion regions in a JFET. (a) For very small drain voltage, the channel is nearly equipotential and the dimensions of the depletion regions are uniform. (b) When V_{DS} is increased to $V_{D,\text{sat}}$, the depletion regions on both sides of the channel meet at the pinch-off point at $y = L$. (c) When $V_{DS} > V_{D,\text{sat}}$, the pinch-off point at $y = L'$ moves slightly closer to the source. (*From Muller and Kamins [3]. Copyright © 1977 by John Wiley & Sons, Inc. Used by permission.*)

When this occurs, Eq. (3-6) becomes indeterminate ($W \to d$). The equations are, therefore, valid only for V_{DS} below the drain voltage that *pinches off* the channel. Current continues to flow when the channel has been pinched off because there is no barrier to the transfer of electrons traveling down the channel toward the drain. As they arrive at the edge of the pinched-off region, they are pulled across it by the field directed from the drain toward the source. If the drain bias is increased further, any additional voltage is dropped across a depleted, high-field region near the drain electrode, and the point at which the channel is entirely depleted moves slightly toward the source (Fig. 3-4c) [3].

If this slight movement is neglected, the drain current remains constant (saturates) as the drain voltage is increased further, and the bias condition is referred to as *saturation*. The drain voltage at which the channel is entirely depleted near the drain electrode is found from Eq. (3-7) to be

$$V_{D,\text{sat}} = \frac{qN_D d^2}{2\varepsilon_s} - (\phi_0 - V_{GS})$$

$$= V_P - \phi_0 + V_{GS} = V_{T0} + V_{GS} \qquad (3\text{-}10)$$

where $V_P = qN_D d^2/2\varepsilon_s$ is referred to as the *pinch-off voltage* and $V_{T0} = V_P - \phi_0$ is referred to as the *threshold voltage*.

From Eqs. (3-9) and (3-10) the drain saturation current is found to be

$$I_{D,\text{sat}} = G_0 \left\{ \frac{qN_D d^2}{6\varepsilon_s} - (\phi_0 - V_{GS}) \left[1 - \frac{2}{3}\sqrt{\frac{2\varepsilon_s(\phi_0 - V_{GS})}{qN_D d^2}} \right] \right\}$$

$$= \frac{G_0 V_P}{3} \left[1 - 3\frac{\phi_0 - V_{GS}}{V_P} + 2\sqrt{\left(\frac{\phi_0 - V_{GS}}{V_P}\right)^3} \right] \qquad (3\text{-}11)$$

The maximum value of $I_{D,\text{sat}}$ (designated I_{DSS}) occurs for $V_{GS} = 0$: if $I_{D,\text{sat}}$ is normalized to I_{DSS} and plotted as a function of V_{GS}/V_P, the curve shown in Fig. 3-5 results. Also plotted in Fig. 3-5 is a square-law transfer characteristic given by

$$I_{D,\text{sat}} = I_{DSS} \left(1 - \frac{V_{GS}}{V_P} \right)^2 \qquad (3\text{-}12)$$

The two curves agree quite closely, and Eq. (3-12) is commonly used as an approximation of the JFET characteristic in the saturation region.

The I_D vs. V_{DS} characteristic may be divided into three regions (see Fig. 3-6): (1) the linear region at low drain voltages, (2) a region with

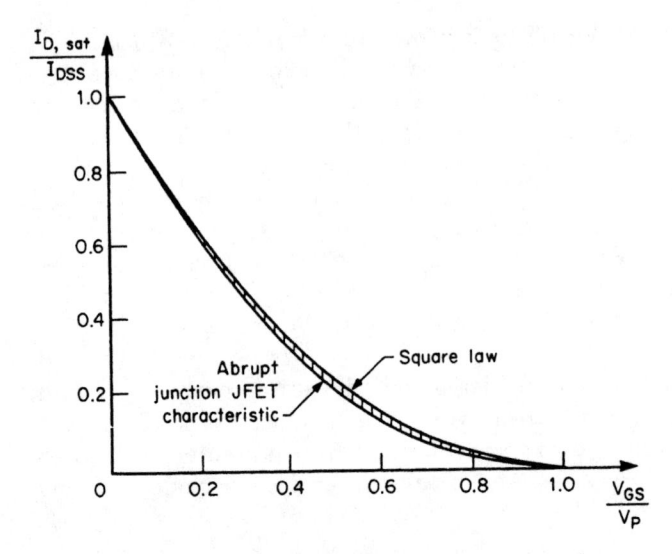

Figure 3-5 Normalized transfer characteristics of an abrupt-junction JFET compared with a square-law characteristic.

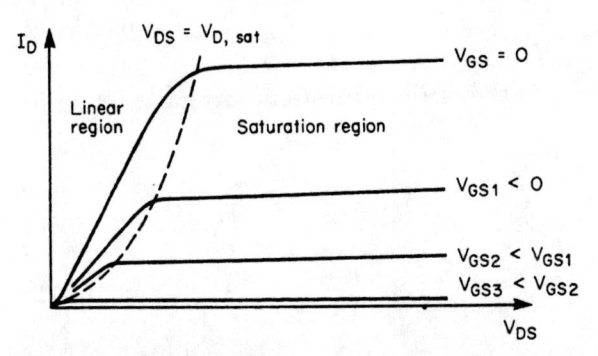

Figure 3-6 The output I_D vs. V_{DS} characteristics of an n-channel JFET as a function of the gate voltage.

less than linear increase of current with drain voltage, and (3) a saturation region where the current remains relatively constant as the drain voltage is increased further.

Because of the physics of the device, Eq. (3-11) predicts the current to be maximum for zero gate bias and to decrease as negative gate voltage is applied [3].

As the gate voltage becomes more negative, the drain saturation voltage and the corresponding current decrease. At a sufficiently negative value of gate voltage, the saturation drain current becomes zero. This *turn-off* voltage V_{T0} is found from Eq. (3-11) to be

$$V_{T0} = \phi_0 - \frac{qN_D d^2}{2\varepsilon_s} \quad (3\text{-}13)$$

Field-effect transistors are often operated in the saturation region where the output current is not appreciably affected by the output (drain) voltage but only by the input (gate) voltage. For this bias condition, the JFET is almost an ideal current source controlled by an input voltage. The *transconductance* g_m of the transistor expresses the effectiveness of the control of the drain current by the gate voltage; it is defined as

$$g_m = \frac{dI_D}{dV_{GS}}\bigg|_{\text{op}} \quad (3\text{-}14)$$

The subscript op denotes that the independent variables assume the values at operating-point bias.

3.1.2 Limitations of the ideal theory

Previous analysis has included several simplifying assumptions. In practical devices, however, some of these assumptions may not be sufficiently valid to obtain a good match between theory and experiment. This section considers some deviations from the simple theory.

a. Control of depletion-region width. One assumption affecting the theory in the previous section is that the depletion-region width is controlled by the gate-channel junction and not by the channel-substrate junction. There will be a variation of the potential across the channel-substrate junction along the channel, with the maximum potential and depletion-region thickness near the drain. Consequently, the channel becomes completely depleted at lower drain voltage than is indicated by Eq. (3-10) [3].

b. Channel-length modulation. It has already been seen in the previous sections that in the saturation region the potential at the end of the channel, at point L of Fig. 3-4b, will be fixed at precisely the value of $V_{D,\text{sat}}$ corresponding to the applied gate voltage. This is so because point L itself is where the two depletion regions just touch. Hence, the reverse bias across the gate junctions at this point is fixed by the condition that $W = d$ there.

As the drain voltage is increased further, the reverse bias between the gate and the drain regions is also increased; hence, the width of the depletion region near the drain will also increase. As a result, the point will move toward the source as indicated in Fig. 3-4c. The voltage at point L' remains at the same value, but the length from the source to point L' shortens. It is evident that the drain current will increase at a given gate voltage as the drain voltage is increased; the end point for the integration in Eq. (3-8) now becomes L' rather than L, where L' is the point at which the channel becomes completely depleted $[V(L') = V_{D,\text{sat}}]$.

Then for $V_{DS} > V_{D,\text{sat}}$, the expression for $I_{D,\text{sat}}$ in Eq. (3-11) should be multiplied by the factor L/L'.

This phenomenon, which is particularly important in devices with short channel lengths, is quite analogous to the Early effect discussed in connection with BJTs. In both cases, the increase in current takes place because the current path is shortened by the widening of a reverse-biased depletion region [5].

Thus λ can be defined as a *channel-length modulation parameter* which is a measure of the JFET output conductance in saturation. If we specify this parameter, the JFET will have a finite but constant output conductance in saturation. λ can be defined as

$$\lambda = \frac{L'}{LV_{DS}} \qquad (3\text{-}15)$$

c. Effect of series resistance.* In the preceding calculation, we considered only the resistance of that portion of the channel which can be modulated by the application of a reverse bias to the gate. In reality, there are series resistances present, both near the source and near the drain, which interpose a voltage drop between the source and drain contacts and the channel.

The effect of these series resistances on the channel conductance in the linear region can be readily calculated by noting that

$$\frac{1}{g(\text{obs})} = \frac{1}{g} + r_S + r_D \qquad (3\text{-}16)$$

where g is the true channel conductance while $g(\text{obs})$ is the conductance observed experimentally; r_S and r_D are the series resistances near the source and the drain, respectively. Thus,

$$g(\text{obs}) = \frac{g}{1 + (r_S + r_D)g} \qquad (3\text{-}17)$$

---

*The material in this section is taken from A. S. Grove, *Physics and Technology of Semiconductor Devices*, copyright © 1967 by John Wiley & Sons, Inc. Reprinted by permission.

which shows that the observed conductance will be reduced due to the two series resistances.

The effect of the series resistance near the source region, r_S, on the transconductance in the saturation region is now considered. Because of this resistance, the potential at the beginning of the channel will not be zero but will have some finite value V_s. Thus the *effective* gate voltage will be

$$V_{GS} = V_{GS,\text{appl}} - V_s \qquad (3\text{-}18)$$

As a result, the observed transconductance is given by

$$g_m(\text{obs}) = \frac{dI_D}{dV_{GS,\text{appl}}} = \frac{dI_D}{d(V_{GS} + V_s)} \qquad (3\text{-}19)$$

which, in turn, yields

$$g_m(\text{obs}) = \frac{1}{dV_{GS}/dI_D + dV_s/dI_D} \qquad (3\text{-}20)$$

and hence

$$g_m(\text{obs}) = \frac{g_m}{1 + r_S g_m} \qquad (3\text{-}21)$$

This last equation shows that the observed transconductance in the saturation region will be reduced due to the presence of a series resistance near the source from that attainable in the absence of such a series resistance.

The series resistance near the drain will act in a different manner. Because of the voltage drop across this resistance, the drain voltage required to bring about saturation of the drain current will be larger than without it. However, since beyond that voltage, i.e., for $V_{DS} > V_{D,\text{sat}}$, the magnitude of V_{DS} has no significant effect on the drain current, the drain series resistance will have no further effect either [6].

d. Breakdown. The *pn* junctions between gate and channel of a JFET are subject to *avalanche breakdown*. Since the JFET is a majority-carrier device whose operation does not depend on minority-carrier concentrations, the breakdown process is quite straightforward. Breakdown occurs when the voltage between gate and channel exceeds a critical value at which an abrupt increase of the drain current occurs.

The breakdown characteristic of the JFET can be described by specifying the breakdown voltage from drain to gate with the source open. This process, however, is not taken into account in SPICE2.

3.1.3 Static model and its implementation in SPICE2

The SPICE2 JFET *static* model follows from the quadratic FET model of Shichman and Hodges [7].

The dc characteristics are defined by the parameters V_{T0} and β, which determine the variation of drain current with gate voltage, λ, which determines the output conductance, and I_S, the saturation current of the two gate junctions.

The results derived in the previous sections can be used to define the static model of the JFET.

The SPICE2 model for an n-channel JFET is shown in Fig. 3-7.

For a p-channel device, the polarities of the terminal voltages V_{GD}, V_{GS}, and V_{DS}, the direction of the two gate junctions, and the direction of the nonlinear current source I_D must be reversed.

The ohmic resistances of the drain and source regions of the JFET are modeled by the two linear resistors r_D and r_S, respectively. The dc characteristics of the JFET are represented by the nonlinear current source I_D. The value of I_D is determined by a set of equations simpler than those derived in Sec. 3.1.1, which are obtained by expanding

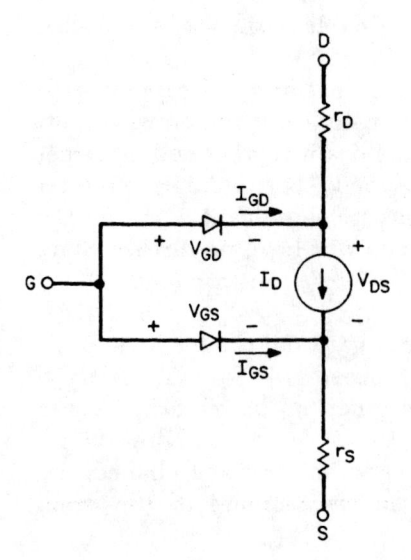

Figure 3-7 SPICE2 n-channel JFET static model.

the square-root terms as a binomial series. Making the expansions implies that $|V_{T0} - V_{GS}|/V_P < 1$.

a. Normal mode. The normal mode operation is characterized in SPICE2 by the following relations (for $V_{DS} \geq 0$):

$$I_D = \begin{cases} 0 & \text{for } V_{GS} - V_{T0} \leq 0 \text{ (channel pinched off)} \\ \beta(V_{GS} - V_{T0})^2(1 + \lambda V_{DS}) & \text{for } 0 < V_{GS} - V_{T0} \leq V_{DS} \text{ (saturated region)} \\ \beta V_{DS}[2(V_{GS} - V_{T0}) - V_{DS}](1 + \lambda V_{DS}) & \text{for } 0 < V_{DS} < V_{GS} - V_{T0} \text{ (linear region)} \end{cases}$$

(3-22)

b. Inverted mode. The inverted mode operation is characterized in SPICE2 by the following relations (for $V_{DS} < 0$):

$$I_D = \begin{cases} 0 & \text{for } V_{GD} - V_{T0} \leq 0 \text{ (channel pinched off)} \\ -\beta(V_{GD} - V_{T0})^2(1 - \lambda V_{DS}) & \text{for } 0 < V_{GD} - V_{T0} \leq -V_{DS} \text{ (saturated region)} \\ \beta V_{DS}[2(V_{GD} - V_{T0}) + V_{DS}](1 - \lambda V_{DS}) & \text{for } 0 < -V_{DS} < V_{GD} - V_{T0} \text{ (linear region)} \end{cases}$$

(3-23)

In SPICE2 the JFET drain current is then modeled by a simple square-law characteristic that is determined by the parameters β and V_{T0} [4]. The convention in SPICE2 is that V_{T0} is negative for all JFETs regardless of polarity. The parameters V_{T0} and β are usually determined from a graph of $\sqrt{I_D}$ vs. V_{GS}. An example for an n-channel device is shown in Fig. 3-8. The parameter V_{T0} is the x-axis intercept of this graph, whereas the parameter β, or its square root, is the slope of the $\sqrt{I_D}$ vs. V_{GS} characteristic.

The parameter λ (which has typical values in the range 0.1 to 0.01 V^{-1}) determines the effect of *channel-length modulation* (see Sec. 3.1.2b) on the JFET characteristic. As the output conductance of the JFET, in the forward saturation region, is given by

$$g_{D,\text{sat}} \equiv \frac{dI_D}{dV_{DS}} = \beta\lambda(V_{GS} - V_{T0})^2 \simeq \lambda I_D \qquad (3\text{-}24)$$

the saturation conductance is directly proportional to the drain current.

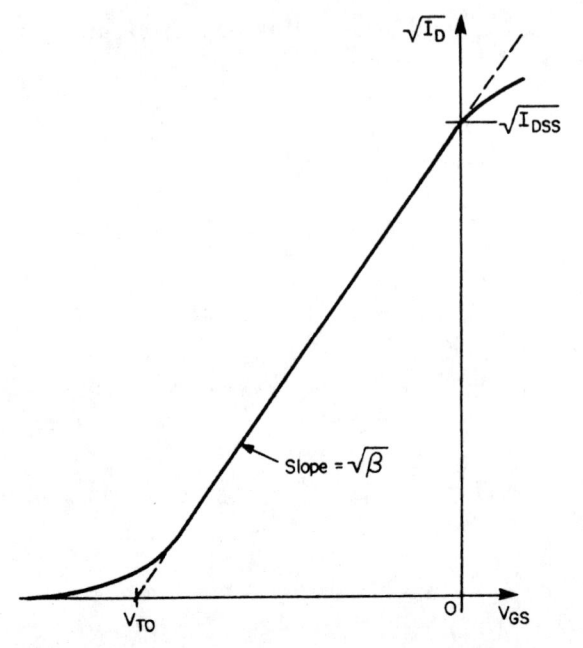

Figure 3-8 How to determine V_{TO} and β parameters.

The two diodes shown in Fig. 3-7 are modeled by the following equations:

$$I_{GD} = \begin{cases} -I_S + V_{GD}\,\text{GMIN} & \text{for } V_{GD} \leq -5\dfrac{kT}{q} \\ I_S(e^{qV_{GD}/kT} - 1) + V_{GD}\,\text{GMIN} & \text{for } V_{GD} > -5\dfrac{kT}{q} \end{cases} \quad (3\text{-}25)$$

$$I_{GS} = \begin{cases} I_S(e^{qV_{GS}/kT} - 1) + V_{GS}\,\text{GMIN} & \text{for } V_{GS} > -5\dfrac{kT}{q} \\ -I_S + V_{GS}\,\text{GMIN} & \text{for } V_{GS} \leq -5\dfrac{kT}{q} \end{cases} \quad (3\text{-}26)$$

where I_S is the gate junction saturation current.

To aid convergence, a small conductance, GMIN, is inserted in SPICE2 in parallel with every junction. The value of this conductance is a program parameter that can be set by the user by selection of the .OPTIONS statement. The default value for GMIN is 10^{-12} mho; moreover, the user cannot set GMIN to zero.

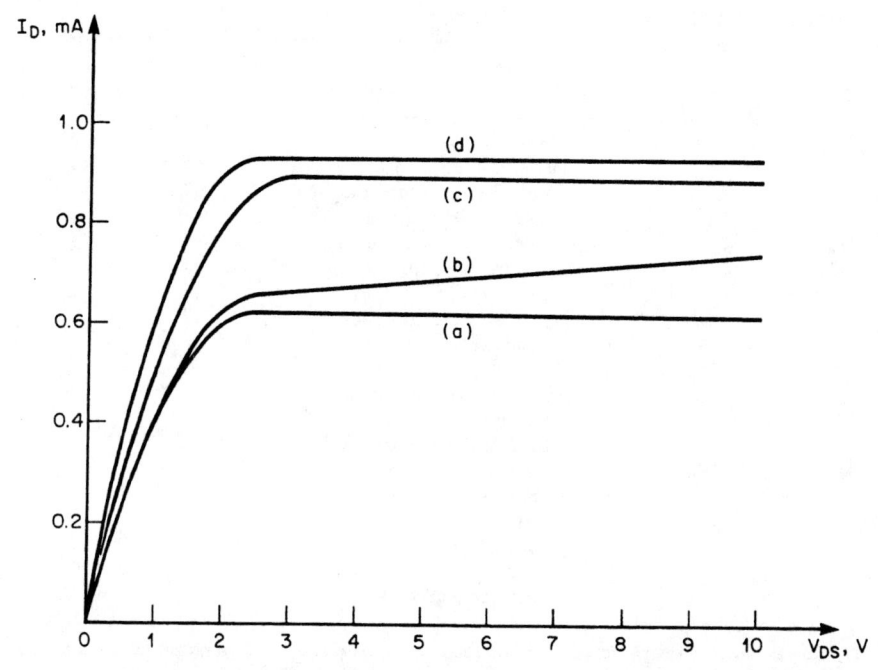

Figure 3-9 SPICE2 I_D vs. V_{DS} characteristics: (a) parameter default values, (b) $\lambda = 0.02$ V^{-1}, (c) $V_{T0} = -2.5$ V, and (d) $\beta = 0.15 \times 10^{-3}$ A/V^2. The characteristics have been derived for $V_{GS} = -0.5$ V.

Summarizing, the SPICE2 JFET static model is characterized by the following parameters which the user can specify in the .MODEL statement [8]:

VTO	Threshold voltage (V_{T0})
BETA	Transconductance parameter (β)
LAMBDA	Channel-length modulation (λ)
IS	Gate-junction saturation current (I_S)
RD	Drain ohmic resistance (r_D)
RS	Source ohmic resistance (r_S)

The keyname used in the text is indicated in parentheses.

Figure 3-9 shows the drain current vs. drain voltage of an n-channel JFET using SPICE2 parameter default values (see Table 3-1) for a given value of gate voltage. This figure also shows curves obtained by varying the λ, β, and V_{T0} parameters one at a time (keeping the default values for the other model parameters).

TABLE 3-1 SPICE2 JFET Model Parameters

Symbol	SPICE2G keyword	Parameter name	Default value	Unit
V_{T0}	VTO	Threshold voltage	-2	V
β	BETA	Transconductance parameter	10^{-4}	A/V^2
λ	LAMBDA	Channel-length modulation	0	V^{-1}
r_D	RD	Drain ohmic resistance	0	Ω
r_S	RS	Source ohmic resistance	0	Ω
C_{GS}	CGS	Zero-bias gate-source junction capacitance	0	F
C_{GD}	CGD	Zero-bias gate-drain junction capacitance	0	F
ϕ_0	PB	Gate-junction potential	1	V
I_S	IS	Gate-junction saturation current	10^{-14}	A
FC	FC	Coefficient for forward-bias depletion capacitance formula	0.5	
k_f	KF	Flicker-noise coefficient	0	
a_f	AF	Flicker-noise exponent	1	
T	T*	Nominal temperature for simulation and at which all input data are assumed to have been measured	27	°C

*Remember, the temperature T is regarded as an operating condition and not as a model parameter.

3.2 Large-Signal Model and Its Implementation in SPICE2

Charge storage in a JFET occurs in the two gate junctions. Since neither gate junction normally is forward-biased, there is no minority-carrier storage, so the capacitances involved are just those due to the ionic space charge in the depletion regions. This charge storage is modeled by the nonlinear elements Q_{GS} and Q_{GD}.

The elements Q_{GS} and Q_{GD} are implemented in SPICE2 as follows:

$$Q_{GS} = \begin{cases} C_{GS}(0) \int_0^{V_{GS}} \left(1 - \frac{V}{\phi_0}\right)^{-m} dV & \text{for } V_{GS} < \text{FC} \times \phi_0 \\ C_{GS}(0)F_1 + \frac{C_{GS}(0)}{F_2} \int_{\text{FC} \times \phi_0}^{V_{GS}} \left(F_3 + \frac{mV}{\phi_0}\right) dV & \text{for } V_{GS} \geq \text{FC} \times \phi_0 \end{cases}$$
(3-27)

$$Q_{GD} = \begin{cases} C_{GD}(0) \int_0^{V_{GD}} \left(1 - \frac{V}{\phi_0}\right)^{-m} dV & \text{for } V_{GD} < \text{FC} \times \phi_0 \\ C_{GD}(0)F_1 + \frac{C_{GD}(0)}{F_2} \int_{\text{FC} \times \phi_0}^{V_{GD}} \left(F_3 + \frac{mV}{\phi_0}\right) dV & \text{for } V_{GD} \geq \text{FC} \times \phi_0 \end{cases}$$
(3-28)

Alternatively, these two charges can be expressed as voltage-dependent capacitors with values determined by the following expressions:

$$C_{GS} = \begin{cases} C_{GS}(0)\left(1-\dfrac{V_{GS}}{\phi_0}\right)^{-m} & \text{for } V_{GS} < \text{FC} \times \phi_0 \\ \dfrac{C_{GS}(0)}{F_2}\left(F_3+\dfrac{mV_{GS}}{\phi_0}\right) & \text{for } V_{GS} \geq \text{FC} \times \phi_0 \end{cases} \quad (3\text{-}29)$$

$$C_{GD} = \begin{cases} C_{GD}(0)\left(1-\dfrac{V_{GD}}{\phi_0}\right)^{-m} & \text{for } V_{GD} < \text{FC} \times \phi_0 \\ \dfrac{C_{GD}(0)}{F_2}\left(F_3+\dfrac{mV_{GD}}{\phi_0}\right) & \text{for } V_{GD} \geq \text{FC} \times \phi_0 \end{cases} \quad (3\text{-}30)$$

where F_1, F_2, and F_3 are program variables defined as follows:

$$F_1 = \dfrac{\phi_0}{1-m}[1-(1-\text{FC})^{1-m}]$$

$$F_2 = (1-\text{FC})^{1+m} \quad (3\text{-}31)$$

$$F_3 = 1 - \text{FC}(1+m)$$

From the above equations, it follows that the SPICE2 JFET large-signal model can be modeled by the following parameters in the .MODEL statement:

CGS, CGD   Zero-bias gate-source and gate-drain junction capacitances, respectively [$C_{GS}(0)$, $C_{GD}(0)$]

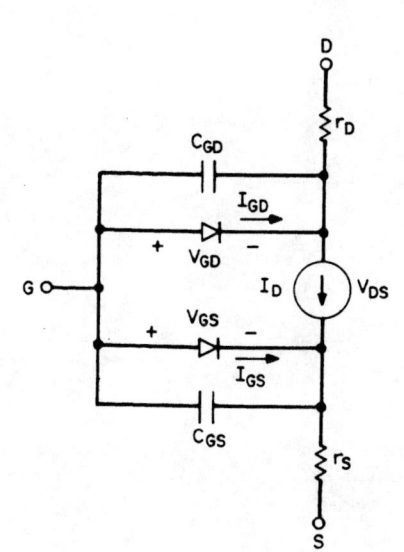

Figure 3-10 SPICE2 n-channel JFET large-signal model.

PB Gate junction potential (ϕ_0)

FC Forward-bias depletion capacitance coefficient (FC); along with ϕ_0, it is used in matching the transition point and voltage characteristics beyond which the junction capacitance is approximated by a linear extrapolation (see Fig. 1-16)

m Junction grading coefficient (m); in SPICE2 it is set to 0.5 and it cannot be varied as a model parameter

The keyname used in the text is indicated in parentheses. The SPICE2 JFET equivalent large-signal model is shown in Fig. 3-10.

3.3 Small-Signal Model and Its Implementation in SPICE2

The SPICE2 small-signal linearized model for the JFET is shown in Fig. 3-11. The conductances g_{GS} and g_{GD} are the conductances of the two gate junctions. Since neither gate junction is forward-biased in normal operation, both g_{GS} and g_{GD} usually are very small. The transconductance g_m and the output conductance g_{DS} are defined by the equations

$$g_m = \frac{dI_D}{dV_{GS}}\bigg|_{op} \quad (3\text{-}32)$$

$$g_{DS} = \frac{dI_D}{dV_{DS}}\bigg|_{op} \quad (3\text{-}33)$$

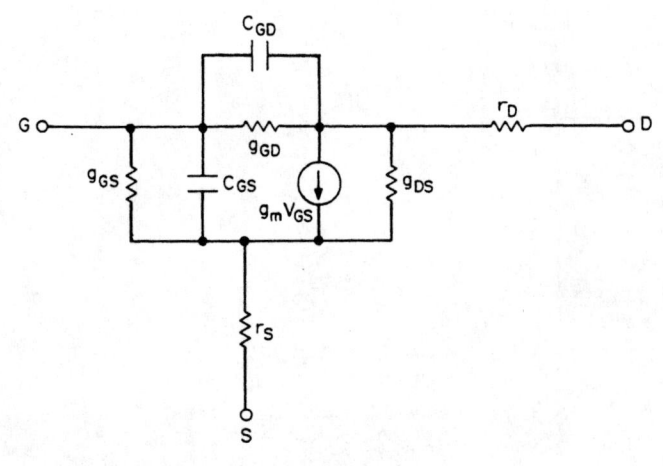

Figure 3-11 SPICE2 JFET small-signal model.

while the conductances g_{GS} and g_{GD} are given by the relations

$$g_{GS} = \left.\frac{dI_{GS}}{dV_{GS}}\right|_{op} \tag{3-34}$$

$$g_{GD} = \left.\frac{dI_{GD}}{dV_{GD}}\right|_{op} \tag{3-35}$$

The values of the preceding equations are determined in SPICE2 by the following equations for the two mode operations [taking into account Eqs. (3-22) to (3-26)].

Normal mode operation ($V_{DS} \geq 0$)

$$g_m = \begin{cases} 0 & \text{for } V_{GS} - V_{T0} \leq 0 \\ 2\beta(1 + \lambda V_{DS})(V_{GS} - V_{T0}) & \text{for } 0 < V_{GS} - V_{T0} \leq V_{DS} \\ 2\beta(1 + \lambda V_{DS})V_{DS} & \text{for } 0 < V_{DS} < V_{GS} - V_{T0} \end{cases} \tag{3-36}$$

$$g_{DS} = \begin{cases} 0 & \text{for } V_{GS} - V_{T0} \leq 0 \\ \lambda\beta(V_{GS} - V_{T0})^2 & \text{for } 0 < V_{GS} - V_{T0} \leq V_{DS} \\ 2\beta(1 + \lambda V_{DS})(V_{GS} - V_{T0} - V_{DS}) \\ \quad + \lambda\beta V_{DS}[2(V_{GS} - V_{T0}) - V_{DS}] & \text{for } 0 < V_{DS} < V_{GS} - V_{T0} \end{cases}$$

$$\tag{3-37}$$

Inverted mode operation ($V_{DS} < 0$)

$$g_m = \begin{cases} 0 & \text{for } V_{GD} - V_{T0} \leq 0 \\ -2\beta(1 - \lambda V_{DS})(V_{GD} - V_{T0}) & \text{for } 0 < V_{GD} - V_{T0} \leq -V_{DS} \\ 2\beta(1 - \lambda V_{DS})V_{DS} & \text{for } 0 < -V_{DS} \leq V_{GD} - V_{T0} \end{cases} \tag{3-38}$$

$$g_{DS} = \begin{cases} 0 & \text{for } V_{GD} - V_{T0} \leq 0 \\ \lambda\beta(V_{GD} - V_{T0})^2 - g_m & \text{for } 0 < V_{GD} - V_{T0} \leq -V_{DS} \\ 2\beta(1 - \lambda V_{DS})(V_{GD} - V_{T0}) \\ \quad - \lambda\beta V_{DS}[2(V_{GD} - V_{T0}) + V_{DS}] & \text{for } 0 < -V_{DS} < V_{GD} - V_{T0} \end{cases}$$

$$\tag{3-39}$$

When I_{GS} and I_{GD} are considered, we can write

$$g_{GS} = \begin{cases} \dfrac{q}{kT} I_S\, e^{qV_{GS}/kT} + \text{GMIN} & \text{for } V_{GS} > -5\dfrac{kT}{q} \\ -\dfrac{I_S}{V_{GS}} + \text{GMIN} & \text{for } V_{GS} \leq -5\dfrac{kT}{q} \end{cases} \tag{3-40}$$

$$g_{GD} = \begin{cases} \dfrac{q}{kT} I_S e^{qV_{GD}/kT} + \text{GMIN} & \text{for } V_{GD} > -5\dfrac{kT}{q} \\ -\dfrac{I_S}{V_{GD}} + \text{GMIN} & \text{for } V_{GD} \le -5\dfrac{kT}{q} \end{cases} \quad (3\text{-}41)$$

The capacitances C_{GS} and C_{GD} are defined in Eqs. (3-29) and (3-30), respectively.

3.4 Temperature and Area Effects on the JFET Model Parameters

3.4.1 Temperature dependence of SPICE2 JFET model parameters

All input data for SPICE2 are assumed to have been measured at 27°C (300 K). The simulation also assumes a nominal temperature (TNOM) of 27°C, which can be changed with the TNOM option of the .OPTIONS statement. The circuits can be simulated at temperatures different from TNOM by using a .TEMP statement.

Temperature appears explicitly in the exponential terms of the JFET model equations. In addition, saturation current I_S, gate-junction potential ϕ_0, zero-bias gate-source and gate-drain junction capacitances C_{GS} and C_{GD}, and the coefficient for forward-bias depletion capacitance formula FC have built-in temperature dependence. The temperature dependence of I_S in the JFET model is determined by

$$I_S(T_2) = I_S(T_1) \exp\left[-\frac{qE_g(300)}{kT_2}\left(1 - \frac{T_2}{T_1}\right)\right] \quad (3\text{-}42)$$

where $E_g(300) = 1.11$ eV for Si.

The effect of temperature on ϕ_0 is modeled by the formula

$$\phi_0(T_2) = \frac{T_2}{T_1}\phi_0(T_1) - 2\frac{kT_2}{q}\ln\left(\frac{T_2}{T_1}\right)^{1.5} - \left[\frac{T_2}{T_1}E_g(T_1) - E_g(T_2)\right] \quad (3\text{-}43)$$

where $E_g(T_1)$ and $E_g(T_2)$ are described by the relations [1]

$$E_g(T) = E_g(0) - \frac{\alpha T^2}{\beta + T} \quad (3\text{-}44)$$

where experimental results give, for Si, $\alpha = 7.02 \times 10^{-4}$, $\beta = 1108$, and $E_g(0) = 1.16$ eV.

The temperature dependence of C_{GS} and C_{GD} is determined by

$$C_{GS}(T_2) = C_{GS}(T_1)\left\{1 + m\left[400 \times 10^{-6}(T_2 - T_1) - \frac{\phi_0(T_2) - \phi_0(T_1)}{\phi_0(T_1)}\right]\right\}$$

(3-45)

$$C_{GD}(T_2) = C_{GD}(T_1)\left\{1 + m\left[400 \times 10^{-6}(T_2 - T_1) - \frac{\phi_0(T_2) - \phi_0(T_1)}{\phi_0(T_1)}\right]\right\}$$

where m is set to 0.5.

The temperature dependence of FC is determined by

$$\text{FCPB}(T_2) = \text{FC} \times \phi_0 = \text{FCPB}(T_1)\frac{\phi_0(T_2)}{\phi_0(T_1)}$$

$$F_1(T_2) = F_1(T_1)\frac{\phi_0(T_2)}{\phi_0(T_1)}$$

(3-46)

In the preceding equations, T_1 and T_2 must be considered as follows:

- If TNOM in the .OPTIONS statement is not specified, then $T_1 =$ TNOM = TREF = 27°C (300 K); this is the default value for TNOM.
- If a TNOM$_{\text{new}}$ value is specified, then $T_1 =$ TNOM = 27°C; $T_2 =$ TNOM$_{\text{new}}$.

These two cases are valid only if one request for temperature is made.

If more than one temperature is requested (using the .TEMP statement), then $T_1 =$ TNOM = 27°C and T_2 is the first temperature in the .TEMP statement. Afterward, T_1 is the last working temperature (T_2) and T_2 is the next temperature in the .TEMP statement to be calculated.

3.4.2 Area dependence of SPICE2 JFET model parameters

The AREA factor used in SPICE2 for the JFET model determines the number of equivalent parallel devices of a specified model. The *JFET* model parameters affected by the AREA factor and specified in the device statement are β, r_D, r_S, C_{GS}, C_{GD}, and I_S; i.e.,

$$\beta = \beta \times \text{AREA}$$

$$r_D = r_D/\text{AREA}$$

$$r_S = r_S/\text{AREA}$$

$$C_{GS}(0) = C_{GS}(0) \times \text{AREA}$$

$$C_{GD}(0) = C_{GD}(0) \times \text{AREA}$$

$$I_S = I_S \times \text{AREA}$$

(3-47)

3.5 SPICE3 Models

SPICE3 [9,10] JFET models are based directly on SPICE2 JFET models. Thus, the *static model* is defined by the parameters VTO, BETA which determine the variation of drain current with gate voltage, LAMBDA, which determines the output conductance, and IS, the saturation current of the two gate junctions. Two ohmic resistances, RD and RS, are included. The *large-signal model* is defined by nonlinear depletion-region capacitances for both gate junctions which vary as the $-1/2$ power of junction voltage and are defined by the parameters CGS, CGD, and PB. The *small-signal model* is automatically computed inside the program by the linearization of the static and large-signal model equations (see Sec. 3.3). The *temperature dependence* and the updating relations for the parameters follow the ones described in Sec. 3.4.1. The values specified for the parameters are assumed to have been measured at the temperature TNOM, which can be specified in the .OPTIONS statement. This value of temperature can be overridden by the specification of the new (in comparison with SPICE2) .MODEL statement parameter TNOM (in units °C) or of the device element statement option TEMP. The *noise* model is described in Sec. 7.1.3c. The parameters affected by the AREA factor follow the same relations as indicated in Sec. 3.4.2. The equivalent circuits for the static, large-signal, small-signal, and noise models are shown in Figs. 3-7, 3-10, 3-11, and 7-5, respectively.

3.6 HSPICE Models

HSPICE [11] contains three JFET model levels. The LEVEL parameter in the .MODEL statement selects either the JFET (LEVEL = 1,2) or the MESFET (LEVEL = 3). The MESFET is dealt with in Chap. 9. In the following, HSPICE JFET *static, large-signal, small-signal,* and *thermal* models are taken into account. The *noise* model is considered in Chap. 7.

3.6.1 Static model (LEVEL = 1,2)

When the user of HSPICE sets LEVEL = 1 in the .MODEL statement automatically the SPICE2 JFET model is selected. Thus, the *static model* equivalent circuit of Fig. 3-7 and Eqs. (3-22) and (3-23) which define the nonlinear current source, I_D, are also valid in HSPICE. However, Eqs. (3-25) and (3-26) which account for the two junction diodes (see Fig. 3-7) are modified by adding the emission coefficient n (N) for gate-drain and gate-source diodes as a multiplier of the thermal voltage in the exponential term. In addition, the limit between the forward- and reverse-bias regions of the two junction diodes is set to $-10nkT/q$, while in SPICE2 it was $-5kT/q$. The LEVEL = 1 static model is then defined by the .MODEL statement parameters V_{T0}(VTO) and β(BETA) which determine the variation of drain current with gate voltage; by λ(LAMBDA), which determines the output conductance; by I_S(IS) the saturation current of the two gate junctions; and by n(N), the emission coefficient of the two gate junctions; two ohmic resistances, r_D(RD) and r_S(RS), are also included. The .MODEL keyword is indicated in parentheses.

At a given V_{GS}, the parameter λ can be determined from a pair of drain current and drain voltage points measured in the saturation region ($V_{GS} - V_{T0} < V_{DS}$), thus [11]

$$\lambda = \frac{I_{D2} - I_{D1}}{I_{D1}V_{DS2} - I_{D2}V_{DS1}} \tag{3-48}$$

When the user of HSPICE sets LEVEL = 2 in the .MODEL statement, a modified SPICE2 JFET static model which accounts for the gate modulation of λ is selected. The equivalent circuit of Fig. 3-7 also represents the JFET static model, but the nonlinear current source I_D is now determined by the following equations [11]:

$$I_D = \begin{cases} 0 & \text{for } V_{GS} - V_{T0} < 0 \text{ (channel pinched off)} \\ \beta(V_{GS} - V_{T0})^2[1 + \lambda(V_{DS} - V_{GS} + V_{T0})(1 + \lambda_1 V_{GS})] \\ & \text{for } 0 < V_{GS} - V_{T0} \leq V_{DS} \text{ (saturation, } V_{GS} > 0) \\ \beta(V_{GS} - V_{T0})^2\left[1 - \lambda(V_{DS} - V_{GS} + V_{T0})\frac{V_{GS} - V_{T0}}{V_{T0}}\right] \\ & \text{for } 0 < V_{GS} - V_{T0} < V_{DS} \text{ (saturation, } V_{GS} < 0) \\ \beta V_{DS}[2(V_{GS} - V_{T0}) - V_{DS}] \\ & \text{for } 0 < V_{DS} < V_{GS} - V_{T0} \text{ (linear region)} \end{cases}$$

(3-49)

The LEVEL = 2 static model is defined by the .MODEL statement parameter λ_1 (LAM1), the channel-length modulation gate voltage parameter, measured in V^{-1}, in addition to the parameters common to the LEVEL = 1 static model.

3.6.2 Large-signal model (LEVEL = 1,2)

The HSPICE JFET large-signal model (Fig. 3-10) is characterized by the same equations [Eqs. (3-29) and (3-30)] as in SPICE2 when the reverse-bias operating mode is considered. However, in the forward-bias operation mode the SPICE2 equations used in calculating the capacitances are modified in HSPICE by considering also the diffusion capacitance component. Thus, it can be written [11],

$$C_{GS} = \tau_D \frac{dI_{GS}}{dV_{GS}} + \frac{C_{GS}(0)}{F_2}\left(F_3 + \frac{mV_{GS}}{\phi_0}\right)^{-m} \quad \text{for } V_{GS} \geq \text{FC} \times \phi_0 \quad (3\text{-}50)$$

and

$$C_{GD} = \tau_D \frac{dI_{GD}}{dV_{GD}} + \frac{C_{GD}(0)}{F_2}\left(F_3 + \frac{mV_{GD}}{\phi_0}\right)^{-m} \quad \text{for } V_{GD} \geq \text{FC} \times \phi_0 \quad (3\text{-}51)$$

It must be taken into account that in HSPICE, the coefficient FC is set to zero when the DCAP = 2 option [11] in the .OPTIONS statement is selected (default). This induces changes in the equations used in calculating the depletion capacitance and in the limits beyond which the capacitance is modeled by a linear extrapolation.

In Eqs. (3-50) and (3-51), two new parameters for the .MODEL statement have been introduced: the transit time τ_D(TT), whose default value is 0 s, and the grading coefficient m(M) whose default value is 0.5. Thus the charge-storage effects are modeled by two nonlinear capacitances which vary as the $-m$ power of junction voltage and are defined by the parameters $C_{GS}(0)$(CGS), $C_{GD}(0)$(CGD), ϕ_0(PB), FC, and τ_D(TT). The .MODEL keyword is indicated in parentheses.

In addition, in HSPICE, the grading coefficient m for gate-drain and gate-source junctions is not fixed as in SPICE2 ($m = 0.5$), but it can vary according to the junction (e.g., $m = 0.5$ for a step junction, or $m = 0.33$ for a linear graded junction).

The described JFET large-signal model uses a diode-like capacitance between source and gate, where the depletion-region thickness (and therefore the capacitance) is determined by the gate-source voltage. A similar diode model is often used to describe the normally much smaller gate-drain capacitance. This model is selected in the .MODEL statement by the CAPOP parameter when it is set to zero (default). The case of CAPOP = 1, which selects the Statz charge-conserving model, is mainly used for the MESFET model (LEVEL = 3) and it is described in Chap. 9.

3.6.3 Small-signal model (LEVEL = 1,2)

As in SPICE2, the HSPICE JFET *small-signal model* is automatically computed inside the program by the linearization of the static and large-signal model equations; thus no new model parameters need to be specified by the user of HSPICE in the .MODEL statement. The small-signal model equivalent circuit is the same as in Fig. 3-11.

3.6.4 Temperature effects on the JFET model parameters (LEVEL = 1,2)

There are several sets of temperature equations for the HSPICE JFET model parameters which may be selected by setting the temperature equation selectors TLEV and TLEVC in the .MODEL statement. In the following, we consider only the equations selected by TLEV and TLEVC set to their default values (zero). Other relations for some of the temperature-dependent parameters can be found in [11].

As in SPICE2, also in HSPICE, the temperature dependence of the energy gap E_g(EG) follows Eq. (3-44). On the other hand, the temperature dependence of C_{GS}(CGS), C_{GD}(CGD), and ϕ_0(PB) follow Eqs. (3-45) and (3-43), respectively.

The effect of temperature on the gate junction saturation current I_S(IS) is described by the relation

$$I_S(T_2) = I_S(T_1) \left(\frac{T_2}{T_1}\right)^{X_{TI}/n} \exp\left[-\frac{qE_g(300)}{nkT_2}\left(1 - \frac{T_2}{T_1}\right)\right] \quad (3\text{-}52)$$

where X_{TI} (XTI) is the saturation current temperature exponent (defaults = 0), and n (N) is the gate-drain and gate-source junction emission coefficient (defaults = 1) which the user can specify in the .MODEL statement.

In addition to the temperature dependence of the above parameters common to SPICE2, HSPICE considers the temperature dependence of the threshold voltage V_{T0}(VTO) and the drain, source, and gate resistances r_D(RD), r_S(RS), r_G(RG), respectively. In particular, the threshold voltage, V_{T0}, varies with temperature according to the equation [11]

$$V_{T0}(T_2) = V_{T0}(T_1) - \text{TCV}(T_2 - T_1) \quad (3\text{-}53)$$

where TCV (default = 0), the temperature coefficient measured in °C^{-1}, is a .MODEL statement parameter.

The mobility temperature compensation equation is updated as follows:

$$\beta(T_2) = \beta(T_1) \left(\frac{T_2}{T_1}\right)^{\text{BEX}} \quad (3\text{-}54)$$

where BEX (default = 0), the mobility temperature exponent, is a .MODEL statement parameter.

The resistances r_D, r_S, r_G vary with temperature according to the equations

$$r_D(T_2) = r_D(T_1)[1 + \text{TRD}(T_2 - T_1)]$$
$$r_S(T_2) = r_S(T_1)[1 + \text{TRS}(T_2 - T_1)] \qquad (3\text{-}55)$$
$$r_G(T_2) = r_G(T_1)[1 + \text{TRG}(T_2 - T_1)]$$

where TRD, TRS, TRG are the temperature coefficients for each resistor, measured in $°C^{-1}$; they are .MODEL statement parameters and their default values are zero.

For the meaning of T_1 and T_2, refer to Sec. 1.8.4, which also explains the role of the DTEMP parameter that allows the user of HSPICE to simulate some devices (here JFETs) at a different temperature in comparison with the circuit temperature (TNOM).

3.6.5 Scaling and area dependence of the JFET model parameters (LEVEL = 1,2)

As for the LEVEL = 3 diode model, scaling is controlled by the SCALE and SCALM options [11] in the .OPTIONS statement. More details about these parameters are found in Sec. 1.8.5.

The area dependence of the parameters is affected by AREA, M, and ACM factors. For more details, the user can directly refer to Ref. [11].

3.7 PSPICE Models

PSPICE [12–14] JFET models are based on the JFET models implemented in SPICE2. In the following, PSPICE JFET *static*, *large-signal*, *small-signal*, and *thermal* models are considered. The *noise* model is described in Chap. 7.

3.7.1 Static model

The PSPICE JFET static model is represented by the equivalent circuit of Fig. 3-7 provided that an added ohmic resistance r_G is considered in series to the gate.

The equations which define the nonlinear current source, I_D, in SPICE2, are also valid in PSPICE. However, Eqs. (3-25) and (3-26), which account for the two junction diodes (see Fig. 3-7), are modified in PSPICE as follows [12].

For the gate-source diode,

$$I_{GS} = I_F + K_{\text{gen}} I_{\text{rec}} + I_I \qquad (3\text{-}56)$$

where I_F is the normal current given by

$$I_F = I_S(e^{qV_{GS}/nkT} - 1) \tag{3-57}$$

I_{rec} is the recombination current given by

$$I_{rec} = I_{SR}(e^{qV_{GS}/n_R kT} - 1) \tag{3-58}$$

K_{gen} is the generation factor given by

$$K_{gen} = \sqrt{\left[\left(1 - \frac{V_{GS}}{\phi_0}\right)^2 + 0.005\right]^m} \tag{3-59}$$

and I_I is the impact ionization current given by

$$I_I = \begin{cases} \alpha I_D [V_{DS} - (V_{GS} - V_{T0})]\, e^{-V_k/[V_{DS} - (V_{GS} - V_{T0})]} \\ \qquad\qquad\qquad\qquad \text{for } 0 < V_{GS} - V_{T0} < V_{DS} \\ 0 \qquad\qquad\qquad\qquad \text{otherwise} \end{cases} \tag{3-60}$$

An analogous expression holds for the gate-drain current I_{GD}, provided that source and drain subscripts are switched in the above I_{GS} equations.

From the above equations, it can be seen that PSPICE introduces, in comparison with SPICE2, five new parameters for the .MODEL statement: N(n), the gate pn emission coefficient (default =1); ISR(I_{SR}), the gate pn recombination current parameter (default = 0 A); NR(n_R), the emission coefficient for ISR (default = 2); ALPHA(α), the ionization coefficient (default = 0 V^{-1}); VK(V_k), the ionization knee voltage (default = 0 V). The other parameters maintain their meaning and keywords encountered in SPICE2.

In particular, VTO < 0 means the device is a depletion-mode JFET (for both n-channel and p-channel) and VTO > 0 means the device is an enhancement-mode JFET. The keyname used in the text is indicated in parentheses.

3.7.2 Large-signal model

The PSPICE JFET *large-signal model* is based directly on SPICE2 JFET large-signal model, and it is defined by the capacitances C_{GS} and C_{GD} in Fig. 3-10. Thus, to take the charge-storage effects into account, the user of PSPICE has to specify, in the .MODEL statement, the parameters as described in Sec. 3.2.

3.7.3 Small-signal model

As in SPICE2, the PSPICE small-signal model is automatically computed inside the program by the linearization of the static and large-signal model equations. Thus the user of PSPICE has to specify no new parameters in the .MODEL statement.

3.7.4 Temperature and area factor effects on the JFET model parameters

The parameters C_{GS}, C_{GD}, and ϕ_0 have the same temperature-dependence equations as in SPICE2. In addition, PSPICE provides new formulations for I_S(IS), I_{SR}(ISR), V_{T0}(VTO), and β(BETA). The temperature dependence of the gate pn saturation current I_S follows Eq. (3-52).

The gate pn recombination current parameter I_{SR} depends on temperature through the relation

$$I_{SR}(T_2) = I_{SR}(T_1) \left(\frac{T_2}{T_1}\right)^{X_{TI}/n_R} \exp\left[-\frac{qE_g(300)}{n_R kT}\left(1 - \frac{T_2}{T_1}\right)\right] \quad (3\text{-}61)$$

The threshold voltage V_{T0} varies with temperature according to the equation

$$V_{T0}(T_2) = V_{T0}(T_1) + \text{VTOTC}(T_2 - T_1) \quad (3\text{-}62)$$

The transconductance coefficient temperature compensation equation is updated as follows

$$\beta(T_2) = \beta(T_1) 1.01^{\text{BETATCE}(T_2 - T_1)} \quad (3\text{-}63)$$

In the above equations, three new .MODEL statement parameters have been introduced: the saturation current temperature coefficient XTI (default = 3); the threshold voltage temperature coefficient VTOTC (default = 0 V °C^{-1}); the transconductance exponential temperature coefficient, BETATCE (default = 0).

Scaling is controlled by the element parameter AREA. All the currents and capacitances are multiplied by AREA, and all the resistances are divided by AREA.

References

1. S. M. Sze, *Physics of Semiconductor Devices*, Wiley, New York, 1969.
2. J. Millman, *Microelectronics*, McGraw-Hill, New York, 1979.
3. R. S. Muller and T. I. Kamins, *Device Electronics for Integrated Circuits*, Wiley, New York, 1977.
4. L. W. Nagel, SPICE2: A Computer Program to Simulate Semiconductor Circuits, Electronics Research Laboratory, Rep. No. ERL-M520, University of California, Berkeley, 1975.
5. P. R. Gray and R. G. Meyer, *Analysis and Design of Analog Integrated Circuits*, Wiley, New York, 1977.
6. A. S. Grove, *Physics and Technology of Semiconductor Devices*, Wiley, New York, 1967.

7. H. Shichman and D. A. Hodges, Modeling and Simulation of Insulated-Gate Field-Effect Transistor Switching Circuits, *IEEE J. Solid-State Circuits*, **SC-3**, 1968.
8. A. Vladimirescu, K. Zhang, A. R. Newton, D. O. Pederson, and Sangiovanni-Vincentelli, *SPICE version 2G User's Guide*, Dept. EECS, University of California, Berkeley, 1981.
9. T. I. Quarles, The SPICE3 Implementation Guide, Electronics Research Laboratory, Rep. No. ERL-M44, University of California, Berkeley, 1989.
10. T. I. Quarles, SPICE3 Version 3C1 Users Guide, Electronics Research Laboratory, Rep. No. ERL-M46, University of California, Berkeley, 1989.
11. Meta-Software Inc., *HSPICE User's Manual*, Campbell, Calif., 1991.
12. MicroSim Corporation, *PSPICE User's Manual*, Irvine, Calif., 1989.
13. P. W. Tuinenga, *SPICE: A Guide to Circuit Simulation and Analysis Using PSPICE*, Prentice-Hall, Englewood Cliffs, N.J., 1988.
14. M. H. Rashid, *SPICE for Circuits and Electronics Using PSPICE*, Prentice-Hall, Englewood Cliffs, N.J., 1990.

Chapter

4

Metal-Oxide-Semiconductor Transistor (MOST)

This chapter establishes *static, large-signal, small-signal,* and *thermal* physical-mathematical models of the MOST, and how the models are implemented in SPICE2 [1] and other SPICE2-based programs. The *noise* model is described in Chap. 7. SPICE2 MOST models are directly regulated by the subroutines *READIN* (it reads and processes the .MODEL statement), *MODCHK* (it performs one-time processing of the model parameters and assigns the parameter default values), *MOSFET* (it processes the device for dc and transient analysis), *MOSCAP* (it computes the equivalent conductances, divides up the channel charge to source and drain, and computes the total terminal charges), *MOSEQ1, MOSEQ2, MOSEQ3* (they evaluate the drain current, its derivates, and the charges associated with the gate, channel, and bulk for the LEVEL1, LEVEL2, and LEVEL3 model, respectively), *MOSQ2, MOSQ3, MQSPOF* (they are related to the previous subroutines), *CMEYER* (it computes the overlap capacitances as functions of the device terminal voltages), and *TMPUPD* (it updates the temperature-dependent parameters of the models), in addition to the subroutines concerning each type of analysis. The noise model is directly regulated by the subroutine *NOISE* (see Chap. 7).

4.1 Structure and Operating Regions of the MOST

Before presenting the equations and the parameters of the models implemented in SPICE2, SPICE3, HSPICE, and PSPICE, it is useful

---

Note: This chapter was written with the contribution of Enrico Profumo, who is with SGS-THOMSON Microelectronics, Milan, Italy.

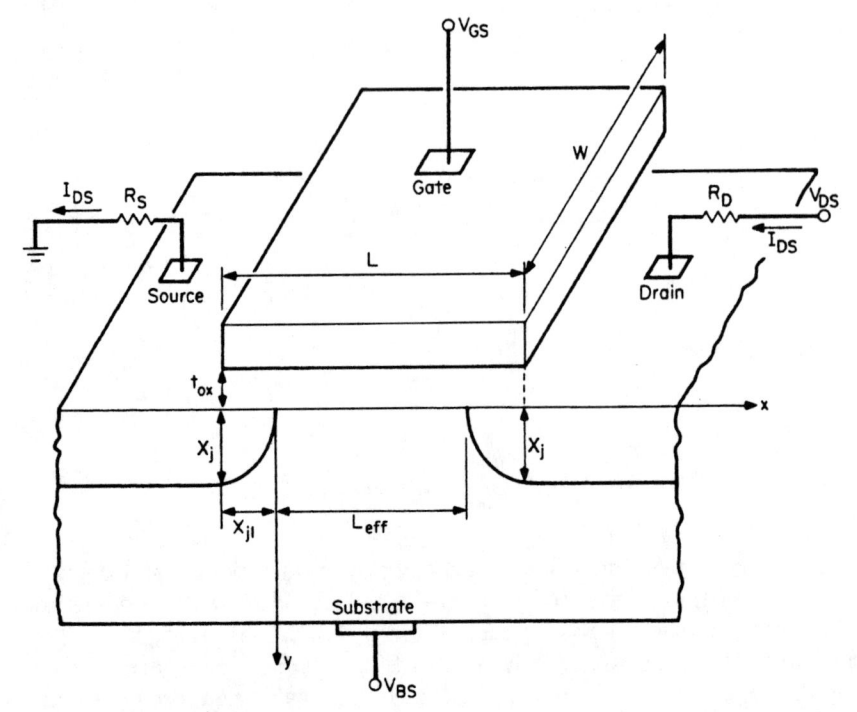

Figure 4-1 Structure of the MOST.

to explain the meaning of the terms used in this chapter as indicated in Figs. 4-1 and 4-2. See Table 4-1 for the list of the model parameters of the MOST.

4.1.1 Introduction to MOST fabrication

Figures 4-1 and 4-2 represent the structure of the MOST; a brief description of its construction follows. In the case shown in Fig. 4-2a, the *substrate* (B) is a *p*-type silicon wafer on which a layer of thermal oxide (t_{oxf}) about 1 μm thick is built up; this oxide is removed (see mask 1 in Fig. 4-2a) from the regions where the MOST will be made. Another layer of thermal oxide t_{ox}, in this case an extremely thin oxide (100 nm or less), is then built up. Afterward, a layer of conducting material, polysilicon in the example, is deposited; this acts as a control electrode and is called the *gate* (G).

In the old processes, this layer was made of aluminum, and the structure formed in this way was called *MOS* (*Metal-Oxide Semiconductor*).

After the second mask has been used, the remaining polysilicon covers part of the thin layer of oxide, which prevents it from being

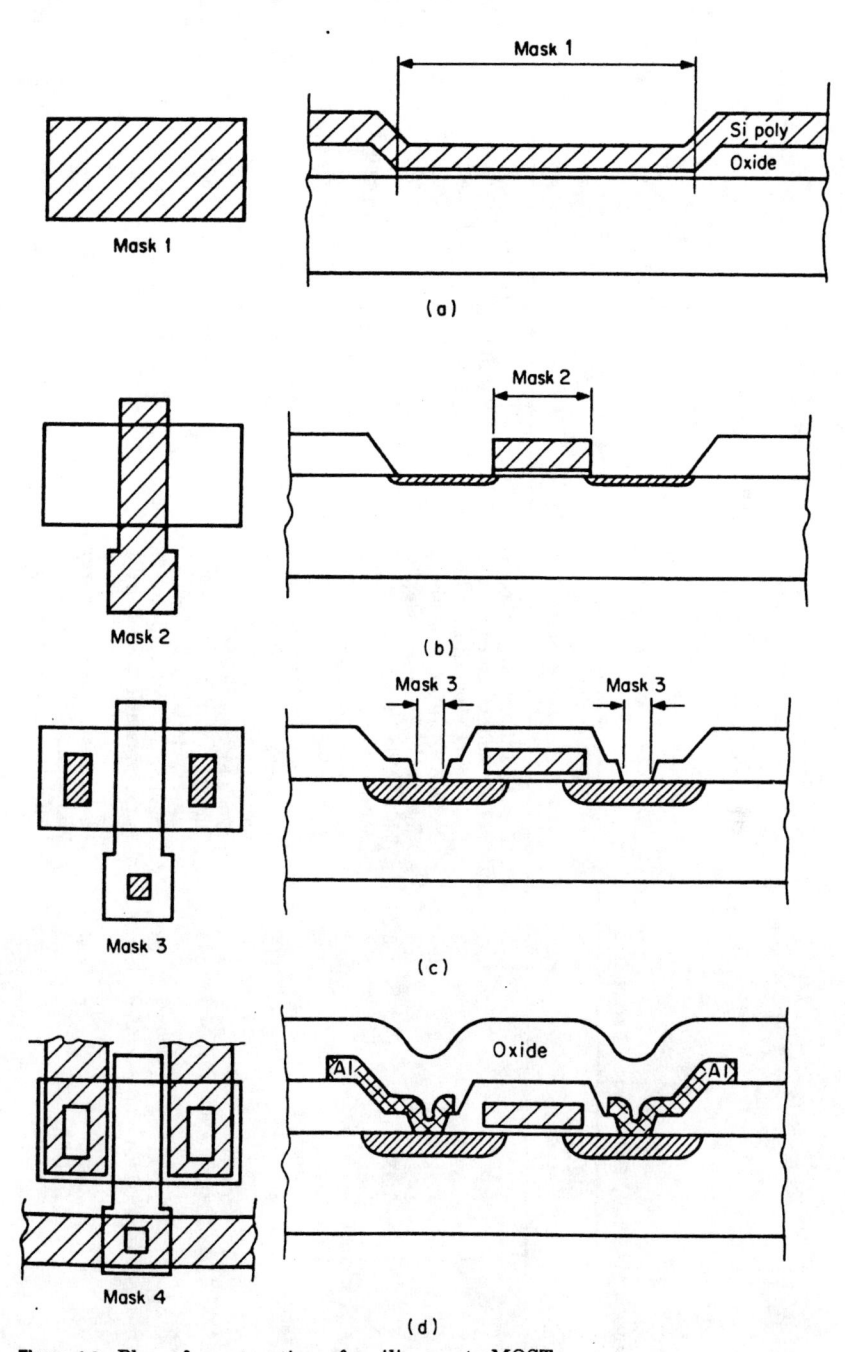

Figure 4-2 Plan of construction of a silicon-gate MOST.

TABLE 4-1 SPICE2 MOST Model Parameters

Symbol	SPICE2G keyword	LEVEL	Parameter name	Default value	Typical value	Units
V_{TO}	VTO	1–3	Zero-bias threshold voltage	1.0	1.0	V
KP	KP	1–3	Transconductance parameter	2×10^{-5}	3×10^{-5}	A/V^2
γ	GAMMA	1–3	Body-effect parameter	0.0	0.35	$V^{1/2}$
$2\phi_p$	PHI	1–3	Surface inversion potential	0.6	0.65	V
λ	LAMBDA	1, 2	Channel-length modulation	0.0	0.02	V^{-1}
t_{ox}	TOX	1–3	Thin oxide thickness	1×10^{-7}	1×10^{-7}	m
N_A	NSUB	1–3	Substrate doping	0.0	1×10^{15}	cm^{-3}
N_{SS}	NSS	1–3	Surface state density	0.0	1×10^{10}	cm^{-2}
N_{FS}	NFS	2, 3	Surface-fast state density	0.0	1×10^{10}	cm^{-2}
N_{eff}	NEFF	2	Total channel charge coefficient	1	5	
X_j	XJ	2, 3	Metallurgical junction depth	0.0	1×10^{-6}	m
X_{jl}	LD	1–3	Lateral diffusion	0.0	0.8×10^{-6}	m
T_{PG}	TPG	1–3	Type of gate material	1	1	
μ_0	UO	1–3	Surface mobility	600	700	$cm^2/(V \cdot s)$
U_c	UCRIT	2	Critical electric field for mobility	1×10^4	1×10^4	V/cm
U_e	UEXP	2	Exponential coefficient for mobility	0.0	0.1	
U_t	UTRA	2	Transverse field coefficient	0.0	0.5	
v_{max}	VMAX	2, 3	Maximum drift velocity of carriers	0.0	5×10^4	m/s
X_{QC}	XQC	2, 3	Coefficient of channel charge share	0.1	0.4	
δ	DELTA	2, 3	Width effect on threshold voltage	0.0	1.0	
K	KAPPA	3	Saturation field factor	0.2	1.0	

Symbol	Name	Description	Range	Value	Unit
η	ETA	Static feedback on threshold voltage	3	0.0	
θ	THETA	Mobility modulation	3	0.0	V^{-1}
a_F	AF	Flicker-noise exponent	1–3	1.0	
k_F	KF	Flicker-noise coefficient	1–3	0.0	
I_S	IS	Bulk junction saturation current	1–3	1×10^{-14}	1×10^{-26}	A
J_S	JS	Bulk junction saturation current per square meter	1–3	0.0	1×10^{-8}	A/m^2
ϕ_j	PB	Bulk junction potential	1–3	0.80	0.75	V
C_j	CJ	Zero-bias bulk junction capacitance per square meter	1–3	0.0	2×10^{-4}	F/m^2
M_j	MJ	Bulk junction grading coefficient	1–3	0.5	
C_{jsw}	CJSW	Zero-bias perimeter capacitance per meter	1–3	0.0	1×10^{-9}	F/m
M_{jsw}	MJSW	Perimeter capacitance grading coefficient	1–3	0.33	0.33	
FC	FC	Forward-bias depletion capacitance coefficient	1–3	0.5	0.5	
C_{GBO}	CGBO	Gate-bulk overlap capacitance per meter	1–3	0.0	2×10^{-10}	F/m
C_{GDO}	CGDO	Gate-drain overlap capacitance per meter	1–3	0.0	4×10^{-11}	F/m
C_{GSO}	CGSO	Gate source overlap capacitance per meter	1–3	0.0	4×10^{-11}	F/m
r_D	RD	Drain ohmic resistance	1–3	0.0	10.0	Ω
r_S	RS	Source ohmic resistance	1–3	0.0	10.0	Ω
R_{sh}	RSH	Source and drain sheet resistance	1–3	0.0	30.0	Ω

removed by a later etching (Fig. 4-2b). Such etching, however, removes the thin oxide from the areas not covered by the polysilicon; in these regions a high concentration of n-type dopant is thermally diffused inside the silicon to obtain the so-called *source* (S) and *drain* (D) regions.

The region covered by the thin layer of oxide and by the gate represents the *channel*, and its length is L_{nom}. This length is indicated in the SPICE2 input description of the MOST and in this chapter as L. L_{eff} is the effective channel length defined as the distance measured on the silicon surface between the two diffused regions (Fig. 4-1). The width of the channel, indicated in SPICE2 as W, is the width of the area covered by the thin oxide.

During diffusion, a new layer of thermal oxide is grown on both the silicon and the polysilicon (Fig. 4-2c); this layer is removed (mask 3) from the regions where the contacts are made. Lastly, the layer of aluminum interconnections is deposited and defined with mask 4 (Fig. 4-2d).

This structure is usually called *MOST* (*MOS Transistor*); an often-used equivalent name is *MOSFET* (*MOS Field-Effect Transistor*).

It has been already seen from the example in Fig. 4-2 that technological developments have brought the use of materials other than aluminum for the gate electrode—e.g., polysilicon. The same has happened with the insulating layer between the gate and the semiconductor, where materials other than silicon oxide, e.g., silicon nitride, have been used. In these processes, the term *IGFET* (*Insulated-Gate Field-Effect Transistor*) has therefore become more appropriate. However, in this text, the transistor will be referred to by the most commonly used name, MOST.

4.1.2 Operating regions

In n-channel devices, here considered, the drain potential is higher than the source one and the voltages are referred to the source (Fig. 4-1). The voltage V_{GS} between gate and source determines the concentration of the carriers in the channel; the gate voltage for which the channel current becomes significant is called the *threshold voltage* V_{TH}. If V_{GS} is greater than V_{TH}, the current I_{DS} flows at the surface region from the drain to the source. The on-off behavior of the MOST is based on the fact that for gate voltages greater than the threshold voltage a current flows, whereas for smaller voltages the current is negligible.

The current vs. voltage output characteristics ($I_{DS} - V_{DS}$) for $V_{GS} > V_{TH}$ can be divided into a *linear region* and a *saturation region*; this is shown in Fig. 4-3. In the linear region the transfer characteris-

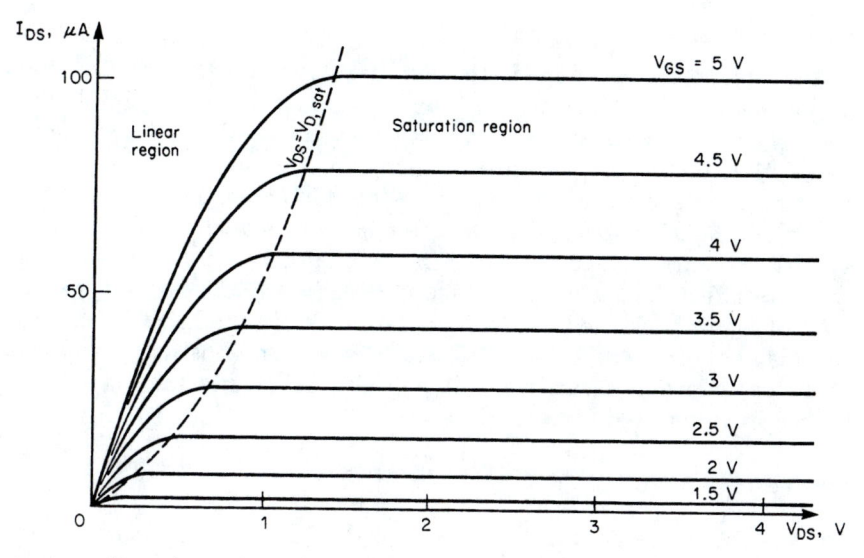

Figure 4-3 Operating regions of the MOST in the plan of the output characteristics.

tics ($I_{DS} - V_{GS}$) are linear (Fig. 4-4), while in the saturation region I_{DS} does not depend on V_{DS}.

There also exists a region near the threshold voltage where the current depends exponentially on V_{GS}; this is the region of *weak inversion* (see Appendix B).

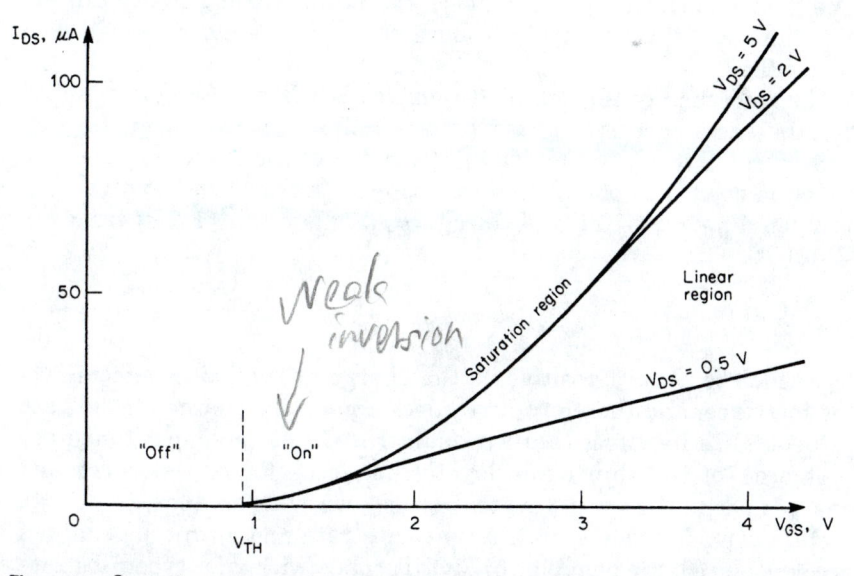

Figure 4-4 Operating regions of the MOST in the plan of the transfer characteristics.

4.1.3 Types of MOSTs

The device described here is the *n*-channel type; i.e., conduction is maintained by the electrons in the channel. In the *p*-channel device the substrate is *n*-type and conduction is due to the holes. In both these MOSTs the carriers are attracted to the surface by the gate voltage; they are called *enhancement MOSTs* (Figs. 4a,c).

In the type of MOST called *buried channel* or *depletion MOST* (Fig. 4-5b,d), the carriers are available in the channel region also without the gate effect, because the surface is weakly doped by an *n*-type (or *p*-type) region, obtained with ion implantation; in the MOST obtained in this way the threshold voltage is negative (positive), and its electrical characteristics make it a useful substitute for load resistances in integrated circuits.

4.2 LEVEL1 Static Model

The first step in studying the theory of the MOST is the behavior of the surface of the semiconductor under the influence of an electric field. This electric field is perpendicular to the oxide-semiconductor interface, and it is produced by the voltage applied between the gate and the substrate. Later, the effect of the source and drain voltages will be introduced.

4.2.1 Surface behavior of the MOS structure

The theory outlined here is concise, the purpose being to explain the equations of SPICE2 models; a more complete theory is described in Appendix B.

The reference condition of the surface is when the semiconductor has the same carrier concentrations at the surface and at the substrate. This state is called the *flat-band* condition; the voltage V_{GB} which is needed to obtain this condition is the *flat-band voltage* V_{FB}.

When $V_{GB} = V_{FB}$, all the voltage V_{GB} falls on the oxide; from Eq. (B-10)

$$V_{FB} = \phi_{MS} - \frac{Q'_0}{C'_{ox}} \qquad (4\text{-}1)$$

Q'_0, which is always positive, is the charge at the oxide-semiconductor interface, and therefore it requires a negative charge on the gate to neutralize its effect (see Appendix B); C'_{ox} is the capacitance per unit area of the thin oxide layer (the terms for capacitances and charges per unit area in this text are shown with an apex, or prime).

Here MOSTs made with a *p*-type substrate and an *n*-type channel are considered; the behavior of MOSTs made with an *n*-type substrate is dual when the sign of the voltages and the current is changed.

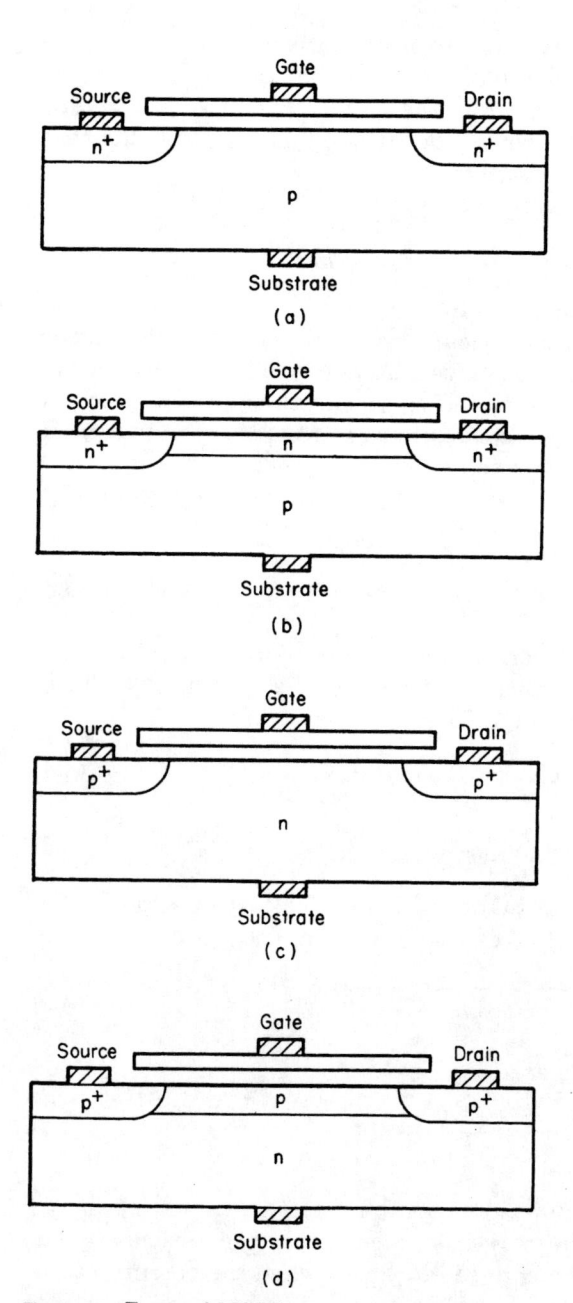

Figure 4-5 Types of MOSTs: (a) n-channel enhancement; (b) n-channel depletion; (c) p-channel enhancement; (d) p-channel depletion.

When $V_{GB} = V_{FB}$, the carrier concentration is constant in the semiconductor, and it is equal to N_A.

The condition $V_{GB} < V_{FB}$ increases the negative charge on the gate and increases the concentration of the holes near the surface in order to balance the gate charge. Thus the concentration of the p-type carriers is greater at the surface than in the substrate; in this case the surface is said to be *accumulated*.

The charge Q'_{SC} in the semiconductor is

$$Q'_{SC} = Q'_G - Q'_0 = (V_{FB} - V_{GB})C'_{ox} \quad (4\text{-}2)$$

where Q'_G is the charge on the gate.

When $V_{GB} > V_{FB}$, the holes (majority carriers) are pushed away from the surface so that the negative charge of the fixed ions restores the balance with the gate charge. The carrier concentration near the surface is less than that in the substrate; in this case the surface is said to be *depleted*. Using the same formulas as in the theory of the step junction (Appendix A), the width X_B of the depleted region (Fig. B-3b) is, from Eq. (B-30),

$$X_B = \sqrt{\frac{2\varepsilon_s}{qN_A}\phi_s} \quad (4\text{-}3)$$

where ϕ_s is the potential across the depleted region. If ϕ_s is small enough to neglect the minority carriers in the channel, the fixed charge Q'_B is equal to $Q'_G - Q'_0$. Therefore

$$Q'_B = N_A q X_B = \sqrt{2\varepsilon_s q N_A \phi_s} \quad (4\text{-}4)$$

and

$$Q'_G - Q'_0 = (V_{GB} - V_{FB} - \phi_s)C'_{ox} \quad (4\text{-}5)$$

By substituting Eq. (4-3) in Eq. (4-4) and making it equal to Eq. (4-5), the surface potential ϕ_s can be obtained. Therefore

$$\phi_s = \tfrac{1}{4}[\sqrt{\gamma^2 + 4(V_{GB} - V_{FB})} - \gamma]^2 \quad (4\text{-}6)$$

where

$$\gamma = \frac{\sqrt{2\varepsilon_s q N_A}}{C'_{ox}} \quad (4\text{-}7)$$

These equations are valid as long as the depletion approximation is valid, i.e., as long as the carrier concentration remains negligible with respect to N_A in the channel. However, when the potential ϕ_s is sufficiently high, the concentration of the electrons at the surface can exceed that of the holes in the substrate. In this case the surface is said to be *inverted*. From Boltzmann's distribution, this condition is reached when $\phi_s = 2\phi_p$, i.e., when

$$n = N_A = n_i e^{\phi_p q/kT} \tag{4-8}$$

or

$$\phi_p = \frac{kT}{q} \ln \frac{N_A}{n_i} \tag{4-9}$$

as it has been defined in Eq. (A-29b). This approximated theory assumes arbitrarily that $\phi_s = 2\phi_p$ as the surface potential needed to change from a condition of depletion to one of inversion. In inversion, the carrier concentration at the surface is greater and of opposite type than the concentration in the substrate; the condition $V_{GB} = V_{TH}$ is therefore reached when $\phi_s = 2\phi_p$.

When $V_{GB} > V_{TH}$, it can be supposed that the surface potential ϕ_s and the fixed charge Q'_B are not altered, because the new charges are supplied by electrons near the surface. In fact, an increase in V_{GB} implies an increase in the charge Q'_I in the inversion region; a slight increase in ϕ_s implies an increase in Q'_I according to the exponential law of Boltzmann's equation, while Q'_B remains almost constant according to Eq. (4-4) (see Appendix B). Therefore the mobile charge in the channel is

$$Q'_I = Q'_G - Q'_0 - Q'_B \tag{4-10}$$

The charge Q'_I determines the conduction of the MOST. Substituting Eqs. (4-4) and (4-5) in Eq. (4-10) and making them equal yields, under the condition $\phi_s = 2\phi_p$

$$V_{TH} = V_{FB} + 2\phi_p + \gamma \sqrt{2\phi_p} \tag{4-11}$$

4.2.2 Equations of the simple model

The first model of the MOST used in SPICE2 is basically the model proposed by Shichman and Hodges [2], and it is called the LEVEL1 model. In order to extend the theory of the MOS capacitor seen in the preceding section to the MOST, the effect of the diffused regions of source and drain must be considered.

In the following analysis the source potential will be used as a reference potential, and the voltage in the channel will be indicated as $V_c(x)$. At equilibrium—that is, when there is a zero voltage between drain and source—in order to make the concentration of minority carriers at the surface equal to that of the majority carriers in the substrate, it is necessary that the surface potential ϕ_s exceeds the inverse bias V_{BS} between source and substrate; i.e.,

$$\phi_s = 2\phi_p - V_{BS} \qquad (4\text{-}12)$$

Substituting Eq. (4-12) in Eq. (4-5), the threshold voltage follows

$$V_{TH} = V_{FB} + 2\phi_p + \gamma\sqrt{2\phi_p - V_{BS}} \qquad (4\text{-}13)$$

The threshold voltage is then a function of the square root of the substrate voltage (or *body bias*); γ is called the *body-effect* parameter, and is defined in Eq. (4-7).

The condition $V_{GS} > V_{TH}$ allows channel formation, and, therefore, by applying a positive voltage to the drain, the electrons in the channel flow by drift from the source to the drain.

The current I_x in a section dx of the channel will be

$$I_x = \frac{dQ_I}{dt} \qquad (4\text{-}14)$$

where dQ_I is the mobile charge present in the dx element, and dt is the time necessary for this charge to cross dx. The value of dQ_I is

$$dQ_I = dQ_G - dQ_0 - dQ_B \qquad (4\text{-}15)$$

where

$$dQ_G - dQ_0 = C'_{ox} W\, dx [V_{GS} - V_{FB} - 2\phi_p - V_c(x)] \qquad (4\text{-}16)$$

$$dQ_B = C'_{ox} W\gamma\, dx \sqrt{2\phi_p - V_{BS}} \qquad (4\text{-}17)$$

In the calculation of dQ_B, no account has been made of the voltage between the channel and the substrate $[V_{BS} + V_c(x)]$, and only V_{BS} has been considered. This approximation is valid only when the value of V_{DS} is small; otherwise the width X_B of the depleted region is noticeably larger near the drain than near the source. The value of dQ_B calculated in this way will be less than the real one, and the value of dQ_I will be overestimated.

By substituting Eqs. (4-16) and (4-17) in Eq. (4-15), and taking into account Eq. (4-13), it follows that

$$dQ_I = C'_{ox} W [V_{GS} - V_c(x) - V_{TH}]\, dx \qquad (4\text{-}18)$$

The speed of the carriers $v(x)$ is linked to the electric field $E_x(x)$ by the transport equation

$$v(x) = \frac{dx}{dt} = -\mu_0 E_x(x) = \mu_0 \frac{dV_c(x)}{dx} \qquad (4\text{-}19)$$

being μ_0 the surface mobility.

From Eq. (4-19)

$$\frac{1}{dt} = \mu_0 \frac{dV_c(x)}{dx^2} \qquad (4\text{-}20)$$

and by substituting Eqs. (4-18) and (4-20) in Eq. (4-14), it follows that

$$I_x = \mu_0 W C'_{ox} [V_{GS} - V_{TH} - V_c(x)] \frac{dV_c(x)}{dx} \qquad (4\text{-}21)$$

The integral between source and drain can be evaluated by using $V_c(x)$ as a variable, which varies from zero for $x=0$ to V_{DS} for $x=L_{\text{eff}}$. Therefore

$$\int_0^{L_{\text{eff}}} I_x \, dx = \mu_0 W C'_{ox} \int_0^{V_{DS}} [V_{GS} - V_{TH} - V_c(x)] \, dV_c(x) \qquad (4\text{-}22)$$

As I_x is constant in every section of the channel, it can be written

$$I_{DS} = \mu_0 C'_{ox} \left(\frac{W}{L_{\text{eff}}}\right) \left[(V_{GS} - V_{TH})V_{DS} - \frac{V_{DS}^2}{2}\right] \qquad (4\text{-}23)$$

The model is valid as long as a continuous channel exists between source and drain. However, when the voltage V_{DS} is such that there is a point in the channel where the voltage between the channel and the gate is equal to the threshold voltage, from this point to the drain, the conditions suitable for channel formation no longer exist; i.e.,

$$V_c(x) < V_{GS} - V_{TH} \qquad (4\text{-}24)$$

In this case the channel is formed only from $x=0$ to $x=L'$, where L' is the point where the channel voltage reaches the *saturation voltage* $V_{D,\text{sat}}$

$$V_{D,\text{sat}} = V_c(L') = V_{GS} - V_{TH} \qquad (4\text{-}25)$$

For $V_{DS} > V_{D,\text{sat}}$, the current I_{DS} is not a function of V_{DS} because the voltage at the end of the channel still remains equal to $V_{D,\text{sat}}$. By substituting Eq. (4-25) in Eq. (4-23), the current for $V_{DS} > V_{D,\text{sat}}$ can be obtained. Thus

$$I_{DS} = \frac{\beta}{2}(V_{GS} - V_{TH})^2 \qquad (4\text{-}26)$$

where

$$\beta = \mu_0 C'_{ox}\left(\frac{W}{L_{\text{eff}}}\right) = KP\left(\frac{W}{L_{\text{eff}}}\right) = KP\left(\frac{W}{L - 2X_{jl}}\right) \qquad (4\text{-}27)$$

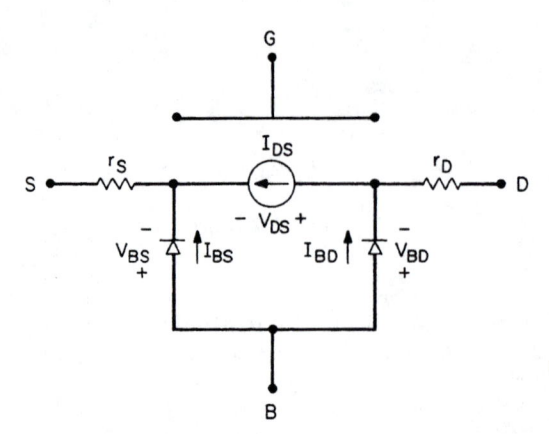

Figure 4-6 MOST static model.

KP and X_{jl} are parameters called the *transconductance* and *lateral diffusion parameter* (see Fig. 4-1). The voltage $V_{DS} - V_{D,\text{sat}}$ falls in the region between $x = L'$ and $x = L_{\text{eff}}$, and the electric field, determined by such voltage difference, transports the carriers from the channel to the drain.

Equations (4-23), (4-26), and (4-27) define the nonlinear current source I_{DS} in the LEVEL1 *static* model of the MOST shown in Fig. 4-6, where the drain and source resistances have also been included. In Fig. 4-6, the two diodes model the source-substrate and the drain-substrate junctions and are defined by expressions following the *pn*-junction law (see Appendix A).

The simplifying assumptions used to formulate the above equations are rather hard, and the precision obtained from this model is limited; however, experience has shown that the simplicity of these equations is useful in many situations.

4.2.3 Implementation of the LEVEL1 static model in SPICE2

The equations used for LEVEL1 MOST model in SPICE2 are as follows.

Linear region

For $V_{GS} > V_{TH}$ and $V_{DS} < V_{GS} - V_{TH}$

$$I_{DS} = \text{KP} \frac{W}{L - 2X_{jl}} \left(V_{GS} - V_{TH} - \frac{V_{DS}}{2} \right) V_{DS}(1 + \lambda V_{DS}) \quad (4\text{-}28)$$

where X_{jl} (LD) is the lateral diffusion, KP is the transconductance, and

$$V_{TH} = V_{T0} + \gamma(\sqrt{2\phi_p - V_{BS}} - \sqrt{2\phi_p}) \qquad (4\text{-}29)$$

is the threshold voltage, V_{T0} (VTO) being the zero-bias threshold voltage, γ (GAMMA) the body-effect parameter, and $2\phi_p$ (PHI) the surface inversion potential. The parameter VTO is positive (negative) for enhancement mode and negative (positive) for depletion mode n-channel (p-channel) devices. The SPICE2 keyword for the .MODEL statement [3] is indicated in parentheses.

Saturation region

For $V_{GS} > V_{TH}$ and $V_{DS} > V_{GS} - V_{TH}$

$$I_{DS} = \frac{KP}{2}\frac{W}{L - 2X_{jl}}(V_{GS} - V_{TH})^2(1 + \lambda V_{DS}) \qquad (4\text{-}30)$$

In the above equations, L and W are the values of the length and width of the channel specified in the SPICE2 input statement of the MOST device.

The term $1 + \lambda V_{DS}$ introduced in the model is an empirical correction of the conductance in the saturation region, by the channel-length modulation parameter λ (LAMBDA).

Equations (4-28) and (4-30) are valid for $V_{DS} > 0$ (normal mode). For $V_{DS} < 0$ (inverted mode) the source and drain in the above equations must be switched.

The two substrate junctions (see Fig. 4-6) are modeled by diode-like equations, i.e.,

$$I_{BS} = \begin{cases} I_{SS}(e^{qV_{BS}/kT} - 1) + \text{GMIN } V_{BS} & \text{for } V_{BS} > 0 \\ \dfrac{qI_{SS}}{kT} V_{BS} + \text{GMIN } V_{BS} & \text{for } V_{BS} \leq 0 \end{cases} \qquad (4\text{-}31)$$

$$I_{BD} = \begin{cases} I_{SD}(e^{qV_{BD}/kT} - 1) + \text{GMIN } V_{BD} & \text{for } V_{BD} > 0 \\ \dfrac{qI_{SD}}{kT} V_{BD} + \text{GMIN } V_{BD} & \text{for } V_{BD} \leq 0 \end{cases} \qquad (4\text{-}32)$$

where I_{SS} and I_{SD} are the substrate-junction saturation currents, both of them identified by IS in the .MODEL statement; GMIN (default = 10^{-12} mho) is a conductance added by default in SPICE2 in parallel with every pn junction. Its value can be set (except GMIN = 0) by the user with the .OPTIONS statement.

The parameters VTO, KP, GAMMA, PHI, IS, which characterize the above equations, are *electrical* parameters, i.e., they refer to the electrical behavior of the MOST. They can be specified directly in the .MODEL statement, or they can be calculated from *geometrical*, *physical*, and *technological* parameters, using the following equations:

$$V_{TO} = \underbrace{\underbrace{tp\, T_{PG}\frac{E_g}{2} - \frac{kT}{q}\ln\frac{N_A}{n_i}}_{\phi_{MS}} - qN_{SS}\frac{t_{ox}}{\varepsilon_{ox}}}_{V_{FB}} + 2\phi_p + \gamma\sqrt{2\phi_p} \qquad (4\text{-}33)$$

$$KP = \mu_0 C'_{ox} = \mu_0 \frac{\varepsilon_{ox}}{t_{ox}} \qquad (4\text{-}34)$$

$$\gamma = \frac{\sqrt{2q\varepsilon_s N_A}}{C'_{ox}} = \frac{t_{ox}}{\varepsilon_{ox}}\sqrt{2q\varepsilon_s N_A} \qquad (4\text{-}35)$$

$$2\phi_p = 2\frac{kT}{q}\ln\frac{N_A}{n_i} \qquad (4\text{-}36)$$

$$I_{SD} = J_S A_D \qquad (4\text{-}37)$$

$$I_{SS} = J_S A_S \qquad (4\text{-}38)$$

The flag *tp* is +1 (−1) for an *n* (*p*)-channel MOST.

In the above equations, T_{PG} (TPG) represents the type of gate and takes a value of 0 for the metal-gate MOST and a value of −1 or +1 for a MOST with the gate made of polysilicon (the value −1 is used if the polysilicon is doped of the same type as the substrate and +1 if it is of the opposite type); N_A (NSUB) is the substrate doping; N_{SS} (NSS) is the surface state density; t_{ox} (TOX) is the oxide thickness; μ_0 (UO) is the surface mobility; J_S (JS) is the substrate-junction saturation current density. A_D (AD) and A_S (AS) are the drain and source diffusion areas, respectively. Thus it is possible to use these parameters instead of VTO, KP, etc., or a combination of the two types of parameters.

In case of conflict (for example, if both μ_0 and KP are present in the .MODEL statement), the value of the electrical parameter is read from the input and not computed from Eq. (4-34). The above parameters can be specified in the .MODEL statement except AD, AS, NRD, and NRS which have to be specified in the SPICE2 input statement of the MOST device.

Simulations of transfer and output characteristics using typical parameters are shown in Figs. 4-7 to 4-10. The values of the electrical parameters of the model used in the simulations are

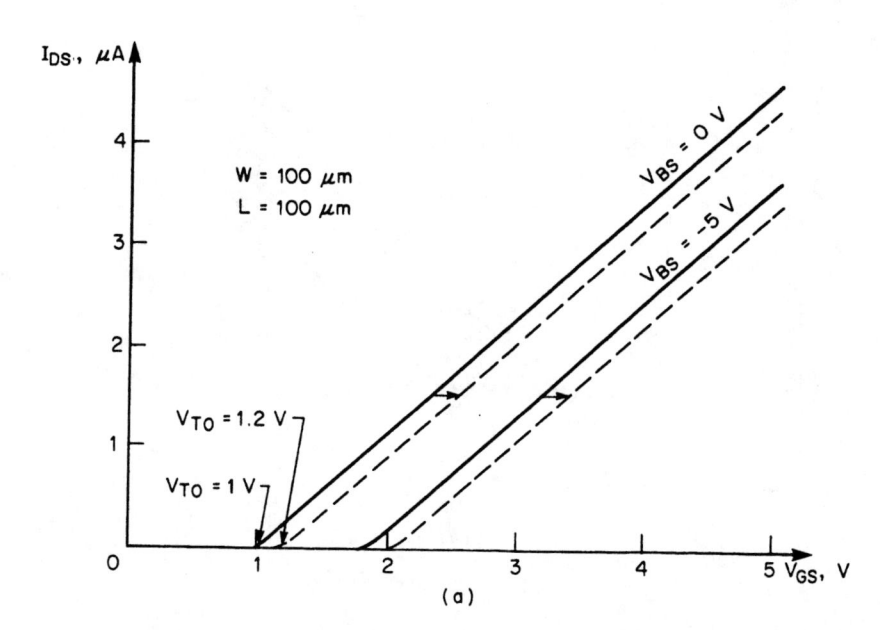

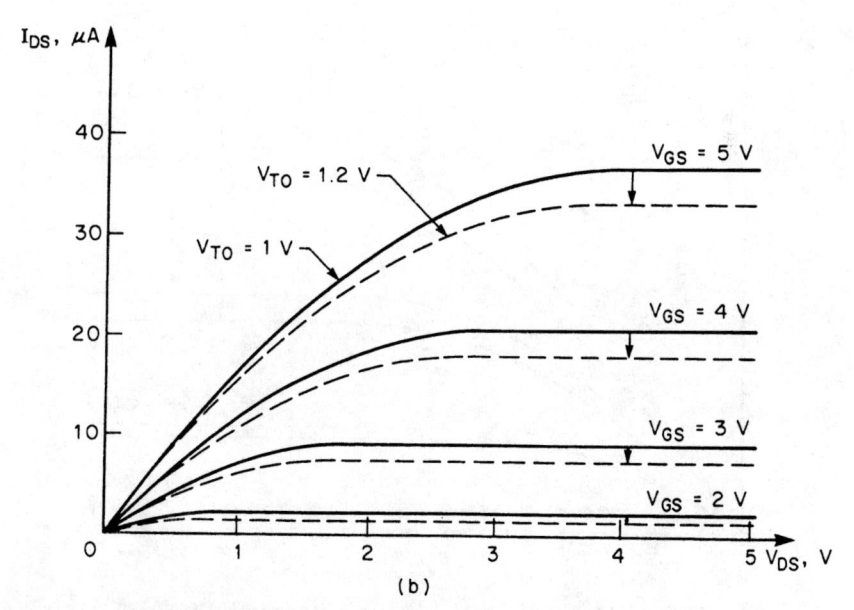

Figure 4-7 Variations of I_{DS} with V_{T0}, for the LEVEL1 model: (a) transfer characteristics and (b) output characteristics.

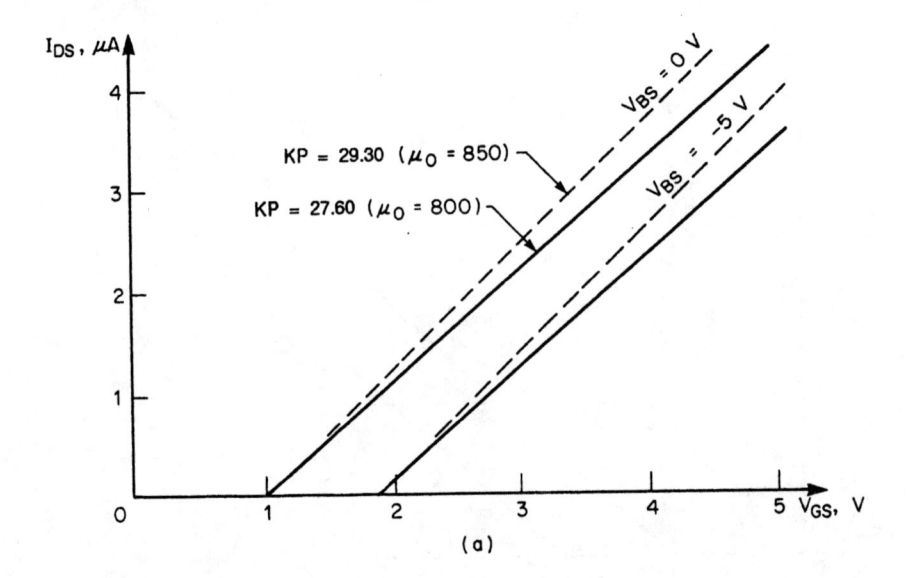

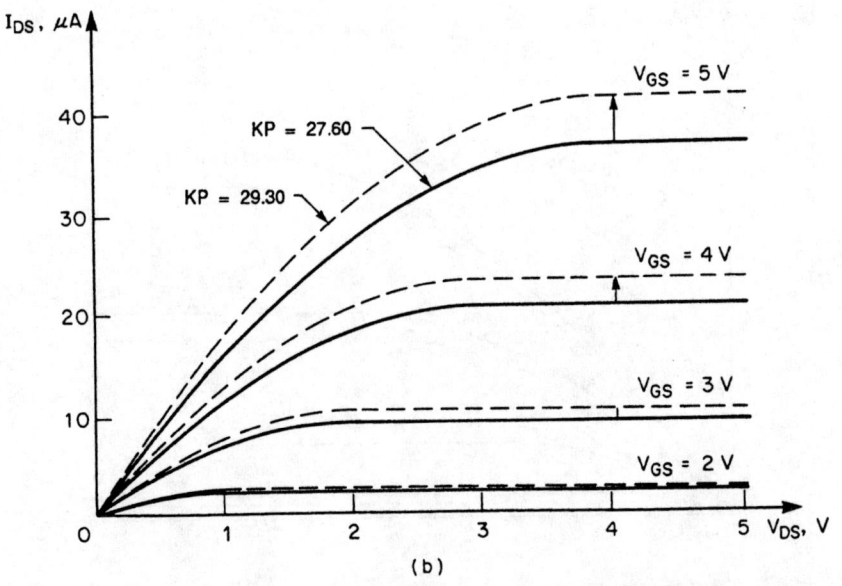

Figure 4-8 Variations of I_{DS} with KP or with μ_0, for the LEVEL1 model: (a) transfer characteristics and (b) output characteristics.

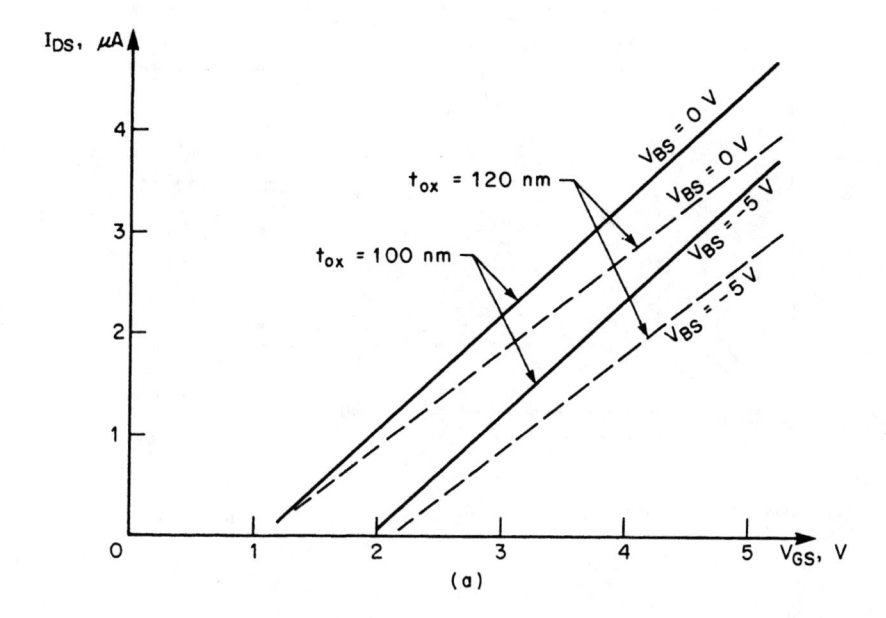

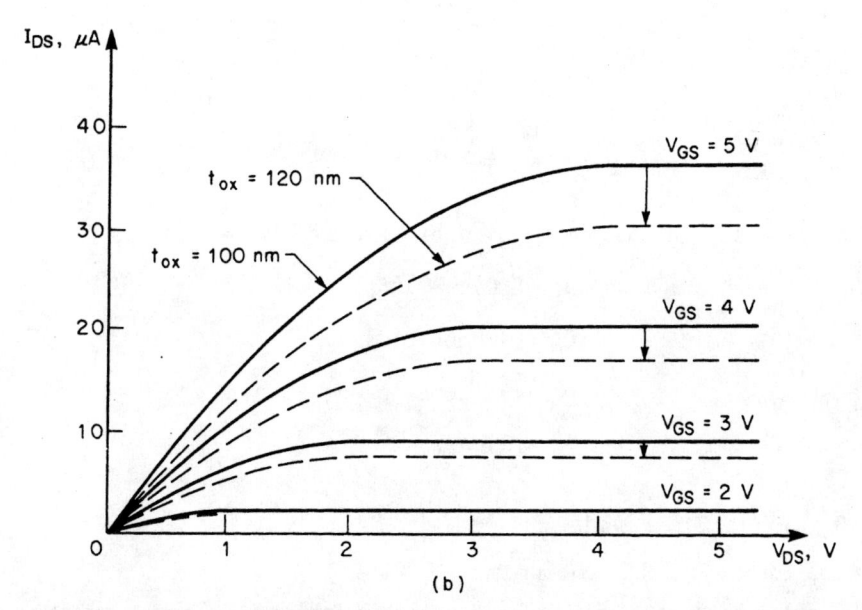

Figure 4-9 Variations of I_{DS} with t_{ox}, for the LEVEL1 model: (a) transfer characteristics and (b) output characteristics.

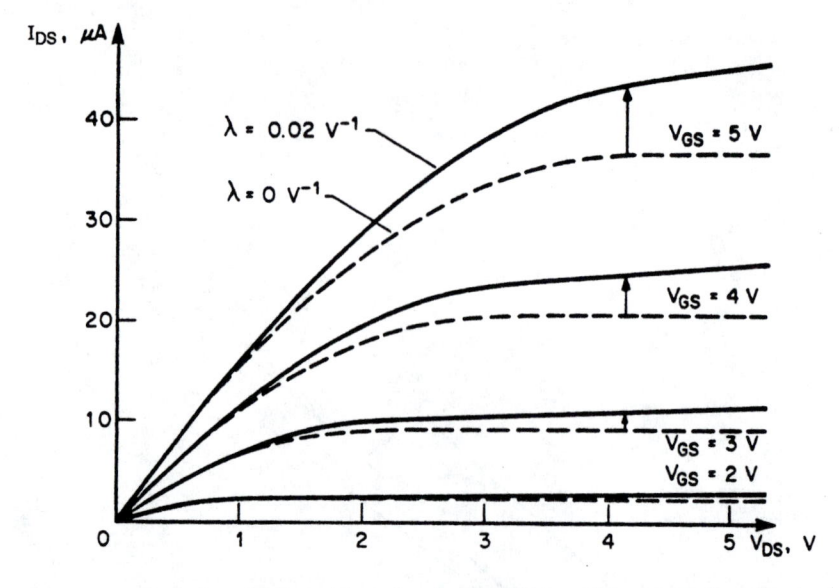

Figure 4-10 Variations of I_{DS} with λ, for the LEVEL1 model: output characteristics.

as follows:

$$KP = 27.6 \ \mu A/V^2$$

$$V_{T0} = 1 \ V$$

$$\gamma = 0.526 \ V^{1/2}$$

$$2\phi_p = 0.58 \ V$$

$$\lambda = 0$$

Typical value

The values of the physical parameters are as follows:

$$\mu_0 = 800 \ cm^2/(V \cdot s)$$

$$t_{ox} = 100 \ nm \ (\text{of } SiO_2)$$

$$N_A = 10^{15} \ cm^{-3}$$

$$X_{jl} = 0.8 \ \mu m$$

4.3 LEVEL2 Static Model

4.3.1 Equations of the basic model

To obtain a better model for I_{DS} it is necessary to eliminate the simplification made in Eq. (4-17) to find the value of Q_B by taking into account the effect of the voltage in the channel on Q_B. This problem was solved by Meyer [4], who started from the charge equation (Eq. 4-15), where the new value of Q_B is

$$dQ_B = W\,dx\,\gamma C'_{ox}\sqrt{2\phi_p - V_{BS} + V_c(x)} \qquad (4\text{-}39)$$

As in the previous theory, a new equation for the current can be obtained. Thus

$$I_{DS} = \beta\left\{\left(V_{GS} - V_{FB} - 2\phi_p - \frac{V_{DS}}{2}\right)V_{DS}\right.$$

linear region

$$\left. - \tfrac{2}{3}\gamma[(V_{DS} - V_{BS} + 2\phi_p)^{1.5} - (-V_{BS} + 2\phi_p)^{1.5}]\right\} \qquad (4\text{-}40)$$

The saturation condition is reached when the charge in the channel for $x = L'$ is zero. By substituting Eqs. (4-16) and (4-39) for $V_{DS} = V_{D,\text{sat}}$ into Eq. (4-15), it follows that

$$Q_I(L') = (V_{GS} - V_{D,\text{sat}} - V_{FB} - 2\phi_p)WL'C'_{ox} - \gamma WL'C'_{ox}$$
$$\times \sqrt{V_{D,\text{sat}} - V_{BS} + 2\phi_p}$$
$$= 0 \qquad (4\text{-}41)$$

from which the value of $V_{D,\text{sat}}$ is

$$V_{D,\text{sat}} = V_{GS} - V_{FB} - 2\phi_p + \frac{\gamma^2}{2}\left[1 - \sqrt{1 + \frac{4}{\gamma^2}(V_{GS} - V_{FB} - V_{BS})}\right] \qquad (4\text{-}42)$$

Also, this model has validity limits that can be easily reached with the present technology. The method chosen to overcome these limitations is to adapt the model by applying semiempirical corrections to the basic equations rather than start again from the basis of the theory each time one needs to develop new and more accurate models.

4.3.2 Implementation of the LEVEL2 static model in SPICE2

This section summarizes the basic equations of Meyer's model as they have been introduced in SPICE2. The corrections made to the model to simulate effects not provided for in the theory of the basic model are described in the following sections.

The current in the linear region is calculated through Eq. (4-40), where the corrective term of the channel-length modulation is also present. Therefore

$$I_{DS} = \frac{KP}{1 - \lambda V_{DS}} \frac{W}{L - 2X_{jl}} \left\{ \left(V_{GS} - V_{FB} - 2\phi_p - \frac{V_{DS}}{2} \right) V_{DS} \right.$$
$$\left. - \tfrac{2}{3}\gamma[(V_{DS} - V_{BS} + 2\phi_p)^{1.5} - (-V_{BS} + 2\phi_p)^{1.5}] \right\} \quad (4\text{-}43)$$

It should be noted that when V_{DS} is very small the values supplied by this model are very close to those of the LEVEL1 model.

In the saturation region the current is

$$I_{DS} = I_{D,\text{sat}} \frac{1}{1 - \lambda V_{DS}} \quad (4\text{-}44)$$

where $I_{D,\text{sat}}$ is calculated from Eq. (4-43) at $V_{DS} = V_{D,\text{sat}}$, and $V_{D,\text{sat}}$ is from Eq. (4-42). These equations give better results than the simple model, but they are still not sufficient for a good agreement with experimental data, even when short- and narrow-channel effects are absent.

The modifications that will improve precision for long and wide channels will be presented first; then the variations introduced in the model for short- and narrow-channel MOSTs will be presented.

4.3.3 Mobility variation

In the calculation of the current in the channel, which has led to the equations of the LEVEL1 and LEVEL2 models, the mobility has been assumed constant with the applied voltage. This approximation is convenient in the calculation of the integral in Eq. (4-22), but the results do not agree with the experimental data; a reduction in mobility with an increase in the gate voltage is observed [5]. In order to simulate this effect, a variation of the parameter KP has been introduced in SPICE2. This new expression is used in Eq. (4-43). Therefore

$$KP' = KP \left(\frac{\varepsilon_s}{\varepsilon_{ox}} \frac{U_c t_{ox}}{V_{GS} - V_{TH} - U_t V_{DS}} \right)^{U_e} \quad (4\text{-}45)$$

The value of the term in parentheses is limited to 1.

The parameter U_c (UCRIT) is the gate-to-channel critical field; above this level, the mobility begins to decrease, while the term $(V_{GS} - V_{TH} - U_t V_{DS})/t_{ox}$ represents the average electric field perpendicular to the channel. The U_t (UTRA) parameter value is chosen between 0 and 0.5, and represents the contribution to the gate-to-channel electric field due to the drain voltage, and U_e (UEXP) is the exponential coefficient for the mobility. The .MODEL statement keyword is indicated in parentheses. The use of this formula allows a good agreement between SPICE2 and experimental data in absence of short- and narrow-channel effects, and in the strong inversion region only.

The effect of U_c and U_e parameters on the characteristics and on KP' is shown in Figs. 4-11 to 4-13.

4.3.4 Conduction in the weak inversion region

The basic model implemented in SPICE2 calculates the drift current when the surface potential is equal to or greater than $2\phi_p$. In reality, as explained in Appendix B, a concentration of electrons near the surface exists also for $V_{GS} < V_{TH}$, and therefore there is a current even when the surface is not in *strong* inversion. This current is due mainly to diffusion between the source and the channel.

The model [6] implemented in SPICE2 introduces an exponential dependence on the current I_{DS} and V_{GS} for the *weak* inversion region. This model has defined a voltage V_{on} which acts as a boundary between the regions of weak and strong inversion. This voltage is defined as

$$V_{on} = V_{TH} + \frac{nkT}{q} \qquad (4\text{-}46)$$

where

$$n = 1 + \frac{qN_{FS}}{C'_{ox}} + \frac{C_B}{C'_{ox}} \qquad (4\text{-}47)$$

C_B is the capacitance associated with the depleted region and is obtained from Eq. (4-17). Therefore

$$C_B = \frac{dQ_B}{dV_{BS}} = \frac{\gamma}{2\sqrt{2\phi_p - V_{BS}}} C_{ox} \qquad (4\text{-}48)$$

N_{FS} (NFS) is a .MODEL statement parameter and is defined as the number of fast superficial states. This parameter determines the slope of $\log I_{DS}$ vs. V_{GS} characteristics. The current in weak inversion is

$$I_{DS} = I_{on} e^{(V_{GS} - V_{on})(q/nkT)} \qquad (4\text{-}49)$$

where I_{on} is the current in strong inversion for $V_{GS} = V_{on}$.

It is clear that the model introduces a discontinuity in the derivative for $V_{GS} = V_{on}$ and therefore the simulation of the transition region between strong and weak inversion is not very precise.

In fact, this model separates two operating regions in which only one mechanism of conduction, diffusion or drift, is considered; there-

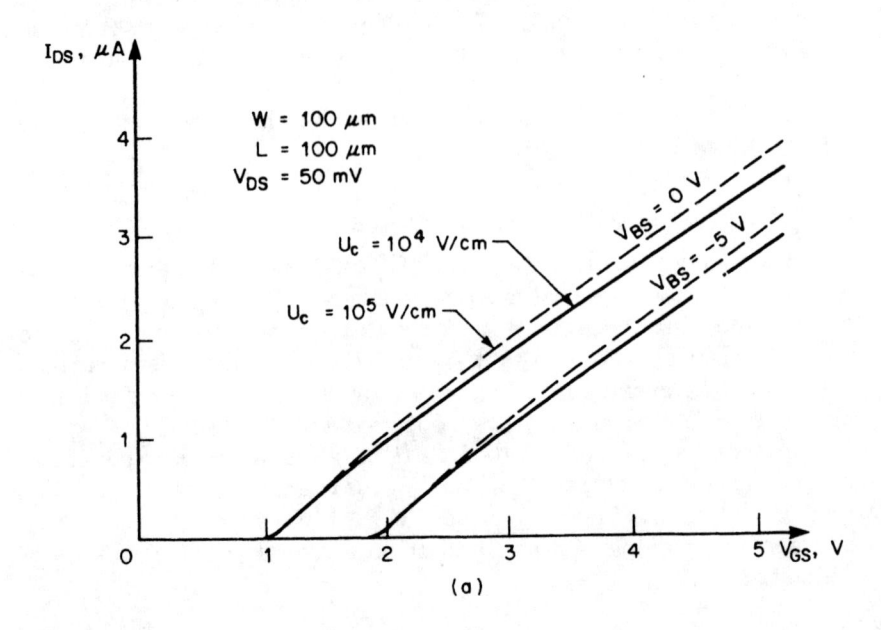

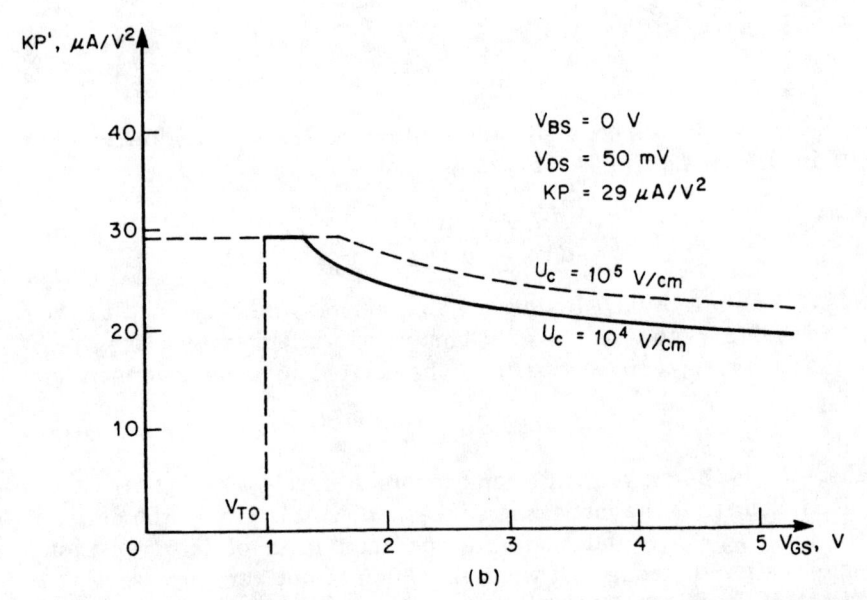

Figure 4-11 Variations of I_{DS} with U_c, for the LEVEL2 model: (a) transfer characteristics and (b) value of KP effective.

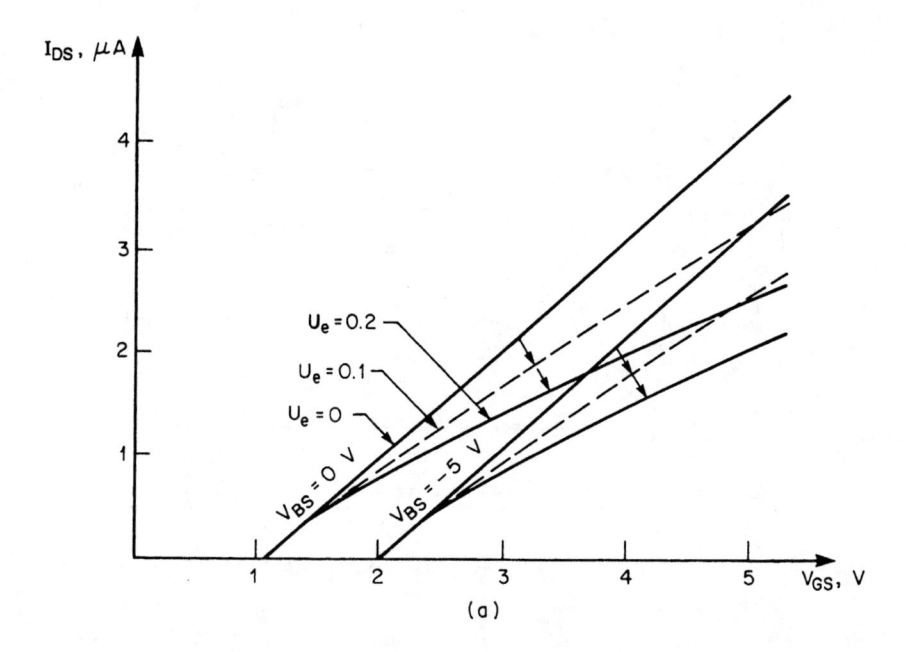

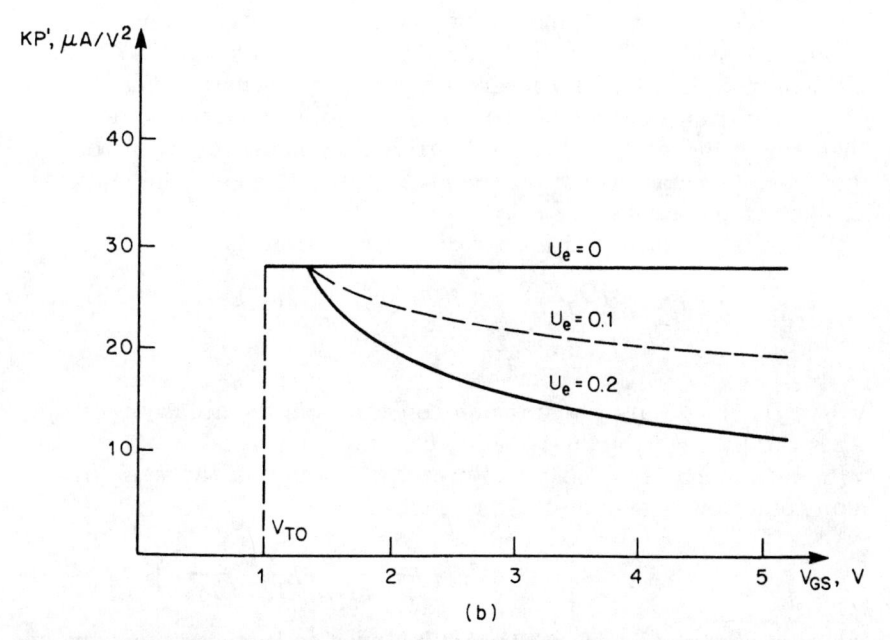

Figure 4-12 Variations of I_{DS} with U_e, for the LEVEL2 model: (a) transfer characteristics and (b) value of KP effective.

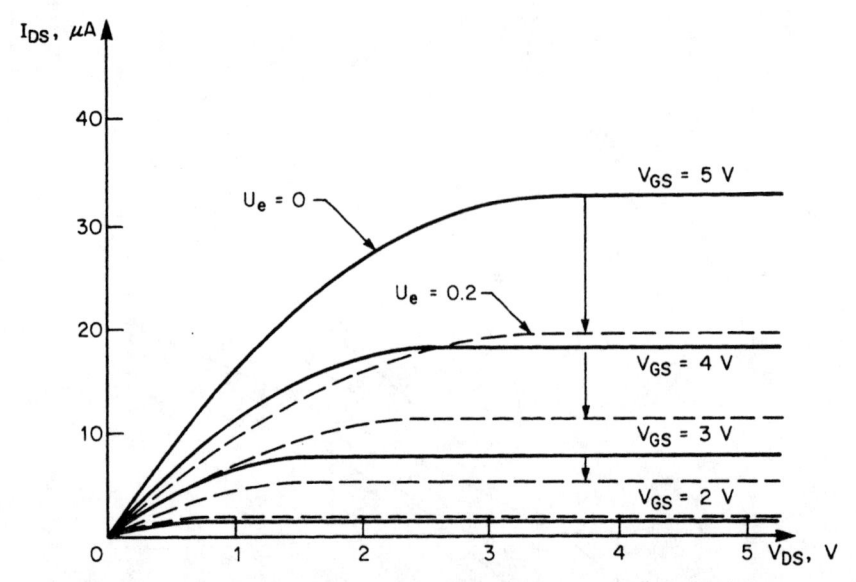

Figure 4-13 Variations of I_{DS} with U_e, LEVEL2 model: output characteristics.

fore, it is not possible, under this assumption, to correctly simulate the intermediate region (or *moderate inversion* region) in which the two conduction mechanisms make comparable contributions.

Another model [7] for weak inversion has been proposed and implemented in SPICE2. This model is based on the separate calculation of the contributions of the drift and diffusion currents, which are then simply added together, without distinguishing between the regions of weak and strong inversion. Figure 4-14 shows the currents I_{drift} and I_{diff} described by this model.

The equation proposed for the diffusion current is

$$I_{\text{diff}} = \frac{qD_n X_c W_{\text{eff}}}{L_n \tanh \dfrac{L_{\text{eff}}}{L_n}} \frac{n_s n_x}{n_s + n_x} (1 - e^{-qV_{DS}/kT}) \qquad (4\text{-}50)$$

where D_n and L_n are the diffusion constant and the diffusion length, X_c is the average thickness of the channel, and n_x is an asymptotic value of the carrier concentration in the channel in the weak inversion condition. The expression for n_s is

$$n_s = N_{\text{surf}} \exp\left[(1 - \alpha V_{BS})\frac{V_{GS} - V_{TH}}{(1 + qN_{FS}/C'_{ox})(kT/q)}\right] \qquad (4\text{-}51)$$

where α is an empirical parameter included to increase the slope of $\log I_{DS}$ vs. V_{GS}, when the body bias is greater than zero.

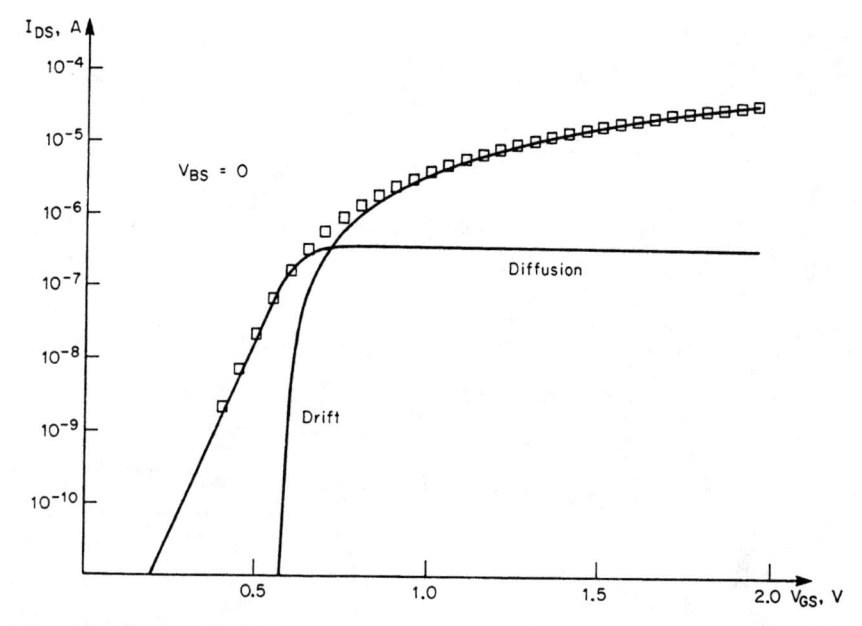

Figure 4-14 Drift and diffusion current components, computed according to the model [7].

4.3.5 Variation of channel length in the saturation region

In Secs. 4.2.3 and 4.3.2, an empirical equation using the parameter λ has been derived for the LEVEL1 and LEVEL2 models to calculate the conductance in the saturation region (see Fig. 4-10).

The LEVEL2 model also offers the possibility of using a physical model to calculate the channel length in saturation on the basis of the same corrective term in Eq. (4-44)

$$L' = L_{\text{eff}}(1 - \lambda V_{DS}) \qquad (4\text{-}52)$$

If the parameter λ is not specified in the .MODEL statement, it is calculated using the equation [8]

$$\lambda = \frac{L_{\text{eff}} - L'}{L_{\text{eff}} V_{DS}} \qquad (4\text{-}53)$$

where

$$L_{\text{eff}} - L' = X_D \left[\frac{V_{DS} - V_{D,\text{sat}}}{4} + \sqrt{1 + \left(\frac{V_{DS} - V_{D,\text{sat}}}{4}\right)^2} \right]^{1/2} \qquad (4\text{-}54)$$

and

$$X_D = \sqrt{\frac{2\varepsilon_s}{qN_A}} \qquad (4\text{-}55)$$

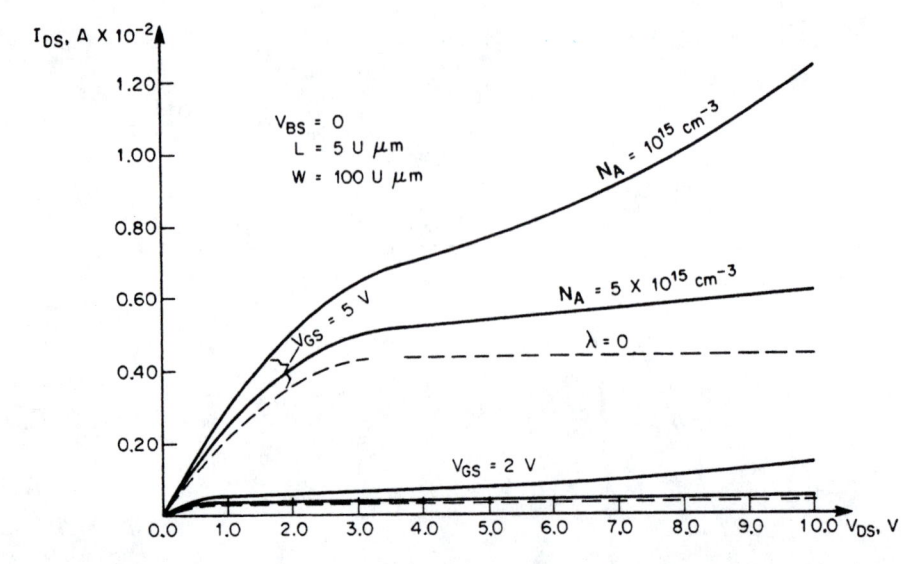

Figure 4-15 Variations of I_{DS} with $\lambda = 0$, for the LEVEL2 model, or without λ in the .MODEL statement, and with several values of N_A: output characteristics.

The value of the slope of $I_{D,\text{sat}}$ vs. V_{DS} calculated with this model, using in Eq. (4-55) the correct substrate doping N_A [the same used in Eq. (4-35) to calculate γ], is usually greater than the measured values (see Fig. 4-15). In this case it is possible to reduce the conductance g_{DS} in saturation by increasing N_A. However, if N_A is used to fit g_{DS}, it is not possible to use the same value to calculate $2\phi_p$ and γ through Eqs. (4-36) and (4-35); it is then necessary to specify these parameters in the .MODEL statement.

This model shows a correct dependence of the conductance in the saturation region on L_{eff}, while if the model with constant λ is used, g_{DS} does not vary with L_{eff} when $V_{DS} > V_{D,\text{sat}}$.

4.3.6 Effect of channel length on the threshold voltage (short-channel)

Equation (4-29) has been obtained from a theory that does not take two-dimensional effects into account. Therefore, this equation does not include a link between the threshold voltage and the channel dimensions W and L_{eff}. Experimental data, on the other hand, show that when the channel length is small enough to be comparable with the width of the depleted region (*short-channel condition*), this relationship exists; it is negligible when L_{eff} is large.

The models presented in the literature explain this phenomenon in two ways: a reduction in the charge Q_B due to the source and drain depleted regions [9–11] or an increase in the channel surface potential ϕ_s due to the effect of the voltage V_{DS} [12–15].

The model used in SPICE2 and described in this section is based on the first of the two hypotheses considered above [9].

The charge Q_B that contributes to the threshold voltage, according to this model, is shown in Fig. 4-16. This effect is introduced in Eq. (4-29) by modifying the value of γ as

$$\gamma' = \gamma \left[1 - \frac{X_j}{2L_{\text{eff}}} \left(\sqrt{1 + \frac{2W_S}{X_j}} + \sqrt{1 + \frac{2W_D}{X_j}} - 2 \right) \right] \quad (4\text{-}56)$$

where X_j is the metallurgical junction depth, while W_S and W_D are the widths of the depleted regions of source and drain, respectively, and are defined as

$$W_S = X_D \sqrt{2\phi_p - V_{BS}} \quad (4\text{-}57)$$

$$W_D = X_D \sqrt{2\phi_p - V_{BS} + V_{DS}} \quad (4\text{-}58)$$

If the real physical values of N_A (NSUB) and X_j (XJ) are used in the .MODEL statement, the model overestimates the reduction of V_{TH} for short L_{eff} with respect to long L_{eff}. The agreement with the experimental data can be improved by changing the values of X_j or N_A, but it is difficult to obtain satisfactory results over a large range of channel lengths (see Fig. 4-17). Moreover, this model does not adequately explain the dependence of V_{TH} on the drain voltage (see Fig. 4-18).

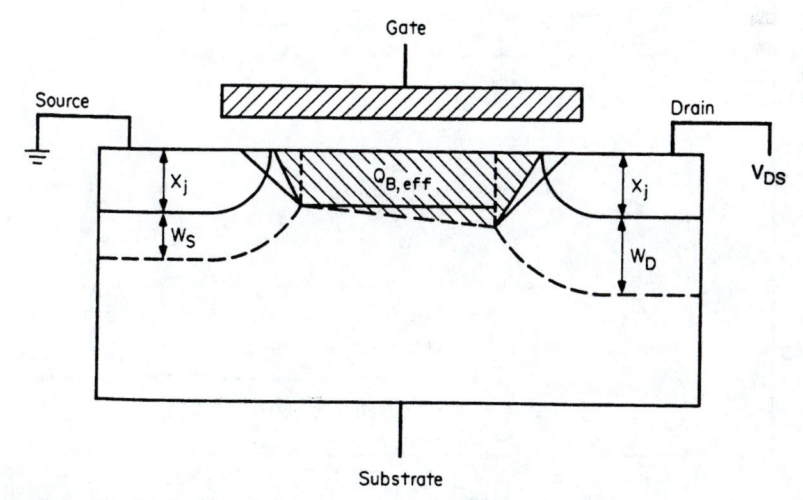

Figure 4-16 Geometrical calculation of the effective substrate charge Q_B with Yau's model [9].

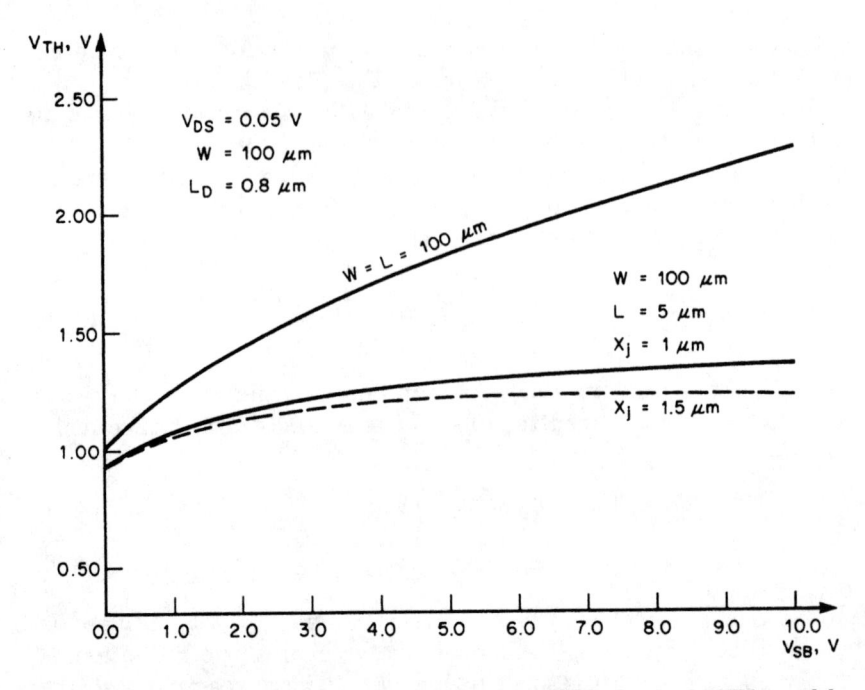

Figure 4-17 Body-effect simulation of short-channel MOSTs for the LEVEL2 model.

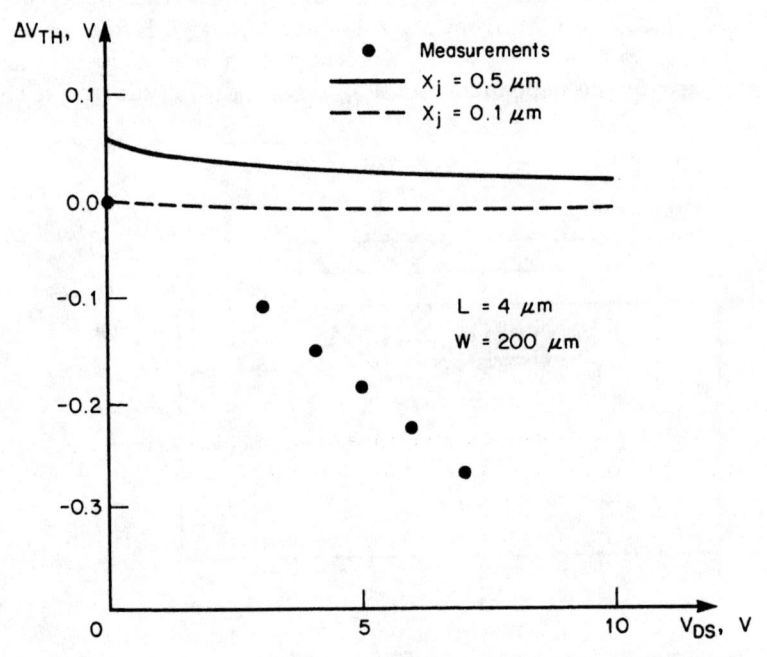

Figure 4-18 Simulation of V_{TH} vs. V_{DS} in a short-channel MOST device, LEVEL2 model.

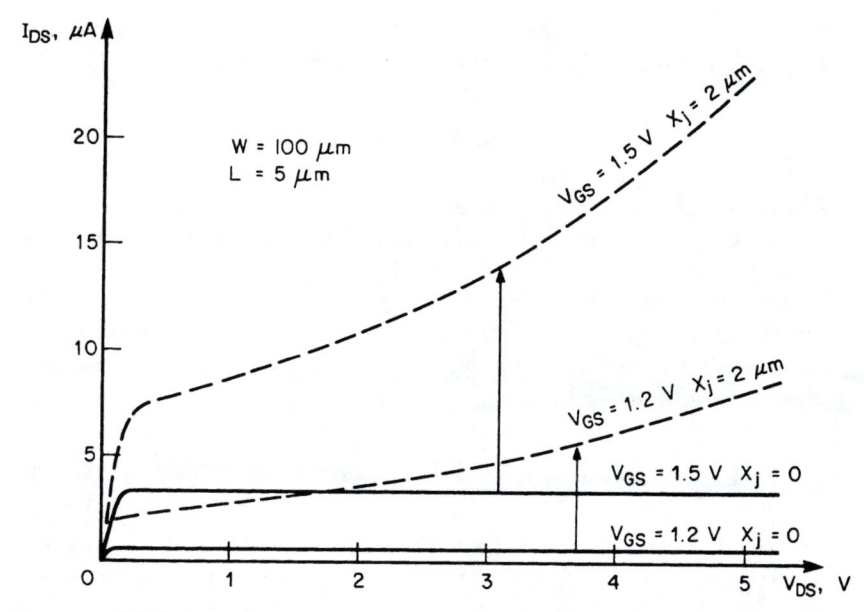

Figure 4-19 Variations of I_{DS} with X_j, for the LEVEL2 model: output characteristics.

The effects of the parameter X_j on the output characteristics are shown in Fig. 4-19.

In conclusion, a correct analytical model of the two-dimensional effects on V_{TH} is in practice too cumbersome. On the other hand, it is possible to obtain a good model for the threshold voltage of a short-channel MOST with a two-dimensional numerical model, which takes into account the spatial configuration of the fields and potentials. This type of analysis is too time-consuming to be implemented in a circuit simulation program; it is used in other programs [16,17], which can be used in the study of physical behavior of short-channel MOSTs.

4.3.7 Effect of speed limit of the carriers

The calculation of the saturation voltage using Eq. (4-42) is based on the hypothesis that the charge in the channel is zero when $x = L'$, i.e., near the drain. This hypothesis is false, because a minimum concentration greater than zero must exist in the channel because of the carriers that sustain the saturation current; this concentration depends on the speed at which the carriers are moving.

Moreover, the electric field between the drain and the channel end ($x = L'$) can be sufficiently high to drift the carriers at the speed limit, which is the maximum speed allowed for the scattering effect within the crystal lattice. This value is indicated as v_{max}, and it is used to calculate the charge Q'_I for $V_{DS} = V_{D,sat}$. Therefore

$$Q'_I = \frac{I_{D,sat}}{W v_{max}} \quad (4\text{-}59)$$

The value of $V_{D,sat}$ is calculated from Eq. (4-59), but the solution of this equation is rather time-consuming. The effect of v_{max} on $V_{D,sat}$ is shown in Fig. 4-20.

If the parameter v_{max} (VMAX) is specified in the .MODEL statement, SPICE2 does not use the model of Sec. 4.3.5 for the modulation of the channel length in the saturation region. In this case the model described by Baum and Beneking [18] is used:

$$L_{eff} - L' = X_D \sqrt{\left(\frac{X_D v_{max}}{2\mu_0}\right)^2 + V_{DS} - V_{D,sat}} - \frac{X_D^2 v_{max}}{2\mu_0} \quad (4\text{-}60)$$

In the calculation of X_D, the value of N_A can be modified through a coefficient N_{eff} (NEFF) that is used as a fitting parameter. Therefore

$$X_D = \sqrt{\frac{2\varepsilon_s}{q N_A N_{eff}}} \quad (4\text{-}61)$$

Figure 4-21 shows the output characteristics simulated with this model.

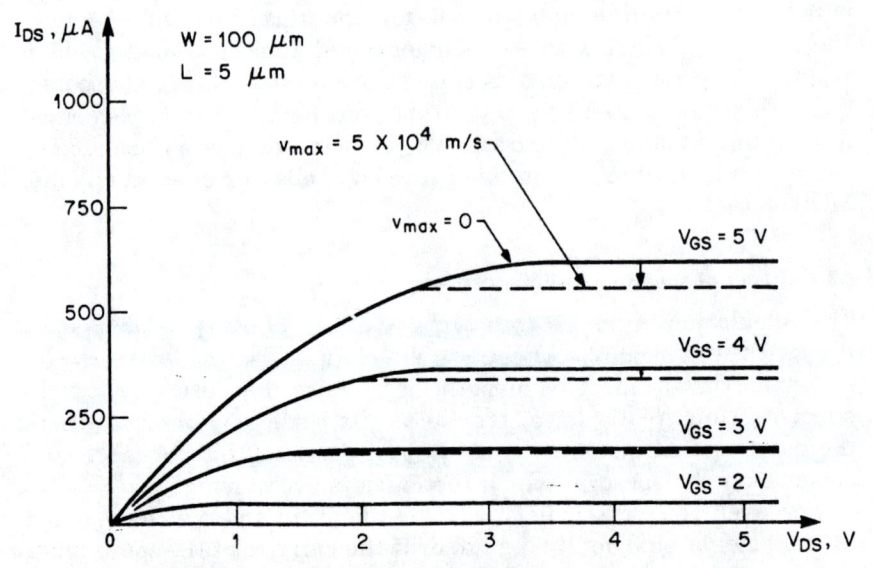

Figure 4-20 Variations of I_{DS} with v_{max}, for the LEVEL2 model: output characteristics.

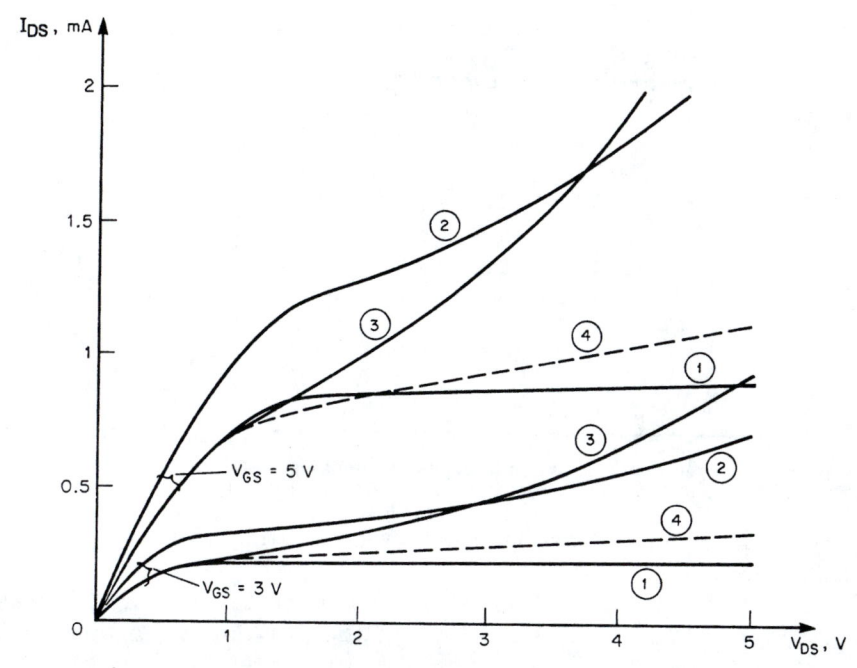

Figure 4-21 Comparison between the models for channel-length modulation in saturation: (1) empirical model (Sec. 4.2.3), $\lambda = 0.05 \text{ V}^{-1}$; (2) flat-junction model (Sec. 4.3.5), $N_A = 10^{15} \text{ cm}^{-3}$; (3) Baum's model [18], $v_{max} = 5 \times 10^4 \text{ m/s}$; (4) Baum's model, $v_{max} = 5 \times 10^4 \text{ m/s}$, $N_A = 10^{15} \text{ cm}^{-3}$, $N_{eff} = 5$.

The model seen so far provides good results in the simulations of MOSTs with a minimum channel length of about 4 to 5 μm. This model, however, introduces a discontinuity in the derivative at the boundary between the saturation and the linear regions; this leads to a less precise calculation of the conductance and is sometimes the cause of difficulties in the convergence of the Newton-Raphson algorithm.

4.3.8 Effect of channel width on the threshold voltage (narrow-channel)

In MOSTs with a small channel width W (e.g., less than 5 or 6 μm), comparable with the width of the depleted region (*narrow-channel condition*), the value of the threshold voltage V_{TH} is greater than the one indicated by the previous theory. This effect is caused by the two-dimensional distribution of the charge Q_B at the edges of the channel (see Fig. 4-22) [10–19].

In the model used in SPICE2, the thickness of the depleted region varies gradually from X_B under the channel to zero under the thick

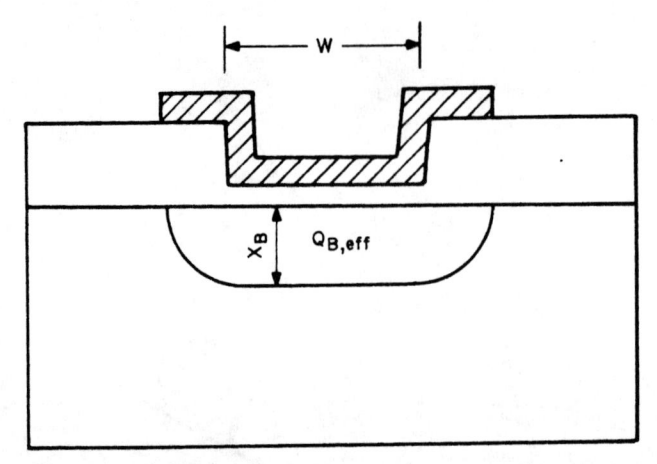

Figure 4-22 Geometrical calculation of the substrate charge Q_B reduced by the depleted regions at the channel edges.

oxide. The empirical parameter δ allows the fitting of the experimental data.

Adding to the value of Q_B the contribution of the boundary regions (see Fig. 4-22), a modified equation for the threshold voltage can be

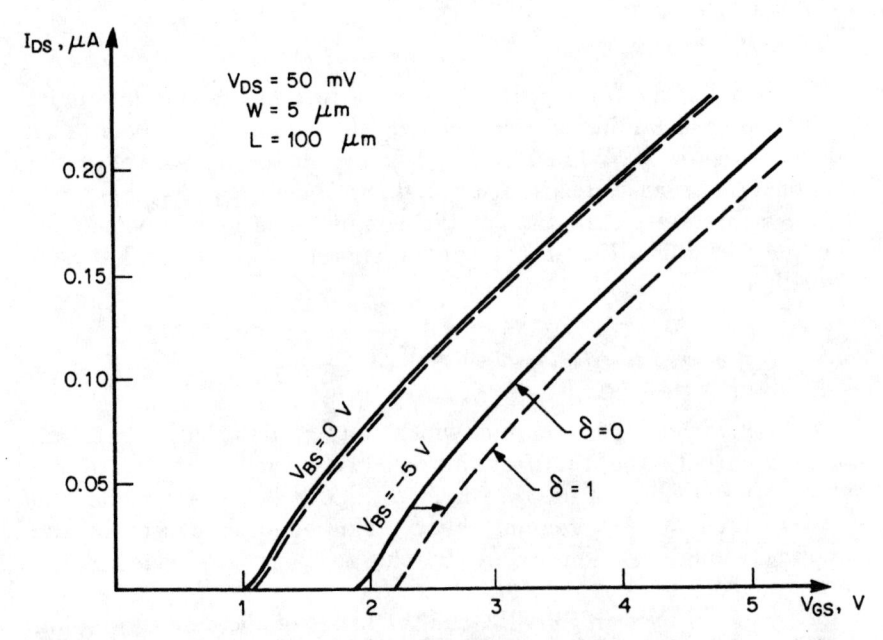

Figure 4-23 Variations of I_{DS} with δ, for the LEVEL2 model: transfer characteristics.

obtained:

$$V_{TH} = V_{FB} + 2\phi_p + \gamma'\sqrt{2\phi_p - V_{BS}} + \frac{\varepsilon_s \delta \pi}{4C'_{ox} W}(2\phi_p - V_{BS}) \quad (4\text{-}62)$$

The calculation of I_{DS} is obviously affected by the increase in the threshold voltage (see Fig. 4-23).

This model is used in SPICE2, even if the parameter δ (DELTA) is not specified in the .MODEL statement; to prevent the program from using this model, $\delta = 0$ must be specified.

4.4 LEVEL1 and LEVEL2 Large-Signal Model

Figure 4-24 shows the large-signal model for the n-channel MOST. In the following discussion each capacitance term will be considered.

4.4.1 Gate capacitance

SPICE2 uses a gate capacitance model similar to that proposed by Meyer [4]. In this simple model, the charge-storage effect is represented by three nonlinear two-terminal capacitors: C_{GB}, C_{GS}, and C_{GD}. The equations of this model are as follows (see Fig. 4-25).

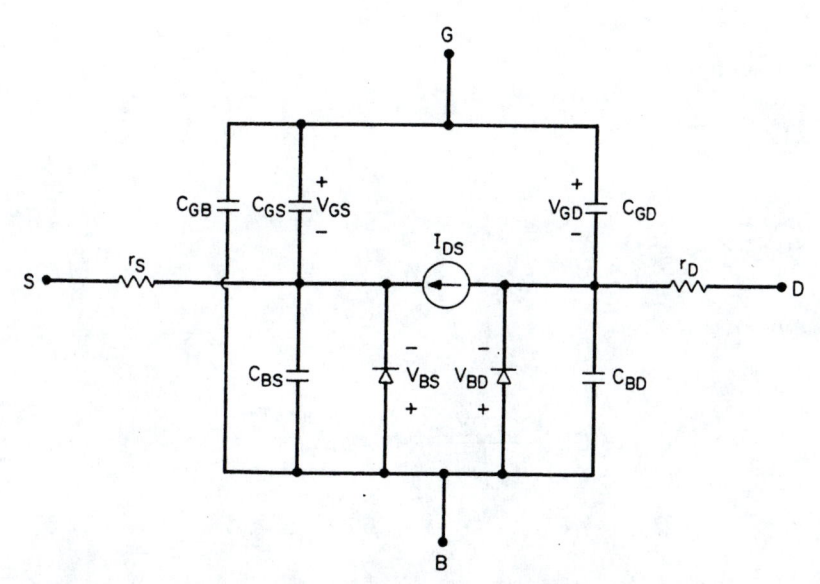

Figure 4-24 Large-signal model for the n-channel MOST.

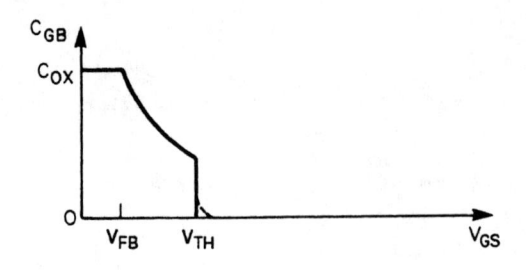

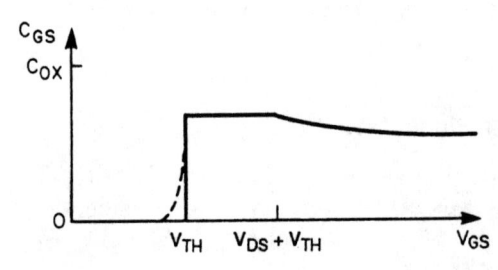

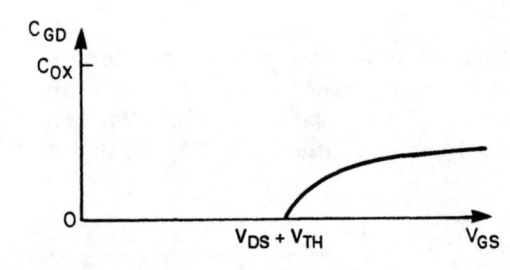

Figure 4-25 Meyer's model of the capacitances.

Accumulation region

For $V_{GS} < V_{on} - 2\phi_p$

$$C_{GB} = C_{ox} + C_{GBO}L_{eff} \tag{4-63}$$

$$C_{GS} = C_{GSO}W \tag{4-64}$$

$$C_{GD} = C_{GDO}W \tag{4-65}$$

Depletion region

For $V_{on} - 2\phi_p < V_{GS} < V_{on}$

$$C_{GB} = C_{ox}\frac{V_{on} - V_{GS}}{2\phi_p} + C_{GBO}L_{eff} \tag{4-66}$$

$$C_{GS} = \tfrac{2}{3}C_{ox}\left(\frac{V_{on} - V_{GS}}{2\phi_p} + 1\right) + C_{GSO}W \tag{4-67}$$

$$C_{GD} = C_{GDO}W \tag{4-68}$$

Saturation region

For $V_{\text{on}} < V_{GS} < V_{\text{on}} + V_{DS}$

$$C_{GB} = C_{GBO} L_{\text{eff}} \qquad (4\text{-}69)$$

$$C_{GS} = \tfrac{2}{3} C_{\text{ox}} + C_{GSO} W \qquad (4\text{-}70)$$

$$C_{GD} = C_{GDO} W \qquad (4\text{-}71)$$

Linear region

For $V_{GS} > V_{\text{on}} + V_{DS}$

$$C_{GB} = C_{GBO} L_{\text{eff}} \qquad (4\text{-}72)$$

$$C_{GS} = C_{\text{ox}} \left\{ 1 - \left[\frac{V_{GS} - V_{DS} - V_{\text{on}}}{2(V_{GS} - V_{\text{on}}) - V_{DS}} \right]^2 \right\} + C_{GSO} W \qquad (4\text{-}73)$$

$$C_{GD} = C_{\text{ox}} \left\{ 1 - \left[\frac{V_{GS} - V_{\text{on}}}{2(V_{GS} - V_{\text{on}}) - V_{DS}} \right]^2 \right\} + C_{GDO} W \qquad (4\text{-}74)$$

where $C_{\text{ox}} = C'_{\text{ox}} W L_{\text{eff}}$.

The voltage V_{on} is calculated from Eq. (4-46) if the parameter N_{FS} is specified in the .MODEL statement; otherwise V_{on} is equal to V_{TH}. C_{GBO} (CGBO), C_{GSO} (CGSO), and C_{GDO} (CGDO) are the overlap capacitances among the gate electrode and the other terminals outside the channel region. The .MODEL statement keyword is indicated in parentheses. The model introduced in SPICE2 differs from that developed by Meyer because discontinuity between the different operating regions causes nonconvergence in the Newton-Raphson algorithm. In particular, Eqs. (4-66) and (4-67) have been introduced as a link between the capacitance with the accumulated surface and with the depleted surface.

The transient current in the MOST terminals is the sum of the current in the dc model and the current of the capacitances.

If Meyer's model is used, the average transient current in each time step $h = t_k - t_{k-1}$ and using trapezoidal integration is

$$i_S(t_k) = -i_S(t_{k-1}) + \frac{2}{h} [V_{GS}(t_k) - V_{GS}(t_{k-1})] \frac{C_{GS}(t_k) + C_{GS}(t_{k-1})}{2}$$

$$(4\text{-}75)$$

$$i_D(t_k) = -i_S(t_{k-1}) + \frac{2}{h} [V_{GD}(t_k) - V_{GD}(t_{k-1})] \frac{C_{GD}(t_k) + C_{GD}(t_{k-1})}{2}$$

$$(4\text{-}76)$$

$$i_B(t_k) = -i_S(t_{k-1}) + \frac{2}{h}[V_{GB}(t_k) - V_{GB}(t_{k-1})]\frac{C_{GB}(t_k) + C_{GB}(t_{k-1})}{2} \tag{4-77}$$

$$I_G = -i_S - i_D - i_B \tag{4-78}$$

where $V(t_k)$ are the tentative voltages in the time step under analysis, and $V(t_{k-1})$ are the voltages in the last converged time step.

The transient currents calculated in Eqs. (4-75) to (4-77) are also functions of the time step, while in the quasi-static operation hypothesis, the charge is only a function of the voltages. This problem causes errors in the simulation of circuits where some nodes in the network cannot change their charge, as node 2 in Fig. 4-26 (see also Fig. 4-27).

A charge-control model for the transient behavior of the MOST has been introduced in SPICE2 to avoid the mistakes due to this model.

If the parameter X_{QC} (XQC) is specified in the .MODEL statement, SPICE2 selects a simplified version of the model proposed by Ward [20,21] instead of that proposed by Meyer [4]. Ward's model calculates analytically the charge in the gate and in the substrate. The charge Q_c in the channel is found by difference, and it is subdivided between the source and the drain with a formula determined by the parameter X_{QC}.

$$Q_c = Q_D + Q_S = -(Q_G + Q_B) \tag{4-79}$$

$$Q_D = X_{QC} Q_c \tag{4-80}$$

$$Q_S = (1 - X_{QC}) Q_c \tag{4-81}$$

This approach is not justified from the physical point of view, and sometimes causes errors and convergence problems (e.g., when V_{DS} is changing sign).

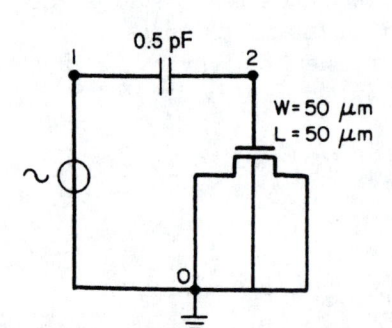

Figure 4-26 An example of a circuit critical for simulation of the capacitance with Meyer's model.

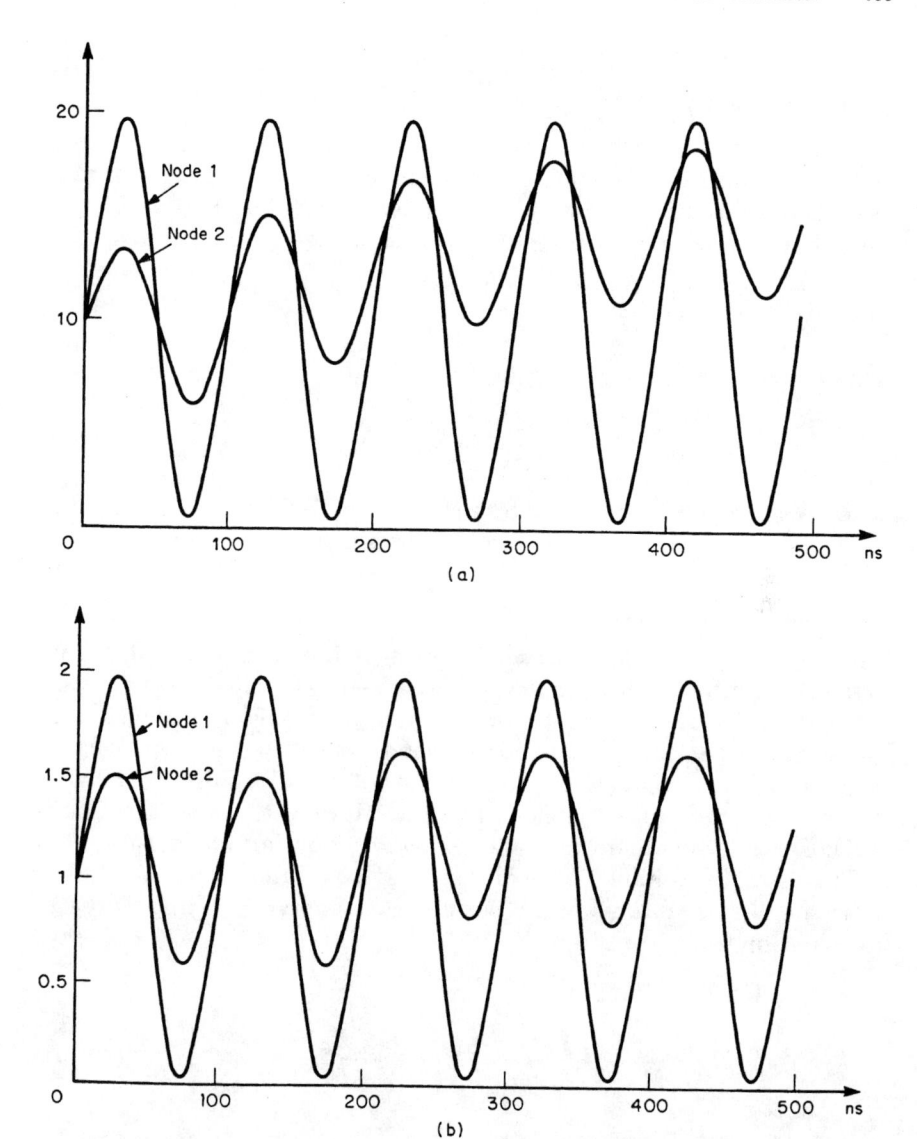

Figure 4-27 Simulation of the circuit seen in Fig. 4-26, with (a) Meyer's model and (b) Ward's model.

The transient currents are calculated from the charges in the following way:

$$i_G = \frac{dQ_G}{dt} \tag{4-82}$$

$$i_B = \frac{dQ_B}{dt} \tag{4-83}$$

$$i_S + i_D = \frac{dQ_c}{dt} \tag{4-84}$$

The capacitances between the MOST terminals are also requested for the definition of the companion model circuit; the model defines a matrix of capacitive terms, and each term is the derivative of the charge at the node x with respect to the voltage at the node y.

$$C_{xy} = \frac{dQ_x}{dV_y} \tag{4-85}$$

This matrix is not reciprocal; i.e.,

$$C_{xy} = \frac{dQ_x}{dV_y} \neq C_{yx} = \frac{dQ_y}{dV_x} \tag{4-86}$$

where x and y are two generic terminals of the MOST. In Fig. 4-28 the capacitance terms of the gate and of the substrate are shown.

4.4.2 Junction capacitance

The capacitance of the diffused regions of the source and drain is simulated with the pn-junction model. A separate capacitance model is defined for the periphery of the junction; this is because the capacitance per unit area and its dependence on the reverse-bias voltage in the boundary regions of the diffusion are different from those associated with the flat junction. Moreover, below the thick oxide region, the doping is usually increased by ion implantation.

The total capacitance of a diffused region is therefore calculated from the sum of an area and a perimeter capacitance. Using SPICE2 formulation,

For $V_{BS}, V_{BD} < (\text{FC} \times \phi_j)$

$$C_{BS} = \frac{C_j A_S}{(1 - V_{BS}/\phi_j)^{m_j}} + \frac{C_{jsw} P_S}{(1 - V_{BS}/\phi_j)^{m_{jsw}}} \tag{4-87a}$$

$$C_{BD} = \frac{C_j A_D}{(1 - V_{BD}/\phi_j)^{m_j}} + \frac{C_{jsw} P_D}{(1 - V_{BD}/\phi_j)^{m_{jsw}}} \tag{4-87b}$$

For $V_{BS}, V_{BD} \geq (\text{FC} \times \phi_j)$

$$C_{BS} = F_3 \left(\frac{C_j A_S}{F_2} + \frac{C_{jsw} P_S}{F_2} \right) + \frac{V_{BS}}{\phi_j} \left(\frac{C_j A_S}{F_2} m_j + \frac{C_{jsw} P_S}{F_2} m_{jsw} \right) \tag{4-88a}$$

$$C_{BD} = F_3 \left(\frac{C_j A_D}{F_2} + \frac{C_{jsw} P_D}{F_2} \right) + \frac{V_{BD}}{\phi_j} \left(\frac{C_j A_D}{F_2} m_j + \frac{C_{jsw} P_D}{F_2} m_{jsw} \right) \tag{4-88b}$$

where F_2 and F_3 have the same formulation of Eqs. (3-31), and

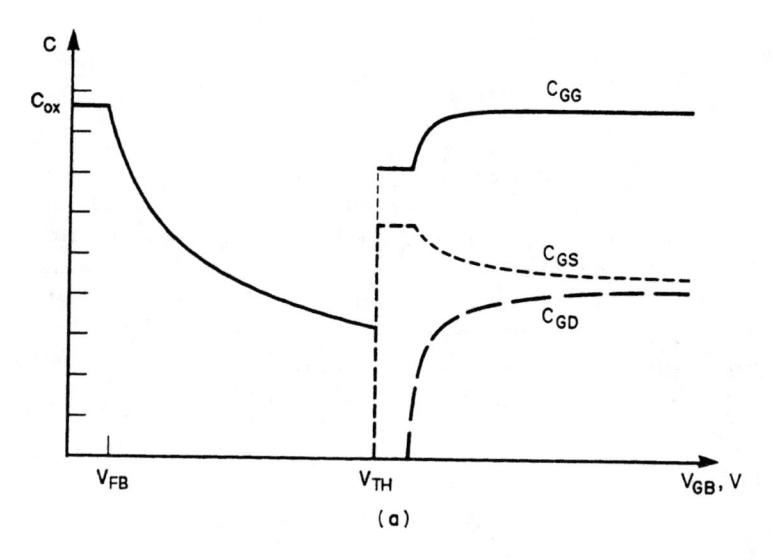

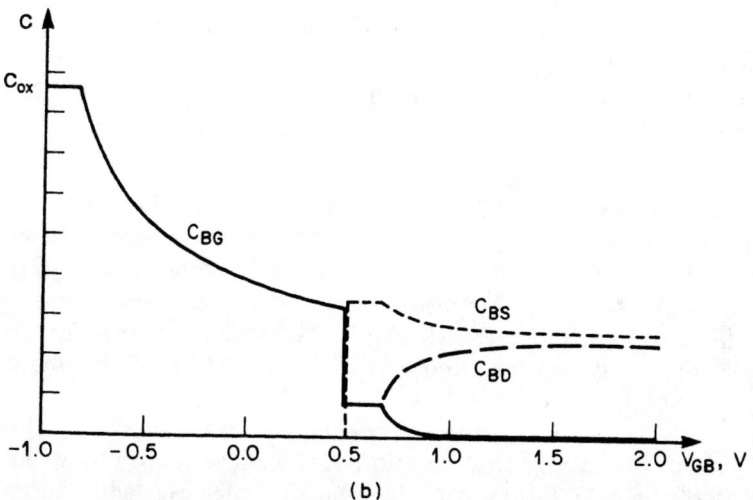

Figure 4-28 Ward's model of the capacitance: (a) gate capacitance and (b) substrate capacitance.

C_j (CJ) and C_{jsw} (CJSW) are the capacitances at zero-bias voltage, for square meter of area and for meter of perimeter; m_j (MJ) and m_{jsw} (MJSW) are the substrate-junction and perimeter capacitance grading coefficient; FC is the substrate-junction forward-bias coefficient; and ϕ_j (PB) is the junction potential that can be specified in the .MODEL statement or calculated from the physical parameters as

$$\phi_j = \frac{E_g}{2} + \frac{kT}{q} \ln \frac{N_A}{n_i} \qquad (4\text{-}89)$$

The junction capacitances are very important not only for the correct description of the circuit to be simulated but also because the convergence of the algorithm is easier if all the nodes are connected to ground with a capacitance. For this reason, it is useful to define the areas of the source and drain diffusions even if the layout is not done when the designer is writing down the input for SPICE2; the use of the default values DEFAD and DEFAS in the .OPTIONS statement to avoid dangerous floating nodes is recommended.

4.5 LEVEL3 Static Model

The LEVEL3 model has been developed to simulate short-channel MOSTs; it simulates quite precisely the characteristics of MOSTs that have a channel length up to 2 μm. The basic equations have been proposed by Dang [22]. The model for the weak inversion region is the same as that for the LEVEL2 model.

4.5.1 The basic LEVEL3 model

The equations for the LEVEL3 model are formulated in the same way as for the LEVEL2 model; however, a simplification of the current equation in the linear region has been obtained with a Taylor series expansion of Eq. (4-43). This approximation allows the development of more manageable basic equations than the LEVEL2 model; the short-channel effects are introduced in the calculation of threshold voltage and mobility.

Many of the equations used in the model are empirical. The purpose is both to improve the precision of the model and to limit the complexity of the calculations and the resulting time needed to carry out the program. The basic equations are very simple. The current in the linear region is

$$I_{DS} = \beta \left(V_{GS} - V_{TH} - \frac{1 + F_B}{2} V_{DS} \right) V_{DS} \qquad (4\text{-}90)$$

where

$$F_B = \frac{\gamma F_s}{4\sqrt{2\phi_p - V_{BS}}} + F_n \qquad (4\text{-}91)$$

is the Taylor series expansion coefficient of the substrate charge, while F_s and F_n will be defined later.

The short-channel effects influence the parameters V_{TH}, F_s, and β, while the narrow-channel effects influence the term F_n.

The dependence of mobility on the gate electric field is simulated with a simpler equation than that used for the LEVEL2 model, without any appreciable loss in precision [23,24]. Thus

$$\mu_s = \frac{\mu_0}{1 + \theta(V_{GS} - V_{TH})} \qquad (4\text{-}92)$$

where θ (THETA) is the mobility modulation parameter that the user can specify in the .MODEL statement.

In fact, as the oxide thickness has been greatly reduced with modern processes, it is practically impossible to find a region where the surface electric field E_x is less than the critical field E_c, as hypothesized in Eq. (4-45).

The effects of the parameters KP (or μ_0) and γ (or N_A) on the characteristics simulated with this model are analogous to those seen for the other models. In Fig. 4-29 the effect of parameter θ is shown.

One suggestion to improve the model is to use V_{T0} instead of V_{TH} in the denominator of Eq. (4-92); in this way it is possible to introduce a mobility variation with V_{BS} very close to its actual measurements without adding any new parameter in the .MODEL statement (see Fig. 4-30).

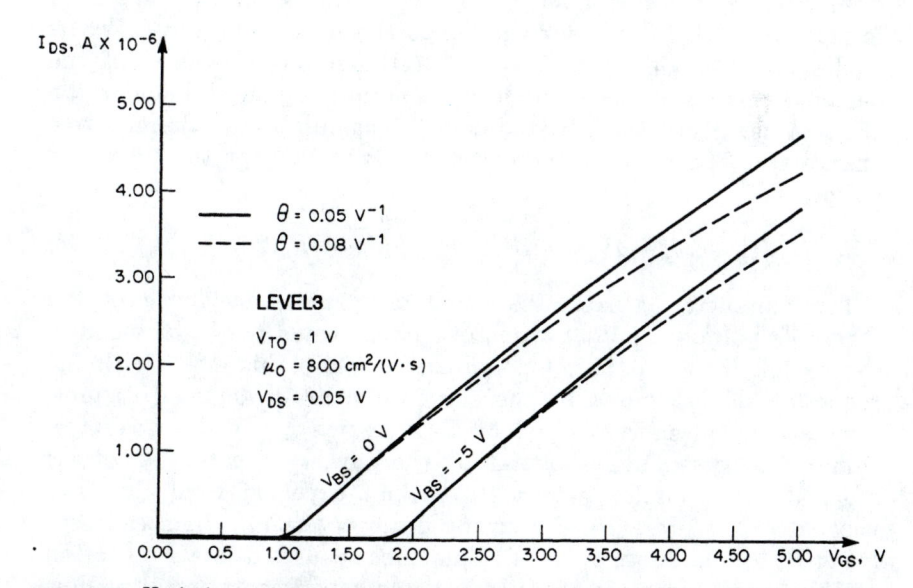

Figure 4-29 Variations of I_{DS} in the linear region with θ, LEVEL3 model: transfer characteristics.

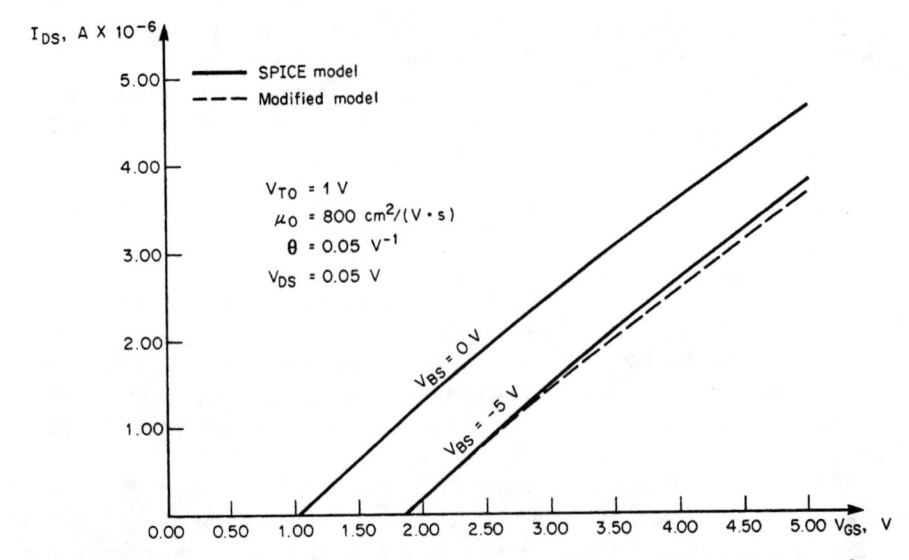

Figure 4-30 Variations of I_{DS} in the linear region with θ, modified model: transfer characteristics.

4.5.2 Effect of channel length on the threshold voltage (short-channel)

The more complex parts of the equations of this model are employed in calculating the threshold voltage V_{TH}. The equations have been formulated with hypotheses similar to those used for the LEVEL2 model (see Sec. 4.3.6); the effect of the different widths of the depleted regions in the cylindrical junction region and under the channel has also been introduced in the calculation of the effective charge Q_B. The equation proposed by Dang [22] for the threshold voltage is

$$V_{TH} = V_{FB} + 2\phi_p - \sigma V_{DS} + \gamma F_s \sqrt{2\phi_p - V_{BS}} + F_n(2\phi_p - V_{BS}) \quad (4\text{-}93)$$

The parameter σ expresses empirically the dependence of the threshold voltage on V_{DS}, which otherwise would not be included in the model. In fact it is not possible to explain this effect with the reduction of Q_B caused by the source and drain depleted regions. Simulations on short-channel MOSTs carried out with two-dimensional programs have shown that the inversion potential $2\phi_p$ is reached at a smaller gate voltage than expected from Eq. (4-6) because of the presence of a strong drain potential, which causes a lower threshold voltage [13,14]. This effect causes a linear variation of V_{TH} vs. V_{DS} (see Fig. 4-31) [25]. These experimental results allow the introduction of the term σV_{DS} seen in Eq. (4-93). The relationship

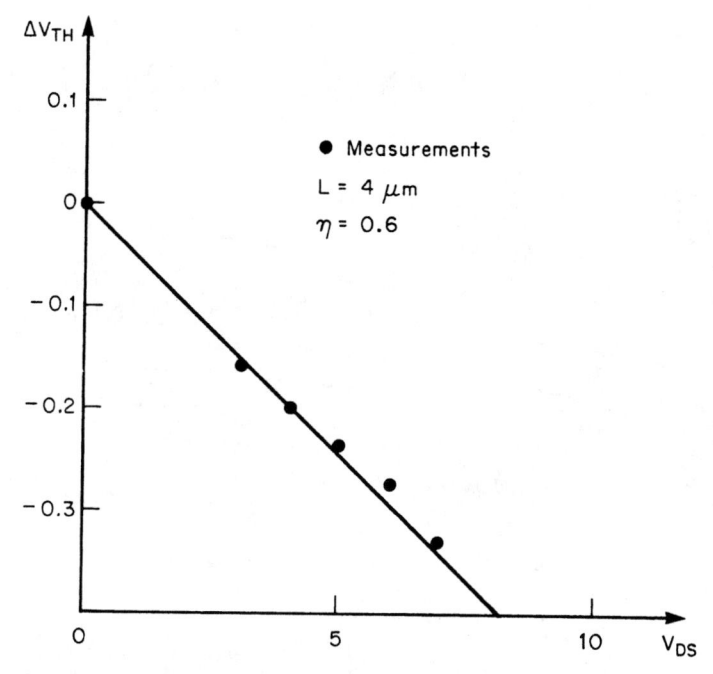

Figure 4-31 Variations of V_{TH} vs. V_{DS}.

between σ and L_{eff} is also empirical [26]:

$$\sigma = \eta \frac{8.15 \times 10^{-22}}{C'_{ox} L^3_{\text{eff}}} \qquad (4\text{-}94)$$

The parameter η (ETA) is a .MODEL statement parameter, and its typical value is 1; its effect on the current I_{DS} is shown in Fig. 4-32.

The term F_B in Eq. (4-90) expresses the dependence of Q_B on the three-dimensional geometry of the MOST. The correction factor of short-channel effect F_s expresses the effect of the short channel and is calculated as

$$F_s = 1 - \frac{X_j}{L_{\text{eff}}} \left(\frac{X_{jl} + W_c}{X_j} \sqrt{1 - \left(\frac{W_p}{X_j + W_p} \right)^2} - \frac{X_{jl}}{X_j} \right) \qquad (4\text{-}95)$$

W_p is the width of the depleted region on the flat source junction (see Fig. 4-33); i.e.,

$$W_p = X_D \sqrt{\phi_j - V_{BS}} \qquad (4\text{-}96)$$

where

$$X_D = \sqrt{\frac{2\varepsilon_s}{qN_A}} \qquad (4\text{-}97)$$

is the coefficient of the depletion region width.

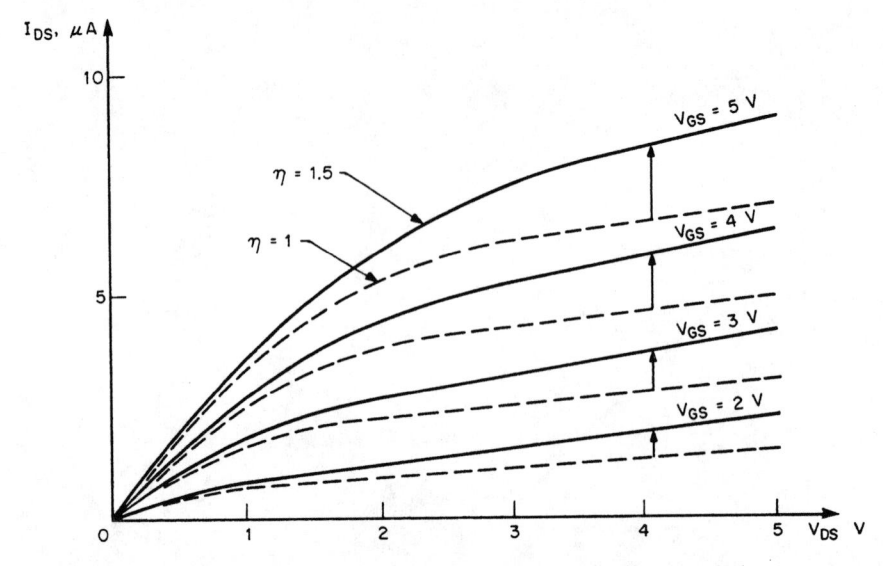

Figure 4-32 Variations of I_{DS} with η, LEVEL3 model: output characteristics.

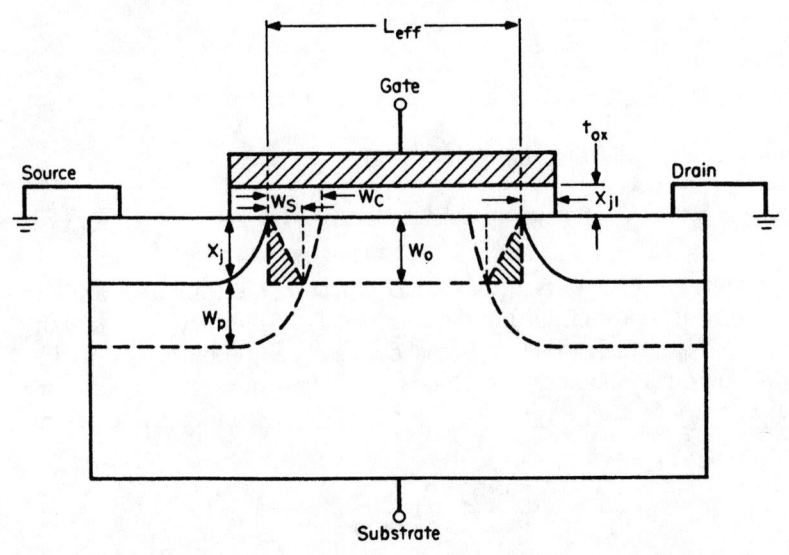

Figure 4-33 Geometrical calculation of the substrate charge Q_B reduced by the depleted regions at the channel edges.

W_c is the width of the depleted cylindrical region of the source-substrate junction; an empirical formula has been used again in this case, i.e.,

$$\frac{W_c}{X_j} = 0.0631353 + 0.8013292 \frac{W_p}{X_j} - 0.01110777 \left(\frac{W_p}{X_j}\right)^2 \quad (4\text{-}98)$$

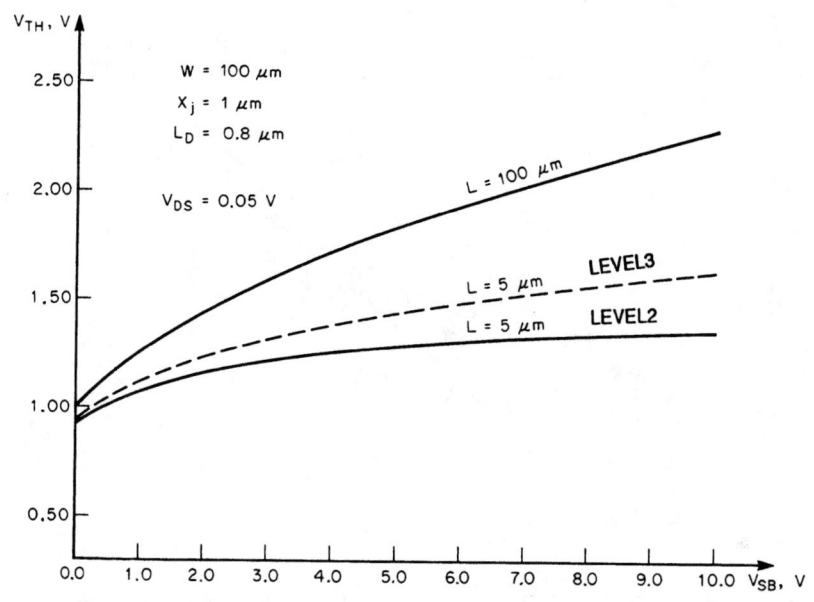

Figure 4-34 Comparison between the models of the body effect with the LEVEL2 and LEVEL3 models.

The junction depth X_j (XJ) and the lateral diffusion X_{jl} (LD) are the parameters for the .MODEL statement to take into account for this effect.

The effects of the parameter X_j on the value of the threshold voltage and on the current are shown in Figs. 4-34 and 4-35.

4.5.3 Effect of channel width on the threshold voltage (narrow-channel)

The correction term F_n in Eq. (4-93) introduces the narrow-channel effect in the model and is calculated with the same equation used in the LEVEL2 model, i.e.,

$$F_n = \frac{\varepsilon_s \delta \pi}{2 C'_{ox} W} \quad (4\text{-}99)$$

where δ (DELTA) is the parameter to be specified in the .MODEL statement.

4.5.4 Effect of channel length on the mobility

The solution chosen for the LEVEL3 model is simpler and more accurate than that of the LEVEL2 model. This model includes a

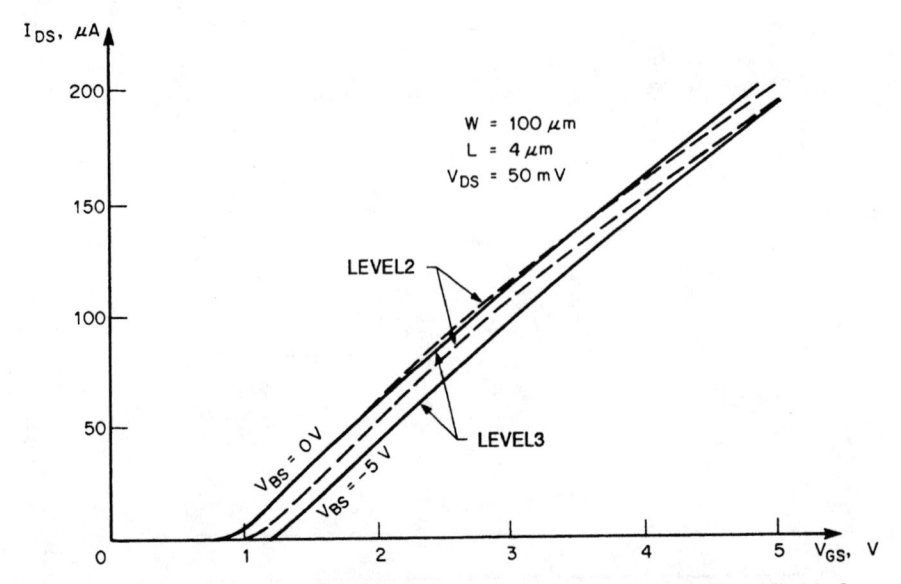

Figure 4-35 Comparison between the transfer characteristics simulated with (a) the LEVEL2 model and (b) the LEVEL3 model. The common parameters are $V_{T0} = 1$ V, $X_j = 1$ μm, $X_{jl} = 0.8$ μm; the parameters of the LEVEL2 model are $\mu_0 = 800$ cm/(V · s), $U_c = 5 \times 10^4$ V/cm, and $U_e = 0.15$; the parameters of the LEVEL3 model are $\mu_0 = 850$ cm (V · s), $\theta = 0.04$ V^{-1}.

decrease in the effective mobility with the average electric field between the source and the drain [23]. Therefore

$$\mu_{\text{eff}} = \frac{\mu_s}{1 + \mu_s V_{DS}/v_{\max} L_{\text{eff}}} \quad (4\text{-}100)$$

where μ_s is the surface mobility calculated from Eq. (4-92). In this model, the calculation of the saturation voltage of the short-channel MOST is modified by the parameter $v_{\max}$; the hypothesis is that the saturation voltage is reached when the carriers reach the limit speed $v_{\max}$. The saturation voltage can be expressed with an equation simpler than that of the LEVEL2 model, i.e.,

$$V_{D,\text{sat}} = V_a + V_b - \sqrt{V_a^2 + V_b^2} \quad (4\text{-}101)$$

where V_a is the saturation voltage if $v_{\max}$ (VMAX) is not included in the .MODEL statement, and V_b modifies $V_{D,\text{sat}}$ if $v_{\max}$ is included. They are expressed as

$$V_a = \frac{V_{GS} - V_{TH}}{1 + F_B} \quad (4\text{-}102)$$

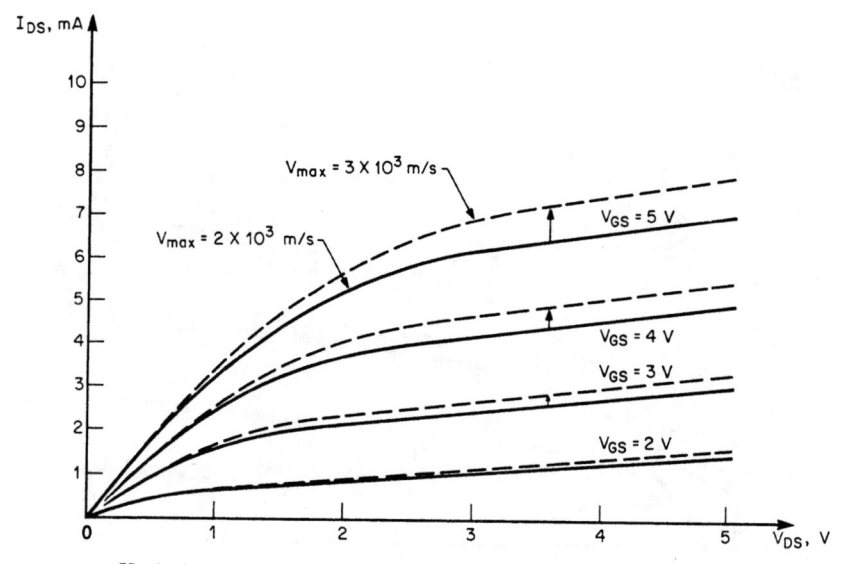

Figure 4-36 Variations of I_{DS} with v_{max}, LEVEL3 model: output characteristics.

$$V_b = \frac{v_{max} L_{eff}}{\mu_s} \quad (4\text{-}103)$$

The use of Eqs. (4-100) and (4-101) together with the model described in the following section allows the continuity of the derivative of the current I_{DS} with respect to V_{DS} at $V_{DS} = V_{D,sat}$ (see Fig. 4-36).

4.5.5 Variation of channel length in the saturation region

The equation of the LEVEL3 model simulates the modulation of the channel length in saturation [18] as

$$L_{eff} - L' = \sqrt{\left(\frac{E_p X_D^2}{2}\right)^2 + K X_D^2 (V_{DS} - V_{D,sat})} - \frac{E_p X_D^2}{2} \quad (4\text{-}104)$$

where

$$E_p = \frac{I_{D,sat}}{g_{D,sat} L_{eff}} \quad (4\text{-}105)$$

is the lateral field at channel pinchoff point.

$I_{D,sat}$ and $g_{D,sat}$ represent the current and the conductance for $V_{DS} = V_{D,sat}$, calculated with the equation derived from the linear region; K (KAPPA) is an empirical fitting parameter for the .MODEL statement, with a typical value of 1, whose effect is shown in Fig. 4-37.

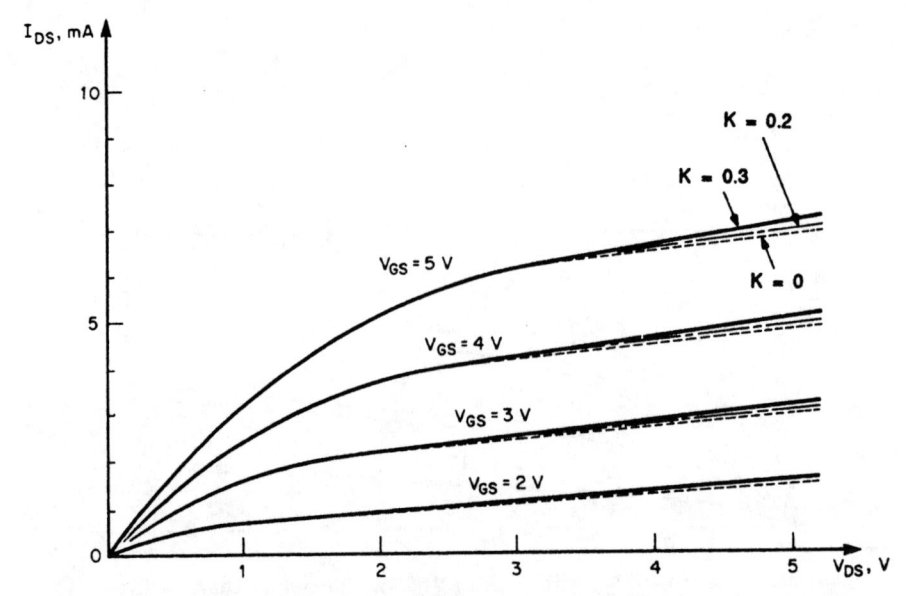

Figure 4-37 Variations of I_{DS} with K, LEVEL3 model: output characteristics.

With Eqs. (4-104) and (4-105), the derivative of I_{DS} vs. V_{DS} for $V_{DS} = V_{D,\text{sat}}$ in the saturation region ($g_{D,\text{sat}}$) should be equal to g_{DS} calculated in the linear region; to obtain this continuity it is necessary to correct Eq. (4-105) according to

$$E_p = \frac{I_{D,\text{sat}} K}{g_{D,\text{sat}} L_{\text{eff}}} \quad (4\text{-}106)$$

4.6 LEVEL3 Large-Signal Model

4.6.1 Model of the gate capacitance

It is possible with the LEVEL3 model to choose between the simple Meyer's model and the charge-controlled Ward's model. Moreover, the equations of the LEVEL3 model allow simpler equations also for Q_G and Q_B. Thus

$$Q_G = C'_{\text{ox}} W L_{\text{eff}} \left(V_{GS} - V_{FB} - 2\phi_p + \sigma V_{DS} - \frac{V_{DS}}{2} + \frac{1+F_B}{12 F_i} V_{DS}^2 \right) \quad (4\text{-}107)$$

where

$$F_i = V_{GS} - V_{TH} - \frac{1+F_B}{2} V_{DS} \quad (4\text{-}108)$$

The charge in the substrate is

$$Q_B = C'_{ox} WL_{eff} \left[\gamma F_s \sqrt{2\phi_p - V_{BS}} + F_n(2\phi_p - V_{BS}) \right.$$
$$\left. + \frac{F_B}{2} V_{DS} - \frac{F_B(1+F_B)}{12F_i} V_{DS}^2 \right] \quad (4\text{-}109)$$

where F_s, F_n, and F_B have been defined in the previous sections.

The charge in the channel is

$$Q_c = -Q_G - Q_B \quad (4\text{-}110)$$

4.7 Comments on the Three Models

At this point, if someone needed to choose the best model for a circuit simulation, it would be useful to reexamine the main differences between the three models.

Usually the LEVEL1 model is not sufficiently precise because the theory is too approximated and the number of fitting parameters too small; its usefulness is in a quick and rough estimate of circuit performances.

The LEVEL2 model can be used with differing complexity by adding the parameters related to the effects needed to simulate with this model. However, if all the parameters are used, this model

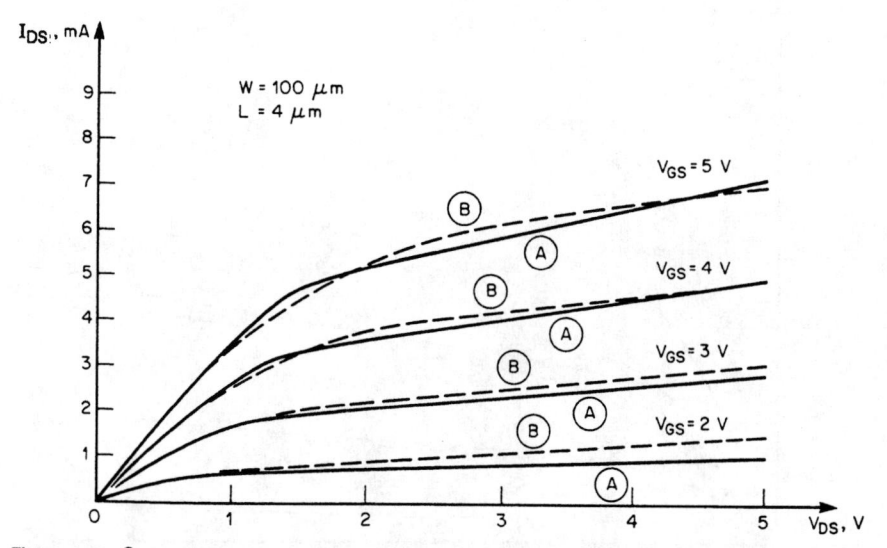

Figure 4-38 Output characteristics calculated with (a) the LEVEL2 model and (b) the LEVEL3 model. The parameters are the same as those specified in Fig. 4-35, with $v_{max} = 5 \times 10^4$ m/s and $K = 0.1$ for the LEVEL3 model.

requires a great amount of CPU (Central Processing Unit) time for the calculations, and it often causes trouble with the convergence of the Newton-Raphson algorithm.

A comparison between the LEVEL2 and LEVEL3 models is interesting. In Fig. 4-38, the LEVEL3 model shows an average quadratic error just less than that obtained with LEVEL2, but the CPU time for every model evaluation is 25 percent less and the iteration number is lower.

In conclusion, the only disadvantage of the LEVEL3 model is the complexity in the calculation of some of its parameters (see Chap. 6). It is best to use the LEVEL3 model if possible, and the LEVEL1 model if great precision is not required.

4.8 The Effect of Series Resistances

The presence of series resistances at the source and drain regions causes a degradation of the electrical characteristics of the MOST, because the effective voltages at the source and the drain are less than those applied at the external terminals (see Fig. 4-39).

In SPICE2 it is possible to insert in the equivalent circuit of the MOST (Fig. 4-24) two resistances r_S and r_D in series to the source and the drain. These resistances are specified in the .MODEL statement, and they are equal for all MOSTs with the same model parameters. It is also possible to specify the resistance for the square of the diffused regions R_{sh}; in this case the value of r_D (RD) and r_S (RS) is

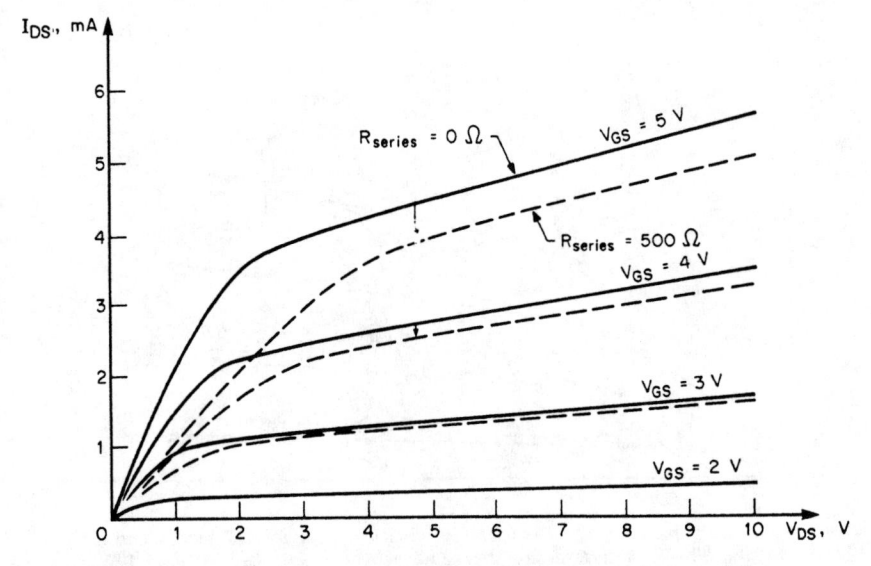

Figure 4-39 The effect of the series resistance on I_{DS}: output characteristics.

calculated for every MOST using the parameters N_{RD} (NRD) and N_{RS} (NRS), which are the number of squares of diffused region in series to the drain and the source.

$$r_D = R_{sh} N_{RD} \qquad (4\text{-}111)$$

$$r_S = R_{sh} N_{RS} \qquad (4\text{-}112)$$

The .MODEL statement keyword is indicated in parentheses.

This method makes it possible to specify different values of series resistance for each MOST, and consequently it gives a better description of the circuit.

These models, however, do not make it possible to describe adequately the resistance contribution of the diffused region near the channel, where the resistance is linearly proportional to W^{-1} [27].

4.9 Small-Signal Models

The *small-signal* linearized model for the MOST is shown in Fig. 4-40. Considering, for simplicity, the LEVEL1 model equations (for

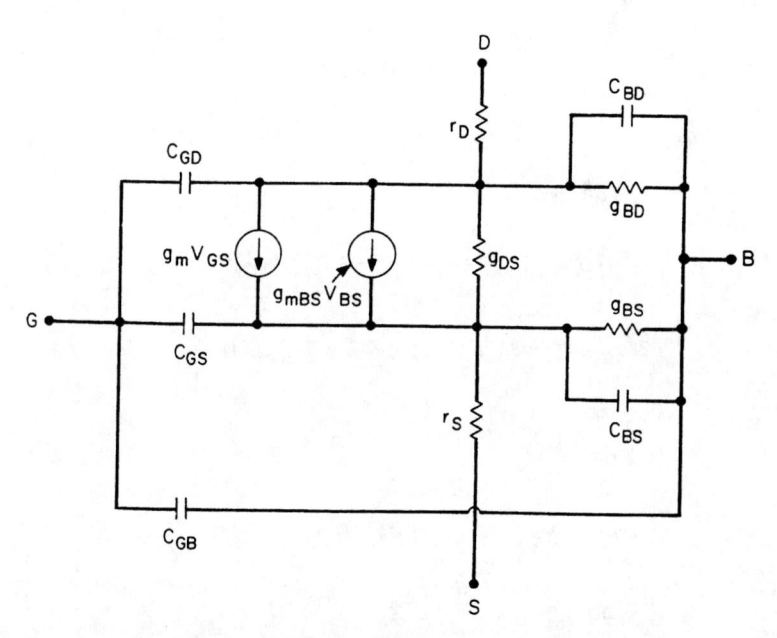

Figure 4-40 Small-signal model of the MOST.

LEVEL2 and LEVEL3 models, analogous equations can be obtained provided the appropriate relations are taken into account), it can be written

$$g_m = \frac{dI_{DS}}{dV_{GS}}\bigg|_{op} = \begin{cases} 0 & \text{for } (V_{GS} - V_{TH}) \leq 0 \\ \beta(V_{GS} - V_{TH})(1 + \lambda V_{DS}) & \text{for } (V_{GS} - V_{TH}) \leq V_{DS} \\ \beta V_{DS}(1 + \lambda V_{DS}) & \text{for } (V_{GS} - V_{TH}) > V_{DS} \end{cases}$$

(4-113)

$$g_{mBS} = \frac{dI_{DS}}{dV_{BS}}\bigg|_{op} = \begin{cases} 0 & \text{for } (V_{GS} - V_{TH}) \leq 0 \\ g_m \text{ ARG} & \text{for } (V_{GS} - V_{TH}) \leq V_{DS} \\ g_m \text{ ARG} & \text{for } (V_{GS} - V_{TH}) > V_{DS} \end{cases}$$

(4-114)

where

$$\text{ARG} = \frac{\gamma}{2\sqrt{2\phi_p - V_{BS}}} \quad \text{for } V_{BS} \leq 0 \tag{4-115a}$$

$$\text{ARG} = \frac{\gamma}{2\left(\sqrt{2\phi_p} - \dfrac{V_{BS}}{2\sqrt{2\phi_p}}\right)} \quad \text{for } V_{BS} > 0 \tag{4-115b}$$

$$g_{DS} = \frac{dI_{DS}}{dV_{DS}}\bigg|_{op} =$$

$$\begin{cases} 0 & \text{for } (V_{GS} - V_{TH}) \leq 0 \\ \dfrac{\beta}{2}\lambda(V_{GS} - V_{TH})^2 & \text{for } (V_{GS} - V_{TH}) \leq V_{DS} \\ \beta(V_{GS} - V_{TH} - V_{DS})(1 + \lambda V_{DS}) \\ \quad + \lambda\beta V_{DS}\left(V_{GS} - V_{TH} - \dfrac{V_{DS}}{2}\right) & \text{for } (V_{GS} - V_{TH}) > V_{DS} \end{cases}$$

(4-116)

$$g_{BS} = \frac{dI_{BS}}{dV_{BS}}\bigg|_{op} = \begin{cases} \dfrac{qI_{SB}}{kT} e^{qV_{BS}/kT} + \text{GMIN} & \text{for } V_{BS} > 0 \\ \dfrac{qI_{SB}}{kT} + \text{GMIN} & \text{for } V_{BS} \leq 0 \end{cases}$$

(4-117)

$$g_{BD} = \frac{dI_{BD}}{dV_{BD}}\bigg|_{op} = \begin{cases} \dfrac{qI_{SD}}{kT} e^{qV_{BD}/kT} + \text{GMIN} & \text{for } V_{DB} > 0 \\ \dfrac{qI_{SD}}{kT} + \text{GMIN} & \text{for } V_{BD} \leq 0 \end{cases}$$

(4-118)

4.10 The Effect of Temperature on the MOST Model Parameters

All input data for SPICE2 is assumed to have been measured at 27°C (300 K). The simulation also assumes a nominal temperature (TNOM) of 27°C that can be changed with the TNOM option of the .OPTIONS control statement. The circuits can be simulated at temperature other than TNOM by using a .TEMP control statement.

Temperature affects some MOST model parameters which are automatically updated in SPICE according to the new temperature.

The temperature dependence of K_p and μ_0 is determined by

$$K_P(T_2) = K_P(T_1) \left(\frac{T_1}{T_2}\right)^{1.5} \tag{4-119}$$

$$\mu_0(T_2) = \mu_0(T_1) \left(\frac{T_1}{T_2}\right)^{1.5} \tag{4-120}$$

The effect of temperature on ϕ_j and $2\phi_p$ (generically indicated with ϕ) is modeled by

$$\phi(T_2) = \frac{T_2}{T_1} \phi(T_1) - \frac{2kT_2}{q} \ln\left(\frac{T_2}{T_1}\right)^{1.5} - \left[\frac{T_2}{T_1} E_g(T_1) - E_g(T_2)\right] \tag{4-121}$$

where

$$E_g(T) = E_g(0) - \frac{\alpha T^2}{\beta + T} \tag{4-122}$$

Experimental data for silicon gives [28] $\alpha = 7.02 \times 10^{-4}$, $\beta = 1108$, $E_g(0) = 1.16$ eV.

The temperature dependence of C_{BD}, C_{BS}, C_j, and C_{jsw} is described by the following relations:

$$C_{BD}(T_2) = C_{BD}(T_1) \left\{1 + m_j \left[400 \times 10^{-6}(T_2 - T_1) - \frac{\phi(T_2) - \phi(T_1)}{\phi(T_1)}\right]\right\} \tag{4-123}$$

$$C_{BS}(T_2) = C_{BS}(T_1) \left\{1 + m_j \left[400 \times 10^{-6}(T_2 - T_1) - \frac{\phi(T_2) - \phi(T_1)}{\phi(T_1)}\right]\right\} \tag{4-124}$$

$$C_j(T_2) = C_j(T_1) \left\{1 + m_j \left[400 \times 10^{-6}(T_2 - T_1) - \frac{\phi(T_2) - \phi(T_1)}{\phi(T_1)}\right]\right\} \tag{4-125}$$

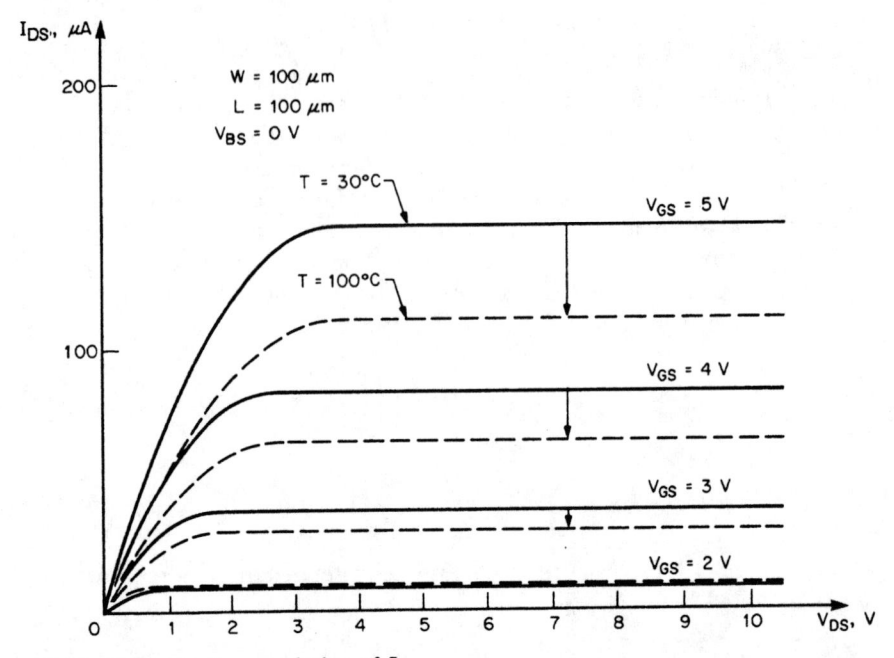

[Figure 4-41: Temperature variation of I_{DS}. Graph showing I_{DS} (μA) vs V_{DS} (V) for W = 100 μm, L = 100 μm, V_{BS} = 0 V, with solid curves at T = 30°C and dashed curves at T = 100°C, for V_{GS} = 5 V, 4 V, 3 V, 2 V.]

$$C_{jsw}(T_2) = C_{jsw}(T_1)\left\{1 + m_{jsw}\left[400 \times 10^{-6}(T_2 - T_1) - \frac{\phi(T_2) - \phi(T_1)}{\phi(T_1)}\right]\right\}$$
(4-126)

The temperature dependence of I_S is given by the relation

$$I_s(T_2) = I_s(T_1)\exp\left[-\frac{q}{kT_2}\left(\frac{T_2}{T_1}E_g(T_1) - E_g(T_2)\right)\right] \quad (4\text{-}127)$$

An analogous dependence holds for J_S.

The variations of the current I_{DS} (see Fig. 4-41), of the threshold voltage (see Fig. 4-42), and of the mobility (see Fig. 4-43) vs. the temperature are shown in the indicated figures.

4.11 BSIM1 Model

The *Berkeley Short-Channel IGFET Model* (*BSIM*) [29–32] is based on the device physics of small-geometry MOSTs.

The special effects included are (1) the vertical field dependence of carrier mobility, (2) the carrier-velocity saturation, (3) the drain-induced barrier lowering, (4) the depletion-charge sharing by the

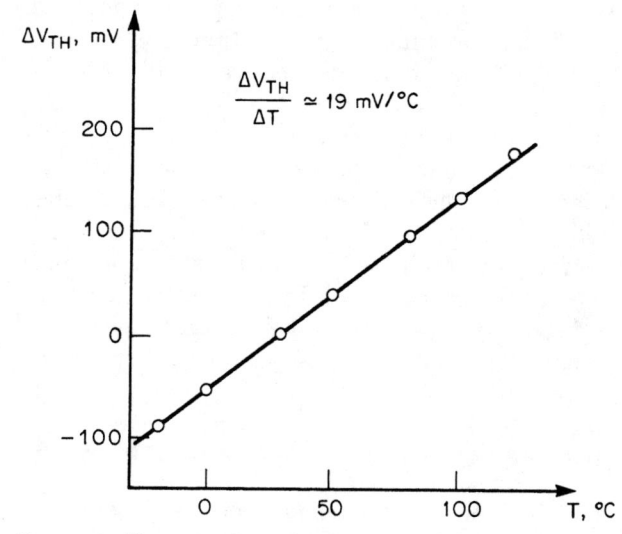

Figure 4-42 Temperature variation of V_{TH}.

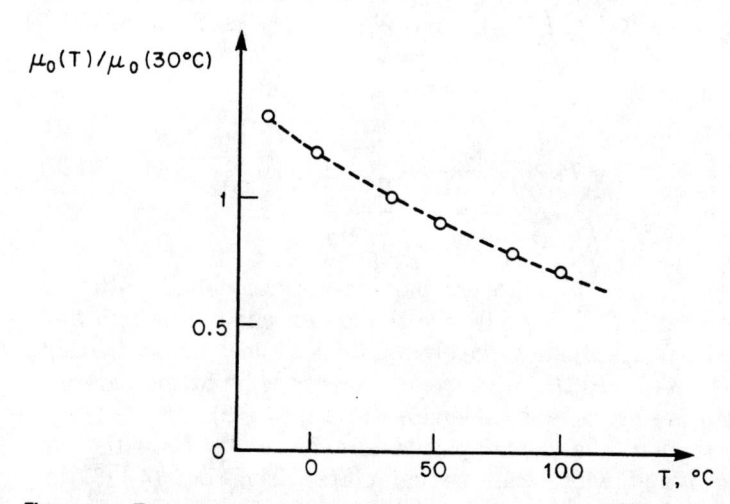

$$\frac{\Delta V_{TH}}{\Delta T} \simeq 19 \text{ mV/°C}$$

Figure 4-43 Temperature variation of μ_0.

source and the drain, (5) the nonuniform doping profile for ion-implanted devices, (6) the channel-length modulation, (7) the subthreshold conduction, and (8) geometric dependencies. In order to speed up the circuit-simulation execution time, the dependence of the drain current on the substrate bias has been modeled with a numerical approximation. The charge model has been derived from its drain-current counterpart to preserve consistency of device physics. Charge conservation is also guaranteed by this model. The model parameters are extracted by an automated parameter-extraction program [33] which generates a process file which contains a set of parameter values for circuit analysis. Circuit designers need only describe the layout geometries of MOSTs and parasitic elements to execute circuit simulation.

4.11.1 Parameter listing

Before describing the BSIM1 model, a summary of the parameters involved in the model and their corresponding SPICE2 keywords is listed as a reference for user's convenience.

The names of the process parameters are defined by appending the letters l and w to the names of the BSIM1 electrical parameters to denote the dependence of that electrical parameter on the device length and width, respectively. In addition, the parameters that account for the sensitivity to V_{BS} are prefixed by X2, and the sensitivity to V_{DS} are prefixed by X3. The names of the process parameters used in the input of SPICE2 are enclosed in parentheses. For example, the electrical parameter V_{FB} has three corresponding process parameters: V_{FB} (VFB), V_{FBl} (LVFB), and V_{FFBw} (WVFB) and the keywords VFB, LVFB, and WVFB are used in SPICE2 input.

The value of an electrical parameter is obtained by

$$P = P_0 + \underbrace{\frac{P_L}{L_i - \text{DL}}}_{L_{\text{eff}}} + \underbrace{\frac{P_W}{W_i - \text{DW}}}_{W_{\text{eff}}} \tag{4-128}$$

where P_0, P_L, P_W are the process parameters associated with the electrical parameter P; L_i and W_i are the drawn channel length and width; DL and DW are the net size changes due to various fabrication steps. The parameters P_0, P_L, and P_W are generated by fitting parameter files with different device dimensions to Eq. (4-128). Once P_0, P_L, and P_W are known, the model parameters for a device with any channel length and width can be calculated from Eq. (4-128) by replacing L_{eff} and W_{eff} with the desired dimensions.

a. Format of the process parameters

	MOST		
Name	L sens. factor	W sens. factor	Units of basic parameter
V_{FB} (VFB)	V_{FBl} (LVFB)	V_{FBw} (WVFB)	V
ϕ_s (PHI)	ϕ_{sl} (LPHI)	ϕ_{sw} (WPHI)	V
K_1 (K1)	K_{1l} (LK1)	K_{1w} (WK1)	$V^{1/2}$
K_2 (K2)	K_{2l} (LK2)	K_{2w} (WK2)	–
η_0 (ETA)	η_{0l} (LETA)	η_{0w} (WETA)	–
μ_Z (MUZ)	δ_l (DL)	δ_w (DW)	$cm^2/V \cdot s \; (\mu m \cdot \mu m)$
U_{0Z} (U0)	U_{0Zl} (LU0)	U_{0Zw} (WU0)	V^{-1}
U_{1Z} (U1)	U_{1Zl} (LU1)	U_{1Zw} (WU1)	$\mu m \, V^{-1}$
μ_{ZB} (X2MZ)	μ_{ZBl} (LX2MZ)	μ_{ZBw} (WX2MZ)	$cm^2/V^2 \cdot s$
η_B (X2E)	η_{Bl} (LX2E)	η_{Bw} (WX2E)	V^{-1}
η_D (X3E)	η_{Dl} (LX3E)	η_{Dw} (WX3E)	V^{-1}
U_{0B} (X2U0)	U_{0Bl} (LX2U0)	U_{0Bw} (WX2U0)	V^{-2}
U_{1B} (X2U1)	U_{1Bl} (LX2U1)	U_{1Bw} (WX2U1)	$\mu m \, V^{-2}$
μ_s (MUS)	μ_{sl} (LMS)	μ_{sw} (WMS)	$cm^2/V^2 \cdot s$
μ_{sB} (X2MS)	μ_{sBl} (LX2MS)	μ_{sBw} (WX2MS)	$cm^2/V^2 \cdot s$
μ_{sD} (X3MS)	μ_{sDl} (LX3MS)	μ_{sDw} (WX3MS)	$cm^2/V^2 \cdot s$
U_{1D} (X3U1)	U_{1Dl} (LX3U1)	U_{1Dw} (WX3U1)	$\mu m \, V^{-2}$
t_{ox} (TOX)	T_{emp} (TEMP)	V_{DD} (VDD)	$\mu m \, (°C \cdot V)$
CGDO	CGSO	CGBO	F/m
XPART	DUM1	DUM2	–
n_0 (N0)	LN0	WN0	–
n_B (NB)	LNB	WNB	–
n_D (ND)	LND	WND	–

	INTERCONNECTS					
R_{sh} (RSH)	C_j (CJ)	C_{jw} (CJW)	I_{js} (IJS)		P_j (PJ)	
P_{jw} (PJW)	M_j (MJ)	M_{jw} (MJW)	W_{df} (WDF)		δ_l (DL)	

b. Names of the process parameters

V_{FB}	Flat-band voltage
ϕ_s	Surface inversion potential
K_1	Body effect coefficient
K_2	Drain/source depletion charge sharing coefficient
η_0	Zero-bias drain-induced barrier lowering coefficient
μ_Z	Zero-bias mobility
U_{0Z}	Zero-bias transverse-field mobility degradation coefficient
U_{1Z}	Zero-bias velocity saturation coefficient
μ_{ZB}	Sensitivity of mobility to substrate bias at $V_{DS} = 0$

η_B	Sensitivity of drain-induced barrier lowering effect to substrate bias
η_D	Sensitivity of drain-induced barrier lowering effect to drain bias, at $V_{DS} = V_{DD}$
U_{0B}	Sensitivity of transverse-field mobility degradation effect to substrate bias
U_{1B}	Sensitivity of velocity saturation effect to substrate bias
μ_s	Mobility at zero substrate bias and at $V_{DS} = V_D$
μ_{sB}	Sensitivity of mobility to substrate bias at $V_{DS} = V_{DD}$
μ_{sD}	Sensitivity of mobility to drain bias at $V_{DS} = V_{DD}$
U_{1D}	Sensitivity of velocity saturation effect to drain bias at $V_{DS} = V_{DD}$
t_{ox}	Gate-oxide thickness
T_{emp}	Temperature at which the process parameters are measured
V_{DD}	Measurement bias range
n_0	Zero-bias subthreshold slope coefficient
n_B	Sensitivity of subthreshold slope to substrate bias
n_D	Sensitivity of subthreshold slope to drain bias
CGDO	Gate-drain overlap capacitance per meter channel width
CGSO	Gate-source overlap capacitance per meter channel width
CGBO	Gate-bulk overlap capacitance per meter channel length
XPART	Gate-oxide capacitance model flag. XPART = 0 selects 40/60 drain/source charge-partitioning in the saturation region, while XPART = 1 and XPART = 0.5 select 0/100 and 50/50 drain/source charge-partitioning in the saturation region, respectively.

c. Names of the process parameters of diffusion, poly, and metal layers

R_{sh}	Sheet resistance/square	Ω/square
C_j	Zero-bias bulk junction bottom capacitance/unit area	F/m^2
C_{jw}	Zero-bias bulk junction sidewall capacitance/unit length	F/m
I_{js}	Bulk junction saturation current/unit area	A/m^2
P_j	Bulk junction bottom potential	V
P_{jw}	Bulk junction sidewall potential	V
M_j	Bulk junction bottom grading coefficient	–
M_{jw}	Bulk junction sidewall grading coefficient	–
W_{df}	Default width of the layer	m
δ_l	Average variation of size due to side etching or mask compensation	m

4.11.2 BSIM1 static model

In this section, the BSIM1 equations for the *static model* are summarized; for the users who are interested, the derivation of these equations is given in the next section. Among the needed BSIM1 parameters, five parameters model the threshold voltage, three parameters model the drain current and one parameter models the subthreshold current.

a. Threshold voltage. The threshold voltage can be expressed as [29–32]

$$V_{TH} = V_{FB} + \phi_s + K_1\sqrt{\phi_s - V_{BS}} - K_2(\phi_s - V_{BS}) - \eta V_{DS} \quad (4\text{-}129)$$

where V_{FB} is the flat-band voltage, ϕ_s is the surface-inversion potential ($\phi_s = 2\phi_p$), K_1 is the body effect coefficient (equivalent to the parameter γ), K_2 is the source-drain depletion-charge-sharing effect coefficient, and η accounts for the drain-induced barrier-lowering effect.

The parameters V_{FB}, ϕ_s, K_1, and K_2 are bias-independent, while η can further be modeled by three parameters.

$$\eta(V_{DS}, V_{BS}) = \eta_0 + \eta_B V_{BS} + \eta_D(V_{DS} - V_{DD}) \quad (4\text{-}130)$$

where V_{DD} is the measurement bias range, and the other parameters are defined in Sec. 4.11.1b.

b. Drain current. In the *cutoff region* ($V_{GS} \leq V_{TH}$), the drain current is

$$I_{DS} = 0 \quad (4\text{-}131)$$

In the *linear region* ($V_{GS} > V_{TH}$ and $0 < V_{DS} < V_{D,\text{sat}}$), the drain current is expressed as

$$I_{DS} = \frac{\mu_0}{[1 + U_0(V_{GS} - V_{TH})]} \frac{C'_{ox}\dfrac{W_{\text{eff}}}{L_{\text{eff}}}}{\left(1 + \dfrac{U_1}{L_{\text{eff}}} V_{DS}\right)} \left[(V_{GS} - V_{TH})V_{DS} - \frac{a}{2} V_{DS}^2\right]$$

$$(4\text{-}132)$$

where

$$a = 1 + \frac{gK_1}{2\sqrt{\phi_s - V_{BS}}} \quad (4\text{-}133)$$

and

$$g = 1 - \frac{1}{1.744 + 0.8364(\phi_s - V_{BS})} \quad (4\text{-}134)$$

In the *saturation region* ($V_{GS} > V_{TH}$ and $V_{DS} \geq V_{D,\text{sat}}$), the drain current is expressed as

$$I_{D,\text{sat}} = \frac{\mu_0}{[1 + U_0(V_{GS} - V_{TH})]} \frac{C'_{\text{ox}} \frac{W_{\text{eff}}}{L_{\text{eff}}}}{2aK} (V_{GS} - V_{TH})^2 \quad (4\text{-}135)$$

where

$$K = \frac{1 + v_c + \sqrt{1 + 2v_c}}{2} \quad (4\text{-}136)$$

and

$$v_c = \frac{U_1}{L_{\text{eff}}} \frac{(V_{GS} - V_{TH})}{a} \quad (4\text{-}137)$$

The above equations are determined by the carrier mobility μ_0 and by the parameters U_0 and U_1 which account for the mobility degradation effects due to the vertical and horizontal electric fields, respectively. The factor a represents the bulk (body) doping effect, while the factor g is the coefficient of average body effect on drain current.

The parameters U_0 and U_1 are bias-dependent and are modeled as follows [29–32]:

$$U_0(V_{DS}, V_{BS}) = U_{0Z} + U_{0B} V_{BS} \quad (4\text{-}138)$$

$$U_1(V_{DS}, V_{BS}) = U_{1Z} + U_{1B} V_{BS} + U_{1D}(V_{DS} - V_{DD}) \quad (4\text{-}139)$$

The parameter μ_0 is obtained by quadratic interpolation through three data points: μ_0 at $V_{DS} = 0$, μ_0 at $V_{DS} = V_{DD}$, and the sensitivity of μ_0 to the drain bias at $V_{DS} = V_{DD}$, with

$$\mu_0(V_{DS} = 0) = \mu_Z + \mu_{ZB} V_{BS} \quad (4\text{-}140)$$

$$\mu_0(V_{DS} = V_{DD}) = \mu_s + \mu_{sB} V_{BS} \quad (4\text{-}141)$$

c. Subthreshold current. Previous SPICE2 MOST models include subthreshold conduction by matching the strong inversion component with the weak inversion component at a transition point close to the threshold voltage. Discontinuity of drain-current derivatives exists, as it has been pointed out by Antognetti et al. [7]. In BSIM1, the total drain current is modeled as the linear sum of a *strong inversion* component $I_{DS,S}$ (as from b above) and a *weak inversion* component $I_{DS,W}$, i.e.,

$$I_{DS,\text{tot}} = I_{DS,S} + I_{DS,W} \quad (4\text{-}142)$$

The weak inversion component can be expressed as

$$I_{DS,W} = \frac{I_{\text{exp}} I_{\text{limit}}}{I_{\text{exp}} + I_{\text{limit}}} \quad (4\text{-}143)$$

where

$$I_{\exp} = \mu_0 C'_{ox} \frac{W_{\text{eff}}}{L_{\text{eff}}} \left(\frac{kT}{q}\right)^2 e^{1.8}\, e^{q(V_{GS} - V_{TH})/nkT}(1 - e^{-qV_{DS}/kT}) \qquad (4\text{-}144)$$

and

$$I_{\text{limit}} = \frac{\mu_0 C'_{ox}}{2} \frac{W_{\text{eff}}}{L_{\text{eff}}} \left(\frac{3kT}{q}\right)^2 \qquad (4\text{-}145)$$

The factor $e^{1.8}$ is empirically chosen to achieve best fits in the subthreshold characteristics with minimum effect on the strong inversion characteristics. The subthreshold parameter n is modeled by

$$n(V_{DS}, V_{BS}) = n_0 + n_B V_{BS} + n_D V_{DS}$$

where n_0, n_B, and n_D are three BSIM1 subthreshold parameters whose meaning can be found in Sec. 4.11.1b.

This approach avoids discontinuity in the drain current expression and first derivatives, and thus does not cause problems of convergence during circuit simulation.

4.11.3 Model derivation

Even if a detailed derivation of the BSIM1 static model equations can be found in Ref. [29], the main steps in obtaining these equations are described in the following discussion.

The relationship between the electric field E_{ox} in the oxide, the gate-substrate voltage V_{GB}, and the potential at the oxide-semiconductor interface ϕ_s can be obtained by summing up the voltages on both sides of the gate and semiconductor. Thus one obtains

$$\phi_{ox} = t_{ox} E_{ox} = V_{GB} - \phi_s - \phi_{MS} \qquad (4\text{-}147)$$

where t_{ox} is the gate oxide thickness, ϕ_{ox} is the potential across the oxide, and ϕ_{MS} is the gate-semiconductor work function difference.

In addition, asuming (a) $t_{ox} \ll L$ which implies that E_{ox} is perpendicular to the gate plane and (b) the horizontal component E_x of the electric field in the semiconductor at the oxide-semiconductor interface is much smaller than the vertical component E_y, by Gauss's law, one obtains

$$\varepsilon_s E_y = \varepsilon_{ox} E_{ox} + Q_{SS} \qquad (4\text{-}148)$$

where ε_s and ε_{ox} are the dielectric constants of the semiconductor and oxide, respectively, and Q_{SS} is the surface charge.

The vertical electric field E_y consists of three components [29], i.e.,

$$\varepsilon_s E_y = Q_c + Q_B + \varepsilon_s E_{dm} \qquad (4\text{-}149)$$

where Q_B is the depletion region charge in the semiconductor given, for the abrupt junction approximation, by Eq. (4-4); Q_c is the channel charge; E_{dm} is the E_y component which originates from the drain (as explained later).

For short-channel MOSTs, Eq. (4-4) must be modified to take into account the influence of the source and drain junction fields. In fact, consider the structure of Fig. 4-44 where the drain, source, and substrate are at the same potential, and a positive voltage is applied to the gate. The depletion charge shown in the dashed area has to balance both the electric field E_{ox} coming in from the oxide-semiconductor interface and E_j due to the source or drain to the substrate junction. Let a fraction of the dashed charge be used to balance E_j. Since this fraction of the charge is not available to balance E_{ox}, one could say that the substrate doping N_A is effectively lowered. Thus it can be shown that Eq. (4-4) may be written as [29]

$$Q_B = f(L_{\text{eff}}, N_A)\sqrt{2q\varepsilon_s N_A \phi_s} = \left(1 - \frac{K_f}{L_{\text{eff}}\sqrt{N_A}}\right)\sqrt{2q\varepsilon_s N_A \phi_s} \qquad (4\text{-}150)$$

where $f(L_{\text{eff}}, N_A)$ represents the effective fraction of the charge available to balance E_{ox}, and K_f is a constant.

Now, let us consider the component E_{dm} of the electric field E_y. When V_{DS} is zero but V_{GB} and V_{BS} are not zero, the substrate is depleted according to the solution of the Poisson equation. When a not zero V_{DS} bias is applied, an additional potential appears across the region already depleted. Since no additional charge appears in the Poisson equation solution for zero drain bias, this additional potential will satisfy the Laplace equation whose solution gives the additional electric fields. These electric fields induce additional chan-

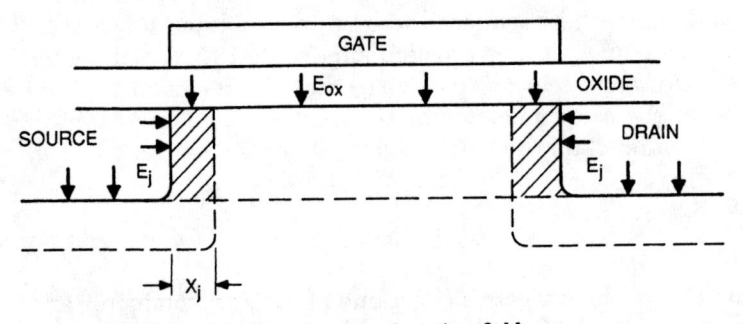

Figure 4-44 Effect of drain and source junction fields.

nel charge which enhances the drain current and results in a finite output conductance in the saturation region. It can be shown [29] that

$$E_{dm} = -\frac{V_{DS}}{\pi L_{\text{eff}}} \quad (4\text{-}151)$$

Taking into account Eqs. (4-147) to (4-151) and assuming to be in the *inversion* condition, the channel charge Q_c can be expressed as [29]

$$Q_c = C'_{\text{ox}}\{[V_{GS} - V_{FB} - \phi_s - V_c(x)] - K_1\sqrt{\phi_s - V_{BS} + V_c(x)}\} - \varepsilon_s E_{dm} \quad (4\text{-}152)$$

where $V_{GS} = (V_{GB} + V_{BS})$ is the gate-substrate voltage, V_{FB} is the flat-band voltage, $V_c(x)$ is the voltage in the channel with respect to the source and varies from zero at $x = 0$ to V_{DS} at $x = L_{\text{eff}}$, V_{BS} is the substrate-source voltage, and K_1 is given by [29]

$$K_1 = f(L_{\text{eff}}, N_A)\frac{\sqrt{2q\varepsilon_s N_A}}{C'_{\text{ox}}} \quad (4\text{-}153)$$

The drain current, assuming it flows parallel to the x-axis, is given by

$$I_{DS}\int_0^{L_{\text{eff}}} dx = \mu_0 W \int_0^{V_{DS}} Q_c\, dV_c \quad (4\text{-}154)$$

By combining Eqs. (4-152) and (4-154) and integrating with respect to V_c in the linear region, one obtains

$$I_{DS} = \beta\{(V_{GS} - \phi_s - V_{FB})V_{DS} + (\eta - \tfrac{1}{2})V_{DS}^2$$
$$- \tfrac{2}{3}K_1[(V_{DS} + \phi_s - V_{BS})^{3/2} - (\phi_s - V_{BS})^{3/2}]\} \quad (4\text{-}155)$$

where $\beta = \mu_0 C'_{\text{ox}} W/L_{\text{eff}}[1 + U_0(V_{GS} - V_{TH})]$ is the conductance coefficient and $\eta = \varepsilon_s/\pi L_{\text{eff}} C'_{\text{ox}}$ is the drain-induced barrier-lowering coefficient.

a. Expansion of the bulk-doping term. In order to speed up the circuit-simulation execution time, the dependence of the drain current on the substrate bias, specified by the function

$$F(V_{DS}, \phi_s - V_{BS}) = \tfrac{2}{3}[(V_{DS} + \phi_s - V_{BS})^{3/2} - (\phi_s - V_{BS})^{3/2}] \quad (4\text{-}156)$$

is expanded in series. But as such an expansion is not valid when $(\phi_s - V_{BS}) \gg V_{DS}$, it is approximated as [29,32]

$$F(V_{DS}, \phi_s - V_{BS}) = \sqrt{\phi_s - V_{BS}}\, V_{DS} + \frac{gV_{DS}^2}{4\sqrt{\phi_s - V_{BS}}} \quad (4\text{-}157)$$

where $g(\phi_s - V_{BS})$ is determined by requiring the Eq. (4-157) to give the best fit to $F(V_{DS}, \phi_s - V_{BS})$ in the desired voltage range. It is found that for $0.7 \leq (\phi_s - V_{BS}) \leq 20.7$ over a range of $0 \leq V_{DS} \leq 10$, the factor g is [29,32]

$$g = 1 - \frac{1}{1.744 + 0.8364(\phi_s - V_{BS})} \quad (4\text{-}158)$$

The drain current in the *linear region* can then be expressed as

$$I_{DS} = \beta \left[(V_{GS} - V_{TH})V_{DS} - \frac{a}{2} V_{DS}^2 \right] \quad (4\text{-}159)$$

where

$$V_{TH} = V_{FB} + \phi_s + K_1\sqrt{\phi_s - V_{BS}} - K_2(\phi_s - V_{BS}) - \eta V_{DS} \quad (4\text{-}160)$$

and

$$a = 1 + \frac{gK_1}{2\sqrt{\phi_s - V_{BS}}} \quad (4\text{-}161)$$

b. Channel-charge modeling. Using Eqs. (4-160) and (4-161) into Eq. (4-152), the channel-charge expression becomes [29,32]

$$Q_c = -C'_{ox}(V_{GS} - V_{TH} - aV_c) \quad (4\text{-}162)$$

If the continuous velocity-saturation characteristics (see Fig. A-1) defined by the relation

$$v_d = \frac{\mu_0 E_x}{\left(1 + \dfrac{E_x}{E_c}\right)} \quad (4\text{-}163)$$

with $v_d \to v_{\text{sat}} \equiv \mu_0 E_c$ when $E_x \to \infty$ and E_c the critical field for carrier velocity saturation is used in the integration of Q_c [see Eq. (4-154)], the drain current expression in the *linear region* becomes

$$I_{DS} = \beta \left[(V_{GS} - V_{TH} - I_{DS}R_{\text{sat}})V_{DS} - \frac{a}{2} V_{DS}^2 \right] \quad (4\text{-}164)$$

where

$$R_{\text{sat}} = \frac{1}{W_{\text{eff}} C'_{ox} v_{\text{sat}}} \quad (4\text{-}165)$$

Equation (4-164) can be rearranged as

$$I_{DS} = \frac{\beta}{\left(1 + \dfrac{U_1}{L_{\text{eff}}} V_{DS}\right)} \left[(V_{GS} - V_{TH})V_{DS} - \frac{a}{2} V_{DS}^2 \right] \quad (4\text{-}166)$$

where

$$\frac{U_1}{L_{\text{eff}}} = \beta R_{\text{sat}} \qquad (4\text{-}167)$$

The saturation voltage $V_{D,\text{sat}}$ is obtained from Eq. (4-162) setting $Q_c = 0$. This condition is not realistic for actual short-channel MOSTs. A more realistic assumption is that at the point in the channel where V_c reaches $V_{D,\text{sat}}$ the channel current is limited by carrier velocity saturation, i.e. [29],

$$Q_c = -\frac{I_{D,\text{sat}}}{W_{\text{eff}} v_{\text{sat}}} = -C'_{\text{ox}}(V_{GS} - V_{TH} - aV_{D,\text{sat}}) \qquad (4\text{-}168)$$

Substitution of Eq. (4-168) into Eq. (4-162) gives an upper limit for the integral in Eq. (4-154), as

$$V_c(x) \to V_{D,\text{sat}} = \frac{1}{a}[(V_{GS} - V_{TH}) - I_{D,\text{sat}} R_{\text{sat}}] \qquad (4\text{-}169)$$

By integrating with this upper limit, the drain current in the *saturation region* is

$$I_{D,\text{sat}} = \beta\left[(V_{GS} - V_{TH} - I_{D,\text{sat}} R_{\text{sat}})V_{D,\text{sat}} - \frac{aV_{D,\text{sat}}^2}{2}\right] \qquad (4\text{-}170)$$

Equation (4-170) can be rearranged as

$$I_{D,\text{sat}} = \frac{\beta}{2a}(V_{GS} - V_{TH} - I_{D,\text{sat}} R_{\text{sat}})^2 \qquad (4\text{-}171)$$

which is quadratic for $I_{D,\text{sat}}$. To make Eq. (4-171) to be compared with the well-known Eq. (4-26), that equation is written as

$$I_{D,\text{sat}} = \frac{\beta}{2aK}(V_{GS} - V_{TH})^2 \qquad (4\text{-}172)$$

where K is obtained by equating Eqs. (4-171) and (4-172), i.e. [29,32],

$$K^2 - K\left[1 + \frac{U_1}{L_{\text{eff}}}\frac{(V_{GS} - V_{TH})}{a}\right] + \left(\frac{U_1}{L_{\text{eff}}}\right)^2\frac{(V_{GS} - V_{TH})^2}{4a^2} = 0 \qquad (4\text{-}173)$$

or

$$K = \frac{1 + v_c + \sqrt{1 + 2v_c}}{2} \qquad (4\text{-}174)$$

where

$$v_c = \frac{U_1}{L_{\text{eff}}}\frac{(V_{GS} - V_{TH})}{a} \qquad (4\text{-}175)$$

If $v_c \ll 1$, then
$$K \to 1 + \frac{U_1}{L_{\text{eff}}} \frac{(V_{GS} - V_{TH})}{a} \tag{4-176}$$

If $v_c \gg 1$, then
$$K \to \frac{U_1}{L_{\text{eff}}} \frac{(V_{GS} - V_{TH})}{2a} \tag{4-177}$$

and
$$I_{D,\text{sat}} \to C'_{\text{ox}} v_{\text{sat}} W_{\text{eff}} (V_{GS} - V_{TH}) \tag{4-178}$$

Equation (4-178) states that in the limit of carrier velocity being saturated at v_{sat}, the saturation current is linear instead of squared with respect to $(V_{GS} - V_{TH})$ and its value is independent of the channel length. The saturation drain voltage can be obtained from (4-169) and (4-172) as [29]

$$V_{D,\text{sat}} = \frac{(V_{GS} - V_{TH})}{a \sqrt{K}} \tag{4-179}$$

If $U_1/L_{\text{eff}}(V_{GS} - V_{TH}) \ll 1$
$$V_{D,\text{sat}} \to \frac{(V_{GS} - V_{TH})}{a} \tag{4-180}$$

If $U_1/L_{\text{eff}}(V_{GS} - V_{TH}) \gg 1$
$$V_{D,\text{sat}} \to \sqrt{\frac{2(V_{GS} - V_{TH})}{a} \frac{L_{\text{eff}}}{U_1}} \tag{4-181}$$

4.11.4 BSIM1 Large-Signal (Charge) Model

The issue of modeling MOST charge storage has attracted integrated-circuit designers' attention since Ward and Dutton [20] pointed out that charge nonconservation is a particular problem for the Meyer model [4] in simulating circuits sensitive to capacitive components of the MOST currents. Although various solutions have been proposed in the literature, there is unanimous agreement that the solution to the charge nonconservation problem is to use a model based on the terminal charges instead on terminal voltages as state variables. The terminal charges Q_G, Q_B, Q_S, and Q_D are the charges associated with the gate, bulk (body), source, and drain, and the capacitive current components of I_G, I_B, I_S, and I_D are, respectively, the derivatives of Q_G, Q_B, Q_S, and Q_D with respect to time.

In addition, the partitioning of channel charge into source and drain components is given significant attention. A number of channel-charge partitioning methods have been attempted. Ward [34] has developed a partitioning method based on the following expressions:

$$Q_S = W_{\text{eff}} \int_0^{L_{\text{eff}}} \left(1 - \frac{x}{L_{\text{eff}}}\right) q_c(x)\, dx \qquad (4\text{-}182)$$

$$Q_D = W_{\text{eff}} \int_0^{L_{\text{eff}}} \frac{x}{L_{\text{eff}}} q_c(x)\, dx = Q_c - Q_S \qquad (4\text{-}183)$$

where $q_c(x)$ is the mobile channel charge per unit length along the channel, x is the position along the channel from 0 to L_{eff} and Q_c is the total charge stored in the channel.

The point of partition, according to this method, is at 3/5 of the channel from the source end. This partitioning method leads to an admittance matrix for the MOST whose capacitance terms are all nonzero and nonreciprocal, and the capacitances between the source and the drain (C_{SD} and C_{DS}) are negative (remember that Meyer's model [4] asserts that MOST capacitances are reciprocal and no ac coupling exists between the source and the drain, i.e., C_{SD} and C_{DS} are zero).

Charge conservation (guaranteed by using terminal charges as state variables) and channel-charge partitioning (which gives non-reciprocal property) characterize the MOST BSIM1 *large-signal model* which, with its static characteristic counterpart, derived from the same considerations of MOST physics, form a unified model previously unavailable.

The charge equations in the different regions are given below [29].

a. **Accumulation region**

$$Q_G = W_{\text{eff}} L_{\text{eff}} C'_{\text{ox}} (V_{GS} - V_{FB} - V_{BS}) \qquad (4\text{-}184)$$

$$Q_B = -Q_G \qquad (4\text{-}185)$$

$$Q_S = 0 \qquad (4\text{-}186)$$

$$Q_D = 0 \qquad (4\text{-}187)$$

b. **Subthreshold region**

$$Q_G = W_{\text{eff}} L_{\text{eff}} C'_{\text{ox}} \frac{K_1^2}{2} \left\{ -1 + \left[1 + \frac{4(V_{GS} - V_{FB} - V_{BS})}{K_1^2} \right]^{1/2} \right\} \qquad (4\text{-}188)$$

$$Q_B = -Q_G \qquad (4\text{-}189)$$

$$Q_S = 0 \qquad (4\text{-}190)$$

$$Q_D = 0 \qquad (4\text{-}191)$$

When a MOST is biased in the linear region, the distributed charge densities of the gate, channel, and bulk can be expressed as [29]

$$q_G(x) = C'_{ox}[V_{GS} - V_{FB} - \phi_s - V_c(x)] \tag{4-192}$$

$$q_c(x) = -C'_{ox}[V_{GS} - V_{TH} - \alpha_x V_c(x)] \tag{4-193}$$

$$q_B(x) = -C'_{ox}[V_{TH} - V_{FB} - \phi_s - (1-\alpha_x)V_c(x)] \tag{4-194}$$

where

$$\alpha_x = a\left[1 + \frac{U_1}{L_{eff}}(V_{GS} - V_{TH})\right] \tag{4-195}$$

The total charge stored in the gate, bulk, and channel regions can be obtained by integrating the distributed charge densities over the area of the active gate region (from $x=0$ to $x=L_{eff}$ and $z=0$ to $z=W_{eff}$). Thus taking into account Eqs. (4-192) to (4-194) and replacing the differential channel length dx with the corresponding differential potential dV, the following expressions for the total gate, bulk, and channel charges are obtained:

$$Q_G = W_{eff}\int_0^{L_{eff}} q_G(x)\,dx = W_{eff}L_{eff}C'_{ox}$$

$$\times \left[V_{GS} - V_{FB} - \phi_s - \frac{V_{DS}}{2} + \frac{V_{DS}}{12}\frac{\alpha_x V_{DS}}{\left(V_{GS} - V_{TH} - \frac{\alpha_x}{2}V_{DS}\right)}\right] \tag{4-196}$$

$$Q_c = W_{eff}\int_0^{L_{eff}} q_c(x)\,dx = -W_{eff}L_{eff}C'_{ox}$$

$$\times \left[V_{GS} - V_{TH} - \frac{\alpha_x}{2}V_{DS} + \frac{\alpha_x V_{DS}}{12}\frac{\alpha_x V_{DS}}{\left(V_{GS} - V_{TH} - \frac{\alpha_x}{2}V_{DS}\right)}\right] \tag{4-197}$$

$$Q_B = W_{eff}\int_0^{L_{eff}} q_B(x)\,dx$$

$$= W_{eff}L_{eff}C'_{ox}\left[-V_{TH} + V_{FB} + \phi_s + \frac{(1-\alpha_x)}{2}V_{DS}\right.$$

$$\left. - \frac{(1-\alpha_x)V_{DS}}{12}\frac{\alpha_x V_{DS}}{\left(V_{GS} - V_{TH} - \frac{\alpha_x}{2}V_{DS}\right)}\right] \tag{4-198}$$

It is necessary to partition the channel charge into the drain component Q_D and the source component Q_S. In BSIM1, the drain/source partitioning of the channel charge changes from 40/60 or (0/100) in the saturation region smoothly to 50/50 in the linear region.

4.11.4.1 BSIM1 40/60 channel-charge partitioning method.

The channel-charge partitioning method proposed by Ward et al. [34] is used. Carrying out the integrations in Eqs. (4-182) and (4-183), with $q_c(x)$ replaced by Eq. (4-193), the total charges associated with the source and the drain are obtained.

a. Linear region. Expressions for the gate and bulk charges are the same as those in Eqs. (4-196) and (4-198), and

$$Q_S = -W_{eff}L_{eff}C'_{ox}\left\{\frac{V_{GS}-V_{TH}}{2}+\frac{\alpha_x V_{DS}}{12}\right.$$

$$\times \frac{\alpha_x V_{DS}}{\left(V_{GS}-V_{TH}-\frac{\alpha_x}{2}V_{DS}\right)} - \frac{\alpha_x V_{DS}}{\left(V_{GS}-V_{TH}-\frac{\alpha_x}{2}V_{DS}\right)^2}$$

$$\left.\times \left[\frac{(V_{GS}-V_{TH})^2}{6}-\frac{\alpha_x V_{DS}(V_{GS}-V_{TH})}{8}+\frac{(\alpha_x V_{DS})^2}{40}\right]\right\} \quad (4\text{-}199)$$

$$Q_D = -W_{eff}L_{eff}C'_{ox}\left\{\frac{V_{GS}-V_{TH}}{2}-\frac{\alpha_x V_{DS}}{2}+\frac{\alpha_x V_{DS}}{\left(V_{GS}-V_{TH}-\frac{\alpha_x}{2}V_{DS}\right)^2}\right.$$

$$\left.\times \left[\frac{(V_{GS}-V_{TH})^2}{6}-\frac{\alpha_x V_{DS}(V_{GS}-V_{TH})}{8}+\frac{(\alpha_x V_{DS})^2}{40}\right]\right\} \quad (4\text{-}200)$$

b. Saturation region

$$Q_G = W_{eff}L_{eff}C'_{ox}\left(V_{GS}-V_{FB}-\phi_s-\frac{V_{GS}-V_{TH}}{3\alpha_x}\right) \quad (4\text{-}201)$$

$$Q_B = W_{eff}L_{eff}C'_{ox}\left[V_{FB}+\phi_s-V_{TH}+\frac{(1-\alpha_x)(V_{GS}-V_{TH})}{3\alpha_x}\right] \quad (4\text{-}202)$$

$$Q_S = -\tfrac{2}{5}W_{eff}L_{eff}C'_{ox}(V_{GS}-V_{TH}) \quad (4\text{-}203)$$

$$Q_D = -\tfrac{4}{15} W_{\text{eff}} L_{\text{eff}} C'_{\text{ox}} (V_{GS} - V_{TH}) \qquad (4\text{-}204)$$

4.11.4.2 BSIM1 0/100 channel-charge partitioning method. The 0/100 channel-charge partitioning method can be derived using the following boundary conditions:

- Expressions for all terminal charges and capacitances are continuous from the saturation region through the linear region.
- The charges Q_S and Q_D are equal and the capacitances C_{SG} and C_{DG} are equal when V_{DS} is zero.
- In the saturation region, all the channel mobile charge is only associated with the source, and Q_D is zero [35].
- For the simplicity of model equations, partitioning is done using the existing terms in the Q_c expression. No additional terms are introduced.

a. Linear region. Expressions for the gate and bulk charges are the same as those in Eqs. (4-196) and (4-198), and

$$Q_S = -W_{\text{eff}} L_{\text{eff}} C'_{\text{ox}} \left[\frac{V_{GS} - V_{TH}}{2} + \frac{\alpha_x V_{DS}}{4} - \frac{(\alpha_x V_{DS})^2}{24\left(V_{GS} - V_{TH} - \frac{\alpha_x V_{DS}}{2}\right)} \right] \qquad (4\text{-}205)$$

$$Q_D = -W_{\text{eff}} L_{\text{eff}} C'_{\text{ox}} \left[\frac{V_{GS} - V_{TH}}{2} - \frac{3\alpha_x V_{DS}}{4} + \frac{(\alpha_x V_{DS})^2}{8\left(V_{GS} - V_{TH} - \frac{\alpha_x V_{DS}}{2}\right)} \right] \qquad (4\text{-}206)$$

b. Saturation region. Expressions for the gate and bulk charges are the same as those in Eqs. (4-201) and (4-202), and

$$Q_S = -\tfrac{2}{3} W_{\text{eff}} L_{\text{eff}} C'_{\text{ox}} (V_{GS} - V_{TH}) \qquad (4\text{-}207)$$

$$Q_D = 0 \qquad (4\text{-}208)$$

4.11.4.3 BSIM1 50/50 channel-charge partitioning method. The charge expressions for the 50/50 partitioning method are listed below.

a. **Linear region.** Expressions for the gate and bulk charges are the same as those in Eqs. (4-196) and (4-198), and

$$Q_S = -\tfrac{1}{2} W_{\text{eff}} L_{\text{eff}} C'_{\text{ox}} \left[V_{GS} - V_{TH} - \frac{\alpha_x V_{DS}}{2} + \frac{(\alpha_x V_{DS})^2}{12\left(V_{GS} - V_{TH} - \dfrac{\alpha_x V_{DS}}{2}\right)} \right]$$

(4-209)

$$Q_D = Q_S \tag{4-210}$$

b. **Saturation region.** Expressions for the gate and bulk charges are the same as those in Eqs. (4-201) and (4-202), and

$$Q_D = -\tfrac{1}{3} W_{\text{eff}} L_{\text{eff}} C'_{\text{ox}} (V_{GS} - V_{TH}) \tag{4-211}$$

$$Q_S = Q_D \tag{4-212}$$

The user of SPICE can select 40/60, 0/100, or 50/50 drain/source charge-partitioning setting the flag XPART to 0, 1, or 0,5 respectively.

4.11.5 BSIM1 Large-Signal (Capacitance) Model

The charge elements can be equivalently represented by capacitances expressed as [34]

$$C_{ij} = \delta_{ij} \frac{\partial Q_i}{\partial V_i}, \quad \text{and} \quad \delta_{ij} = \begin{cases} 1 & \text{if } i=j \\ -1 & \text{if } i \neq j \end{cases} \tag{4-213}$$

where i and j stand for G, B, D, and S. The factors δ_{ij} are defined so that the C_{ij} will normally be positive.

These sixteen MOST capacitances have the following properties. First, if all the terminal voltages are changed by an equal amount, the terminal charges remain the same. It can be expressed as

$$\sum_{j=G,B,D,S} C_{ij} = 0 \quad \text{for} \quad i = G,D,S,B \tag{4-214}$$

Second, charge neutrality implies that the sum of the terminal capacitive currents is zero. Therefore, the capacitances obey the following relations:

$$C_{GG} + C_{BG} + C_{DG} + C_{SG} = 0 \tag{4-215}$$

$$C_{GB} + C_{BB} + C_{DB} + C_{SB} = 0 \tag{4-216}$$

$$C_{GD} + C_{BD} + C_{DD} + C_{SD} = 0 \tag{4-217}$$

$$C_{GS} + C_{BS} + C_{DS} + C_{SS} = 0 \tag{4-218}$$

Nine of the capacitances are independent and must be calculated explicitly. Since 12 capacitances are required in the assembly of nodal admittance matrix, the remaining three may be obtained through Eqs. (4-214) to (4-218).

Expressions for the drain and source junction capacitances are the same as those in Eqs. (4-87) and (4-88) when the value of FC is set to zero.

4.11.6 BSIM1 Small-Signal Model

The BSIM1 *small-signal model* is obtained by the linearization of the static and large-signal model equations described in the previous sections. As our goal is the SPICE model, and as in SPICE the small-signal model is automatically computed without any more parameters specification by the user, we think it is not interesting to give in the following a list of the small-signal model equations. Readers interested in such equations can refer to Appendix C of Ref. [29].

4.12 BSIM2 Model

The BSIM1 model has been shown to be adequate in modeling MOSTs with dimensions limited to $L \simeq 1\,\mu\text{m}$ and $t_{ox} \simeq 15\,\text{nm}$. The channel length modulation effect (or the output resistance) is empirically modeled by the β parameter. This approach is in general valid in modeling the drain current but does not correctly predict the output resistance. In addition, the transition from the subthreshold region to the strong inversion region is given by summing up the currents obtained in both regions. This approach leads to a constant current offset (given by I_{limit} term) in the strong inversion region. When the MOST operates at low currents, this current offset may cause large errors in simulation results. Also, an empirical constant $e^{1.8}$ is used in Eq. (4-144) to account for the slight difference in threshold voltage between strong inversion and subthreshold regions. It has been found that this threshold voltage offset is technology-dependent, varies with process, and is a function of device dimensions.

With these problems of BSIM1 in mind, Jeng [36] developed a BSIM1-based deep-submicrometer model (BSIM2) which has been successfully used to model the drain current and output resistance of MOSTs with gate oxide thicknesses as thin as 3.6 nm and channel length as small as $0.2\,\mu\text{m}$. The drain current equations and their first derivatives are continuous throughout the bias ranges. An improved parameter extraction algorithm and system for BSIM2 were also developed [36].

In the following, the effects (1)–(8) listed in Sec. 4.11, in addition to (9) source/drain parasitic resistance, (10) hot-electron induced output resistance reduction, and (11) inversion-region capacitance effects, are reexamined, based on the BSIM2 model.

4.12.1 BSIM2 Static Model

a. Strong inversion region. The *threshold voltage* expression [Eq. (4-129)] used in BSIM1 which includes (3), (4), and (5) effects is retained in BSIM2. The only change concerns the drain-induced barrier-lowering coefficient η. In fact, while in BSIM1 η is empirically expressed as a linear function of V_{DS} and V_{BS}, in BSIM2 the dependence of η on V_{DS} is removed [term η_D in Eq. (4-130)] because it does not agree with physical principles and may cause negative output resistance at low currents if η_D has a wrong sign [36].

As far as the *velocity saturation* [Eq. (4-163), effect (2)] is concerned, the BSIM2 model uses a new expression for the critical field E_c in order to have a high accuracy and to avoid discontinuity problems encountered in some proposed models.

Thus [36]

$$E_c = E_{c0}\left[1 + \frac{E_{cD}(V_{DS} - V_{D,\text{sat}})^2}{V_{D,\text{sat}}^2}\right] \quad \text{for } V_{DS} \leq V_{D,\text{sat}} \quad (4\text{-}219a)$$

$$E_c = E_{c0} \quad \text{for } V_{DS} > V_{D,\text{sat}} \quad (4\text{-}219b)$$

where E_c starts from a large value of $E_{c0}(1 + E_{cD})$ in the low-field region and smoothly changes to E_{c0} at $V_{DS} = V_{D,\text{sat}}$. In Eq. (4-219a), E_{cD} is a fitting parameter. The quadratic function used in Eq. (4-219a) is to keep the first derivative of the drain current equation continuous at $V_{D,\text{sat}}$.

The vertical field dependence of carrier *mobility* [effect (1)] is expressed in BSIM1 by the relation [see also LEVEL2 Eq. (4-92)]

$$\mu = \frac{\mu_0}{1 + U_0(V_{GS} - V_{TH})} \quad (4\text{-}220)$$

which can be considered as a first-order approximation of the well-known function

$$\mu = \frac{\mu_0}{1 + \left(\dfrac{E_{\text{eff}}}{E_0}\right)^n} \quad (4\text{-}221)$$

where μ_0, E_0, and n are constants, and E_{eff} is the effective vertical field in the inversion region given by

$$E_{\text{eff}} = \frac{2Q_{BU} + Q_I}{2\varepsilon_s} \quad (4\text{-}222)$$

In Eq. (4-222), Q_{BU} and Q_I are the bulk and inversion charge density, respectively. For deep-submicrometer MOSTs with thin gate oxides, the vertical field is too large and Eq. (4-220) does not fit well with measured data. An improvement to Eq. (4-220) consists in adding a second-order term to the denominator of this equation. Thus BSIM2 uses the relation [36]

$$\mu = \frac{\mu_0}{1 + U_0(V_{GS} - V_{TH}) + U_2(V_{GS} - V_{TH})^2} \quad (4\text{-}223)$$

where U_2 is the second-order parameter of vertical field effect, while the other parameters maintain the meaning indicated in the previous sections.

In BSIM2, the effect of the *parasitic resistance* [effect (9)] is included in Eq. (4-223). Therefore, when the mobility parameters are extracted from test devices, the source/drain parasitic resistance effect is automatically incorporated in the model.

BSIM1 *drain current* accounts for the velocity saturation and mobility reduction due to vertical field effects with the product of Eqs. (4-163) and (4-220) in the denominator of Eq. (4-132). Such a relation implies that when V_{GS} increases, the saturation velocity decreases, which does not agree with the observation that v_d is constant at a given temperature, independent of the vertical field [28].

The approach adopted in BSIM2 results in the following drain current expression (for the *linear region*)

$$I_{DS} = \frac{\mu_0 C'_{ox} \dfrac{W_{\text{eff}}}{L_{\text{eff}}} \left(V_{GS} - V_{TH} - \dfrac{aV_{DS}}{2}\right) V_{DS}}{[1 + U_0(V_{GS} - V_{TH}) + U_2(V_{GS} - V_{TH})^2] + \dfrac{U_1}{L_{\text{eff}}} V_{DS}} \quad (4\text{-}224)$$

Referring to Eq. (4-224), it has been shown [36] that if the channel length is longer than 1 μm, either using a product term or a summation term in the denominator, it produces very little difference, as far as accuracy is concerned. However, if the channel length is reduced, velocity saturation effects become more important and simulations obtained from the summation method results more accurate than the multiplication method as used in Eq. (4-132). For consistency with the modification in the denominator of Eq. (4-224), the expression for v_c in Eq. (4-137) should be also changed to

$$v_c = \frac{U_1}{L_{\text{eff}}[1 + U_0(V_{GS} - V_{TH}) + U_2(V_{GS} - V_{TH})^2]} \frac{V_{GS} - V_{TH}}{a} \quad (4\text{-}225)$$

b. Subthreshold region. The drain current in the subthreshold region is dominated by the diffusion current [28]. Using the charge-sheet approximation (i.e., the inversion layer is compressed into a conducting sheet of zero thickness) [37], the subthreshold current of a MOST can be expressed as

$$I_{\text{subth}} = \frac{3}{2}\mu_0 C'_{\text{ox}} \frac{W_{\text{eff}}}{L_{\text{eff}}} \left(\frac{kT}{q}\right)^2 \frac{C_B}{C'_{\text{ox}}} \frac{q\, e^{(\phi_s - 2\phi_p)}}{kT} (1 - e^{qV_{DS}/kT}) \quad (4\text{-}226)$$

where ϕ_p is given by Eq. (4-9) and C_B is the depletion-region capacitance given, through Eq. (B-31), as

$$C_B = -\frac{dQ_B}{d\phi_s} = \sqrt{\frac{q\varepsilon_s N_A}{2\phi_s}} \quad (4\text{-}227)$$

In order to use Eq. (4-226), an explicit relation between ϕ_s and V_{GS} must be known. It is found that ϕ_s in the subthreshold region can be approximated as a linear function of V_{GS} with slope nkT/q and y intercept $-V_0$. Thus [36]

$$\phi_s \simeq \frac{q(V_{GS} - V_{FB})}{nkT} - V_0 \quad (4\text{-}228)$$

where n is the subthreshold swing coefficient given by

$$n = 1 + \frac{C_B}{C'_{\text{ox}}} \quad (4\text{-}229)$$

and

$$V_0 \simeq \frac{2\phi_p C_B}{C_B + C'_{\text{ox}}} \quad (4\text{-}230)$$

Readers interested in a detailed derivation of the above equations are referred to Appendix C of Ref. [36].

Combining Eq. (4-228) and Eq. (4-226), yields [36]

$$I_{\text{subth}} = \mu_0 C'_{\text{ox}} \frac{W_{\text{eff}}}{L_{\text{eff}}} \left(\frac{kT}{q}\right)^2 \frac{q\, e^{(V_{GS} - V_{TH} - V_{\text{offset}})}}{nkT} (1 - e^{qV_{DS}/kT}) \quad (4\text{-}231)$$

where the fitting parameter V_{offset}, which contains all the missing terms in Eq. (4-231) coming from the substitution of Eq. (4-228) into Eq. (4-226), accounts for the threshold voltage offset between strong inversion and subthreshold regions.

4.12.2 BSIM2 Large-Signal Model (inversion-region capacitance)

When the gate oxide thickness is comparable to the depletion-region thickness as in deep-submicrometer MOSTs, the inversion-region capacitance effect is no longer negligible. The effect of the inversion-region capacitance [effect (11)], makes the inversion charge density to deviate from its linear approximation when the gate voltage is near the threshold voltage. This deviation increases as the gate oxide thickness decreases. Although the drain current equations in both the strong inversion and the subthreshold regions can be derived, there is no simple analytical expression for I_{DS} near the threshold voltage, or better in the *transition region* as shown in Fig. 4-45.

To account for the inversion-region capacitance effect, and to define the transition region, BSIM2 uses a cubic spline function of V_{GS} through the definition of a gate capacitance C_{ox}^*, given by

$$C_{ox}^* = \frac{C'_{ox} C_I}{C'_{ox} + C_I} \quad (4\text{-}232)$$

where

$$C_I = C_B \, e^{q(\phi_s - 2\phi_p)/kT} \quad (4\text{-}233)$$

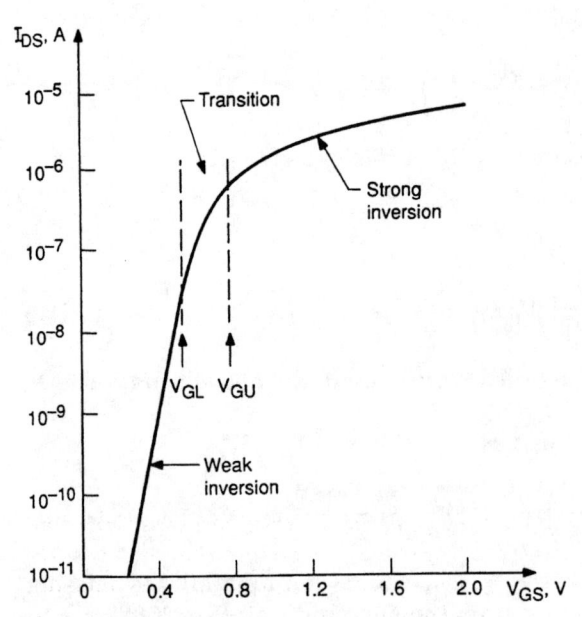

Figure 4-45 Input characteristics of a MOST showing the smooth transition from subthreshold region to the strong-inversion region [36].

is the inversion-region capacitance, and C_B is the depletion region capacitance. Using the same approach used in obtaining Eq. (4-231), C_I can be approximated as [36]

$$C_I \simeq C_B \, e^{q(V_{GS} - V_{TH} - V_{\text{offset}})/nkT} \tag{4-234}$$

In the strong inversion region, C_I is much larger than C'_{ox} therefore, $C^*_{\text{ox}} \simeq C'_{\text{ox}}$ (linear approximation). In the subthreshold region, C'_{ox} is much larger than C_I (diffusion current dominates). By comparing C'_{ox} and C_I along V_{GS}, the lower bound V_{GL} and upper bound V_{GU} of the transition region can be determined. For example, if we define the lower bound to be at which $C'_{\text{ox}} = 100 C_I$, then V_{GL} will be [36]

$$V_{GL} = V_{TH} - V_{\text{offset}} + n\frac{kT}{q} \ln\left(\frac{C'_{\text{ox}}}{100 C_B}\right) \tag{4-235}$$

Similarly, the upper bound, defined at $C_I = 100 C'_{\text{ox}}$, can also be calculated as [36]

$$V_{GU} = V_{TH} + n\frac{kT}{q} \ln\left(\frac{100 C'_{\text{ox}}}{C_B}\right) \tag{4-236}$$

The width of the transition region ($V_{GU} - V_{GL}$) is about 0.2–0.3 V depending on the oxide thickness.

A gate voltage V'_{GS} is introduced in BSIM2 to describe the transition region by using a cubic spline function of V_{GS}. Thus

$$V'_{GS} = C_0 + C_1 V_{GS} + C_2 V_{GS}^2 + C_3 V_{GS}^3 \tag{4-237}$$

where the coefficients C_i's are to be determined from the boundary conditions. The drain current equation used in the transition region is the same as that in strong inversion region except V'_{GS} is used instead of V_{GS}. Equation (4-237) leads to a smooth transition from the substrate region to the strong inversion region. This approach avoids the need to determine whether the linear region or the saturation region drain current equations should be used at the upper bound. Readers interested in a detailed derivation of the coefficients of Eq. (4-237) are referred to Appendix D of Ref. [36].

4.12.3 BSIM2 output resistance model

The *output resistance* (R_{out}) in the saturation region can be attributed to three mechanisms: drain-induced barrier lowering, channel length modulation, and hot-electron-induced output resistance reduction [effect (10)]. In BSIM2 the drain-induced barrier-lowering effect is modeled again by the η parameter, therefore only the other two effects will be considered.

a. Channel-length modulation. Channel-length modulation (see Sec. 4.2.3) is the dominant mechanism affecting R_{out} when V_{DS} is near $V_{D,\text{sat}}$. This effect is implemented in BSIM2 through an empirical combination of a hyperbolic tangent function and a quadratic function given by [36]

$$\beta'_0 = \beta_0 + \beta_1 \tanh\left(\frac{\beta_2 V_{DS}}{V_{D,\text{sat}}}\right) + \beta_3 V_{DS} - \beta_4 V_{DS}^2 \qquad (4\text{-}238)$$

where β_0 is the conductance coefficient extracted at the linear region, and the other coefficients are fitting parameters. Equation (4-238) eliminates the discontinuity problem at $V_{D,\text{sat}}$ encountered in other models.

b. Hot-electron-induced output resistance reduction. As the drain increases beyond $V_{D,\text{sat}}$, electron-hole pairs are generated due to impact ionization. The holes generated are collected by the substrate and originate a substrate current I_{sub} that, when flowing through the substrate, slightly forward biases the source junction with respect to the substrate because of the ohmic voltage drop $V'_{BS} = I_{\text{sub}} R_{\text{sub}}$, R_{sub} being the resistance of the substrate. This positive body bias (for n-type MOST) reduces the threshold voltage causing the drain current to increase and in turn degrades the output resistance.

The resultant drain saturation current $I'_{D,\text{sat}}$ due to the hot-electron effect can be expressed as [36]

$$I'_{D,\text{sat}} = I_{D,\text{sat}} [1 + A_i\, e^{-B_i/(V_{DS} - V_{D,\text{sat}})}] \qquad (4\text{-}239)$$

where A_i and B_i are impact ionization coefficients and $I_{D,\text{sat}}$ is the drain saturation current without the hot-electron effect given by Eq. (4-135). Under normal bias conditions, the second term in the brackets is much smaller than one, hence Eq. (4-239) has very little effect on the magnitude of the calculated drain current, but has significant effect on the output resistance.

4.13 SPICE3 Models

SPICE3 [38,39] provides four basic MOST models, which differ in the formulation of the current-voltage characteristic. The LEVEL (parameter) specifies the model to be used: LEVEL1 defines the Shichman-Hodges model, LEVEL2 defines the geometry-based analytical MOS2 model, LEVEL3 defines the semiempirical short-channel MOS3 model, LEVEL4 defines the BSIM1 model. LEVEL1, LEVEL2, and LEVEL3 MOST models are based directly on the SPICE2 analogous models and have been described in the previous sections.

Thus, briefly, the *static models* of the LEVEL1 through LEVEL3 are defined by the .MODEL statement parameters VTO, KP, LAMBDA, PHI, and GAMMA. These parameters are computed by SPICE3 if process parameters (NSUB, TOX, ...) are given, but user-specified values always override. The parameter VTO is positive (negative) for enhanced mode and negative (positive) for depletion mode n-channel (p-channel) MOSTs. Two ohmic resistances, RD and RS, are also included.

The *large-signal models* are defined by three constant capacitors, (.MODEL statement parameters), CGSO, CGDO, and CGBO which represent overlap capacitances, by the nonlinear thin-oxide capacitances which are distributed among the gate, source, drain, and bulk regions, and by the nonlinear depletion-region capacitances for both substrate junctions divided into bottom and periphery, which vary as the MJ and MJSW power of junction voltage, respectively, and are determined by the .MODEL statement parameters CBD, CBS, CJ, CJSW, MJ, MJSW, and PB.

Charge-storage effects are modeled by the piecewise linear voltage-dependent capacitance model proposed by Meyer [4]. The thin-oxide charge-storage effects are treated slightly different for the LEVEL1 model. These voltage-dependent capacitances are included only if TOX is specified in the .MODEL statement and they are represented using Meyer's model.

There is an overlap among the parameters describing the junctions, e.g., the reverse current can be input either as IS (in A) or as JS (in A/m^2) .MODEL statement parameters. Whereas the first is an absolute value, the second is multiplied by AD and AS to give the reverse current of the drain and source junctions, respectively.

The same approach applies also to the zero-bias junction capacitances CBD and CBS (in F) on one hand, and CJ (in F/m^2) on the other. The parasitic drain and source series resistance can be expressed as either RD and RS (in Ω) or RSH (in Ω/sq), the latter being multiplied by the number of squares NRD and NRS input on the device statement.

The *small-signal models* are automatically computed inside the program by the linearization of the static and large-signal model equations (see Sec. 4.9).

The *temperature* dependence and the updating relations for the parameters follow the ones described in Sec. 4.10.

The values specified for the parameters are assumed to have been measured at the temperature TNOM (default = 27°C), which can be specified in the .OPTIONS statement. This value of temperature can be overridden by the specification of the new (in comparison with SPICE2) .MODEL statement parameter TNOM (in units °C) or of the device element statement option TEMP.

The *noise model* is described in Sec. 7.1.3d.

The equivalent circuits for the static, large-signal, small-signal, and noise models are shown in Figs. 4-6, 4-24, 4-40, and 7-6, respectively.

The BSIM1 model parameters (LEVEL4) are all values obtained from process characterization, and can be generated automatically [33] and converted into a sequence of .MODEL statements suitable for inclusion in a SPICE3 input. Note that unlike the other models in SPICE, the BSIM1 model is designed for use with a process characterization system that provides all the parameters, thus there are no defaults for the parameters.

SPICE3 BSIM1 models are described in Sec. 4.11. The BSIM2 models (see Sec. 4.12) are also implemented in some versions of SPICE3.

4.14 HSPICE Models

HSPICE [40] MOST models consist of public and client private models selected by the .MODEL statement parameter LEVEL. New models are constantly being added by Meta-Software. By dwelling upon the public models, a partial list of the available models includes: LEVEL1 (Shichman-Hodges model), LEVEL2 (MOS2 Grove-Frohman model used in SPICE2G), LEVEL3 (MOS3 empirical model, used in SPICE2G), LEVEL4 (Grove-Frohman model, LEVEL2 model derived from SPICE2E.3), LEVEL5 (Taylor-Huang model, AMI-ASPEC depletion and enhancement), LEVEL13 (BSIM model), and LEVEL27 (SOSFET model).

As each model is described exhaustively in Ref. [40], this section points out only the main characteristics of the above (mostly used) models and, eventually, the most important differences in equations with respect to SPICE2. General notes valid for all the models are first presented.

HSPICE MOST is defined by parameters specified in the .MODEL statement and in the element (device) statement, and two submodels selected by the CAPOP and ACM .MODEL statement parameters.

4.14.1 MOST diode model

The Area Calculation Method (ACM) parameter allows for the precise control of modeling bulk-source and bulk-drain diodes within MOST models. The ACM .MODEL parameter is used to select one of three different modeling schemes for the MOST bulk diodes [40]. If ACM = 0, the *pn* bulk junctions of the MOST are modeled in the SPICE2 style. The ACM = 1 diode model is the original ASPEC model. The ACM = 2 specifies the HSPICE improved diode model, which is based on a model similar to the ASPEC MOST diode model. The ACM = 3 diode model is a further HSPICE improvement that deals with capacitances of shared sources, and drains and gate edge source/drain-bulk periphery capacitance. If the ACM is not set, the diode model defaults to the ACM = 0 SPICE2 model.

The user interested in the model parameters and equations used for the different MOST diode models can see Ref. [40]. For the SPICE2-style diodes the user can also refer to the previous sections of this chapter.

4.14.2 MOST impact ionization

The impact ionization current for MOSTs is available for all LEVELs. The controlling parameters are the impact ionization current coefficient α (ALPHA) and the critical voltage V_{CR} (VCR), in addition to the parameters which take account the length and width sensitivity of the controlling parameters. Another parameter, I_{RAT} (IIRAT) controls the splitting of the impact ionization current between source and bulk. The current due to impact ionization effect is then calculated by the expression

$$I_{\text{impact}} = \alpha_{\text{eff}} I_{DS}(V_{DS} - V_{D,\text{sat}}) e^{-V_{CR,\text{eff}}/(V_{DS} - V_{D,\text{sat}})} \qquad (4\text{-}240)$$

which, multiplied by I_{RAT} and summed up to I_{DS}, gives the effective I_{DS} current.

4.14.3 MOST capacitance models

The HSPICE algorithm used for calculating nonlinear, voltage-variable MOST capacitance depends on the value of the .MODEL statement parameter CAPOP. The selection of the CAPOP model depends on the MOST model LEVEL being used. Specific CAPOP models can only be used with certain MOST model LEVELs [40]. The selected CAPOP model affects the accuracy of the simulation results and the simulation time. HSPICE offers nine gate capacitance models [40]: CAPOP = 0 (selects the SPICE2G.5 Meyer model); CAPOP = 1 (MSINC, modified Meyer model); CAPOP = 2 (selects the ASPEC modified MSINC Meyer model); CAPOP = 3 (uses the same set of equations and parameters as the CAPOP = 2 model, but the charges are obtained by Simpson numeric integration instead of the box integration found in CAPOP models 1 and 2); CAPOP = 4 (selects a modified charge conservation model based on the Ward-Dutton model not implemented correctly in SPICE2G.6. The channel-charge partitioning between drain and source is also taken into account); CAPOP = 5 (gate capacitances are not calculated); CAPOP = 6 (selects the AMI model); CAPOP = 7 (selects SOSFET model); CAPOP = 9 (selects an improved CAPOP = 4 model for LEVEL3 model).

4.14.4 Temperature dependence of the model parameters

There are several sets of temperature equations for the HSPICE MOST model parameters that may be selected by setting the temperature equation selectors TLEV and TLEVC in the .MODEL statement as explained in Sec. 3.6.4. The HSPICE user can find these equations in Ref. [40].

4.14.5 HSPICE LEVELx models

In this section a quick survey on the most common HSPICE MOST models is presented.

1. *LEVEL1* model (Shichman-Hodges model [2]). It uses the same set of equations and parameters as the SPICE2 MOST LEVEL1 model which has been described in Sec. 4.2. The only difference consists in the definition of the effective channel length and width which in HSPICE is

$$L_{eff} = L \cdot LMLT + XL - 2(LD + DEL)$$
$$W_{eff} = M(W \cdot WMLT + XW - 2WD)$$
(4-241)

where LMLT and WMLT are, respectively, the length and width shrink factors; XL and XW account for masking and etching effects; DEL is the channel-length reduction on each side; LD and WD are the lateral diffusions into the channel from source and drain diffusions or bulk along the width; M is the multiple device option which multiplies the channel width. The above parameters, except for LMLT and WMLT, can be scaled according to the rules which are a function of the parameter's units [40].

2. *LEVEL2* model (Grove-Frohman model). It is based on the SPICE2 MOST LEVEL2 model (see Sec. 4.3). In addition, HSPICE provides a mobility equation selector (MOB). If MOB = 0 (default), Eq. (4-45) is used; otherwise if MOB is set to 7, HSPICE invokes the channel-length modulation and mobility equations of MOST LEVEL3 model. Another .MODEL statement parameter, WIC, operates as a subthreshold current model selector. Thus if WIC = 0 (default) the SPICE2 LEVEL2 model is invoked; otherwise if WIC = 3, the subthreshold current is calculated differently [40]. As usual, the effective channel length and width in HSPICE are calculated from the drawn length and width as from Eq. (4-241) and the parameters involved follow the rules that control the parameter's units [40].

3. *LEVEL3* model (empirical model). It uses the same equations and parameters as the SPICE2 MOST LEVEL3 model which has been described in Sec. 4.5. The subthreshold current model selector WIC is also present as in HSPICE LEVEL2 model. The same considerations on scaling and effective measurements of the parameters hold as in LEVEL2.

4. *LEVEL4* model (modified Grove-Frohman model). It is the same as the LEVEL2 model with the following main exceptions: (a) no narrow width effects ($\eta = 1$), (b) no short-channel effects (γ = GAMMA), (c) TPG parameter defaults to zero, i.e., Al gate.

5. *LEVEL5* model (Taylor-Huang model [41]). This model, which uses micrometer units rather than the usual meter units, has been expanded to include two models of operation: enhancement and depletion. These two modes of operation are accessed by the .MODEL parameter ZENH (ZENH = 1 selects enhancement model, default; ZENH = 0 selects depletion model). The HSPICE enhancement and depletion models are basically identical to AMI models; however, certain aspects have been revised to enhance performance. The depletion mode MOST model was formulated assuming a silicon gate construction with an ion implant used to obtain the depletion characteristics. This implanted layer also causes the formation of an additional channel, offering a conductive pathway through the bulk silicon, as well as through the surface channel.

The physical model for a depletion MOST is basically the same as an enhancement model, except that the depletion implant is approximated by a one-step profile with a depth DP. Due to the implant profile, the drain current equation must be calculated by region. MOST model LEVEL5 has three regions; depletion, enhancement, and partial enhancement.

6. *LEVEL13* model (BSIM model [29]). The HSPICE LEVEL13 MOST model is an adaptation of BSIM model from SPICE2G.6 and described in Sec. 4.11.

7. *LEVEL27* model (SOSFET model). Three-terminal Silicon-On-Sapphire (SOS) FET transistor models have been integrated into HSPICE. This SOSFET model is based on a sapphire insulator which isolates the substrate and is able to model the behavior of SOS devices more accurately than standard MOST models with physically unreal parameter values. The SOSFET model also includes a charge conservation model based upon that of Ward and Dutton [20]. Because the defaults of the SOSFET model parameters are channel-length-dependent, the model parameter SOSLEV must be specified to select either the 5 μm or 3 μm processing model.

The parameter ACM is not available for this model because no junction diodes are included.

4.15 PSPICE Models

PSPICE [42–44] provides four MOST models which differ in the formulation of the current-voltage characteristic. The LEVEL parameter selects the model to be used: LEVEL1 defines the Shichman-Hodges model, LEVEL2 defines the geometry-based analytical MOS2 model, LEVEL3 defines the semiempirical short-channel MOS3 model, and LEVEL4 defines the BSIM1 model. LEVEL1, LEVEL2, and LEVEL3 MOST models are based directly on the SPICE2 analogous models and have been developed in the previous sections, while the LEVEL4 MOST model is based on the BSIM1 model, described in Sec. 4.11.

Thus in the following discussion, only the differences encountered in PSPICE MOST models with respect to SPICE2 analogous models are described. As the BSIM model is the same as that described in Sec. 4.11, it is not taken into account.

4.15.1 Static model

The PSPICE MOST *static model* is represented by the equivalent circuit of Fig. 4-6, provided that a shunt resistance r_{DS} in parallel with the drain-source channel, and the gate and bulk series resistances (r_G and r_B) are considered. In conjunction with the gate and bulk resistances, as in the other versions of SPICE, PSPICE introduces the relative resistivities of the gate and bulk in squares (NRG and NRB) which multiply the diffusion sheet resistance RSH to obtain the absolute values r_G and r_B.

The equations that define the nonlinear current source I_{DS} in SPICE2 are also valid in PSPICE. However, Eqs. (4-31) and (4-32) which account for the two junction diodes (see Fig. 4-6) are modified in the term kT/q (thermal voltage) which is multiplied by the bulk emission coefficient n.

In addition, Eqs. (4-37) and (4-38) which define the drain and source junction saturation currents are modified in PSPICE as follows:

$$I_{DS} = A_D J_S + P_D J_{SSW} \qquad (4\text{-}242)$$

$$I_{SS} = A_S J_S + P_S J_{SSW} \qquad (4\text{-}243)$$

where J_{SSW} is a new parameter that represents the bulk *pn* saturation sidewall current/length.

From the above equations it can be seen that PSPICE introduces, with respect to SPICE2, new parameters for the .MODEL statement, i.e., N(n), the bulk *pn* emission coefficient (default = 1); RG(r_G), the gate ohmic resistance (default = 0 Ω), RB(r_B), the bulk ohmic resistance (default = 0 Ω); RDS(r_{DS}), the drain-source shunt resistance (default = ∞Ω); JSSW(J_{SSW}), the bulk pn saturation sidewall current/length (default = 0 A/m). Another .MODEL parameter, not present in SPICE2, is WD(W_D), the lateral diffusion width (default = 0 m) which is used to evaluate the effective device width. The other parameters maintain their meaning and keywords encountered in SPICE2. The keyname used in the text is indicated in parentheses.

4.15.2 Large-signal model

The PSPICE MOST *large-signal model* is based directly on the SPICE2 MOST large-signal model and it is represented by the equivalent circuit of Fig. 4-24. However, Eqs. (4-87) and (4-88) which account for the bulk-source capacitance C_{BS} and the bulk-drain capacitance C_{BD}, are modified by adding a transit-time capacitance term $\tau_D g_{BS}$ and $\tau_D g_{BD}$, being $g_{BS} = dI_{BS}/dV_{BS}$ and $g_{BD} = dI_{BD}/dV_{BD}$, respectively.

The effect of this new term is taken into account by PSPICE by specifying, in the .MODEL statement, the bulk pn transit time TT (τ_D) (default = 0 s).

4.15.3 Small-signal model

As in SPICE2, the PSPICE *small-signal model* is automatically computed inside the program by the linearization of the static and large-signal model equations. Thus the user of PSPICE has to specify no new parameters in the .MODEL statement.

4.15.4 Temperature and area factor effects on the model parameters

PSPICE parameters have the same temperature dependence as in SPICE2. In addition, PSPICE provides a relation for J_{SSW} which is equal in form to Eq. (4-127).

As far as the area effect on the model parameters are concerned, PSPICE provides a multiplier parameter M (default = 1) which simulates the effect of multiple MOSTs in parallel. This parameter is specified in the element statement (keyword M, see Ref. [42]). The effective width, overlap and junction capacitances, and junction currents of the MOST are multiplied by M; the resistances are divided by M.

References

1. L. W. Nagel, SPICE2: A Computer Program to Simulate Semiconductor Circuits, Electronics Research Laboratory, Rep. No ERL-M520, University of California, Berkeley, 1975.

2. H. Shichman and D. A. Hodges, Modeling and Simulation of Insulated-Gate Field-Effect Transistor Switching Circuits, *IEEE J. Solid-State Circuits*, **SC-3**, 1968.
3. A. Vladimirescu, K. Zhang, A. R. Newton, D. O. Pederson, and A. Sangiovanni-Vincentelli, *SPICE Version 2G User's Guide*, Dept. EECS, University of California, Berkeley, 1981.
4. J. E. Meyer, MOS Models and Circuit Simulation, *RCA Rev.*, **32**, 1971.
5. D. Frohman-Bentchkowsky and A. S. Grove, On the Effect of Mobility Variation on MOS Device Characteristics, *Proc. IEEE*, **56**, 1968.
6. R. M. Swanson and J. D. Meindl, Ion-Implanted Complementary MOS Transistor in Low-Voltage Circuits, *IEEE J. Solid-State Circuits*, **14**, 1979.
7. P. Antognetti, D. Caviglia, and E. Profumo, CAD Model for Threshold and Sub-threshold Conduction in MOSFETs, *IEEE J. Solid-State Circuits*, **17**, 1982.
8. D. Frohman-Bentchkowsky and A. S. Grove, Conductance of MOS Transistors in Saturation, *IEEE Trans. Electron Devices*, **ED-16**, 1969.
9. L. D. Yau, A Simple Theory to Predict the Threshold Voltage of Short-Channel IGFETs, *Solid-State Electron.*, **17**, 1974.
10. L. A. Akers, An Analytical Expression for the Threshold Voltage of a Small Geometry MOSFET, *Solid-State Electron.*, **24**, 1981.
11. O. Jantsch, A Geometrical Model of the Threshold of Short and Narrow Channel MOSFETs, *Solid-State Electron.*, **25**, 1982.
12. W. R. Bandy and D. P. Kokalis, A Simple Approach for Accurately Modeling the Threshold Voltage of Short-Channel MOSTs, *Solid-State Electron.*, **20**, 1977.
13. R. R. Troutman and A. G. Fortino, Simple Model for Threshold Voltage in a Short-Channel IGFET, *IEEE Trans. Electron Devices*, **10**, 1977.
14. Y. Omura and K. Ohwada, Threshold Voltage Theory for a Short-Channel MOSFET Using a Surface-Potential Distribution Model, *Solid-State Electron.*, **22**, 1979.
15. Y. Ohno, Short-Channel MOSFET $V_{TH} - V_{DS}$ Characteristics Model Based on a Point Charge and Its Mirror Images, *IEEE Trans. Electron Devices*, **2**, 1982.
16. S. Selberherr, A. Schutz, and H. W. Potzl, MINIMOS—A Two-Dimensional MOS Transistor Analyzer, *IEEE Trans. Electron Devices*, **ED-27**, 1980.
17. J. A. Greenfield and R. W. Dutton, Nonplanar VLSI Device Analysis Using the Solution of Poisson's Equation, *IEEE Trans. Electron Devices*, **ED-27**, 1980.
18. G. Baum and H. Beneking, Drift Velocity Saturation in MOS Transistors, *IEEE Trans. Electron Devices*, **17**, 1970.
19. K. O. Jeppson, Influence of the Channel Width on the Threshold Voltage Modulation in MOSFETs, *Electron. Lett.*, **EDL-11**, 1975.
20. D. E. Ward and R. W. Dutton, A Charge-Oriented Model for MOS Transistors Capacitances, *IEEE J. Solid-State Circuits*, **SC-13**, 1978.
21. S. Y. Oh, D. E. Ward, and R. W. Dutton, Transient Analysis of MOS Transistors, *IEEE Trans. Electron Devices*, **ED-27**, 1980.
22. L. M. Dang, A Simple Current Model for Short Channel IGFET and Its Application to Circuit Simulation, *IEEE J. Solid-State Circuits*, **14**, 1979.
23. G. Merkel, J. Borel, and N. Z. Cupcea, An Accurate Large Signal MOS Transistor Model for Use in Computer-Aided Design, *IEEE Trans. Electron Devices*, **ED-19**, 1972.
24. K. Y Fu, Mobility Degradation Due to the Gate Field in the Inversion Layer of MOSFET's *Electron Lett.*, **EDL-3**, 1982.
25. R. R. Troutman, VLSI Limitations from Drain-Induced Barrier Lowering, *IEEE J. Solid-State Circuits*, **SC-14**, 1979.
26. H. Masuda, M. Nakai, and M. Kubo, Characteristics and Limitations of Scaled-Down MOSFET's Due to Two-Dimensional Field-Effects, *IEEE Trans. Electron Devices*, **ED-14**, 1979.
27. P. Antognetti, C. Lombardi, and D. Antoniadis, Use of Process and 2-D MOS Simulation in the Study of Doping Profile Influence on S/D Resistance in Short Channel MOSFET's, *IEDM*, Washington D.C., 1981.
28. S. M. Sze, *Physics of Semiconductor Devices*, Wiley, New York, 1969.

29. B. J. Sheu, MOS Transistor Modeling and Characterization for Circuit Simulation, Electronics Research Laboratory, Rep. No. ERL-M85/85, University of California, Berkeley, 1985.
30. M.-C. Jeng, B. J. Sheu, and P. K. Ko, BSIM Parameter Extraction, Algorithms and User's Guide, Electronics Research Laboratory, Rep. No. ERL-M85/79, University of California, Berkeley, 1985.
31. B. J. Sheu, D. L. Scharfetter, and P. K. Ko, SPICE2 Implementation of BSIM, Electronics Research Laboratory, Rep. No. ERL-M85/42, University of California, Berkeley, 1985.
32. B. J. Sheu, D. L. Scharfetter, P. K. Ko, and M.-C. Jeng, BSIM: Berkeley Short-Channel IGFET Model for MOS Transistors, *IEEE J. Solid-State Circuits*, SC-22, 1987.
33. J. R. Pierret, A MOS Parameter Extraction Program for the BSIM Model, Electronics Research Laboratory, Rep. No. ERL-M84/99, University of California, Berkeley, 1984.
34. D. E. Ward, Charge-Based Modeling of Capacitance in MOS Transistor, Ph.D. Dissertation, Rep. G201-11, Integrated Circuit Laboratory, Stanford University, Stanford, California, 1981.
35. P. Yang, B. D. Epler, and P. K. Chatterjee, An Investigation of the Charge Conservation Problem for MOSFET Circuit Simulation, *IEEE J. Solid-State Circuits*, SC-18, 1983.
36. M.-C. Jeng, Design and Modeling of Deep-Submicrometer MOSFETS, Electronics Research Laboratory, Rep. No. ERL-M90/90, University of California, Berkeley, 1990.
37. J. R. Brews, A Charge-Sheet Model of the MOSFET, *Solid-State Electronics*, 21, 1978.
38. T. J. Quarles, SPICE3 Version 3C1 User's Guide, Electronics Research Laboratory, Rep. No. ERL-M46, University of California, Berkeley, 1989.
39. T. J. Quarles, The SPICE3 Implementation Guide, Electronics Research Laboratory, Rep. No. ERL-M44, University of California, Berkeley, 1989.
40. Meta-Software Inc., *HSPICE User's Manual*, Campbell, Calif., 1991.
41. J. S. Huang and G. W. Taylor, Modeling of an Ion-implanted Silicon Gate Depletion-Mode IGFET, *IEEE Trans. Electron Devices*, ED-22, 1975.
42. MicroSim Corporation, *PSPICE User's Manual*, Irvine, Calif., 1989.
43. P. W. Tuinenga, *SPICE: A Guide to Circuit Simulation and Analysis Using PSPICE*, Prentice-Hall, Englewood Cliffs, N.J., 1988.
44. M. H. Rashid, *SPICE for Circuit and Electronics Using PSPICE*, Prentice-Hall, Englewood Cliffs, N.J., 1990.

Chapter 5

BJT Parameter Measurements

Accurate circuit simulation is possible only by specifying accurate and meaningful parameter values for each model. This chapter discusses how the BJT model parameters for the .MODEL statement of SPICE2 can be can be extracted from the measurement data.

The described measurement techniques are by no means meant to encompass all possible ones; there are other techniques, some of which may be just as good or even better.

5.1 Input and Model Parameters

To completely specify a transistor model, a program such as SPICE2, requires three types of information [1]: (a) fundamental physical constants, (b) operating conditions, and (c) model parameters.

The *fundamental physical constants*, such as Boltzmann's constant k and electronic charge q, are normally defined inside the program.

The *operating conditions* define the circumstances under which the model equations are to be used. In a nodal analysis program, for example, the operating conditions are normally the transistor's bias voltages, say V_{BE} and V_{BC}. These bias voltages are determined internally as the computer iterates to the solution. That is, the program assumes a set of bias voltages, solves the equations, and

---

Note: The material contained in this chapter is based in part upon the book *Modeling the Bipolar Transistor*, by Ian E. Getreu, copyright © Tektronix, Inc., 1976. All rights reserved. Reproduced with permission.

then selects new and better values until it converges to an adequate solution; this is all done internally.

In the previous chapters, it has been assumed that the operating conditions consist of not only the bias voltages but also the temperature T at which the analysis is to be performed. The value of T is normally required as input, and it will be assumed throughout that T has been specified.

The third type of information required is the set of *model parameters* for each different device in the circuit. The measurement of the model parameters is the subject of this chapter.

The values of the model parameters must be supplied by the user in a manner predetermined by the program. Some programs are very flexible and allow some model parameters to be specified indirectly. For example, τ_F (the total transit time in the normal active region) is normally determined through the measurement of f_T (the unity-gain bandwidth).

In some programs, the user can specify either τ_F or the f_T value at a bias condition. If the f_T value is provided, the program determines the value of τ_F internally, taking into account the effects of junction capacitances. It must be noted that in this particular case, τ_F is the model parameter, yet f_T is the input. Therefore, a distinction must be made between the *model parameters* and the program's *input parameters* [1]. This distinction is illustrated in Fig. 5-1.

The model parameters are those parameters used in the model equations to describe the device for a given set of operating conditions. The input parameters are the data required by the program to specify the model parameters. Some or all of the input parameters may be model parameters, depending on the program.

Tables 5-1 and 5-2 list the parameters (which may be specified in the .MODEL statement) required to model the BJT. Included in the tables are the keywords and the default values used in SPICE2 version G.

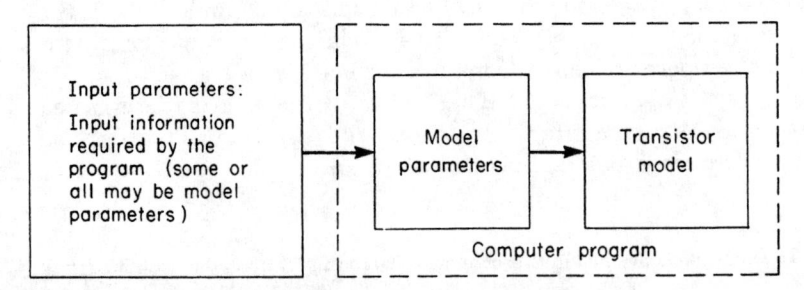

Figure 5-1 Distinction between input parameters and model parameters [1].

TABLE 5-1 SPICE2 BJT Parameters: Ebers-Moll Model

Symbol	SPICE2G keyword	Affected by area	Parameter name	Default value	Unit
\multicolumn{6}{c}{Static model parameters}					
I_S	IS	X	Saturation current	10^{-16}	A
β_F	BF		Ideal maximum forward current gain	100	
β_R	BR		Ideal maximum reverse current gain	1	
r_B	RB	X	Zero-bias base resistance	0	Ω
r_C	RC	X	Collector resistance	0	Ω
r_E	RE	X	Emitter resistance	0	Ω
V_A	VAF		Forward Early voltage	∞	V
E_g	EG		Energy gap (silicon)	1.11	eV
\multicolumn{6}{c}{Large-signal model parameters}					
C_{JE}	CJE	X	Zero-bias base-emitter junction capacitance	0	F
ϕ_E	VJE		Base-emitter built-in potential	0.75	V
C_{JC}	CJC	X	Zero-bias base-collector junction capacitance	0	F
ϕ_C	VJC		Base-collector built-in potential	0.75	V
C_{JS}	CJS	X	Zero-bias collector-substrate capacitance	0	F
ϕ_S	VJS		Substrate-junction built-in potential	0.75	V
τ_F	TF		Ideal forward transit time	0	s
τ_R	TR		Ideal reverse transit time	0	s
\multicolumn{6}{c}{Noise parameters}					
k_f	KF		Flicker-noise coefficient	0	
a_f	AF		Flicker-noise exponent	1	

5.2 Parameter Measurements

5.2.1 Static parameter measurement techniques

Measurement techniques to determine SPICE2 forward dc parameters for a BJT are presented in this section. The reverse parameters are calculated from the reverse measurement using the same algorithm. In the reverse mode measurements, the emitter and the collector are interchanged and the transistor is handled as in the forward mode.

As discussed in Chap. 2, $\ln I_C$, $\ln I_B$ vs. V_{BE} curves (see Fig. 5-2) depict all the physical phenomena represented by the static model.

TABLE 5-2 SPICE2 BJT Parameters: Gummel-Poon Model*

Symbol	SPICE2G keyword	Affected by area	Parameter name	Default value	Unit
			Static model parameters		
C_2	ISE = $C_2 I_S$	X	Base-emitter leakage saturation current	0	A
C_4	ISC = $C_4 I_S$	X	Base-collector leakage saturation current	0	A
V_B	VAR		Reverse Early voltage	∞	V
I_{KF}	IKF	X	Corner for forward β high-current roll-off	∞	A
I_{KR}	IKR	X	Corner for reverse β high current roll-off	∞	A
n_{EL}	NE		Base-emitter leakage emission coefficient	1.5	
n_{CL}	NC		Base-collector leakage emission coefficient	2	
r_{BM}	RBM	X	Minimum base resistance at high currents	RB	Ω
I_{rB}	IRB	X	Current where base resistance falls halfway to its minimum value	∞	A
n_F	NF		Forward current emission coefficient	1	
n_R	NR		Reverse current emission coefficient	1	
			Large-signal model parameters		
m_E	MJE		Base-emitter junction grading coefficient	0.33	
m_C	MJC		Base-collector junction grading coefficient	0.33	
m_S	MJS		Substrate-junction exponential factor	0	
X_{CJC}	XCJC		Fraction of base-collector junction capacitance connected to internal base node	1	
FC	FC		Coefficient for forward-bias junction capacitance formula	0.5	
$X_{\tau F}$	XTF		Coefficient for bias dependence of TF	0	
$V_{\tau F}$	VTF		Voltage describing V_{BC} dependence of TF	∞	V
$I_{\tau F}$	ITF	X	High-current parameter for effect on TF	0	A
$P_{\tau F}$	PTF		Excess phase at $f = \frac{1}{2}\pi\tau_F$	0	° (deg)
$X_{T\beta}$	XTB		Forward and reverse β temperature coefficient	0	
X_{TI}	XTI		Saturation current temperature exponent	3	

*Any combination of these parameters may be specified in addition to the Ebers-Moll model parameters.

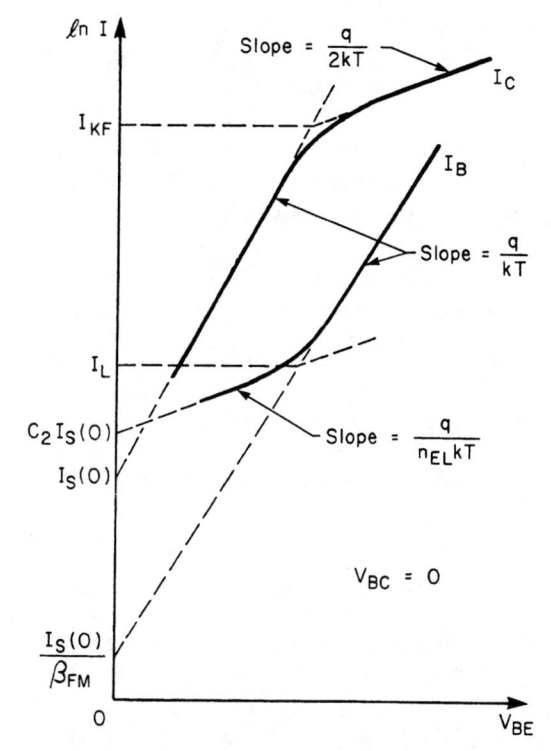

Figure 5-2 Plots of $\ln I_C$ and $\ln I_B$ vs. V_{BE} (for $V_{BC} = 0$).

The parameters for the model can be determined from the curves, as described below. First, the curve for I_C is considered.

a. I_S. I_S is the transistor saturation current. With regard to the curve for I_C, by extrapolating the region with slope q/kT, the value of I_S is obtained. In this region, the collector current is given at low-current levels by

$$I_C = I_S(e^{qV_{BE}/kT} - 1) \qquad (5\text{-}1)$$

I_S is directly proportional to the emitter-base junction area and therefore can vary significantly from device to device. A typical value of I_S for an integrated circuit transistor is 10^{-16} A.

b. I_{KF}. I_{KF} is the knee current; it models the drop in β_F at high collector currents due to high-level injection. As already seen, in regions where high-level injection dominates, the curve has a slope of $q/2kT$. If it were possible to assume that low-level and high-level effects could be distinguished by regions, a value of I_{KF} could be obtained by the intersection of the two asymptotes with slopes q/kT

and $q/2kT$. In most transistors, however, the two effects overlap and some trial and error is required to produce the correct parameter [2]. I_{KF} typically ranges from approximately 0.1 to 10 mA.

c. C_2 and n_{EL}. With regard to the curve for I_B, by extrapolating the region with a slope of $q/n_{EL}kT$ at low-current levels, the value of C_2 can be determined and, by determining this slope, n_{EL} can be found.

C_2 (magnitude) and n_{EL} (variation with V_{BE}) describe the nonideal component of I_B, which is dominant at low currents. Then,

$$I_{B,\text{nonideal}} = C_2 I_S(0)(e^{qV_{BE}/n_{EL}kT} - 1) \quad (5\text{-}2)$$

This component of I_B is responsible for the drop in β_F at low currents.

The value of C_2 is typically on the order of 100 to 1000; n_{EL} is typically between 2 and 4 (most often 2).

d. β_{FM}. β_{FM} is the maximum value of β_F, which is the ratio of I_C to I_B when the BJT is in the normal active region. β_{FM} is determined by the region of the two curves when they are parallel and have a slope of q/kT. In this region, both I_C and I_B have dominant ideal components, and the ratio of I_C and I_B gives β_{FM}, which can be determined from the I_B curve by extrapolating the region with a slope of q/kT, as shown in Fig. 5-2.

Thus β_{FM} defines the magnitude of the ideal component of I_B at a given value of V_{BE}. Then

$$I_{B,\text{ideal}} = \frac{I_S(0)}{\beta_{FM}}(e^{qV_{BE}/kT} - 1) \quad (5\text{-}3)$$

Typical values of β_{FM} range from the order of 10 for an IC lateral *pnp* or a power BJT, to about 100 for a small-signal *npn* BJT, to the order of 1000 for a super-β *npn* BJT.

e. V_A and V_B. V_A is the Early voltage that models the effect on the transistor characteristics of base-width modulation due to variations in the collector-base depletion region. It is always a positive number.

Typical values of V_A are 50 to 70 V for *npn* BJTs, 100 V for lateral *pnp* BJTs, and on the order of 1 to 10 V for devices with extremely thin base widths (such as punch-through devices or super-β BJTs).

The simplest method of obtaining V_A is from the slope of the I_C vs. V_{CE} curve in the linear region. The base-emitter voltage should be constant and the BJT should be biased in the forward active region. The collector current can be simplified as

$$I_C = \frac{I_S}{q_b} e^{qV_{BE}/kT} \qquad (5\text{-}4)$$

where q_b is given by Eq. (2-88).

By keeping the base-emitter voltage constant, the collector current would be proportional to $1/q_b$. Therefore, in the linear region, the collector current is proportional to $(1 - V_{BC}/V_A - V_{BE}/V_B)$. If V_{CE} is much larger than V_{BE}, the above term simplifies to $(1 + V_{CE}/V_A)$. The parameter V_A can now be calculated as the intersection of the extrapolated I_C curve with the V_{CE} axis, as shown in Fig. 5-3.

The extrapolation can be performed graphically directly from the curve-tracer display. However, the experimental error associated with this approach can be very high.

A more accurate technique is the use of $\ln I_C$ vs. V_{BE} curves. This technique [1] for finding V_A is the determination of the $\ln I_C$ vs. V_{BE} characteristic at two different values of V_{BC}; this theoretically results in two parallel lines. The Early voltage is then determined from the ratio of the (extrapolated) I_S values or the ratio of the value of two I_C values at the same V_{BE}.

$$\frac{I_S(V_{BC1})}{I_S(V_{BC2})} = \frac{I_C(V_{BC1})}{I_C(V_{BC2})} = \frac{1 + V_{BC2}/V_A}{1 + V_{BC1}/V_A} \qquad (5\text{-}5)$$

Equation (5-5), which is used to find V_A, can be simplified if one of the V_{BC} values is zero. To simplify or eliminate the need for correcting for finite r_B, r_E, and r_C, the curves should be measured at currents as low as possible and with V_{BC} as close to zero as practical [1].

When V_A is large, all of the preceding techniques for finding V_A become inaccurate. However, this is acceptable because the larger V_A

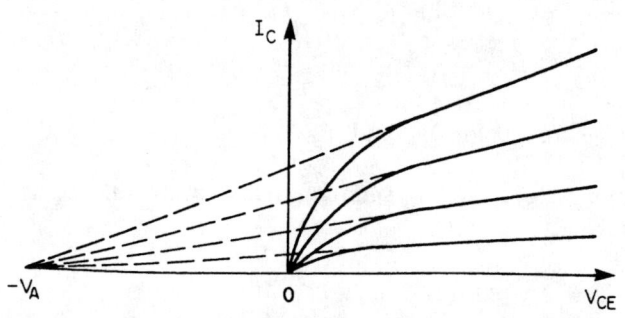

Figure 5-3 Approximate definition of V_A from I_C vs. V_{CE} characteristics.

the less important the effect of base-width modulation becomes and, therefore, the need to accurately determine V_A diminishes.

The above measurement techniques assume that base-width modulation due to variation in the width of the emitter-base depletion region is negligible (normally a valid assumption). To test for this, the above measurements must be repeated at different V_{BE} values. If the change in V_{BE} significantly affects the value of V_A obtained from the described techniques, the above assumption is not valid. In this case the method described for the measurement of V_B can be used (see the following discussion).

V_B is the reverse Early voltage that models the effect of base-width modulation due to variations in the width of the emitter-base depletion region. It is always a positive number. V_B is typically on the order of 10 to 50 V.

There is a complicating factor in measuring V_B in an analogous way to that used for measuring V_A. The measurement of V_A assumed that the variation in the width of the emitter-base depletion region had a negligible effect on the BJT characteristics in the normal active region, that is, that V_{BE}/V_B is much less than unity. The equivalent assumption that would be necessary if the methods for measuring V_A were to be used directly for measuring V_B is that V_{BC}/V_A is much less than unity. However, the technique for measuring V_B given in the following assumes that V_{BC}/V_A is *not* negligible compared with unity [1].

From the $\ln I_C$ vs. V_{BE} characteristics in the active region and from the $\ln I_E$ vs. V_{BC} characteristics in the inverse region, it follows that

$$I_C = \frac{I_S(e^{qV_{BE}/kT} - 1)}{1 + V_{BE}/V_B + V_{BC}/V_A} \to \frac{I_C(0)}{I_C(V_{BC1})}$$
$$= \frac{1 + V_{BE1}/V_B + V_{BC1}/V_A}{1 + V_{BE1}/V_B}$$
$$I_E = \frac{I_S(e^{qV_{BC}/kT} - 1)}{1 + V_{BE}/V_B + V_{BC}/V_A} \to \frac{I_E(0)}{I_E(V_{BE2})}$$
$$= \frac{1 + V_{BE2}/V_B + V_{BC2}/V_A}{1 + V_{BC2}/V_A}$$

(5-6)

Equations (5-6) can be solved for V_A and V_B.

f. E_g. E_g is the energy gap of the semiconductor material. E_g is typically at $T = 300$ K

1.11 eV for Si
0.67 eV for Ge
0.69 eV for SBD

E_g is used to model the variation of I_S with temperature and is therefore obtained by curve-fitting the model equation to the I_S vs. T curve. The I_S vs. T curve is determined from measurements of I_S with the device in a controlled-temperature environment, for example, an oven (in making these measurements care must be taken that the power dissipated by the device does not raise the junction temperature significantly above the ambient temperature).

The model equation given in Sec. 2.8.1 [see Eq. (2-143)] assumes that I_S is proportional to $T^{X_{TI}} e^{-E_g/kT}$. When the T-term power before the exponential term is allowed to vary, typical values generated by curve-fitting for Si are 1 to 4 for the power of T, while for E_g (the extrapolated, zero-temperature energy gap), typical values range from 1.206 eV for low doping levels to approximately 1.11 eV for high doping levels [1].

g. r_E. r_E is a constant resistor that models the resistance between the emitter region and the emitter terminal.

The value of r_E can be obtained by observing the base current I_B as a function of collector-emitter voltage V_{CE} when the collector is open ($I_C = 0$). The slope of the resultant graph is approximately $(r_E)^{-1}$.

A typical value of r_E is approximately $1\,\Omega$, of which the metal contact resistance is normally a significant portion.

h. r_C. r_C models the resistance between the BJT's collector region and its collector terminal. r_C is calculated from the slope of the I_C vs. V_{CE} curve when β_F is forced to 1 in the measurement.

Parameter r_C, calculated in saturation mode, $r_{C,\text{sat}}$, should not be confused with $r_{C,\text{lin}}$ found by measuring the slope of the I_C vs. V_{CE} curve in the linear region, which is usually higher. For switching circuits $r_{C,\text{sat}}$ can be used, but for linear applications, the higher value of $r_{C,\text{lin}}$ will be more appropriate [3]. Figure 5-4 shows these two limiting values of r_C.

For a given BJT, r_C actually varies with current level, but for SPICE2 it is considered constant. The value of r_C can vary significantly from a few ohms for discrete and deep-collector integrated devices to hundreds of ohms for standard integrated devices.

i. r_B. r_B models the resistance between the base region and the base terminal. Values of r_B can range from about $10\,\Omega$ (for microwave devices) to several kilohms (for low-frequency devices).

The value obtained for r_B depends strongly on the measurement technique used as well as the BJT's operating conditions.

r_B should be determined by the method closest to the operating conditions under study. If r_B is being measured to verify its effect on

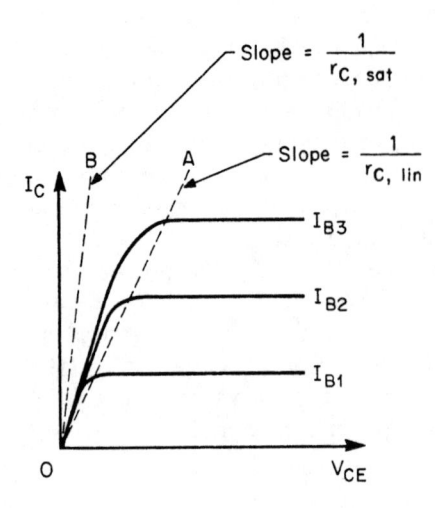

Figure 5-4 I_C vs. V_{CE} characteristics at constant I_B, showing the two limiting values of r_C [1].

noise performance, the noise measurement technique [1] should be used. Similarly, if the BJT is to be used in a switching application, the pulse measurement techniques [1] may provide the most appropriate value. For small-signal analyses, four measurement techniques can be used: the input impedance circle method, the phase-cancellation technique, the two-port network method, and the h-y ratio technique [1].

5.2.2 Large-signal parameter measurement techniques

The parameters for each junction capacitance and the algorithm for extracting the parameters associated with τ_F are described in this section.

a. $C_j(0)$, ϕ, and m. These three parameters describe the junction capacitance due to the fixed charge in the junction depletion region.

In the BJT models, when the appropriate junction voltage V is less than or equal to $\phi/2$, the junction capacitance is modeled by

$$C_J(V) = \frac{C_j(0)}{(1 - V/\phi)^m} \tag{5-7}$$

where $C_j(0)$ is the value of C_J at $V = 0$, ϕ is the built-in potential, and m is the junction gradient factor. The emitter-base junction capacitance C_{JE} is a function of the internal base-emitter voltage V_{BE}, and the parameters are $C_{JE}(0)$, ϕ_E, and m_E. Similarly, for the collector-base junction, C_{JC} is expressed in terms of $C_{JC}(0)$, ϕ_C, and m_C.

In SPICE2 regardless of the type of device (*npn* or *pnp*), V is positive if the junction is forward-biased and negative if the junction is reverse-biased.

Both junction capacitances can be obtained as a function of voltage by means of a bridge. The two junction contacts are connected to the bridge and the third contact is left open. For example, for C_{JE}, the emitter and base leads are connected to the bridge and the collector contact is left open. The measurement frequency is normally low enough that it can be assumed that the ohmic resistances have a negligible effect.

A complicating factor is the extra capacitance C_k, caused mainly by pin capacitance, stray capacitance, and pad capacitance. C_k is normally assumed to be constant. The capacitance measured by the bridge is

$$C_{\text{meas}} = \frac{C_J(0)}{(1 - V/\phi)^m} + C_k \qquad (5\text{-}8)$$

C_k can be determined in four ways: by an estimate (approximately 0.4 to 0.7 pF), by measurement with a dummy can, by a computer parameter optimization procedure, or by graphical techniques [1].

This last technique consists in making an initial guess for ϕ and C_k and then plotting the resultant value of $(C_{\text{meas}} - C_k)$ as a function of $(\phi - V)$ on ln-ln graph paper [1]. If a straight line (with a slope between 0.5 and 0.33) results, the chosen values are assumed correct. If the plotted line is not straight, a second guess is made for C_k and/or ϕ and the plot redone. This process continues until the appropriate straight line is obtained. Since the slope of the straight line is equal to $-m$, the values of ϕ, m, C_k, and $C_J(0)$ can be determined from this plot.

An alternative graphical technique is to plot $(C_{\text{meas}} - C_k)^{-1/m}$ as a function of V. When a straight line is obtained, ϕ is determined by extrapolating the line to the V axis.

Typically C_J varies with V as shown in Fig. 5-5 for V less than or equal to $\phi/2$. The built-in potential ϕ is usually about 0.5 to 0.7 V. The gradient factor m is assumed to be between 0.33 and 0.5, the graded-junction and abrupt-junction values, respectively.

b. C_{JS}. C_{JS} is the epitaxial-layer–substrate capacitance (only in the case of an integrated circuit). It can be measured directly on a capacitance bridge at the bias voltage to be used in the analysis. If the bias voltage will change drastically, an averaging process should be used.

C_{JS} is mainly important for integrated *npn* BJTs and lateral *pnp* BJTs. For *npn* devices, C_{JS} is represented as a constant capacitance,

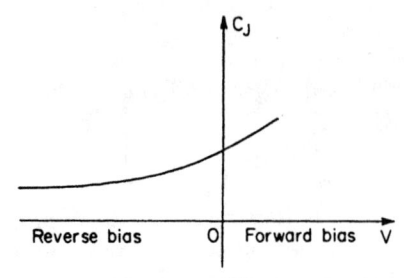

Figure 5-5 Variation of C_J as a function of V [1].

typically 1 to 2 pF, from the collector terminal to ground. Ideally, C_{JS} should be modeled by a junction capacitance distributed across r_C (with its dependence on the epitaxial-layer–substrate voltage) [1].

c. X_{CJC}. X_{CJC} models the split of the base-collector junction capacitance C_{JC} across the base resistance. The capacitance $(X_{CJC} C_{JC})$ is placed between the internal base node and the collector. $(1 - X_{CJC})C_{JC}$ is the capacitance from external base to collector, C_{JC} being the total base-collector capacitance. This parameter is usually important only at very high frequencies [1,3].

X_{CJC} is a difficult parameter to determine from terminal measurements. It can be found with relative ease if the geometry of the device is known, since it is equal to $(1 - A_E/A_B)$, where A_E is the area of the emitter and A_B is the total area of the base, including the emitter area.

It is possible to obtain X_{CJC} via terminal measurements, too, by measuring the product $r_B(1 - X_{CJC})C_{JC}$ [1]. X_{CJC} must lie between 0 and 1. It is typically on the order of 0.8.

d. τ_F. τ_F, the total forward transit time, is used for modeling the excess charge stored in the BJT when its emitter-base junction is forward-biased and $V_{BC} = 0$. It is needed to calculate the BJT's emitter diffusion capacitance.

Generally, τ_F is determined from f_T, the BJT's unity-gain bandwidth, which is defined as the frequency at which the common-emitter, zero-load, small-signal current gain extrapolates to unity.

f_T also varies with the operating point, as well as from device to device. A typical variation of f_T with $\ln I_C$ is sketched in Fig. 5-6.

For discrete devices and integrated npn BJTs, the peak f_T is generally on the order of 600 MHz to 2 GHz. For integrated pnp BJTs, the peak f_T is usually 10 MHz for a substrate pnp device and 1 MHz for a lateral pnp device. The drop in f_T at high currents is caused by the increase in τ_F. The drop in f_T at low currents is caused by junction capacitances C_{JE} and C_{JC}. Since these two capacitances are modeled separately, the drop in f_T at low currents is included inherently in the Gummel-Poon model.

In the region where f_T is constant, τ_F is given by

$$\tau_F = \frac{1}{2\pi f_{T,\max}} - C_{JC} r_C \tag{5-9}$$

where $f_{T,\max}$ is the peak value of f_T and r_C is the $r_{C,\text{lin}}$ value (see r_C measurement). When there is no constant f_T region, τ_F is obtained by plotting $1/f_T$ as a function of $1/I_C$. The resultant curve can then be extrapolated to obtain τ_F. The intercept (noted by $1/f_A$) of the extrapolated straight line at $1/I_C = 0$ is related to τ_F by

$$\tau_F = \frac{1}{2\pi}\frac{1}{f_A} - C_{JC} r_C \tag{5-10}$$

Typically, τ_F varies with I_C as shown in Fig. 5-7. However, for the SPICE2 Ebers-Moll model, τ_F is assumed to be constant. Values of τ_F generally range from 0.3 ns for a standard *npn* BJT to 80 ps for a high-frequency device ($f_T \simeq 2$ GHz).

e. $X_{\tau F}$, $I_{\tau F}$, $V_{\tau F}$, and $P_{\tau F}$. The four parameters associated with the forward transit time τ_F should be calculated from the f_T measurement. A method to determine the parameters $X_{\tau F}$, $I_{\tau F}$, $V_{\tau F}$, and $P_{\tau F}$ is described next; this method finds the approximate values for

Figure 5-6 Variation of f_T with I_C [1].

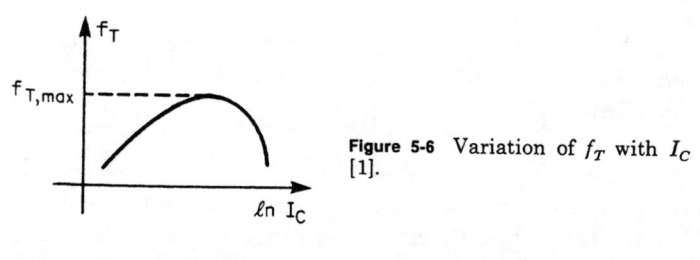

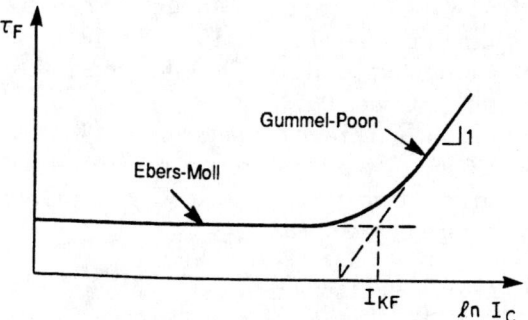

Figure 5-7 Variation of τ_F with I_C.

the parameters. Because of the interaction with other model parameters, a good fit to the measured data can be made only by trial and error.

At a high value of V_{CE}, the peak value of f_T is first found ($f_{T,\max}$). τ_F is then calculated. At a low value of V_{CE} such that $V_{BC} = 0$, I_C is set to a large value such that f_T goes to a minimum. The f_T at this point is called $f_{T,\min}$. $X_{\tau F}$ is then calculated as [3]

$$X_{\tau F} = \frac{f_{T,\max}}{f_{T,\min}} - 1 \qquad (5\text{-}11)$$

With V_{BC} set to almost zero, I_C can be found at the peak of the f_T curve. Assuming the peak value is f_T^*, $I_{\tau F}$ can be calculated as [3]

$$I_{\tau F} = \frac{I_C}{\sqrt{X_{\tau F}(f_{T,\max}/f_T^* - 1)}} - I_C \qquad (5\text{-}12)$$

If $f_T(f_{TX})$ is measured at a high $V_{CE}(V_{CEX})$ and a high I_C and also if $f_T(f_{TY})$ is measured at a low $V_{CE}(V_{CEY})$ and at the same I_C, $V_{\tau F}$ can be calculated as [3]

$$V_{\tau F} = \frac{V_{CEX} - V_{CEY}}{1.44 \ln[(f_{TY}/f_{TX})(f_{T,\max} - f_{TX})/(f_{T,\max} - f_{TY})]} \qquad (5\text{-}13)$$

After the parameters have been approximated, they should be checked with an actual simulation. The parameters can then be trimmed to find a better match.

The excess phase parameter $P_{\tau F}$ is found by measuring the phase at unity-gain bandwidth of the BJT. $P_{\tau F}$ is the difference between the actual phase at f_T and $90°$.

f. TNOM, T_{C1}, T_{C2}, X_{TI}, and X_{TB}. TNOM is the temperature at which all the model parameters are obtained. The simplest technique for measuring TNOM is by means of a thermometer placed near the BJT. As long as the power dissipation of the device is low enough to cause a negligible increase in the junction temperature, then the junction temperature is approximately room temperature. TNOM is considered to be room temperature, about 27°C, or 300 K.

T_{C1} and T_{C2} are the first- and second-order temperature coefficients and are used to model the effect of temperature on resistors (and consequently on r_B, r_C, and r_E if they are specified as *external* resistances and not as model parameters). Thus,

$$r(T) = r(\text{TNOM})[1 + T_{C1}(T - \text{TNOM}) + T_{C2}(T - \text{TNOM})^2] \qquad (5\text{-}14)$$

where r is the resistor being considered, TNOM is the nominal temperature at which the parameter was measured, and T is the temperature at which the analysis is to be performed [1].

The temperature variation of each parameter is obtained by placing the device in a controlled-temperature environment and making the appropriate measurements. The observed temperature variation is then fitted to Eq. (5-14). T_{C1} represents a linear variation with temperature, and T_{C2} represents a nonlinear (square-law) variation with temperature.

Values of T_{C1} and T_{C2} vary from one parameter to the other. Typical values of T_{C1} and T_{C2} for r_C and r_B are $2 \times 10^{-3} \, \text{K}^{-1}$ and $8 \times 10^{-6} \, \text{K}^{-2}$, respectively.

X_{TI} is the saturation current temperature exponent (usually equal to 3) that appears explicitly in the temperature dependence of the saturation current [see Eq. (2-143)].

$X_{T\beta}$ is the forward and reverse β temperature coefficient, according to Eqs. (2-146).

References

1. I. E. Getreu, *Modeling the Bipolar Transistor*, Tektronix Inc., Beaverton, Oreg., 1976.
2. F. Van De Wiele, W. L. Engl, and P. G. Jespers (eds.), *Process and Device Modeling for Integrated Circuit Design*, Nijhoff, The Hague, 1977.
3. E. Khalily, Hewlett-Packard Co., private communication.

Chapter 6

MOST Parameter Measurements

This chapter considers the methods of calculating and measuring the parameters of the three MOST models implemented in SPICE2. For each parameter, the values necessary to simulate n-channel MOSTs will be obtained as examples, with different channel lengths and widths, starting from a material with a doping value of $N_A = 7 \times 10^{14}$ cm^{-3}, with source and drain junctions made with phosphorous diffusions 1 μm deep, and with a gate oxide thickness $t_{ox} = 80$ nm. This example allows us to evaluate the degree of precision reached by the three models by comparing the results of the simulations with the measurements.

6.1 LEVEL1 Model Parameters

The equations of the LEVEL1 model were given in Chap. 4. As an example of the calculation of the parameters for the LEVEL1 model, a MOST with a very large and wide channel will be used, with $W_{nom} = L_{nom} = 100\ \mu$m. This choice allows us to neglect possible differences between the nominal dimensions and the real dimensions W_{eff} and L_{eff}; moreover, this MOST does not present short- and narrow-channel effects, which cannot be simulated with this simple model.

Note: This chapter has been written with the contribution of Enrico Profumo, who is with SGS-THOMSON Microelectronics, Milan, Italy.

6.1.1 Measurements of V_{T0} and KP in the linear region

The parameters V_{T0} and KP can be calculated from the transfer characteristics measured in the linear region or in the saturation region. Figure 6-1a shows the setup for the measurement in the linear region.

In order to account for the condition that the MOST is in the linear region, that is, $V_{DS} < V_{GS} - V_{TH}$, it is necessary to carry out the measurements with a small V_{DS}. In the example in Fig. 6-2, the voltage on the channel is $V_{DS} = 50 \, \text{mV}$; in this condition, all the

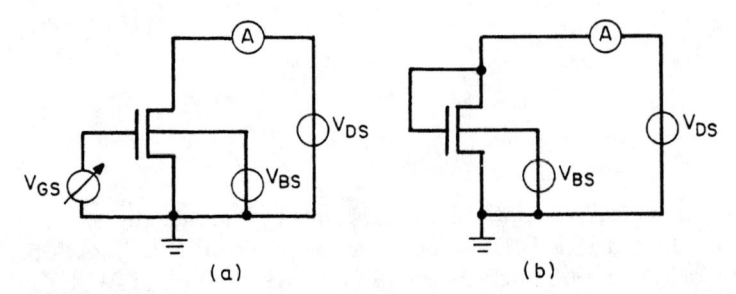

Figure 6-1 Circuit plans for the measurement of the threshold voltage: (a) measurement with V_{DS} constant and (b) measurement with $V_{DS} = V_{GS}$.

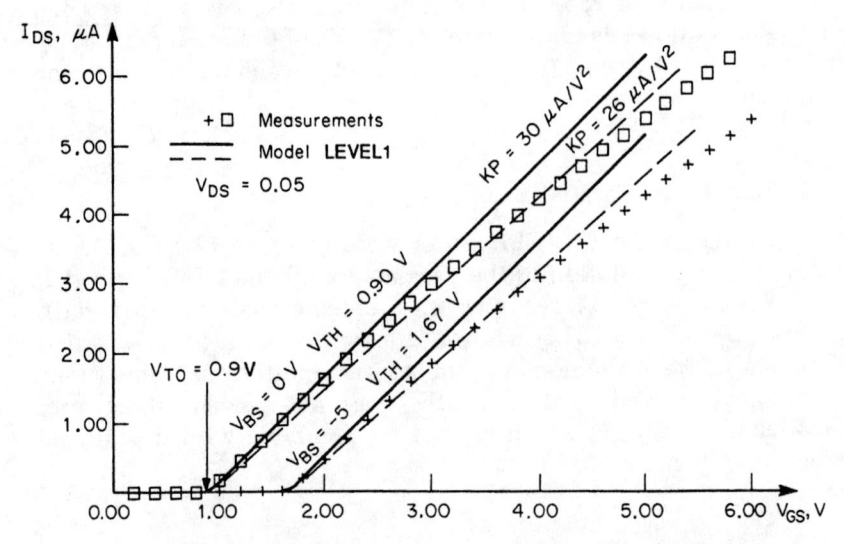

Figure 6-2 Measurements of the transfer characteristics in the linear region; calculation by extrapolation to $I_{DS} = 0$ of V_{TH} and of KP. The value of KP calculated using the lowest current measurements is $\text{KP} = 30 \, \mu\text{A/V}^2$ (continuous lines), and the average value is $\text{KP} = 26 \, \mu\text{A/V}^2$ (dashed lines).

measurements are in the linear region except those with $V_{TH} < V_{GS} < V_{TH} + V_{DS}$.

This means that the measurements of the example, carried out with $V_{GS} - V_{TH} < 50$ mV, are in the saturation region and not in the linear region (see Fig. 4-4).

With this only exception, the current is expressed by Eq. (4-23), in which the term $V_{DS}/2$ can also be eliminated (the term $V_{DS}/2$ is negligible with respect to V_{GS}). Thus

$$I_{DS} \simeq \beta(V_{GS} - V_{TH})V_{DS} \qquad (6\text{-}1a)$$

where

$$\beta = \text{KP}\frac{W_{\text{eff}}}{L_{\text{eff}}} \qquad (6\text{-}1b)$$

The term $W_{\text{eff}}/L_{\text{eff}}$ is the ratio of the effective dimensions of the channel; its method of calculation will be shown later (see Sec. 6.3.1). In this example, the ratio is set to 1.

Equation (6-1a) expresses a linear relationship between I_{DS}, V_{DS}, and $V_{GS} - V_{TH}$, and suggests the calculation of β and V_{T0} by linear extrapolation, as shown in Fig. 6-2.

This model is sufficiently accurate only for a short range of the transfer characteristic. One way to evaluate these parameters is first to find V_{T0} by extrapolation of the measurements at low current and then to find an average value of β in the range of measured V_{GS} voltages.

The extraction of V_{T0} can be performed graphically or with a relation obtained from Eq. (6-1a) applied to the voltages $V_{GS} = V_{GS1}$ and $V_{GS} = V_{GS2}$ which are related to the two measurements I_{DS1} and I_{DS2}. Thus

$$V_{TH} = \frac{V_{GS1} - (I_{DS1}/I_{DS2})V_{GS2}}{1 - I_{DS1}/I_{DS2}} \qquad (6\text{-}2)$$

The input parameter for SPICE2, V_{T0}, is the threshold voltage V_{TH} with zero voltage between source and substrate ($V_{BS} = 0$ V). In order to calculate the input parameter KP from β, it is sufficient to substitute the values obtained from Eq. (6-1a). If the currents I_{DS1} and I_{DS2} are related in such a way that $I_{DS2} = 2I_{DS1}$, Eq. (6-2) becomes simply

$$V_{TH} = 2V_{GS1} - V_{GS2} \qquad (6\text{-}3)$$

The currents I_{DS1} and I_{DS2} are to be chosen so as to avoid both the weak inversion region and the variable transconductance region. In the example in Fig. 6-2, $I_{DS1} = 1\ \mu\text{A}$ and $I_{DS2} = 2\ \mu\text{A}$ have been used.

This equation is very convenient for calculating the threshold voltage with only two measurements, for which a curve tracer can be used. The values calculated from the measurements in Fig. 6-2 are:

1. Using Eq. (6-3) and the experimental points $I_{DS1} = 1\,\mu\text{A}$ and $I_{DS2} = 2\,\mu\text{A}$, the result is $V_{T0} = 0.9\,\text{V}$ and $KP = 30\,\mu\text{A}/\text{V}^2$.
2. Using Eq. (6-2) and $V_{GS1} = 1\,\text{V}$ and $V_{GS2} = 4\,\text{V}$, the result is $V_{T0} = 0.85\,\text{V}$ and $KP = 26\,\mu\text{A}/\text{V}^2$.

The best solution is choosing the voltage V_{T0} calculated at the lowest currents, i.e., $V_{T0} = 0.9\,\text{V}$, and the value of KP calculated at the highest voltages, that is, $KP = 26\,\mu\text{A}/\text{V}^2$.

Finally, to make the best choice of parameters, it is convenient to use a larger number of measurements and calculate V_{TH} and β using a simple algorithm with a linear regression.

6.1.2 Measurements of V_{T0} and KP in the saturation region

As an alternative to the method presented in the preceding section, it is possible to measure the same parameters in the saturation region by referring to Eq. (4-26) and using the setup of Fig. 6-1b. This setup has the advantage of using only two adjustable voltage sources.

Moreover, if the threshold voltage is positive (or negative for the p channel), all the measurements are certainly in the saturation region; this method cannot be applied to MOS depletion transistors, because these transistors are in the linear region for $V_{GS} = V_{DS}$.

In order to again obtain a linear relationship similar to Eq. (4-26), the current in the saturation region can be described in the following way:

$$\sqrt{I_{DS}} = \sqrt{\frac{\beta}{2}}(V_{GS} - V_{TH}) \qquad (6\text{-}4)$$

In this case a graphical calculation can be carried out (where the square root of I_{DS} is on the y axis, Fig. 6-3) before proceeding as in Sec. 6.1.1.

In the same way as in Eq. (6-2), the following equation calculates V_{TH} in an analytical way using only two measurements:

$$V_{TH} = \frac{V_{GS1} - \sqrt{I_{DS1}/I_{DS2}}\,V_{GS2}}{1 - \sqrt{I_{DS1}/I_{DS2}}} \qquad (6\text{-}5)$$

In order to use Eq. (6-3) it is sufficient to make $I_{DS2} = 4I_{DS1}$. From the measurements shown in Fig. 6-3, the following values can be

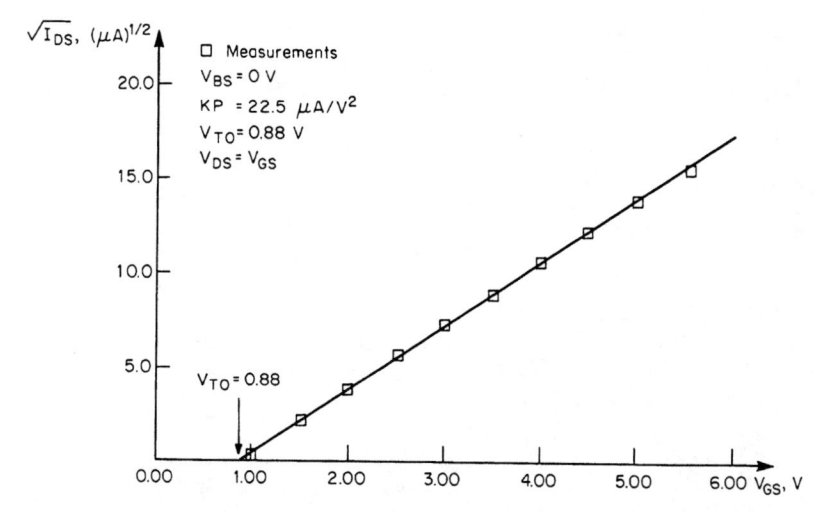

Figure 6-3 Measurements of the transfer characteristics in the saturation region; calculation by extrapolation to $I_{DS} = 0$ of V_{TH} and KP.

calculated:

$$V_{TH} = 0.88 \text{ V} \quad \text{KP} = 22.5 \text{ } \mu\text{A/V}^2$$

Another possible circuit setup for this measurement is shown in Fig. 6-1a, in which V_{DS} is maintained constant at a value sufficiently high to make the condition $V_{GS} < V_{DS} + V_{TH}$ valid; if this condition is not verified, the measurements of the transfer characteristics with $V_{GS} > V_{DS} + V_{TH}$ are in the linear region and not in the saturation region (see Fig. 4-4). This method is not as straightforward as the preceding one; however, it is useful afterward to find the law of variation of V_{TH} with V_{DS} in the short-channel MOSTs.

It is worth noting that the calculation of the same parameters with the two preceding methods does not give equal results. The short transition into the saturation region of the transfer characteristics with $V_{DS} = 50$ mV is the principal reason for the small difference in the threshold voltages when measured with the two methods; it can be demonstrated that the measurements in the linear region exceed by $V_{DS}/2$ the measurements in the saturation region; furthermore, the latter method yields the correct result.

It is now necessary to examine the reason for the much clearer difference in the values calculated for KP. The result in Fig. 6-4, showing the measurements and simulations of the output characteristics, allows us to attribute this difference to the imprecision of the model; in fact, the simulations with the parameters calculated in the linear region are successful only in the linear region with low V_{DS},

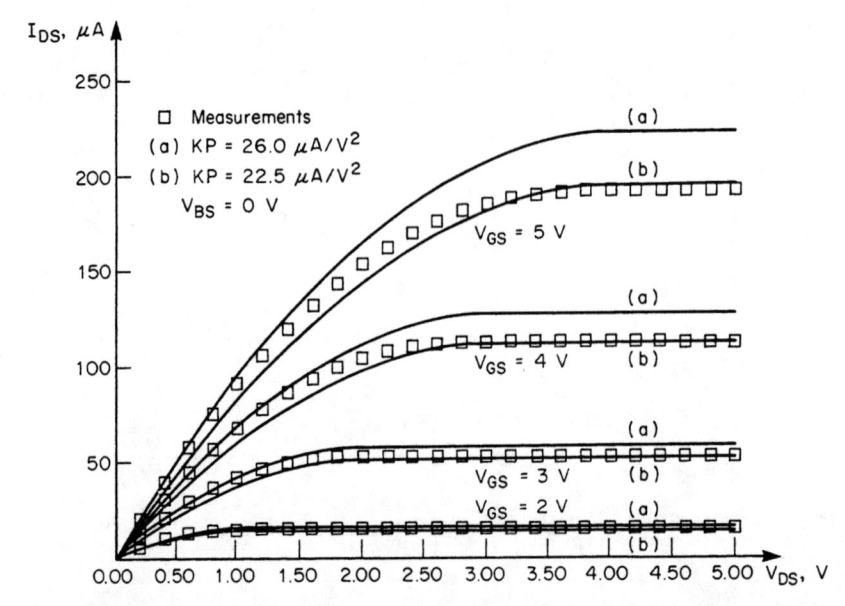

Figure 6-4 Measurements of output characteristics and simulations with parameters calculated in Figs. 6-2 and 6-3.

while those with the parameters calculated in the saturation region successfully simulate the characteristics of the saturation region, as shown in Fig. 6-4.

Given that a good agreement between measurement and simulation is possible only near the region where the parameters have been evaluated, the choice between one or another type of measurement should be based on which working range has to be simulated with greater precision.

6.1.3 Calculation of the body effect

From Eq. (4-29) we can easily calculate the parameter γ, which links the value of V_{TH} to the voltage between source and substrate V_{BS}, using at least two measurements of V_{TH} at different substrate voltages:

$$\gamma = \frac{V_{TH2} - V_{TH1}}{\sqrt{2\phi_p - V_{BS2}} - \sqrt{2\phi_p - V_{BS1}}} \tag{6-6}$$

If more measurements are available, we can resort to a linear interpolation of the measurements of the threshold voltage V_{TH} against $(\sqrt{2\phi_p - V_{BS}} - \sqrt{2\phi_p})$ (see Fig. 6-5).

The value of $2\phi_p$, at this point of the calculations, is not yet known. It is possible to calculate this parameter from the value of the

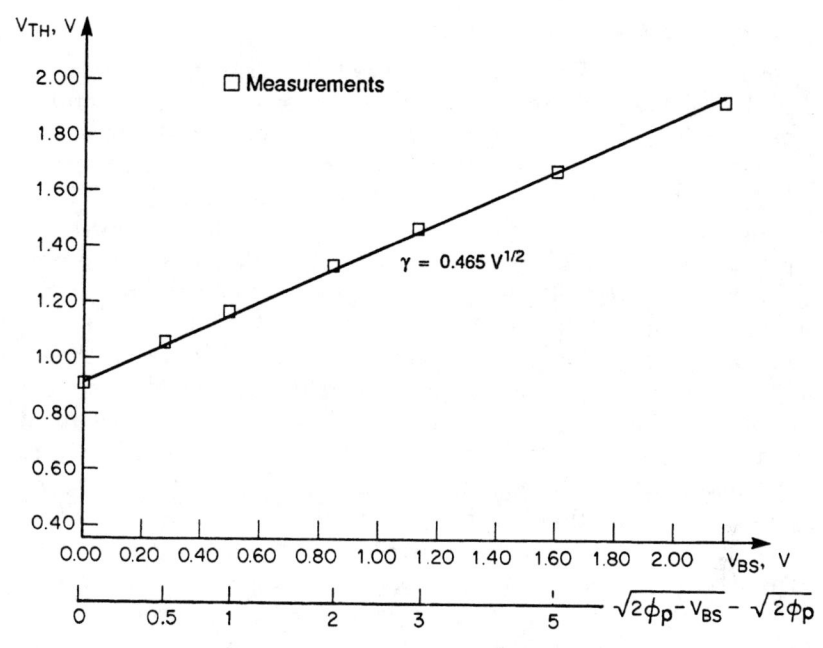

Figure 6-5 Graph calculation of the body-effect constant.

substrate doping, which is obtained from the resistivity of the starting material, or it can be taken to be equal to 0.6 V as a first estimate.

Referring to the example, we find that, with a doping of $N_A = 7 \times 10^{14}$ cm^{-3}, $2\phi_p = 0.56$ V with Eq. (4-36).

Calculating V_{TH1} and V_{TH2} with Eq. (6-2) from the curves in Fig. 6-2, we find that $\gamma = 0.484$ V$^{1/2}$, while with a fitting for all the measurements in Fig. 6-5, we find that $\gamma = 0.465$ V$^{1/2}$; the latter value will be used in future examples.

6.2 LEVEL2 Model (Long-Channel) Parameters

In this and the following sections the methods of calculating the parameters for the equations of the LEVEL2 model seen in Chap. 4 will be presented. The topic has been divided into two sections to distinguish two types of parameters: those of the basic model, for which the example from the preceding chapter will be used (i.e., a long-channel MOST), and the parameters relative to the second-order effects, not included in the basic model. The measurements for this second group of parameters concern almost all short-channel MOSTs.

6.2.1 Measurements in the linear region

The parameters V_{T0}, KP, γ, and $2\phi_p$ have the same significance as in the LEVEL1 model and are calculated in the same way. It is useful to observe that Eq. (4-23) of the current of the simple model with low V_{DS} supplies practically the same values as Eq. (4-40), which relates to the LEVEL2 model; however, the method proposed in Sec. 6.1.1 is still applicable, while Eq. (4-26) for the current in the saturation region used in Sec. 6.1.2 is very different from that of the model now being examined. It is not possible, therefore, to calculate KP in the saturation region with Eq. (6-5) and use the same value in Eq. (4-44); we can, on the other hand, calculate the same parameter with Eq. (6-1a) and use it in Eq. (4-40) without introducing large errors.

Moreover, if the present model is used to simulate a long-channel MOST, the parameters calculated in the linear region are sufficient to simulate quite accurately the current value in the saturation region, too. This good correlation becomes more evident if we use the model indicated in Sec. 4.3.3 instead of using an average constant value for KP. The parameters of Eq. (4-45)—U_c, U_e, and U_t—can be calculated by a graphical method. We can, in fact, rewrite the model in the following way

$$\log \frac{KP'}{KP} = \log \frac{\beta'}{\beta} = U_e \left[\log \frac{t_{ox}\varepsilon_s U_c}{\varepsilon_{ox}} - \log(V_{GS} - V_{TH} - U_t V_{DS}) \right] \quad (6\text{-}7)$$

where β' is the value of β with Eq. (6-1b) for the measurements in Fig. 6-2, with a different V_{GS} and a constant V_{DS}; V_{TH} and β are evaluated by extrapolation at the voltages nearest to the threshold, i.e., in the region where the dependence of I_{DS} on V_{GS} is at a maximum, and then are used in Eq. (6-7).

The values relating to the example are shown in Fig. 6-6a, as KP' vs. $V_{GS} - V_{TH}$, and in Fig. 6-6b as $\log(KP'/KP)$ vs. $\log(V_{GS} - V_{TH} - V_{DS}U_t)$; in this way we can calculate U_e and U_c with Eq. (6-7). The value of the parameter U_t cannot be calculated with this method. However, this parameter does not generally have a great influence on the current calculated from the model; it is best therefore to set $U_t = 0.5$ and to give up the idea of a more precise calculation.

By doing so, we obtain the following values:

$$V_{TH} = 0.9 \text{ V}$$

$$KP = 32.4 \ \mu A/V^2$$

$$U_c = 2.25 \times 10^4 \text{ V/cm}$$

$$U_e = 0.11$$

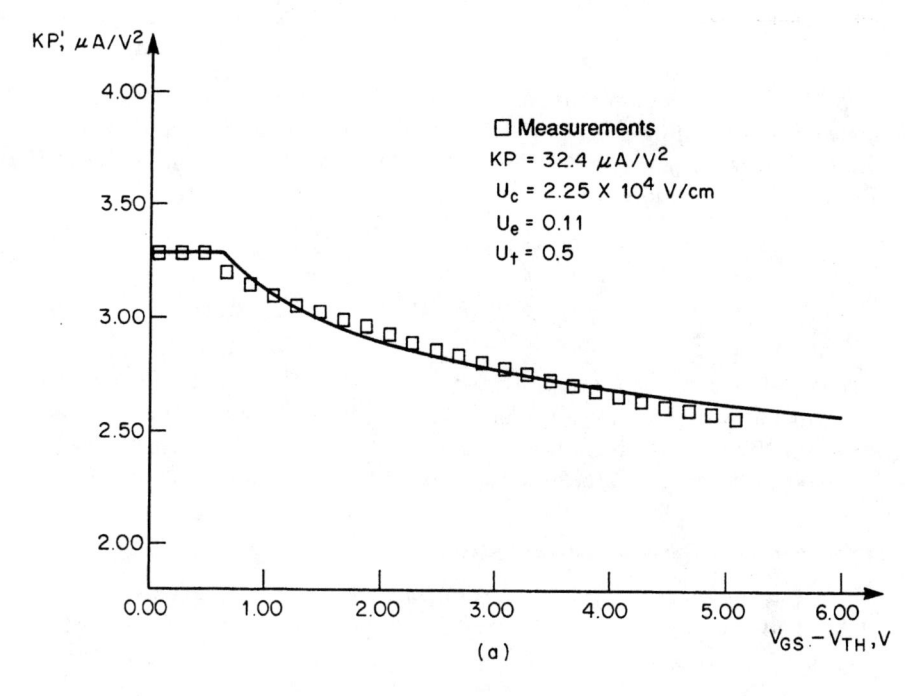

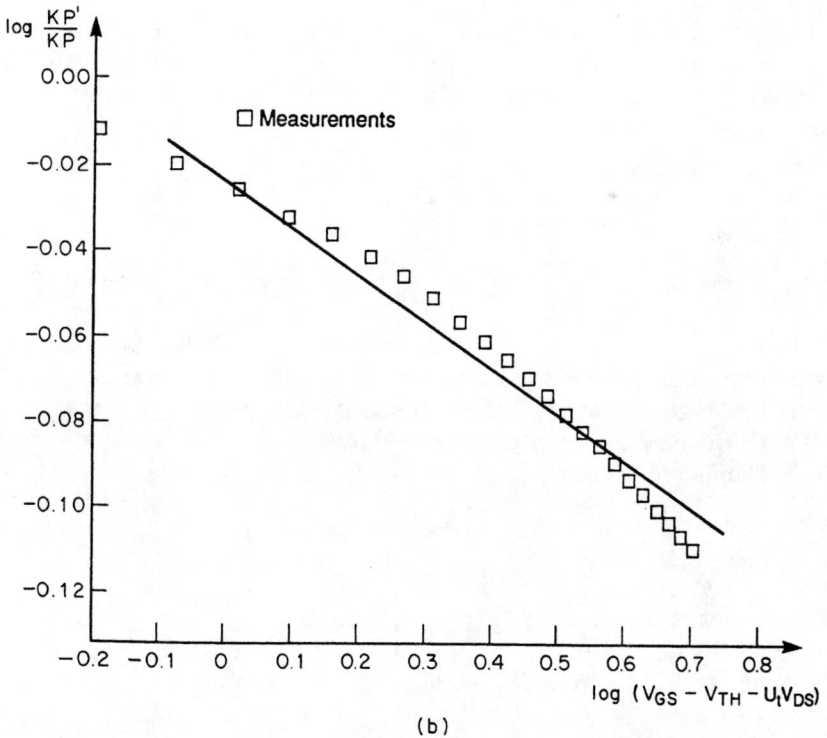

Figure 6-6 (a) Variation of KP′ vs. $V_{GS} - V_{TH}$ and (b) graphic calculation of U_c and U_e.

With a suitable computer, it is best to use a linear regression method that supplies the parameters of the linear equation $y = ax + b$, where $y = \log(KP'/KP)$ and $x = \log(V_{GS} - V_{TH} - V_{DS}/2)$. From this expression we obtain U_e and U_c:

$$U_e = -a \tag{6-8}$$

$$U_c = \frac{\varepsilon_{ox}}{t_{ox}\varepsilon_s} 10^{-b/a} \tag{6-9}$$

The value of γ can be calculated in the way shown in Sec. 6.1.3.

Figure 6-7a and b shows the simulations of the transfer characteristics and the output characteristics with this model; the improvement with respect to the simple model (see Fig. 6-4) is clear.

6.2.2 Calculation of physical parameters

These parameters V_{T0}, KP, $2\phi_p$, and γ are of an *electrical* type and are related to the parameters of a *physical* type—mobility (μ_0), substrate doping (N_A), gate oxide thickness (t_{ox}), and number of surface states (N_{SS}). If the value of t_{ox} is not already known from direct measurements, it can be calculated from a measurement of the gate capacitance, as described in Sec. 6.5.1. Once t_{ox} and C'_{ox} are known, the calculation of the mobility μ_0 and the doping N_A can be obtained by inverting Eqs. (4-34) and (4-35):

$$\mu_0 = \frac{KP}{C'_{ox}} \tag{6-10}$$

$$N_A = \frac{C'^2_{ox}\gamma^2}{2q\varepsilon_s} \tag{6-11}$$

Once N_A is known we can recalculate $2\phi_p$ with Eq. (4-36); if the value found is different from that used to calculate γ (see Sec. 6.1.3), it is better to repeat the calculation with Eq. (6-6).

With the parameters already calculated as in the example and with $t_{ox} = 80$ nm, we obtain

$$C'_{ox} = 4.32 \times 10^{-8} \text{ F/cm}^2$$

$$\mu_0 = 750 \text{ cm}^2/(\text{V} \cdot \text{s})$$

$$N_A = 1.2 \times 10^{15} \text{ cm}^{-3}$$

Recalculating $2\phi_p$ from N_A we obtain

$$2\phi_p = 2\frac{kT}{q}\ln\frac{N_A}{n_i} = 0.58 \text{ V}$$

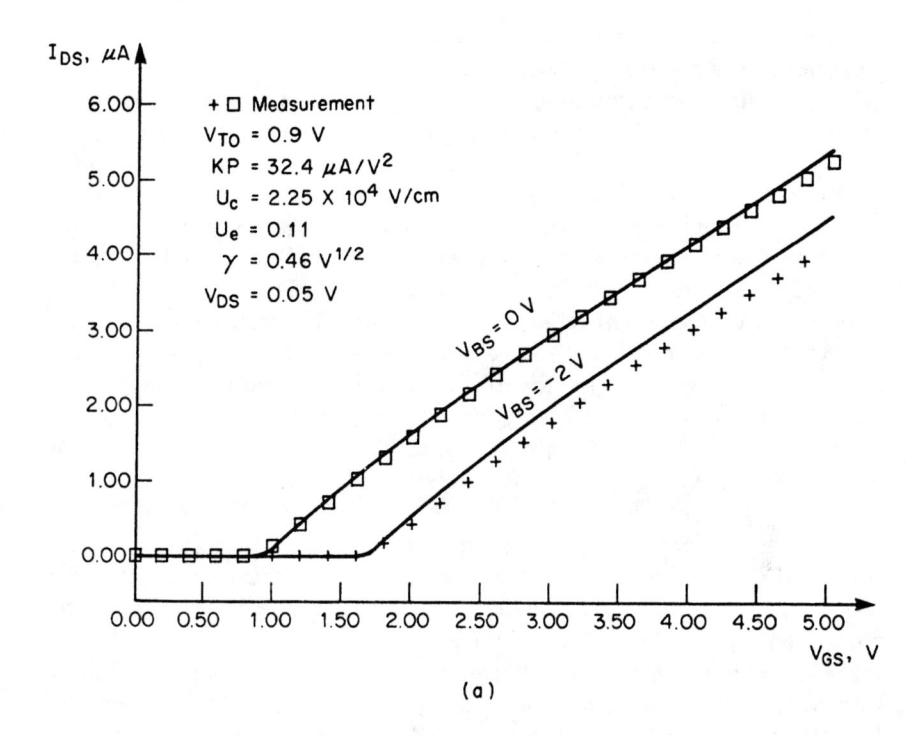

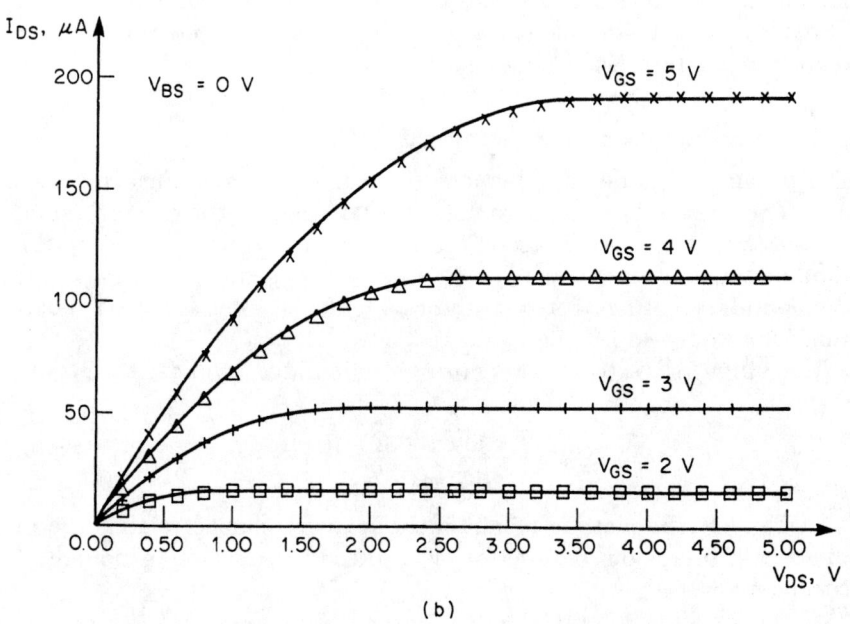

Figure 6-7 Measurements and simulations with the LEVEL2 model of (a) transfer characteristics and (b) output characteristics with a long channel.

Lastly, we can substitute the new value of $2\phi_p$ in Eq. (6-6) and recalculate the body-effect constant $\gamma = 0.463\ V^{1/2}$. This value does not differ significantly from that calculated in Sec. 6.1.3; it is therefore superfluous to calculate the doping N_A from Eq. (6-11) again.

The threshold voltage V_{T0} can also be calculated from physical-type parameters, by using Eq. (4-33).

In our example, the polysilicon gate has a different doping than the substrate (this situation is indicated in the SPICE2 .MODEL statement with the parameter $T_{PG} = 1$); since the equations of the model calculate the same value of threshold voltage measured, that is, $V_{T0} = 0.9\ V$, we need a number of surface states $N_{SS} = -2.5 \times 10^{10}\ cm^{-2}$. This value is physically absurd, due to the fact that in the real MOST the threshold voltage has been obtained with ion implantation; the effect of a doping region different from that of the substrate is not included in the model (see Sec. 4.11.1). Equation (4-33) can therefore be used in the absence of ion implantation.

The use of this group of parameters, as an alternative to the electrical ones, is convenient in order to analyze the sensitivity with respect to the process parameters. As an example, it is possible to predict with reasonable accuracy the currents of a MOST with a gate oxide $t_{ox} = 75$ nm, starting from physical-type parameters calculated with a gate oxide of $t_{ox} = 80$ nm, only by changing the t_{ox} parameter. This is not possible if the values of KP and γ have been given in the SPICE2 .MODEL statement, because these two parameters depend in reality on t_{ox} [see Eqs. (4-34) and (4-35)].

6.2.3 Measurements in the weak inversion region

The parameter to be calculated is the number of fast surface states N_{FS}. The presence of this parameter makes possible the calculation of the subthreshold current (see Sec. 4.3.4), and it is determined by the exponential law that links I_{DS} to V_{GS}, for $V_{GS} < V_{on}$, where V_{on} is the boundary voltage between the regions of weak and strong inversion [see Eq. (4-46)].

The value of N_{FS} can therefore be calculated by Eqs. (4-47) and (4-49):

$$N_{FS} = \left(\frac{\Delta V_{GS}}{\Delta \log I_{DS}} \frac{q}{kT} - 1\right) \frac{C'_{ox}}{q} \qquad (6\text{-}12)$$

In Fig. 6-8, the measurements of the transfer characteristics are in the weak inversion region for $V_{GS} < 0.85$ V, and the exponential coefficient is

$$\frac{\Delta V_{GS}}{\Delta \log I_{DS}} = 116\ mV\ \text{per decade}$$

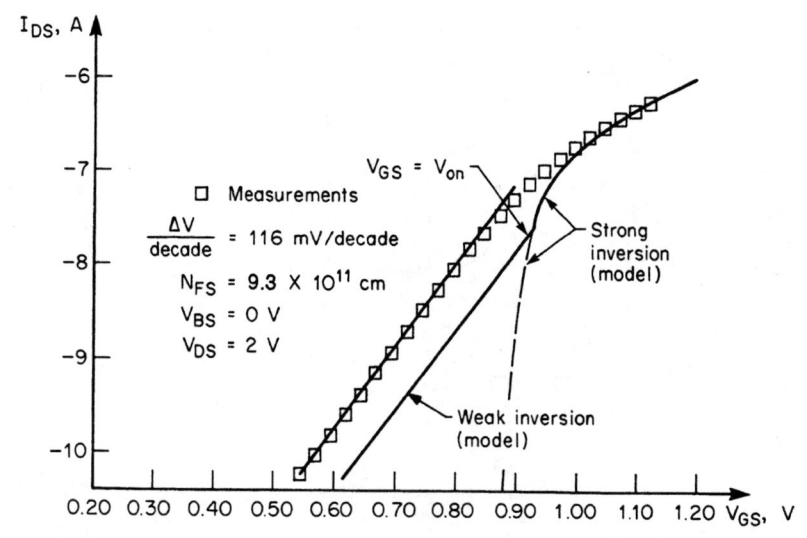

Figure 6-8 Measurements, calculation of N_{FS}, and simulation in the weak inversion region.

From the measurements in Fig. 6-8 we obtain $N_{FS} = 9.3 \times 10^{11}$ cm^{-2}. Unfortunately, the model does not ensure a good correlation with the measurements, even if the slope of the region exponentially growing is calculated correctly; this happens because it is not possible to vary the transition voltage V_{on}.

6.3 LEVEL2 Model (Short-Channel) Parameters

In the preceding section the basic parameters were found for any channel length or width. In this section the parameters not provided by the basic model, which relate to short- and narrow-channel effects, will be calculated. We will therefore use as examples MOSTs with short and narrow channels.

6.3.1 Calculation of the effective dimensions and of the series resistance

All the parameters still to be calculated depend on the length and width of the channel. It is therefore important to determine as precisely as possible the effective values of the length L_{eff} and the width W_{eff}; these values differ from the nominal ones because of constant contributions that are due to the process: the lateral diffusion X_{jl} for the length L_{nom} (Fig. 6-9a) and the step of the thick oxide W_{ox} for the width W_{nom} (Fig. 6.9b). In addition, there are random contributions due to imprecisions of the process in the definition of

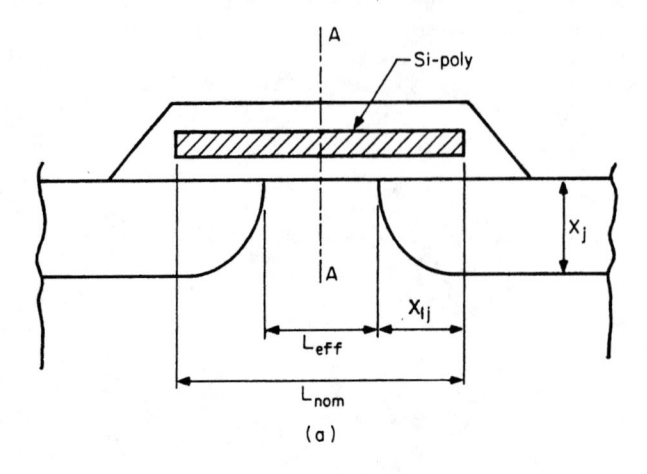

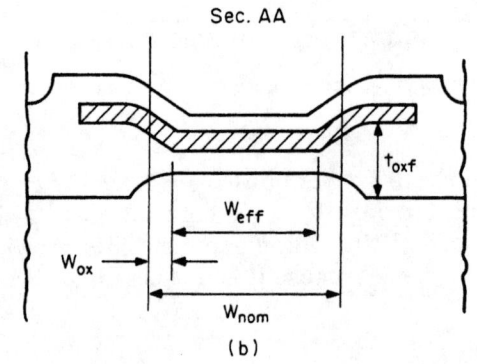

Figure 6-9 Definitions of effective length and width of a MOST.

the gate (L_{err} and W_{err}). Therefore,

$$\Delta L = L_{nom} - L_{eff} = 2X_{jl} + 2L_{err} \qquad (6\text{-}13)$$

$$\Delta W = W_{nom} - W_{eff} = 2W_{ox} + 2W_{err} \qquad (6\text{-}14)$$

The definition of the value of the parameters W and L in the input description for SPICE2 of the MOST device and of X_{jl} in the .MODEL statement must also take into account the value of the parameters L_{err}, W_{ox}, and W_{err}. Moreover, we will see that we also need to know the parasitic series resistance in order to measure L_{eff} precisely.

The effective length and width can be calculated from the electrical measurements, using the formulas relating L_{eff} and W_{eff} to the resistance R_{on} and to the equivalent channel conductance G_{on} in the linear region, obtained from Eq. (6-1):

$$R_{on} = \frac{V_{DS}}{I_{DS}} = \frac{1}{\beta(V_{GS} - V_{TH})} = \frac{L_{eff}}{KP\, W_{eff}(V_{GS} - V_{TH})} \qquad (6\text{-}15)$$

$$G_{on} = \frac{I_{DS}}{V_{DS}} = KP\, \frac{W_{eff}}{L_{eff}}(V_{GS} - V_{TH}) \qquad (6\text{-}16)$$

Repeating the measurements of R_{on} with the same voltages V_{GS} and V_{DS} on MOSTs of equal width and of scaled lengths, we obtain the graph in Fig. 6-10a. With a linear extrapolation to $R_{on} = 0$, we obtain

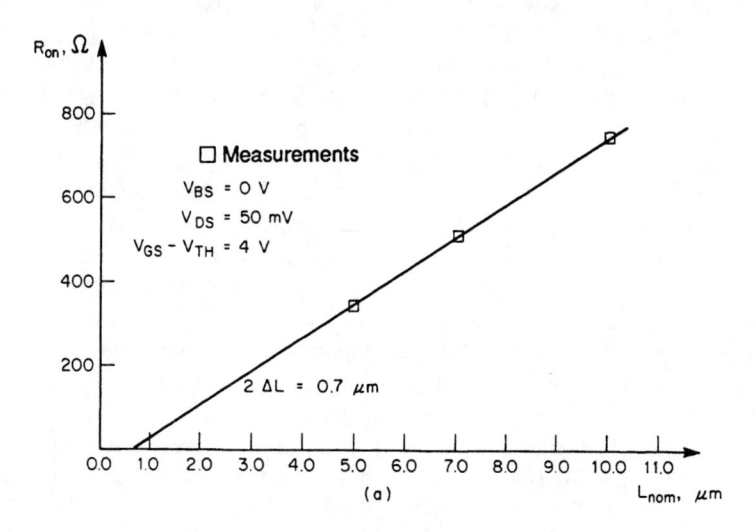

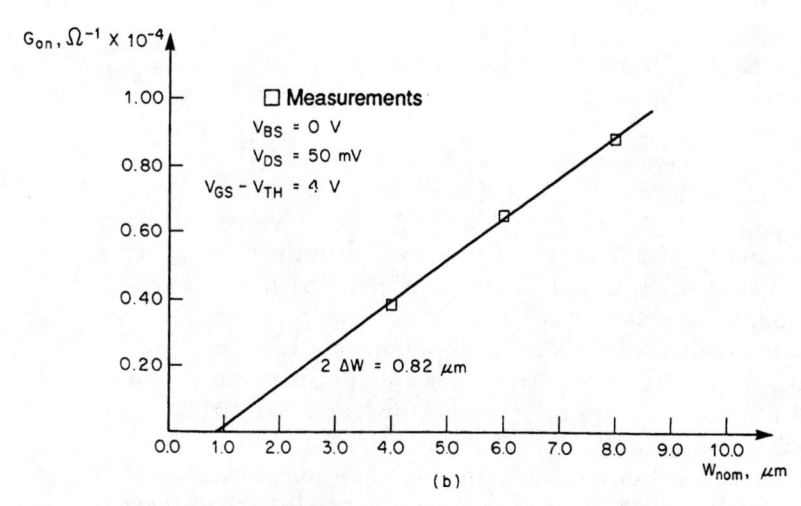

Figure 6-10 Electrical measurement of effective length and width of a MOST, without the effect of series resistance.

the total difference ΔL between the nominal length L_{nom} and the effective length L_{eff}, in the hypothesis that this difference is the same for all the MOSTs measured; if all the MOSTs come from the same chip, this assumption is certainly realistic. Similarly, the graph of G_{on} versus the nominal width W_{nom} in Fig. 6-10b gives ΔW.

To obtain greater precision, we can repeat the measurements with different values of V_{GS} and find the average values of ΔL and ΔW. If the threshold voltage is not the same on all MOSTs with different W or L—i.e., short- and narrow-channel effects are noticeable—it is better to make the measurements at constant $V_{GS} - V_{TH}$ instead of at constant V_{GS}. The values obtained from the example in Fig. 6-10 are $L_{nom} - L_{eff} = 0.70 \, \mu m$ and $W_{nom} - W_{eff} = 0.82 \, \mu m$; in order to separate the contributions indicated in Eq. (6-14) optical measurements are necessary.

If the channel length is small, R_{on} becomes comparable to the parasitic series resistance external to the MOST, which disturbs the measurements of the effective length. Rather complex methods of measurement exist to evaluate at the same time the total resistance series $R_t = r_D + r_S$ and the effective length L_{eff} [1,2]. This explanation will be limited to the simplest case, following the hypothesis that the series resistance is constant and equal for all the MOSTs measured; moreover, the effect of the voltage drop on r_S on the voltage V_{BS} will be neglected, and therefore its influence on the value of V_{TH} will be neglected, too.

In this case, adding the contribution of the total parasitic resistance R_t in Eq. (6-15), we obtain

$$R_{on} + R_t = \frac{L_{nom} - \Delta L}{W_{nom} - \Delta W} \frac{1}{KP(V_{GS} - V_{TH})} \tag{6-17}$$

This means that the linear interpolation of R_{on} against L_{nom}, with V_{GS} constant, intersects not on the axis $R_{on} = 0$ but at the point $R = R_t$ and $L_{nom} = \Delta L$ (see Fig. 6-11), where R_t is the sum of all the parasitic series resistances at the source and at the drain and ΔL is the difference between L_{nom} and L_{eff}.

From the example in Fig. 6-11 we find that the value of the effective length for a MOST with a nominal length of $L_{nom} = 5 \, \mu m$ is $L_{eff} = 3.4 \, \mu m$, which is significantly different from that found in Fig. 6-10a, neglecting the series resistance. We therefore obtain $\Delta L = 1.6 \, \mu m$, equivalent to the double of the process value of X_{jl}; substituting the values of L_{eff} found in this way and the process value of the lateral diffusion X_{jl} in Eq. (6-13), we find that the error L_{err}, due to the real length of the gate, is very small.

The series resistance calculated from the example measurements in Fig. 6-11 is $R_t = 30 \, \Omega$. To insert this value in SPICE2, we can use the parameters for the resistance of the source and of the drain

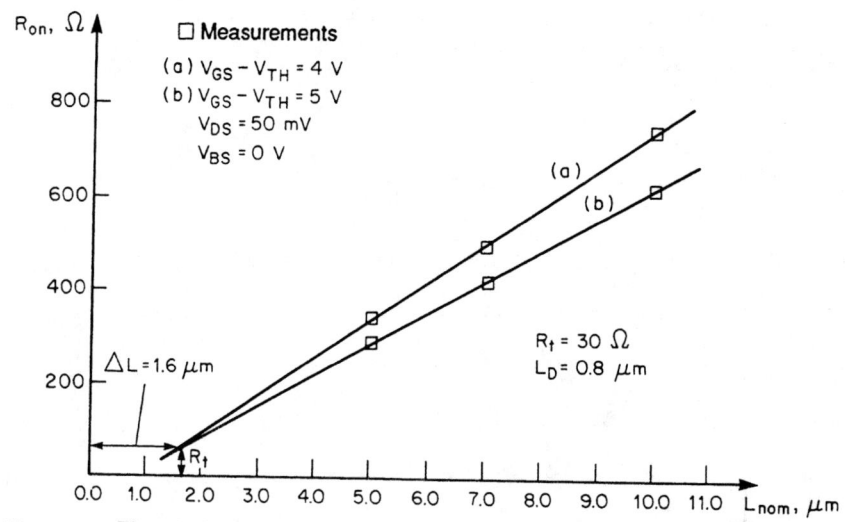

Figure 6-11 Electrical measurement of the effective length and of the resistance series of a MOST.

($r_S = r_D = 15\ \Omega$) or the parameter for the sheet resistance (R_{sh}), which is multiplied by the number of square in series, i.e., by the parameters NRD and NRS in the description of the MOST. Therefore, it is best to use r_D and r_S if the series resistances of the MOSTs in the simulation all have the same width and to use R_{sh} in all other cases.

6.3.2 Parameters for the model of the conductance in the saturation region

Once the effective dimensions are known, we can calculate the dimension-dependent parameters, beginning at the simplest equations of the conductance in the saturation region available in the LEVEL2 model; the parameter used by this model is λ, which relates to the average conductance for $V_{DS} > V_{D,sat}$ as follows:

$$G_{D,sat} = \frac{I_{D,sat}\lambda}{(1 - \lambda V_{DS})^2} \qquad (6\text{-}18)$$

The value of λ that we obtain from Eq. (6-18) can be calculated approximately, neglecting the term λV_{DS}, from the following equation:

$$\lambda = \frac{G_{D,sat}}{I_{D,sat}} \qquad (6\text{-}19)$$

The value of λ calculated in Fig. 6-12 is 0.053 V^{-1}, while the value computed by SPICE2 with Eq. (4-53), starting from the substrate doping, is 0.103 V^{-1} for $V_{GS} = V_{DS} = 5$ V.

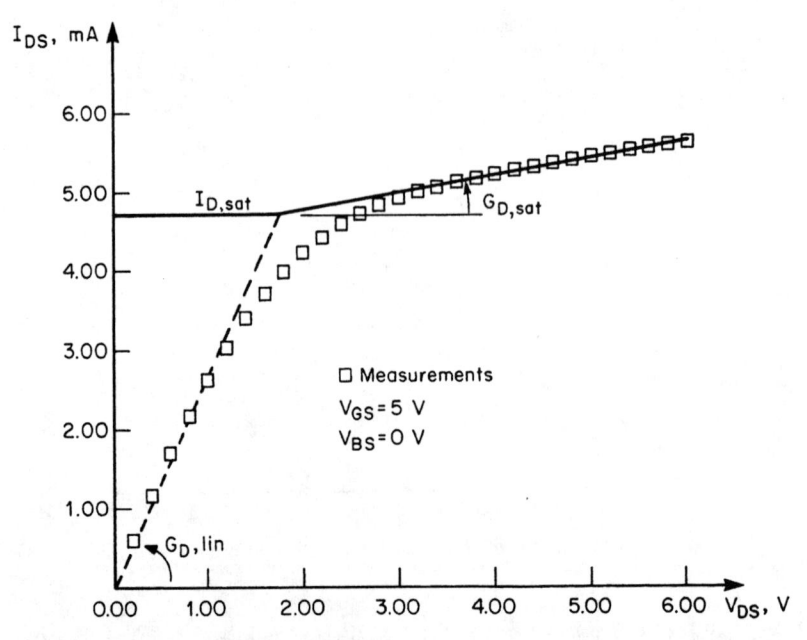

Figure 6-12 Determination of the average conductance in saturation and calculation of the parameter λ.

Remember that the corrective term of the current in the saturation region [see Eq. (4-44)], to avoid discontinuity between the two regions, acts in the linear region as well [see Eq. (4-43)]. This term, however, gives a false reading for the current in the linear region, with respect to that calculated with the model in Sec. 6.2.1; to solve the problem, reduce the value of KP, which has already been calculated, by a term dependent on λ and on a value of V_{DS} to be chosen. Thus

$$\text{KP}' = \text{KP}(1 - \lambda V_{DS}) \qquad (6\text{-}20)$$

The conductance in the saturation region can also be simulated with a more complex model (see Sec. 4.3.7); this model will be described later because the parameters X_j and $v_{\max}$ must be used but have not yet been calculated.

6.3.3 Parameters for the short-channel effect on the threshold voltage

The equations in Sec. 4.3.6 relate the threshold voltage to the channel length, if this is small enough to be compared with the depth of junction X_j or with the thickness of the depleted regions.

The parameter used in SPICE2 to model the effect of the short channel on the threshold voltage is the depth of the source and drain junctions, that is, X_j. We are dealing therefore with a parameter with a precise physical significance; however, due to the limited precision of the model, if we calculate X_j starting from the measurements of the threshold voltages for the short channel, the value obtained is not closely related to the real one.

The model uses the body-effect constant for the long channel, γ, whose value is modified for short-channel MOSTs with Eq. (4-56). The value of the depth of the junction X_j is therefore calculated by solving Eq. (4-56), with the hypothesis that the depleted regions under the source and drain are equal; i.e., with V_{DS} very small, X_j can be calculated with the following explicit equation:

$$X_j = \frac{L_{\text{eff}}^2 (1 - \gamma'/\gamma)^2}{2[W_S - L_{\text{eff}}(1 - \gamma'/\gamma)]} \tag{6-21}$$

where γ' is the body-effect constant calculated for the voltage V_{BS}, and W_S is the thickness of the depleted regions under the source, the drain, and the channel, given by

$$W_S = \sqrt{\frac{2\varepsilon_s}{qN_A}(-V_{BS} + 2\phi_p)} \tag{6-22}$$

Another possibility is to calculate a new value for the substrate doping N_A from the measurements of threshold voltage on a short-channel MOST; this is done by using in the calculations and in the SPICE2 .MODEL statement the real value for X_j. In this case, however, it is necessary to indicate explicitly in the SPICE2 .MODEL statement all the parameters that the program would otherwise calculate from N_A: γ, $2\phi_p$, and ϕ_j. The value of N_A that solves Eq. (4-56) is that for which the thickness of the depleted regions W_S is equal to

$$W_S = \frac{L_{\text{eff}}^2 (1 - \gamma'/\gamma)^2}{2X_j} + L_{\text{eff}}\left(1 - \frac{\gamma'}{\gamma}\right) \tag{6-23}$$

from which

$$N_A = \frac{2\varepsilon_s}{qW_S^2}(-V_{BS} + 2\phi_p) \tag{6-24}$$

The measurements in Fig. 6-13 have been carried out in the linear region of the short-channel MOST, which was chosen as an example.

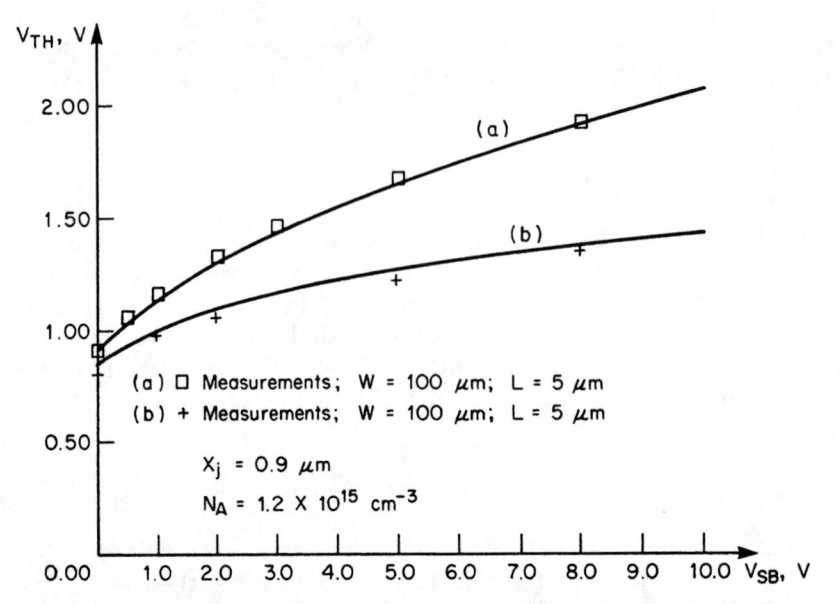

Figure 6-13 Measurements and simulations with the Yau model of the body effect of a short-channel MOST.

The value for the short-channel body effect calculated with $V_{BS} = -5$ V is $\gamma' = 0.269$ V$^{1/2}$. In the example, both methods have been tried: first the value of X_j has been calculated from Eq. (6-21) using the value of N_A calculated in Sec. 6.2.2; then N_A has been calculated from Eqs. (6-23) and (6-24) using the real value of X_j. The results obtained with the two procedures are as follows:

1. From Eq. (6-21):

$$N_A = 1.2 \times 10^{15} \text{ cm}^{-3} \quad X_j = 1.1 \ \mu\text{m}$$

2. From Eq. (6-23);

$$N_A = 1.1 \times 10^{15} \text{ cm}^{-3} \quad X_j = 1.0 \ \mu\text{m}$$

Verification made with a "best-fitting" program, solving Eq. (4-56) without any simplifying hypothesis and using all the measurements in Fig. 6-13, has given

$$N_A = 1.2 \times 10^{15} \text{ cm}^{-3} \quad X_j = 0.9 \ \mu\text{m}$$

The latter two values provide the most precise results.

6.3.4 Parameters for the short-channel effect on the mobility

As seen in Sec. 4.3.7, the reduction of mobility with the transverse electric field E_y is represented in this model by a reduction in the saturation voltage related to the parameter v_{max}.

This parameter is used, together with X_j and N_{eff}, in the calculation of the conductance in the saturation region with a model different from that seen in Sec. 6.3.2.

Given the complexity of the models, there are no methods that are simple and sufficiently precise to evaluate v_{max} and N_{eff}. Thus, we can only try varying v_{max} in order to find the right saturation current (by increasing v_{max}, $I_{D,sat}$ increases) and varying N_{eff} in order to find the conductance (by increasing N_{eff}, $G_{D,sat}$ decreases). Figure 6-14 represents the best result obtainable with the LEVEL2 model in the simulation of the short-channel MOST used as an example, with the parameters $v_{max} = 3 \times 10^4$ cm/s and $N_{eff} = 10$.

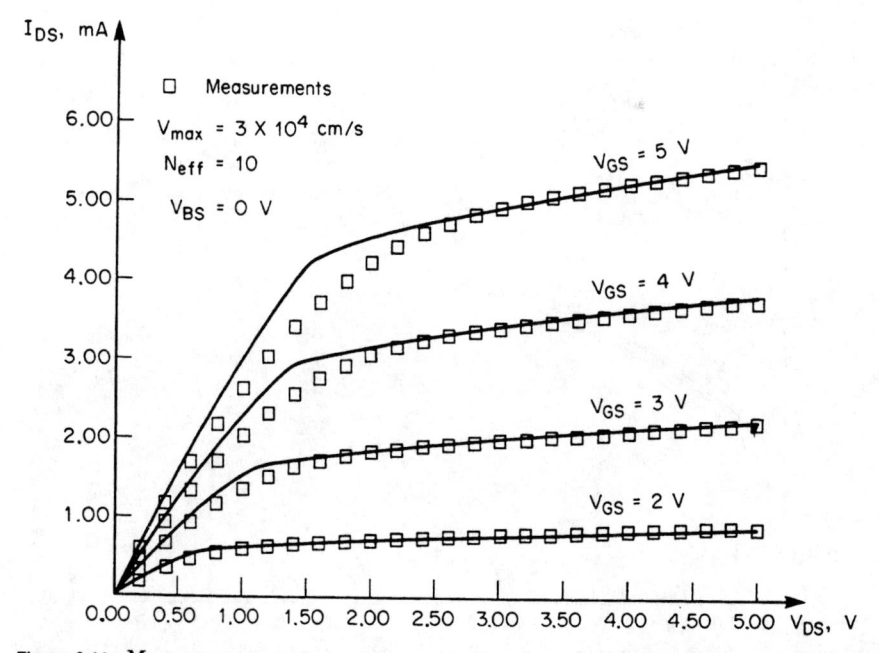

Figure 6-14 Measurements and simulations of the output characteristics of a LEVEL2 model of a short-channel MOST.

6.3.5 Parameters for the narrow-channel effect

As seen in Sec. 4.3.8, the narrowness of the channel influences the threshold voltage as well. The model uses parameter δ in order to calculate the value of the threshold voltage in relation to the width W. The value of δ can be calculated from the fitting of the body-effect curve of a narrow-channel MOST; in our example, $W_{\text{eff}} = 5.2\ \mu\text{m}$. We can find δ with a trial-and-error procedure similar to that used to calculate X_j (see Fig. 6-15) or with a graph evaluating the dependence of the measurements of V_{TH} on $1/W$ by using the formula

$$\delta = \frac{\Delta V_{TH}}{\Delta(1/W)} \frac{4C'_{ox}}{\pi X_D^2} \qquad (6\text{-}25a)$$

where

$$X_D = \sqrt{\frac{2\varepsilon_s}{qN_A}} \qquad (6\text{-}25b)$$

To use this method, however, we must have a sufficient number of MOSTs with different widths, although what we usually need to know is the dependence of the threshold voltage on the substrate voltage V_{BS}.

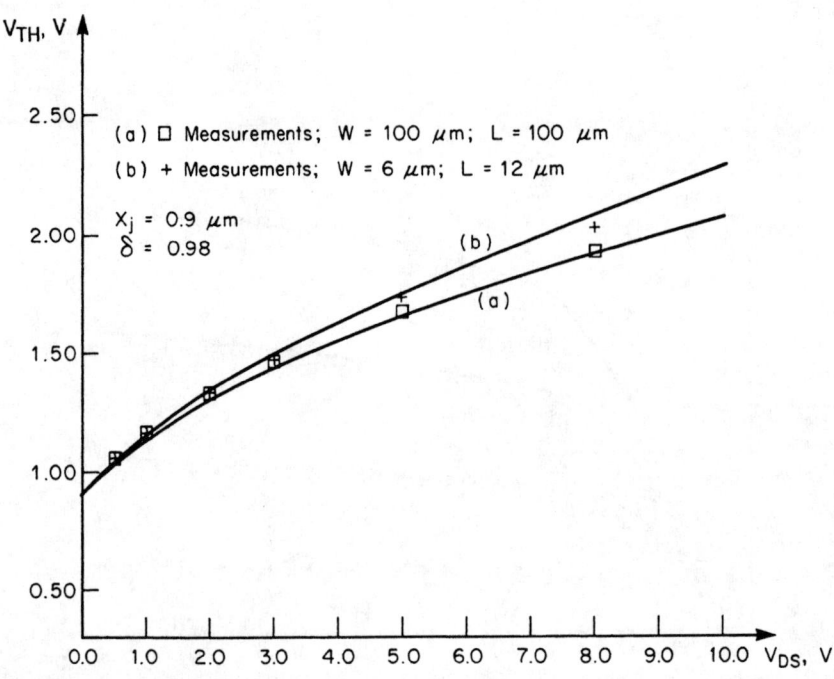

Figure 6-15 Measurements and simulations of the body effect of a narrow-channel MOST.

The MOSTs that are at the same time "short" and "narrow" are worthy of special note. These cannot always be simulated by simply overlapping the short-channel and narrow-channel effects, but often require that we choose the values of the parameters X_j, δ, and $v_{\max}$ which are specific for them; that is due to the difficulty of calculating the effective dimensions of these devices with the methods seen in Sec. 6.3.1.

6.4 LEVEL3 Model Parameters

The equations for the LEVEL3 model have been given in Chap. 4; even if the model has been explicitly developed for short-channel MOSTs, the fundamental parameters V_{T0}, KP (or μ_0) and γ (or N_A) are those of a long-channel MOST and can be calculated as for the other models.

This model is applicable to MOSTs with L_{nom} of $2\,\mu\text{m}$. However, in the example, we will again employ the MOST used to characterize the LEVEL2 model ($L_{\text{nom}} = 5\,\mu\text{m}$), so as to be able to make a comparison between the results obtained by simulating the same MOST with the two models.

6.4.1 Measurements in the linear region

For this model, too, we begin with the transfer characteristics in the linear region, using the equation for the simplest model in the calculations; it is still true that, for a small V_{DS}, the more complex models and the simplest model supply the same current values. Therefore, once V_{T0}, γ, μ_0, N_A, and $2\phi_p$ have been found (in the ways already explained), we must find the parameters for the model of the mobility variation with V_{GS} (see Sec. 4.5.1); this is different from what we saw in Sec. 6.2.1. The simplest method consists of calculating β' with at least two different values for V_{GS} and then finding the values of KP (or μ_0) and of θ to be inserted in the model by solving a system of two equations and two unknowns starting from Eq. (4-92):

$$\theta = \frac{\beta'_1/\beta'_2 - 1}{V_{GS2} - V_{T0} - (\beta'_1/\beta'_2)(V_{GS1} - V_{T0})} \qquad (6\text{-}26)$$

$$\text{KP} = \beta'_1 \frac{L}{W_{\text{eff}}} [1 + \theta(V_{GS1} - V_{T0})] \qquad (6\text{-}27)$$

Using an iterative method, we must calculate V_{T0} and θ, starting from a value of V_{T0} linearly extrapolated from the measurements with V_{GS} nearer to the threshold voltage V_{TH}, provided that it is in a region of strong inversion. Then we have to calculate β and θ with a linear regression method from the equation

$$\frac{\beta}{\beta_i'} = 1 + \theta(V_{GSi} - V_{T0}) \qquad (6\text{-}28)$$

where β_i' is the value of β' calculated for $V_{GS} = V_{GSi}$ (strong inversion). Once β and θ have been obtained, we can recalculate V_{T0} from Eq. (6-28) for each value of V_{GSi} and β_i' and repeat the procedure up to the convergence, which is usually reached very quickly. For the example in Fig. 6-16, the same measurements as in Fig. 6-2 have been used; the values calculated previously (Sec. 6.2) are

$$V_{T0} = 0.90 \text{ V}$$

$$\gamma = 0.46 \text{ V}^{1/2}$$

$$N_A = 1.2 \times 10^{15} \text{ cm}^{-3}$$

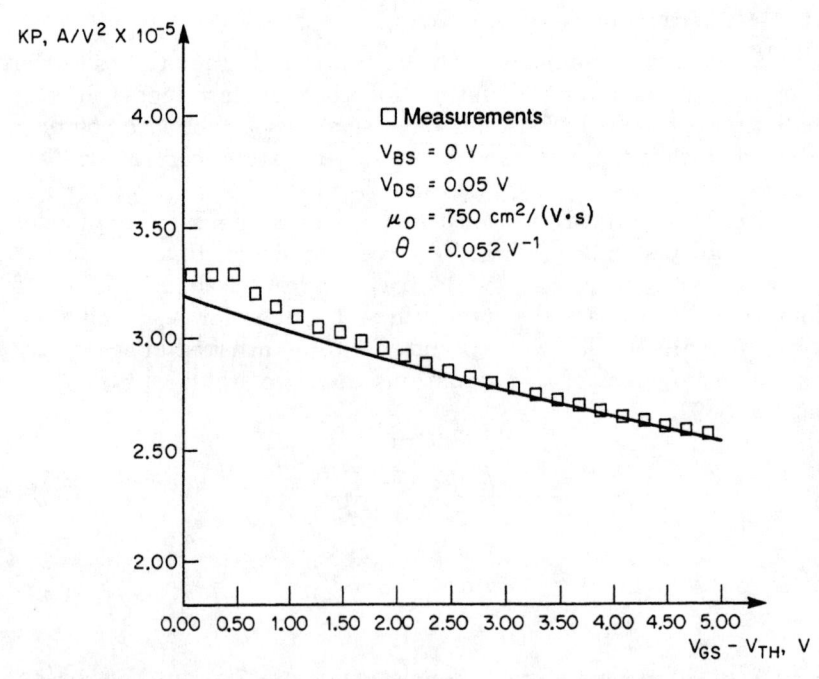

Figure 6-16 Measurements and simulations with the LEVEL3 model of transfer characteristics, calculation of β and θ.

With the simple method, and $V_{GS1} = 2$ V, $V_{GS2} = 5$ V

$$KP = 30.9 \ \mu A/V^2$$

$$\theta = 0.051 \ V^{-1}$$

$$\mu_0 = 715 \ cm^2/(V \cdot s)$$

With the iterative procedure

$$KP = 32.4 \ \mu A/V^2$$

$$\theta = 0.052 \ V^{-1}$$

$$\mu_0 = 750 \ cm^2/(V \cdot s)$$

These latter values provide greater precision; they have been obtained, minimizing the error, by using a least-squares fit starting from a greater number of measurements than in the first method.

6.4.2 Parameters for the short-channel effect on the threshold voltage

The model under examination (see Sec. 4.5.2) uses the two parameters for the effect of the short channel on the threshold voltage: X_j and η. the parameter X_j is used in the model of the body effect similar to, but more accurate than, that used in the equations of the LEVEL2 model.

The value of X_j calculated with Eq. (6-21) can be used only as an initial value for a trial-and-error procedure (see Fig. 6-17).

The parameter η, proportional to the variation of the threshold voltage with V_{DS} (see Sec. 4.5.2), is easily calculable from the threshold measurements in the saturation region and with V_{DS} constant (see Fig. 6-17). We can use the second method in Sec. 6.1.2.

The parameter η can be obtained from the linear dependence of V_{TH} on V_{DS} by inverting Eq. (4-94):

$$\eta = \frac{\Delta V_{TH}}{\Delta V_{DS}} \frac{C'_{ox} L^2_{eff}}{8.15 \times 10^{-22}} \tag{6-29}$$

From the example in Figs. 6-17 and 6-18 we can calculate

$$X_j = 1.1 \ \mu m \quad \eta = 0.98$$

6.4.3 Parameters for the short-channel effect on the mobility

This model also uses the parameter v_{max}, but in a way different from that seen in Sec. 6.3.4. In fact, as well as being part of the calculation

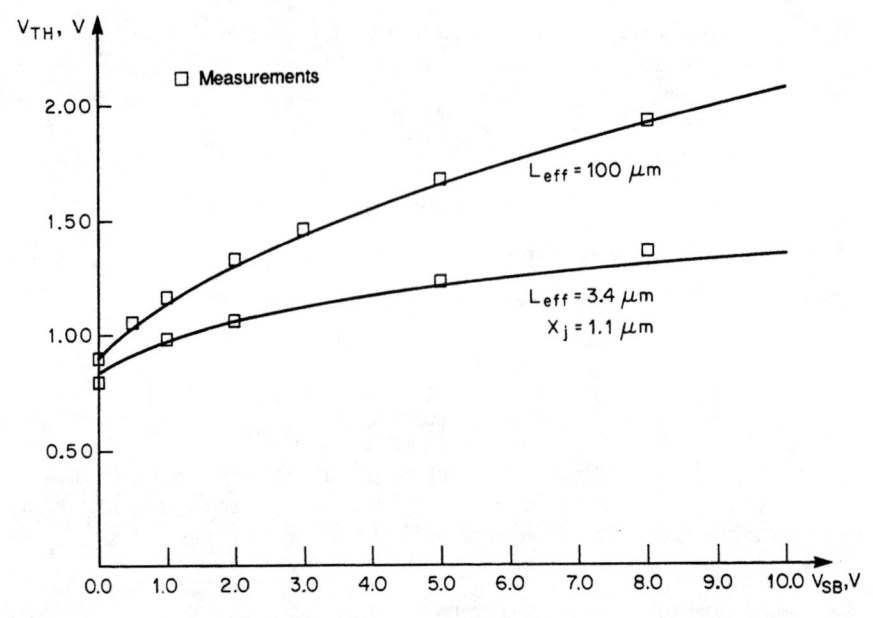

Figure 6-17 Measurements and simulations with the Dang model of body effect of a short-channel MOST.

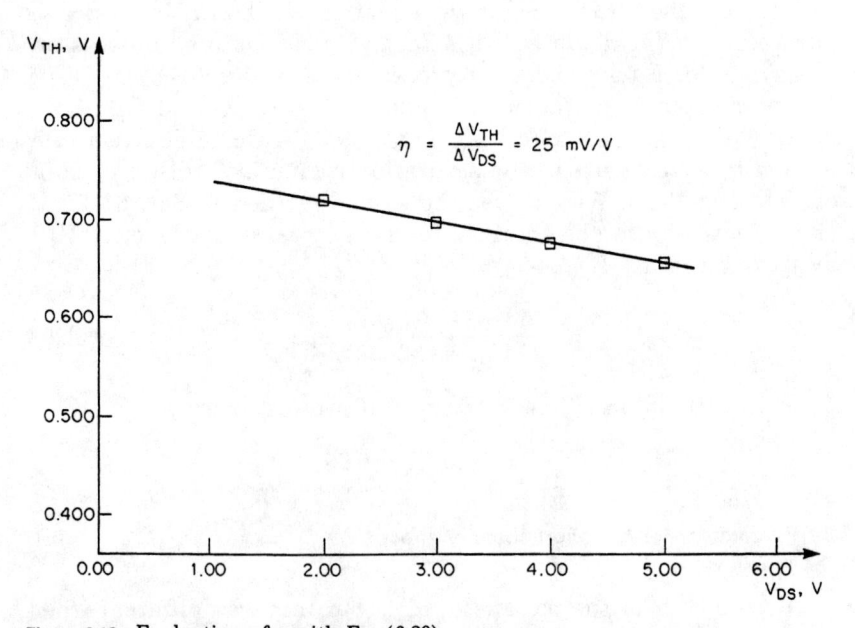

Figure 6-18 Evaluation of η with Eq. (6-29).

of $V_{D,sat}$, the parameter v_{max} reduces the effective mobility μ_{eff} in relation to the average electrical field in the channel $E_y = V_{DS}/L_{eff}$.

The method consists of calculating the mobility with a small V_{DS}, using the method of Sec. 6.4.1, and then calculating, with an increasing V_{DS} and a constant V_{GS}, the values of the effective mobility μ_{eff} from the measured current.

We can then calculate v_{max} also by using a graph based on a relation where μ_s is the mobility calculated with Eq. (4-92) and μ_{eff} is the mobility calculated with Eq. (4-100). Thus

$$\frac{1}{\mu_{eff}} - \frac{1}{\mu_s} = \frac{V_{DS}}{L_{eff} v_{max}} \tag{6-30}$$

The method explained is easily applicable, using a program that obtains μ_{eff} from the measurements; otherwise, it is better to proceed by trial and error.

In the example, the value of v_{max} has been calculated by trial and error in the way shown in Fig. 6-19. Then the value $v_{max} = 1.2 \times 10^5$ cm/s has been found, which is different from that found for the LEVEL2 model. The calculation can be repeated again for the other values of V_{GS}.

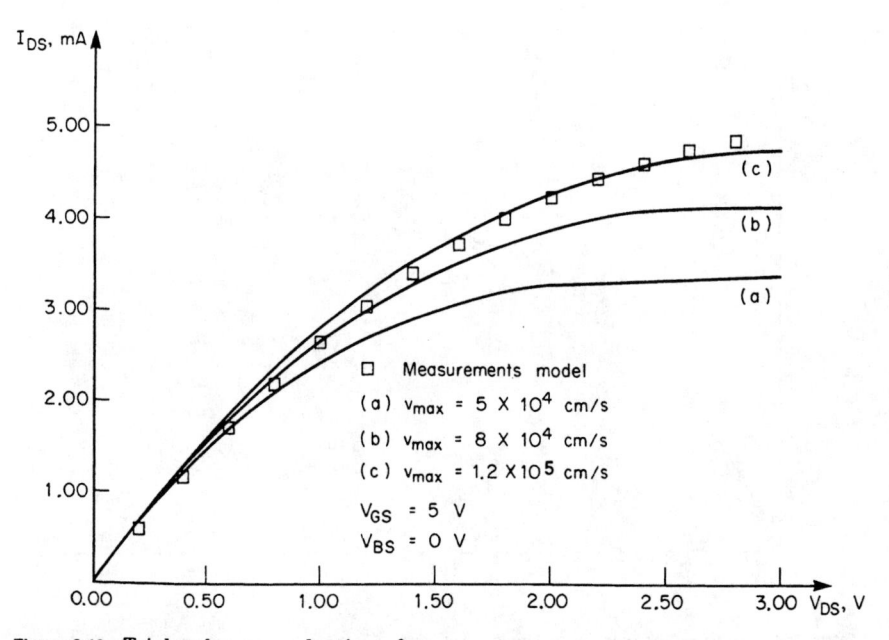

Figure 6-19 Trial-and-error evaluation of v_{max}.

6.4.4 Parameters for the model of the conductance in the saturation region

The parameter used for this model (see Sec. 4.5.5) is K.

Again a graphical method is proposed; it is advisable, however, to carry out some control simulations and, if necessary, to correct the K value by trial and error.

The relationship between the value of K and the shortening of the channel is obtained from Eq. (4-104); from this equation we can get an approximate equation that gives the value of K starting from the shortening of the channel in the saturation region

$$K X_D^2 = \frac{\Delta(L_{\text{eff}} - L')^2}{\Delta V_{DS}} \qquad (6\text{-}31a)$$

where

$$L_{\text{eff}} - L' = L_{\text{eff}}\left(1 - \frac{I_{DS}}{I_{D,\text{sat}}}\right) \qquad (6\text{-}31b)$$

In Fig. 6-20 the measurements of an output characteristic and simulations with different values for K are shown. The simulation shown in Fig. 6-21, in which the value of K is 1.5, represents the best result obtained with the LEVEL3 model for this example.

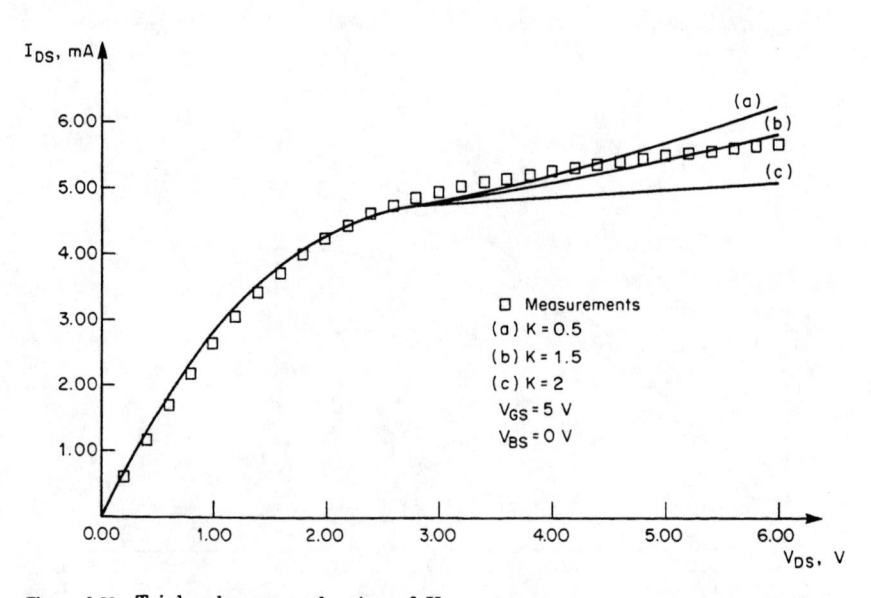

Figure 6-20 Trial-and-error evaluation of K.

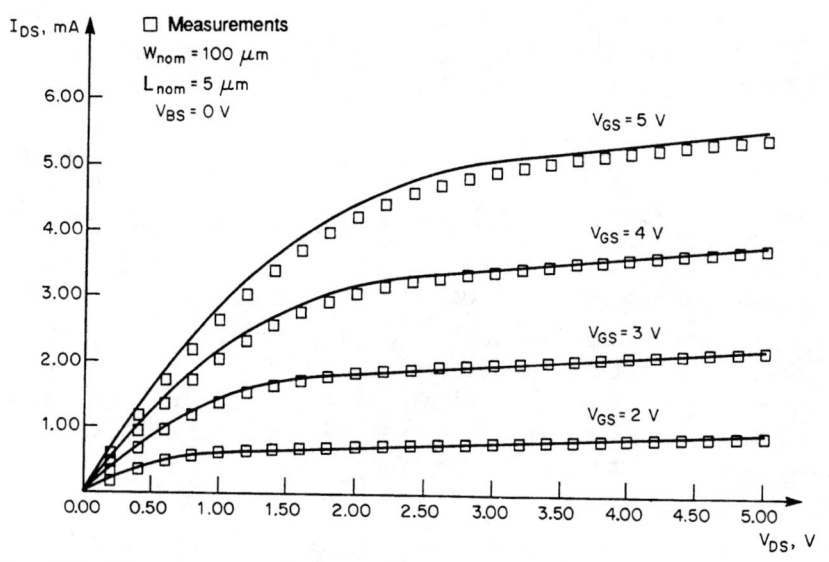

Figure 6-21 Measurements and simulation of the output characteristics of a short-channel MOST with the LEVEL3 model.

6.5 Measurements of Capacitance

6.5.1 Gate capacitance

The value of t_{ox} can be calculated from the measurements of the gate capacitance in the strong inversion region of a MOST with large dimensions (to avoid problems caused by parasitic capacitance) by using the equation

$$t_{ox} = \frac{\varepsilon_s (\text{area})}{C_G} \qquad (6\text{-}32)$$

The overlap capacitance between the gate and the diffusions is difficult to measure with precision, unless there are adequate test chips available. Knowing the lateral diffusion X_{jl} in the processes with a self-aligned gate, or the length of the overlap region for the processes that are not self-aligned, we can calculate the parameters C_{GDO} and C_{GSO} according to the theoretical formula

$$C_{GDO} = C_{GSO} = C'_{ox} X_{jl} \qquad (6\text{-}33)$$

Other, more precise formulas have been proposed [3,4], to take into account the distortion of the electric field at the edge of the gate electrode. Thus

$$C_{GSO} = C'_{ox} X_{jl} \left[1 + \frac{t_{ox}}{X_{jl}} \ln\left(1 + \frac{t_g}{t_{ox}}\right) + \frac{t_g}{X_{jl}} \ln\left(1 + \frac{X_{jl}/2}{t_{ox} + t_g}\right) \right] \qquad (6\text{-}34)$$

where t_g is the thickness of the gate electrode.

The parameter C_{GBO} can be calculated from the thickness of the thick oxide and from the length of the overlap between the gate and the substrate as

$$C_{GBO} = \frac{\varepsilon_{ox}}{t_{oxf}} W_g \qquad (6\text{-}35)$$

In most cases, however, the capacitance C_{GBO} can be neglected.

6.5.2 Capacitance of the diffused regions

To evaluate the values per unit area and perimeter of the junction capacitances, we have to measure the capacitance versus reverse voltage over two diffused regions, designed with different ratios between area and perimeter; these measurements make it possible to solve the following equations, for every voltage:

$$\begin{aligned} A_1 C_{ja} + P_1 C_{jp} &= C_1 \\ A_2 C_{ja} + P_2 C_{jp} &= C_2 \end{aligned} \qquad (6\text{-}36)$$

where the areas and perimeters of the diffused regions (A_1, A_2, P_1, P_2) are known, and C_1 and C_2 are the values of capacitance measured at the reverse voltage V_r.

Once we have determined C_{ja} and C_{jp} as a function of the reverse voltage V_r, the values of CJ and CJSW are those for $V_r = 0$; thus the exponents MJ and MJSW and the junction potential ϕ_j can be calculated by using the graph in Fig. 6-22. The experimental points in

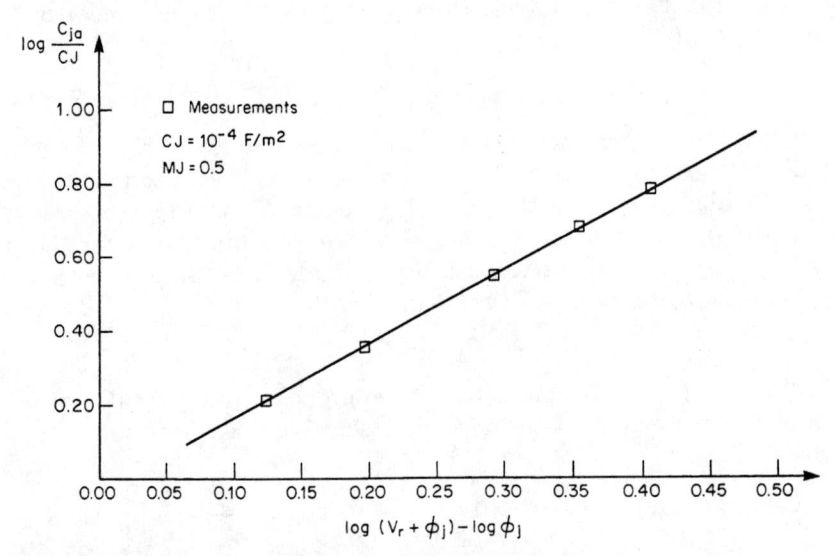

Figure 6-22 Graphic evaluation of the parameters MJ and MJSW.

Fig. 6-23 have been obtained by separating the contributions of the area and of the perimeter of the diffused regions expressed in Eqs. (4-87) and (4-88). On a logarithmic scale they are as follows:

$$\log \frac{C_{ja}}{CJ} = MJ[\log \phi_j - \log(V_r + \phi_j)] \tag{6-37}$$

$$\log \frac{C_{jp}}{CJSW} = MJSW[\log \phi_j - \log(V_r + \phi_j)] \tag{6-38}$$

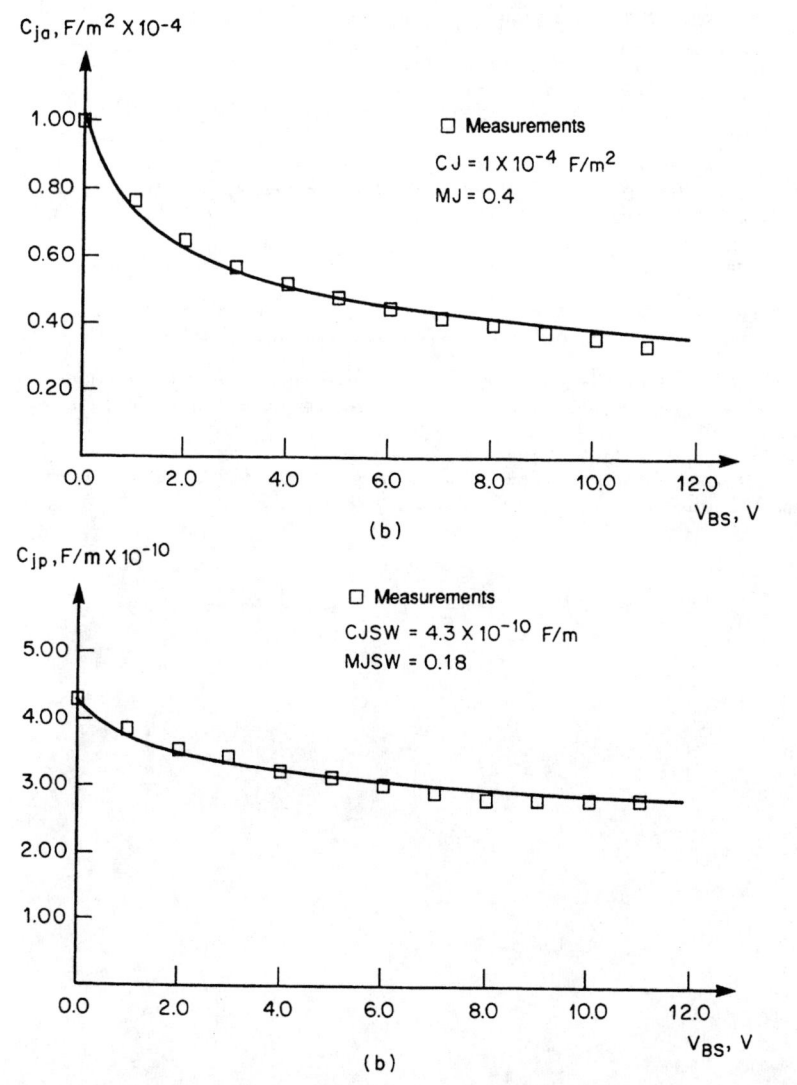

Figure 6-23 Measurements and simulations of (a) C_{ja} and (b) C_{jp}.

The values calculated from the measurements using Eq. (6-36) and the results of the simulation are shown in Fig. 6-23.

6.6 BSIM Model Parameter Extraction

A MOST parameter extraction program has been developed by Pierret [5] using the modified BSIM model (Berkeley Short-Channel IGFET model) and a version of CSIM (Compact Short-Channel IGFET model) parameter extraction program [6].

As the description in Ref. [5] is exhaustive, the user can directly refer to this work.

References

1. P. I. Suciu and R. L. Johnston, Experimental Derivation of the Source and Drain Resistance of MOS Transistors, *IEEE Trans. Electron Devices*, **27**, 1980.
2. K. Terada and H. Muta, A New Method to Determine the Effective Channel Length, *Jpn. J. Appl. Phys.* **18**, 1979.
3. M. I. Elmasry, Capacitance Calculation in MOSFET VLSI, *IEEE Electron Device Lett.*, **EDL-3**, 1982.
4. C. P. Yuan and T. N. Trik, A Simple Formula for the Estimation of the Capacitance of Two-Dimensional Interconnects in VLSI Circuits, *IEEE Electron Device Lett.*, **EDL-3**, 1982.
5. J. R. Pierret, A MOS Parameter Extraction Program for the BSIM Model, Rep. No. ERL-M84/99, Electronics Research Laboratory, University of California, Berkeley, 1984.
6. B. S. Messenger, A Fully Automated MOS Device Characterization System for Process-Oriented Integrated Circuit Design, Rep. No. ERL-M84/18, Electronics Research Laboratory, University of California, Berkeley, 1984.

Chapter 7

Noise and Distortion

7.1 Noise

Noise phenomena are caused by the small current and voltage fluctuations generated within the devices themselves. This section deals with the physical causes that give rise to noise in semiconductor devices and reviews the noise models implemented in SPICE2, SPICE3, HSPICE, and PSPICE.

7.1.1 Noise spectral density

The noise voltage or current $x(t)$ is a quantity with an average value that is often zero. Thus the most significant mathematical characterization is obtained using the mean-square value of $x(t)$:

$$\overline{X^2} = \lim_{T \to \infty} \frac{1}{T} \int_0^T x^2(t)\, dt \qquad (7\text{-}1)$$

or its root-mean-square (rms) value:

$$X = \sqrt{\overline{X^2}} \qquad (7\text{-}2)$$

Moreover, as the variable $x(t)$ might contain frequency components distributed on a large spectrum, it is necessary to correlate $\overline{X^2}$ to its power spectral density $S_x(f)$.

Note: This chapter has been written with the contribution of Ermete Meda, who is with ANSALDO, Genoa, Italy.

It can be shown that $\overline{X^2}$ and $S_x(f)$ are related through the following equation:

$$\overline{X^2} = \int_0^\infty S_x(f)\,df \tag{7-3}$$

The physical meaning of this expression is clear: The mean-square value of the noise, $\overline{X^2}$, is equal to the sum, extended to the whole spectrum, of contributions at different frequencies.

7.1.2 Noise sources

a. Shot noise. *Shot noise* is always associated with a direct-current flow and is present in diodes and bipolar transistors. The origin of shot noise can be understood by considering the *pn* junction and the carrier concentrations in the device in the forward-bias condition (see Appendix A).

The flow of each carrier across the junction is a random event and is dependent on the carrier having sufficient energy and velocity directed toward the junction. Thus external current I, which appears to be a steady current, is, in fact, composed of a large number of random, independent current pulses [1].

The fluctuation in I is termed *shot noise*. It can be shown that the resulting noise current has a mean-square value

$$\overline{i^2} = S_i(f)\,\Delta f = 2qI\,\Delta f \tag{7-4}$$

where Δf is the bandwidth, I is the current through the junction, q is the electron charge, and $S_i(f)$ is the power spectral density. Equation (7-4) is valid until the frequency becomes comparable to $1/\tau$, where τ is the carrier transit time through the depletion region.

The effect of shot noise can be represented in the low-frequency, small-signal equivalent circuit of the diode, as shown in Fig. 7-1.

b. Thermal noise. *Thermal noise* is generated by a mechanism different from that responsible for shot noise. In conventional resistors it is due to the random thermal motion of the electrons and is unaffected by the presence or absence of direct current, since typical electron drift velocities in a conductor are much less than electron thermal velocities. Since this source of noise is due to the thermal motion of electrons, it is related to absolute temperature T. In fact,

---

Note: The material in Sec. 7.1.2 is derived primarily from P. R. Gray and R. G. Meyer, *Analysis and Design of Analog Integrated Circuits*, copyright © 1977 by John Wiley & Sons, Inc. Reprinted by permission.

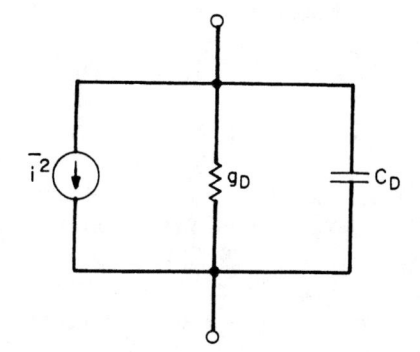

Figure 7-1 Junction diode small-signal equivalent circuit with noise.

thermal noise is *directly proportional* to T (unlike shot noise, which is *independent* of T), and as T approaches zero, thermal noise also approaches zero [1].

In a resistor R, thermal noise can be shown to be represented by a shunt current source $\overline{i^2}$, as in Fig. 7-2. This is equivalent to

$$\overline{i^2} = S_i(f)\,\Delta f = \frac{4kT}{R}\,\Delta f \tag{7-5}$$

where k is Boltzmann's constant. From Eq. (7-5) it can be seen that thermal noise is frequency-independent.

c. Flicker noise, or 1/f noise. *Flicker noise* is a type of noise found in all active devices (as well as in some discrete passive elements, such as carbon resistors). The origins of flicker noise are varied, but in BJTs it is caused mainly by traps associated with contamination and crystal defects in the emitter-base depletion region. These traps capture and release carriers in a random fashion, and the time constants associated with the process give rise to a noise signal with energy concentrated at low frequencies [1].

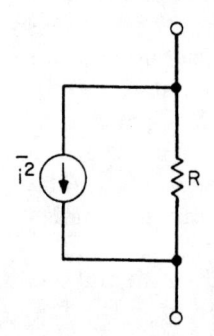

Figure 7-2 Representation of thermal noise.

Flicker noise is always associated with a flow of direct current and displays a spectral density of the form

$$\overline{i^2} = S_i(f)\,\Delta f = k_f \frac{I^{a_f}}{f^b}\,\Delta f \tag{7-6}$$

where Δf is a small bandwidth at frequency f, I is a direct current, k_f is a constant for a particular device (it depends on contamination and crystal imperfections), a_f is a constant in the range 0.5 to 2, and b is a constant, approximately unity. If $b = 1$ in Eq. (7-6), the noise spectral density has $1/f$ frequency dependence, thus the alternative name $1/f$ noise.

d. Burst, or popcorn, noise. *Burst noise* is another type of low-frequency noise found in some integrated circuits and discrete transistors. The source of this noise is not fully understood, although it has been shown to be related to the presence of heavy-metal ion contamination. Gold-doped devices show very high levels of burst noise [1]. The spectral density of burst noise can be shown to be

$$\overline{i^2} = S_i(f)\,\Delta f = k_b \frac{I^c}{1 + (f/f_c)^2}\,\Delta f \tag{7-7}$$

where k_b is a constant for a particular device, I is a direct current, c is a constant in the range 0.5 to 2, and f_c is a particular frequency for a given noise process.

7.1.3 Noise models and their implementation in SPICE2

The *ac* analysis portion of SPICE2 [2] contains the capability of evaluating the noise characteristics of an electronic circuit. Every resistor in the circuit generates thermal noise, and every semiconductor device in the circuit generates both shot and flicker noise, in addition to the thermal noise generated by the ohmic resistances of the device.

The noise generated by a circuit element can be modeled as an electrical excitation to the small-signal circuit. Each noise source in the circuit is statistically uncorrelated to the other noise sources in the circuit. The contribution of each noise source in the circuit to the total noise, at a specified output, is determined separately. The total output noise is then the rms sum of the individual noise contributions.

Noise is specified with respect to a certain noise bandwidth and has the units $V/Hz^{1/2}$ or $A/Hz^{1/2}$. The output noise of a circuit is the rms level of noise at the output divided by the square root of the bandwidth. The equivalent input noise is the output noise divided by the transfer function of the output with respect to the specified input.

In this section, the noise models for the semiconductor devices implemented in SPICE2 and other SPICE2-based versions are presented.

It is important to note that the equations developed in the following pages for each device model are implemented in the program by their rms values [see Eq. (7-2)], as shown in the figures.

a. DIODE noise model. The equivalent circuit for a junction diode was considered briefly in the description of shot noise. The basic equivalent circuit of Fig. 7-1 can be made complete by adding the series resistance r_S as shown in Fig. 7-3. Since r_S is a physical resistor due to the resistivity of the silicon, it exhibits thermal noise. Experimentally it has been found that any flicker noise present can be represented by a current source in shunt with a resistor; this is conveniently combined with the shot-noise source as indicated by

$$\overline{i_{rS}^2} = \frac{4kT}{r_S} \Delta f$$

$$\overline{i_D^2} = 2qI_D \, \Delta f + k_f \frac{I_D^{a_f}}{f} \Delta f \tag{7-8}$$

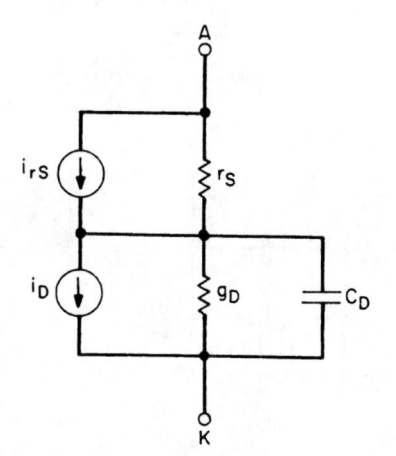

Figure 7-3 Complete diode small-signal equivalent circuit with noise sources.

Note: The material in Secs. 7.1.3*a* to *c* is taken from P. R. Gray and R. G. Meyer, *Analysis and Design of Analog Integrated Circuits*, copyright © 1977 by John Wiley & Sons, Inc. Reprinted by permission.

The parameters, k_f (KF) and a_f (AF) are .MODEL statement parameters. For silicon diodes, typical values for these parameters are $k_f = 10^{-16}$ and $a_f = 1$. The keyword used in the .MODEL statement is indicated in parentheses.

b. BJT noise model. In a bipolar transistor in the forward active region, minority carriers diffuse and drift across the base region to be collected at the collector-base junction. Minority carriers entering the collector-base depletion region are accelerated by the field existing there and swept across this region to the collector. The time of arrival at the collector-base junction of the diffusion (or drifting) carriers is a random process, and thus the transistor collector current consists of a series of random current pulses. Consequently, collector current I_C shows shot noise, and this is represented by a shot-noise current source $\overline{i_C^2}$.

Base current I_B in a transistor is due to recombination in the base and base-emitter depletion regions and also to carrier injection from the base into the emitter. All of these are independent, random processes, and thus I_B also shows shot noise. This is represented by a shot-noise current source $\overline{i_B^2}$ in Fig. 7-4.

Transistor base resistor r_B is a physical resistor and thus has thermal noise. Collector series resistor r_C also shows thermal noise, but since this is in series with the high impedance collector node, this noise is negligible and is usually not included in the model. Note that resistors r_π and r_o in the model are *fictitious* resistors that are used for modeling purposes only, and they do *not* exhibit thermal noise [1].

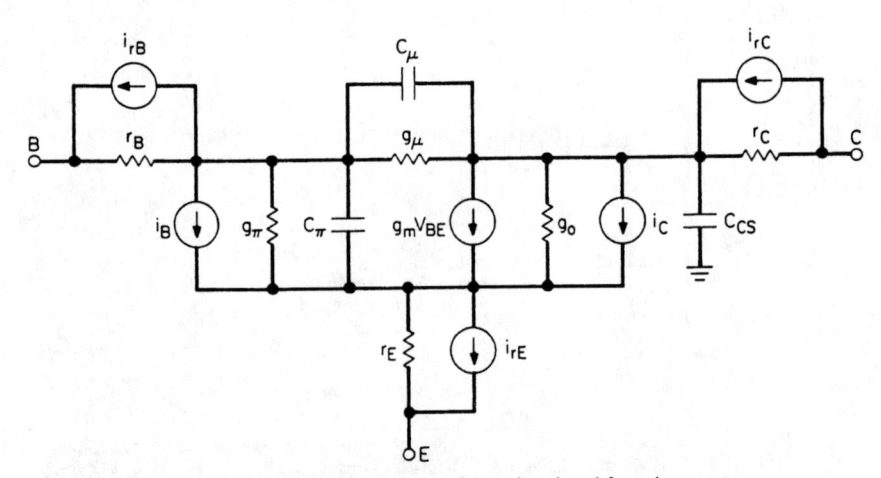

Figure 7-4 Complete BJT small-signal equivalent circuit with noise sources.

Flicker noise and burst noise in a bipolar transistor have been found experimentally to be represented by current sources across the internal base-emitter junction. However, burst noise is not modeled in SPICE2, and flicker noise is conveniently combined with the shot noise sources in $\overline{i_B^2}$.

The full small-signal equivalent circuit including noise for the BJT is shown in Fig. 7-4. Since they arise from separate, independent physical mechanisms, all the noise sources are independent of each other and have mean-square values.

$$\overline{i_{rB}^2} = \frac{4kT}{r_B} \Delta f$$

$$\overline{i_{rC}^2} = \frac{4kT}{r_C} \Delta f$$

$$\overline{i_{rE}^2} = \frac{4kT}{r_E} \Delta f \tag{7-9}$$

$$\overline{i_C^2} = 2qI_C \Delta f$$

$$\overline{i_B^2} = 2qI_B \Delta f + k_f \frac{I_B^{a_f}}{f} \Delta f$$

This equivalent circuit is valid for both *npn* and *pnp* transistors. For *pnp* devices, the magnitudes of I_B and I_C are used in the above equations. The parameters a_f (AF) and k_f (KF) are .MODEL statement parameters. Typical values for KF are 10^{-17} to 10^{-12}; they are calculated as $2qf_k$, where f_k is noise knee frequency and q is the electron charge.

c. JFET noise model. It was shown in Chap. 3 that the resistive channel joining source and drain is modulated by the gate-source voltage so that the drain current is controlled by the gate-source voltage. Since the channel material is resistive, it exhibits *thermal noise*, and this is the major source of noise in JFETs. It can be shown that this noise source can be represented by a noise current source $\overline{i_D^2}$ from drain to source in the JFET small-signal equivalent circuit of Fig. 7-5. Flicker noise in the JFET is also found experimentally to be represented by a drain-source current source, and these two can be lumped into one noise source $\overline{i_D^2}$. The other source of noise in JFETs is shot noise generated by the gate leakage current, but it is not modeled in SPICE2.

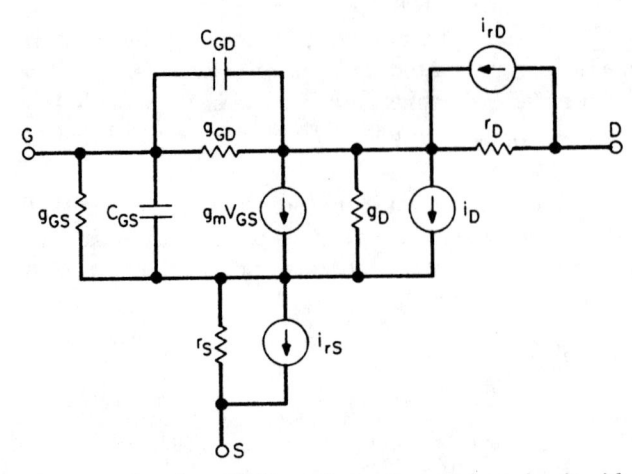

Figure 7-5 Complete JFET small-signal equivalent circuit with noise sources.

The noise sources in Fig. 7-5 have the following values:

$$\overline{i_{rD}^2} = \frac{4kT}{r_D} \Delta f$$

$$\overline{i_{rS}^2} = \frac{4kT}{r_S} \Delta f \tag{7-10}$$

$$\overline{i_D^2} = \frac{8kTg_m}{3} \Delta f + k_f \frac{I_D^{a_f}}{f} \Delta f$$

where I_D is the drain bias current, k_f is a constant for a given device whose values are in the range 10^{-19} to 10^{-25}, a_f is a constant in the range 0.5 to 2, and g_m is the transconductance of the device at the operating point. The parameters a_f (AF) and k_f (KF) are .MODEL statement parameters.

d. MOST noise model. As in the JFET, also in the MOST we can find thermal noise in the drain and source resistances as well as shot and flicker noise. All these noise sources can be represented by noise current sources (see Fig. 7-6) whose values are

$$\overline{i_{rD}^2} = \frac{4kT}{r_D} \Delta f$$

$$\overline{i_{rS}^2} = \frac{4kT}{r_S} \Delta f \tag{7-11}$$

$$\overline{i_D^2} = \frac{8kTg_m}{3} \Delta f + k_f \frac{I_D^{a_f}}{fC_{ox}L_{eff}^2}$$

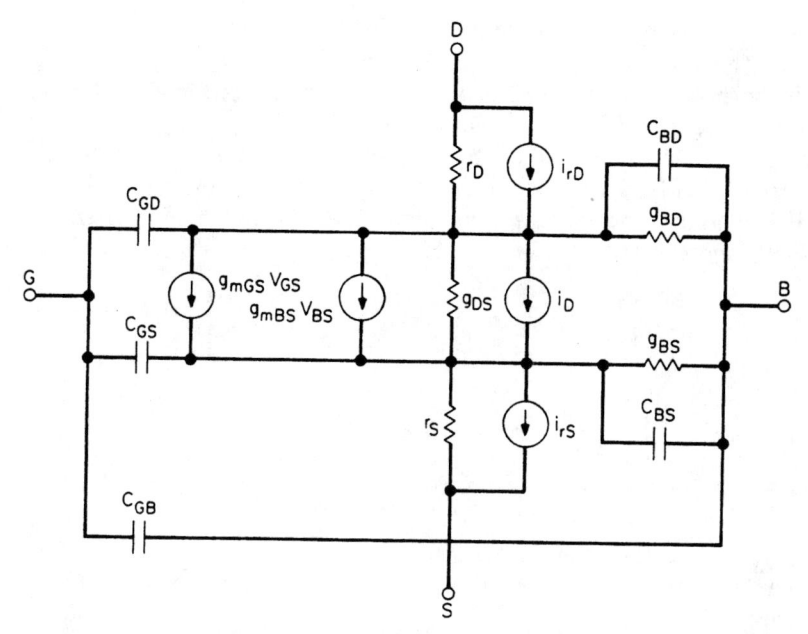

Figure 7-6 Complete MOST small-signal equivalent circuit with noise sources.

where I_D is the drain bias current, a_f (AF) is the flicker noise exponent, k_f (KF) is the flicker noise coefficient, $C'_{ox} = \varepsilon_{ox}/\text{TOX}$ is the oxide capacitance for unit area, $L_{\text{eff}} = (L - 2\text{LD})$ is the channel real length, with L being the channel nominal length (device element statement parameter), LD the lateral diffusion into channel from source and drain diffusion (.MODEL statement parameter), and TOX the oxide thickness (.MODEL statement parameter). Reasonable values for k_f are in the range 10^{-19} to 10^{-25}.

7.1.4 Noise models and their implementation in SPICE2-based versions

According to the SPICE2-based versions considered in this book, as SPICE3 exhibits no significant deviations from the SPICE2 noise models, we focus the attention mainly on HSPICE and PSPICE.

a. HSPICE noise models. DIODE and BJT noise models in HSPICE [3] do not differ from the analogous ones of SPICE2.

The HSPICE JFET noise model is based on the noise model implemented in SPICE2 from which it differs by an added thermal noise component due to the gate resistance r_G [in the form of the thermal terms in Eqs. (7-10)], and by two shot noise equations selected by setting the .MODEL statement parameter NLEV (default = 2) to a proper value. If NLEV < 3, then the JFET shot noise equation defaults to the one of SPICE2. This shot noise formula [Eq. (7-10)] is used in both saturation and linear regions, which can lead to wrong results in the linear region. For example, at $V_{DS} = 0$, shot noise becomes zero, because $g_m = 0$; this is physically impossible. If NLEV is set to 3, HSPICE will use a different equation, which is valid in both linear and saturation regions. Thus the shot noise component in Eq. (7-10) becomes [3]

$$\overline{i^2_{shot}} = \left[\frac{8kT}{3}\beta(V_{GS} - V_{T0})\frac{1+\alpha+\alpha^2}{1+\alpha}g_{DSN}\right]\Delta f \qquad (7\text{-}12)$$

where

$$\alpha = \begin{cases} 1 - \dfrac{V_{DS}}{V_{GS} - V_{T0}} & \text{(linear region)} \\ 0 & \text{(saturation region)} \end{cases} \qquad (7\text{-}13)$$

and g_{DSN} (GDSNOI), the shot noise coefficient, is a .MODEL statement parameter (default = 1), whose HSPICE keyword is indicated in parentheses.

The HSPICE MOST noise model equations have a selector parameter NLEV (default = 2) that the user of HSPICE can specify in the .MODEL statement and that is used to select either the original SPICE2 shot and flicker noise or new equations [3]. The thermal noise generated by the drain and source resistances have the same relations as in SPICE2. If NLEV < 3, then the MOST shot noise equation defaults to the one of SPICE2. This shot noise formula [Eq. (7-11)] is used in both saturation and linear regions, which can lead to wrong results in the linear region. For example, at $V_{DS} = 0$, shot noise becomes zero, because $g_m = 0$. This is physically impossible. If NLEV is set to 3, HSPICE will use a different equation which is valid in both linear and saturation regions. Thus the shot noise component in Eq. (7-11) becomes [3]

$$\overline{i^2_{shot}} = \left[\frac{8kT}{3}\beta(V_{GS} - V_{TH})\frac{1+\alpha+\alpha^2}{1+\alpha}g_{DSN}\right]\Delta f \qquad (7\text{-}14)$$

where

$$\alpha = \begin{cases} 1 - \dfrac{V_{DS}}{V_{D,sat}} & \text{(linear region)} \\ 0 & \text{(saturation region)} \end{cases} \qquad (7\text{-}15)$$

and g_{DSN} (GDSNOI) is the same as for the JFET. As mentioned before, the parameter NLEV also affects the flicker noise. In this case, if NLEV = 0, the flicker noise formula defaults to Eq. (7-11), i.e., the SPICE2 formula. If NLEV = 1, the L_{eff}^2 term in Eq. (7-11) is replaced by $L_{\text{eff}} W_{\text{eff}}$. If NLEV = 2, 3, the flicker noise component in Eq. (7-11) becomes [3]

$$\overline{i_{\text{flicker}}^2} = \frac{k_f g_m^2}{f^{a_f} C_{\text{ox}}' W_{\text{eff}} L_{\text{eff}}} \Delta f \tag{7-16}$$

In Eq. (7-16), neglecting masking, etching, and scaling effects, $W_{\text{eff}} = (W - 2WD)$, W being the channel width (device statement parameter), and WD the lateral diffusion into channel from bulk along width (.MODEL statement parameter).

The HSPICE JFET noise model describes the HSPICE MESFET noise model as well.

b. PSPICE noise models. DIODE and JFET noise models in PSPICE [4] do not differ from the analogous ones of SPICE2.

The PSPICE BJT noise model, on the other hand, is characterized, in comparison with SPICE2, by an added flicker noise component for the noise source generated by the collector current. This component modifies the $\overline{i_C^2}$ term in Eq. (7-9) as follows [4]:

$$\overline{i_C^2} = 2qI_C \Delta f + k_f \frac{I_C^{a_f}}{f} \Delta f \tag{7-17}$$

The PSPICE MOST noise model is based on the noise model implemented in SPICE2, from which it differs by two added thermal noise components due to the gate (r_G) and bulk (r_B) resistances, respectively [in the form of the thermal terms in Eqs. (7-11)].

The PSPICE JFET noise model describes the PSPICE MESFET noise model provided that the thermal noise component generated by the gate resistance r_G is taken into account.

7.2 Distortion

This section deals first with the results necessary for a better understanding of harmonic distortion, at low and high frequency. Subsequently, the methodology used in SPICE2 for the calculation of distortion coefficients is outlined [5].

7.2.1 Harmonic distortion coefficients

The methodology for determining the coefficients of harmonic distortion, at low and high frequency, is now briefly reviewed.

a. Low frequencies. At low frequencies, where reactive effects can be neglected, the output signal $y(t)$ (voltage or current) of a nonlinear circuit can be expressed in terms of its input $x(t)$ by a *Taylor* series [6]:

$$y(t) = a_1 x(t) + a_2 x^2(t) + a_3 x^3(t) + \cdots \qquad (7\text{-}18)$$

Coefficient a_1 represents the linear gain of the circuit, whereas coefficients $a_2, a_3, \ldots$, represent its distortion. It is important to note that, at low frequencies, $a_1, a_2, a_3, \ldots, a_n$ are constants.

Applying a cosine wave of frequency ω and amplitude V_A at the input, the nth harmonic distortion (HD_n) is defined as the ratio of the component at frequency $n\omega$ to the one at the fundamental ω. This is obtained by a trigonometric manipulation [7] so that Eq. (7-18) becomes

$$y(t) = (a_1 + \tfrac{3}{4} a_3 V_A^2) V_A \cos \omega t + \tfrac{1}{2} a_2 V_A^2 \cos 2\omega t$$
$$+ \tfrac{1}{4} a_3 V_A^3 \cos 3\omega t + \cdots \qquad (7\text{-}19)$$

Under low-distortion conditions, only the second- and third-order distortion components are considered, so that harmonic distortion is then specified by

$$HD_2 = \frac{\text{amplitude 2d harmonic}}{\text{amplitude fundamental at the output}} = \frac{a_2 V_A}{2 a_1} \qquad (7\text{-}20a)$$

$$HD_3 = \frac{\text{amplitude 3d harmonic}}{\text{amplitude fundamental at the output}} = \frac{a_3 V_A^2}{4 a_1} \qquad (7\text{-}20b)$$

Applying now the sum of two cosine waves of frequencies ω_1 and ω_2, both of amplitude V_A at the input, gives rise to output signal components at all combinations of ω_1, ω_2, and their multiples. Under low-distortion conditions, the number of terms can be reduced to the ones caused by coefficients a_2 and a_3 only [7].

Second-order intermodulation distortion IM_2 is defined by the ratio of the component at frequency $\omega_1 + \omega_2$ to the one at ω_1 or ω_2, so that

$$IM_2 = \frac{a_2}{a_1} V_A \qquad (7\text{-}21)$$

and by comparison with Eq. (7-20a)

$$IM_2 = 2 HD_2 \qquad (7\text{-}22)$$

Third-order intermodulation distortion IM_3 can be detected at the frequencies $2\omega_1 + \omega_2$ and $2\omega_2 + \omega_1$ and is given by

$$IM_3 = \tfrac{3}{4} \frac{a_3}{a_1} V_A^2 \qquad (7\text{-}23)$$

such that

$$IM_3 = 3HD_3 \tag{7-24}$$

These components can be mapped versus frequency, as shown in Fig. 7-7, by considering their Fourier transforms [6]

$$F(n\omega) = \int_{-\infty}^{\infty} V_A^n \cos n\omega t \, e^{-j\omega t} \, dt \tag{7-25}$$

b. High frequencies. At high frequencies the simple power-series approach cannot be used because reactive elements (linear and nonlinear) are important—for example, C_μ and C_π in the BJT small-signal model. Thus coefficients a_1, a_2, and a_3 are not constant but are functions of frequency [8].

Then, in this case, the linear term in Eq. (7-18), $a_1 x(t)$, is replaced by the convolution integral [$x(t) = 0$, for $t < 0$]:

$$y_1(t) = \int_0^t a_1(t-\tau) x(\tau) \, d\tau \tag{7-26}$$

In the frequency domain, Eq. (7-26) may be written as

$$Y_1(\omega) = A_1(\omega) X(\omega) \tag{7-27}$$

A generalization of the second-degree term $a_2 x^2(t)$ is the double convolution integral

$$y_2(t) = \int_0^t \int_0^t a_2(t-\tau_1, t-\tau_2) x(\tau_1) x(\tau_2) \, d\tau_1 \, d\tau_2 \tag{7-28}$$

In the frequency domain, Eq. (7-28) can be written

$$Y_2(\omega_1, \omega_2) = A_2(\omega_1, \omega_2) X(\omega_1) X(\omega_2) \tag{7-29}$$

The expression in Eq. (7-29) involves a product of the input with itself, thus representing a quadratic system. $A_2(\omega_1, \omega_2)$ is known as the second-degree *Volterra kernel*.

It is obvious that when two sinusoidal signals at frequencies ω_a and ω_b are applied, the output at the harmonic frequencies $\omega_a \pm \omega_b$ is given by

$$|A_2(\omega_a \pm \omega_b)| \cos[(\omega_a \pm \omega_b)t + \varphi_{a \pm b}] \tag{7-30}$$

Note: The material in Sec. 7.2.1b is taken from Narayanan [8]. Reprinted with permission from *The Bell System Technical Journal*. Copyright © 1967 by AT&T.

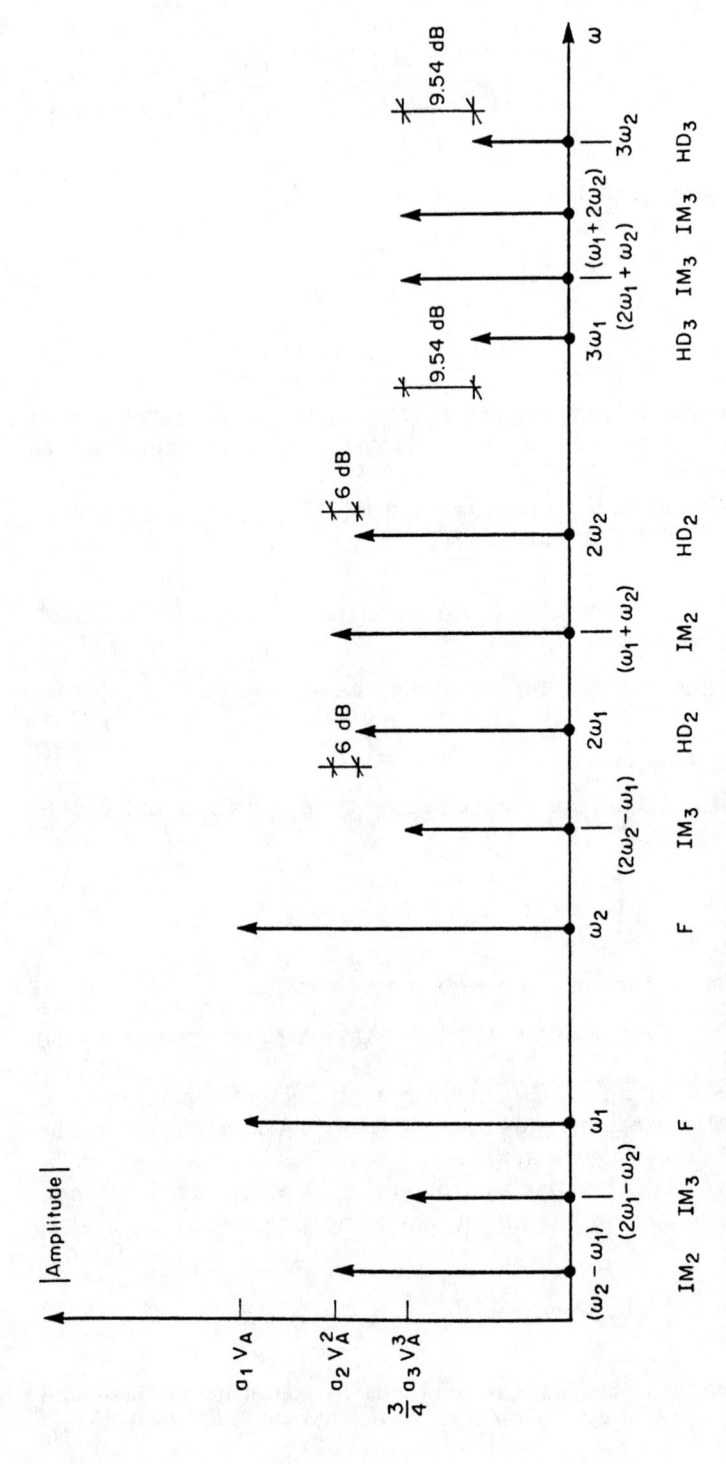

Figure 7-7 Intermodulation components in the output signal caused by coefficients a_1, a_2, and a_3 in the power series expansion. The input signals have frequencies ω_1 and ω_2 and amplitude V_A [6].

Since in general $A_2(\omega_a, \omega_b)$ will not be equal to $A_2(\omega_a, -\omega_b)$, different values of distortion at different harmonic frequencies are directly reflected in the kernel. Moreover, as in the low-frequency case of Eq. (7-22), the 2ω product is less by a factor of 2.

Likewise, the third-degree term $a_3 x^3(t)$ can be generalized to a triple convolution integral

$$y_3(t) = \int_0^t \int_0^t \int_0^t a_3(t-\tau_1, t-\tau_2, t-\tau_3) x(\tau_1) x(\tau_2) x(\tau_3) \, d\tau_1 \, d\tau_2 \, d\tau_3 \quad (7\text{-}31)$$

In the transform domain, Eq. (7-31) can be written as

$$Y_3(\omega_1, \omega_2, \omega_3) = A_3(\omega_1, \omega_2, \omega_3) X(\omega_1) X(\omega_2) X(\omega_3) \quad (7\text{-}32)$$

$A_3(\omega_1, \omega_2, \omega_3)$ is known as the third-degree *Volterra kernel*.

Thus, in this analysis, the output voltage

$$y(t) = y_1(t) + y_2(t) + y_3(t) \quad (7\text{-}33)$$

is expressed in terms of a Volterra series of the input signal $x(t)$, so that the kernels $A_1(\omega)$, $A_2(\omega_1, \omega_2)$, and $A_3(\omega_1, \omega_2, \omega_3)$ are the transfer ratios.

For example, the second and third harmonic distortions are given by

$$\begin{aligned} HD_2 &= \tfrac{1}{2} \frac{A_2(2\omega)}{A_1(\omega)} V_A \\ HD_3 &= \tfrac{1}{4} \frac{A_3(3\omega)}{A_1(\omega)} V_A^2 \end{aligned} \quad (7\text{-}34)$$

7.2.2 Distortion models implemented in SPICE2

The application of the distortion theory to circuit analysis is mathematically difficult and tedious, and it is beyond the aim of this book. This section provides only a brief overview of the methodology, followed by three examples to point out how SPICE2 analyzes a network. Finally, the models implemented in SPICE2 for the distortion anaysis and computation are presented.

a. Methodology description. The distortion algorithm implemented in SPICE2 requires two independent steps: evaluation of the Taylor series coefficients for each nonlinearity and evaluation of the distortion current excitation vector for each distortion component frequency of analysis [9].

Then the Volterra series approach consists of expanding the equations describing the nonlinearities as Taylor series, with the junction voltages as independent variables, and expressing each small-signal node voltage as a Volterra series expansion of the source voltage. In particular, the small-signal nonlinear equations of the circuit are formulated and separated into the usual linear part and a nonlinear part that is a function of the junction voltages expanded in a power series and that can be represented by equivalent current sources.

Thus the analysis of the circuit results in different sets of linear equations, one for each order, that are solved separately and successively beginning with the first order. The first-order solution is then used to calculate the equivalent current sources in order to determine the second-order solution. Finally, both first- and second-order solutions are used to find the third-order solution. The method shows that the equivalent current sources associated with each nonlinearity can be interpreted as the sources that produce distortion at the circuit output. This procedure amounts to evaluating, at each distortion component frequency of the analysis, the admittance matrix Y and finding the equivalent excitation vector in the node equations [10]

$$Y \cdot V = I \qquad (7\text{-}35)$$

b. Methodology examples

Example 1: First, the simple case of a resistive nonlinearity is considered, as shown in Fig. 7-8. At node j, it can be written

$$I_{Dj} = I_D e^{qV_j/kT} \qquad (7\text{-}36)$$

where I_{Dj} is the small-signal current, V_j is the small-signal voltage, and I_D is the bias current.

A useful way to handle this circuit analytically is to represent the nonlinearity as a voltage-controlled current source. Equation (7-36) can then be rewritten as an expansion of the exponential term in a

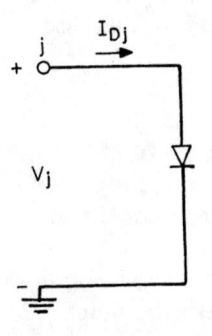

Figure 7-8 Resistive nonlinearity.

Taylor series

$$I_{Dj} = a_1 V_j + a_2 V_j^2 + a_3 V_j^3 \qquad (7\text{-}37)$$

where the coefficients of the series are

$$a_1 = \frac{q}{kT} I_D$$

$$a_2 = \frac{1}{2!} \left(\frac{q}{kT}\right)^2 I_D \qquad (7\text{-}38)$$

$$a_3 = \frac{1}{3!} \left(\frac{q}{kT}\right)^3 I_D$$

Equation (7-37) can be represented by the circuit shown in Fig. 7-9.
The Volterra kernels can now be evaluated by the following procedure.

1. Calculate the linear transfer functions from the sinusoidal input voltage V_{in} to the output voltage V_{out} and to every node voltage V_j of the circuit generating distortion, i.e., every node having nonlinear branches. Thus it can be written as

$$V_{out} = T_1(\omega) V_{in} \qquad (7\text{-}39)$$

$$V_j = H_1(\omega) V_{in} \qquad (7\text{-}40)$$

2. Consider two sinusoidal input signals $V_{in} = V_1(\sin \omega_1 t + \sin \omega_2 t)$ at frequencies ω_1 and ω_2. Since the two linear terms $H_1(\omega_1)$ and $H_1(\omega_2)$ are known from Eq. (7-40), the second-order (and third-order) distortion values can be calculated at the node j as

$$a_2 V_j^2 = a_2 V_{j1}^2 [\cos(\omega_1 t + \varphi_1) + \cos(\omega_2 t + \varphi_2)]^2 \qquad (7\text{-}41)$$

where

$$V_j = H_1(\omega) V_{in} \qquad (7\text{-}42)$$

so that

$$a_2 V_j^2 = a_2 H_1(\omega_1) H_1(\omega_2) V_{in}^2(\omega_1, \omega_2) \qquad (7\text{-}43)$$

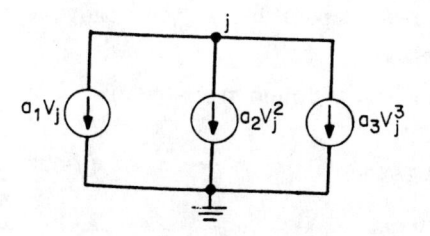

Figure 7-9 Equivalent circuit of resistive nonlinearity.

3. Analyze the circuit with these second-order source values as input signals and calculate the second-order distortion HD_2 at the point of interest. Obviously, the linear transfer function $H_1(\omega)$ from the input voltage V_{in} to the voltage V_j is equivalent to the term $A_1(\omega)$ in Eq. (7-27). Similarly, this calculation can be repeated for the third-order distortion.

Example 2: Now the case of a capacitance nonlinearity, for instance, C_μ in the BJT small-signal model, is considered.

In Chap. 2 it was shown that this capacitance is defined by a relation of the form

$$C_\mu = \frac{K}{(\phi + V_{BC})^{1/n}} \tag{7-44}$$

where

$$C_\mu = \frac{dQ_{BC}}{dV_{BC}}\bigg|_{op} \tag{7-45}$$

If $V_{BC} = V_0 + V_j$, with V_0 the base-collector bias voltage and V_j the small-signal voltage at node j, then

$$C_\mu = \frac{K}{(\phi + V_0 + V_j)^{1/n}}$$

$$= \frac{K}{(\phi + V_0)^{1/n}}\left(1 + \frac{V_j}{\phi + V_0}\right)^{-1/n} \tag{7-46}$$

Equation (7-46) can now be expressed in terms of the power series of V_j, so that

$$C_\mu(V_j) = C_{\mu 0} + C_{\mu 1} V_j + C_{\mu 2} V_j^2 + \cdots \tag{7-47}$$

and the total current flowing in the capacitance C_μ is

$$I = \begin{cases} \dfrac{dQ_{BC}}{dt} = \dfrac{dQ_{BC}}{dV_j}\dfrac{dV_j}{dt} = C_\mu(V_j)\dfrac{dV_j}{dt} & (7\text{-}48a) \\[2mm] C_{\mu 0}\dfrac{dV_j}{dt} + \tfrac{1}{2}C_{\mu 1}\dfrac{dV_j^2}{dt} + \tfrac{1}{3}C_{\mu 2}\dfrac{dV_j^3}{dt} & (7\text{-}48b) \end{cases}$$

Equations (7-48) can be represented by the circuit shown in Fig. 7-10. Thus, for circuits containing nonlinear capacitances, the analysis procedure is as follows.

1. As above, solve the circuit to obtain the linear transfer function in the form

$$V_{out} = T_1(\omega) V_{in} \tag{7-49}$$

$$V_j = H_1(\omega) V_{in} \tag{7-50}$$

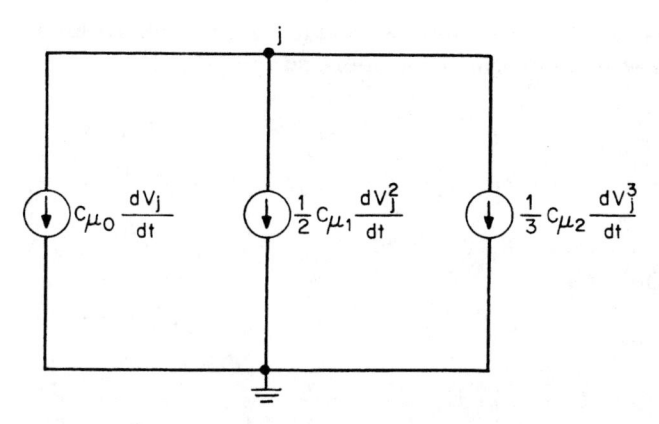

Figure 7-10 Equivalent circuit of capacitive nonlinearity.

2. Insert the second-order distortion sources in the circuit. The preceding capacitive nonlinearity [see Eq. (7-47)] gives

$$\frac{C_{\mu 1}}{2}\frac{dV_j^2}{dt} = \frac{C_{\mu 1}}{2}\frac{d}{dt}[H_1(\omega_1)H_1(\omega_2)V_{in}^2]$$

$$= \frac{C_{\mu 1}}{2}j(\omega_1 + \omega_2)H_1(\omega_1)H_1(\omega_2)V_{in}^2(\omega_1, \omega_2) \quad (7\text{-}51)$$

Note that this case is different from simple resistive nonlinearity; the major difference consists of the presence of the factor $j(\omega_1 + \omega_2)$.

Example 3: In the previous examples, we dealt with factors contributing to distortion for one nonlinear element at a time. Here, we analyze a general network containing two nonlinear elements (a capacitor C_X and a diode D) at the same time in order to evaluate distortion. Look at Fig. 7-11, taking into account that the diode bias current I_Q is not represented and that V_{in} and V_2 are the input and output voltages, respectively.

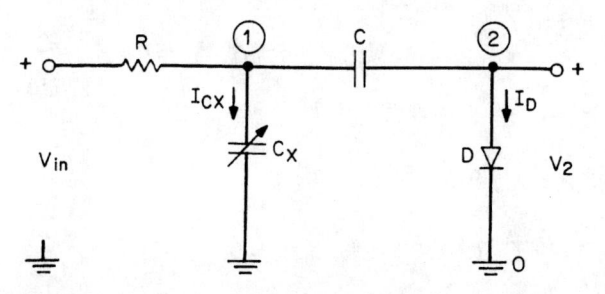

Figure 7-11 General network for evaluating distortion.

As explained earlier, the first step is to express each small-signal node voltage as a Volterra series expansion, so that

$$V_2 = A_1(j\omega)V_{in} + A_2(j\omega_1, j\omega_2)V_{in}^2 + A_3(j\omega_1, j\omega_2, j\omega_3)V_{in}^3$$
$$V_1 = B_1(j\omega)V_{in} + B_2(j\omega_1, j\omega_2)V_{in}^2 + B_3(j\omega_1, j\omega_2, j\omega_3)V_{in}^3$$
(7-52)

Then, the equations describing the nonlinearities are expanded in a Taylor series, as follows.

For diode D (see Example 1):

$$I_D = a_1 V_2 + a_2 V_2^2 + a_3 V_2^3 \qquad (7\text{-}53a)$$

For capacitor C_X (see Example 2):

$$I_{CX} = C_0 \frac{dV_1}{dt} + \frac{C_1}{2}\frac{dV_1^2}{dt} + \frac{C_2}{3}\frac{dV_1^3}{dt} \qquad (7\text{-}53b)$$

Finally, the Kirchhoff nodal equations can be written as follows.

For node 1: $\quad \dfrac{V_1 - V_{in}}{R} + C\dfrac{d}{dt}(V_1 - V_2) + I_{CX} = 0$

For node 2: $\quad -C\dfrac{d}{dt}(V_1 - V_2) + I_D = 0$
(7-54)

where the terms I_{CX} and I_D are obtained by Eqs. (7-53) and V_1 and V_2 are obtained by Eqs. (7-52).

These small-signal nonlinear node equations must now be separated into their linear and nonlinear terms, so that the method of analysis involves three steps (one for each order), which are solved separately and successively, beginning with the first order.

Linear (first-order) terms. Considering only linear terms, Eqs. (7-54) become

$$\frac{B_1 V_{in} - V_{in}}{R} + j\omega C(B_1 V_{in} - A_1 V_{in}) + j\omega C_0 B_1 V_{in} = 0$$
$$-j\omega C(B_1 V_{in} - A_1 V_{in}) + a_1 A_1 V_{in} = 0$$
(7-55)

Dividing Eqs. (7-55) by V_{in} yields

$$\frac{B_1 - 1}{R} + j\omega C(B_1 - A_1) + j\omega C_0 B_1 = 0$$
$$-j\omega C(B_1 - A_1) + a_1 A_1 = 0$$
(7-56)

Solving Eqs. (7-56) with respect to A_1 and B_1,

$$A_1 = \frac{1}{R} \frac{j\omega C}{(1/R + j\omega C + j\omega C_0)(a_1 + j\omega C) + \omega^2 C^2}$$

$$B_1 = \frac{1}{R} \frac{a_1 + j\omega C}{(1/R + j\omega C + j\omega C_0)(a_1 + j\omega C) + \omega^2 C^2}$$

(7-57)

Second-order terms. Considering only second-order terms, Eqs. (7-54) become

$$\frac{B_2 V_{in}^2}{R} + j(\omega_1 + \omega_2)C(B_2 V_{in}^2 - A_2 V_{in}^2) + j(\omega_1 + \omega_2)C_0 B_2 V_{in}^2$$

$$+ \tfrac{1}{2} C_1 j(\omega_1 + \omega_2) B_1(j\omega_1) B_1(j\omega_2) = 0 \qquad (7\text{-}58)$$

$$-j(\omega_1 + \omega_2)C(B_2 V_{in}^2 - A_2 V_{in}^2) + a_1 A_2 V_{in}^2 + a_2 A_1(j\omega_1) A_1(j\omega_2) V_{in}^2 = 0$$

and solving as in the previous case

$$A_2 = \frac{\left[\dfrac{a_2 A_1(j\omega_1) A_1(j\omega_2)}{j(\omega_1 + \omega_2)C}\right]\left[\dfrac{1}{R} + j(\omega_1 + \omega_2)(C + C_0)\right] + \dfrac{C_1}{2} j(\omega_1 + \omega_2) B_1(j\omega_1) B_1(j\omega_2)}{\left[1 + \dfrac{a_1}{j(\omega_1 + \omega_2)C}\right]\left[\dfrac{1}{R} + j(\omega_1 + \omega_2)(C + C_0)\right] - j(\omega_1 + \omega_2)C}$$

(7-59)

$$B_2 = \left[1 + \frac{a_1}{j(\omega_1 + \omega_2)C}\right] A_2 + \frac{a_2 A_1(j\omega_1) A_1(j\omega_2)}{j(\omega_1 + \omega_2)C}$$

where the values of A_1 and B_1 are given by Eqs. (7-57).

Third-order terms. If we consider only third-order terms and second-order interactions, Eqs. (7-54) become

$$\frac{B_3}{R} V_{in}^3 + j(\omega_1 + \omega_2 + \omega_3)C(B_3 V_{in}^3 - A_3 V_{in}^3)$$

$$+ j(\omega_1 + \omega_2 + \omega_3)C_0 B_3 V_{in}^3 + j(\omega_1 + \omega_2 + \omega_3)C_1 B_1 B_2 V_{in}^3$$

$$+ j(\omega_1 + \omega_2 + \omega_3)\frac{C_2}{3} B_1(j\omega_2) B_1(j\omega_3) V_{in}^3 = 0 \qquad (7\text{-}60)$$

$$-j(\omega_1 + \omega_2 + \omega_3)C(B_3 V_{in}^3 - A_3 V_{in}^3) + a_1 A_3 V_{in}^3 + 2 a_2 A_1 A_2 V_{in}^3$$

$$+ a_3 A_1(j\omega_1) A_1(j\omega_2) A_1(j\omega_3) V_{in}^3 = 0$$

Equations (7-60) can now be solved with respect to A_3 and B_3, taking into account the values of A_1, B_1 and A_2, B_2 given by Eqs. (7-57) and (7-59), respectively.

When the coefficients $A_1, B_1, A_2, B_2, A_3, B_3$ of the Volterra series have been calculated, the amount of distortion can be easily evaluated. For example, suppose we wish to calculate HD_2 when an input signal $V_{in} = V_A \cos \omega_1 t$ is applied to the circuit of Fig. 7-11. From Eq. (7-34) it follows that

$$HD_2 = \tfrac{1}{2} \frac{A_2(2\omega_1)}{A_1(\omega_1)} V_A \qquad (7\text{-}61)$$

where Eqs. (7-57) and (7-59) must be used.

c. Diode distortion model. The diode distortion model is shown in Fig. 7-12. Current source I_{DI} has been added to the small-signal diode model to represent the distortion introduced by the diode. The value of the current source is given by the following equations.

For second-order harmonic distortion (HD_2):

$$I_{DI}(2\omega_1) = \left(\tfrac{1}{2} \frac{d^2 I_D}{dV_D^2} \bigg|_{op} + j2\omega_1 \tfrac{1}{2} \frac{dC_D}{dV_D} \bigg|_{op} \right) V_D^2(2\omega_1) \qquad (7\text{-}62)$$

For second-order intermodulation distortion (IM_2):

$$I_{DI}(\omega_1 \pm \omega_2) = \left(\tfrac{1}{2} \frac{d^2 I_D}{dV_D^2} \bigg|_{op} + j(\omega_1 \pm \omega_2) \tfrac{1}{2} \frac{dC_D}{dV_D} \bigg|_{op} \right) V_D(\omega_1) V_D(\omega_2) \qquad (7\text{-}63)$$

where $V_D(\omega_n)$ is the voltage across the diode due to a response at frequency ω_n. All currents and voltages in these equations are expressed as phasors. The value of the second-order harmonic distortion

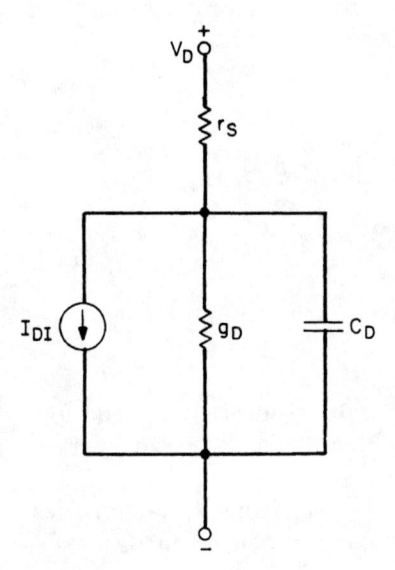

Figure 7-12 Diode distortion model.

current source $I_{DI}(2\omega_1)$ is determined by the response at fundamental frequency ω_1.

It is important to note that the first term in Eq. (7-62) corresponds to a_2 in Eq. (7-37) and the second term in Eq. (7-62) corresponds to the second term in Eq. (7-48b).

Similarly, the value of second-order intermodulation distortion current source $I_{DI}(\omega_1 + \omega_2)$ is determined solely by the response at fundamental frequencies ω_1 and ω_2. The equations used to determine the third-order distortion (HD_3 and IM_3) for a diode are similar to those given previously for second-order effects, but they are more complex.

The third-order harmonic distortion current source $I_{DI}(3\omega_1)$ is determined by the response at fundamental frequency ω_1 plus the response at the second harmonic frequency $2\omega_1$. Similarly, the third-order intermodulation distortion current source $I_{DI}(2\omega_1 - \omega_2)$ is determined by the response at fundamental frequencies ω_1 and ω_2, plus the response at the second-order harmonic frequency $2\omega_1$.

d. **BJT distortion model.** The BJT distortion model is shown in Fig. 7-13. Current sources I_{DI1}, I_{DI2}, and I_{DI3} have been added to the small-signal BJT model to represent the distortion introduced by the transistor. The values of these sources are calculated from the following equations.

For second-order harmonic distortion (HD_2):

$$I_{DI1}(2\omega_1) = (g_{\pi 2} + j2\omega_1 C_{\pi 2}) V_{BE}^2(\omega_1)$$

$$I_{DI2}(2\omega_1) = (g_{\mu 2} + j2\omega_1 C_{\mu 2}) V_{BC}^2(\omega_1) \qquad (7\text{-}64)$$

$$I_{DI3}(2\omega_1) = g_{m2} V_{BE}^2(\omega_1) + g_{o2} V_{CE}^2(\omega_1) + g_{mo2} V_{BE}(\omega_1) V_{CE}(\omega_1)$$

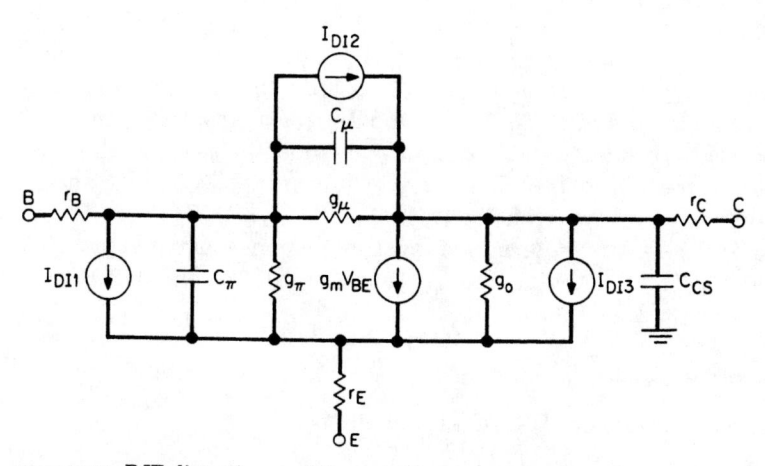

Figure 7-13 BJT distortion model.

with
$$g_{\pi 2} = \tfrac{1}{2}\frac{d^2 I_B}{dV_{BE}^2}\bigg|_{op} \qquad g_{\mu 2} = \tfrac{1}{2}\frac{d^2 I_B}{dV_{BC}^2}\bigg|_{op}$$

$$C_{\pi 2} = \tfrac{1}{2}\frac{d^2 C_{BE}}{dV_{BE}^2}\bigg|_{op} \qquad C_{\mu 2} = \tfrac{1}{2}\frac{d^2 C_{BC}}{dV_{BC}^2}\bigg|_{op} \qquad (7\text{-}65)$$

$$g_{m2} = \tfrac{1}{2}\frac{d^2 I_C}{dV_{BE}^2}\bigg|_{op} - g_{o2} \qquad g_{o2} = \tfrac{1}{2}\frac{d^2 I_C}{dV_{BC}^2}\bigg|_{op} - g_{\mu 2}$$

$$g_{mo2} = \frac{d^2 I_C}{dV_{BE}\, dV_{BC}}\bigg|_{op} - 2g_{o2}$$

For second-order intermodulation distortion (IM_2):

$$I_{DI1}(\omega_1 \pm \omega_2) = [g_{\pi 2} + j(\omega_1 \pm \omega_2)C_{\pi 2}]V_{BE}(\omega_1)V_{BE}(\omega_2)$$

$$I_{DI2}(\omega_1 \pm \omega_2) = [g_{\mu 2} + j(\omega_1 \pm \omega_2)C_{\mu 2}]V_{BC}(\omega_1)V_{BC}(\omega_2) \qquad (7\text{-}66)$$

$$I_{DI3}(\omega_1 \pm \omega_2) = g_{m2}V_{BE}(\omega_1)V_{BE}(\omega_2) + g_{o2}V_{CE}(\omega_1)V_{CE}(\omega_2)$$
$$+ g_{mo2}[V_{BE}(\omega_1)V_{CE}(\omega_2) + V_{BE}(\omega_2)V_{CE}(\omega_1)]$$

where $V_{BC}(\omega_n)$, $V_{BE}(\omega_n)$, and $V_{CE}(\omega_n)$ are, respectively, the base-collector, base-emitter, and collector-emitter voltages due to a response at frequency ω_n. The admittances and capacitances are evaluated at the dc operating point of the network. All voltages and currents in these equations are expressed as phasors.

The values of second-order harmonic distortion current sources are determined by the response at the fundamental frequency ω_1. Similarly, the values of the second-order intermodulation distortion current sources are determined solely by the response at the fundamental frequencies ω_1 and ω_2.

The equations used to determine third-order distortion (HD_3 and IM_3) for a BJT are similar to those given previously for second-order effects, but they are more complex. Third-order harmonic distortion current sources, $I_{DI1}(3\omega_1)$, $I_{DI2}(3\omega_1)$, and $I_{DI3}(3\omega_1)$ are determined by the response at fundamental frequency ω_1 plus the response at the second-order harmonic frequency $2\omega_1$. Similarly, third-order intermodulation distortion current sources $I_{DI1}(2\omega_1 - \omega_2)$, $I_{DI2}(2\omega_1 - \omega_2)$, and $I_{DI3}(2\omega_1 - \omega_2)$ are determined by the response at fundamental frequencies ω_1 and ω_2, plus the response at the second-order harmonic frequency $2\omega_1$ [11].

7.2.3 .DISTO statement

The general form of the .DISTO statement is

.DISTO RLOAD   INTER SKW2 REFPWR SPW2

where RLOAD is the name of the output load resistor into which all distortion power products are computed. Other coefficients are optional.

This statement determines whether SPICE2 will compute the distortion characteristic of the circuit in a small-signal mode as part of the small-signal sinusoidal steady-state analysis [12–17].

The analysis is performed assuming that one or two signal frequencies be ω_1 (the nominal analysis frequency) and $\omega_2 = SKW_2 \cdot \omega_1$. The program then computes the following distortion measures: HD_2, HD_3, IM_2 as $SIM_2(\omega_1 + \omega_2)$ and $DIM_2(\omega_1 - \omega_2)$, and IM_3 as $DIM_3(2\omega_1 - \omega_2)$.

At the conclusion of the distortion analysis, any of these distortion components can be listed or plotted as a function of frequency. The contributions of every distortion source also can be printed at selected frequency points. The correct use of the REFPWR parameter is very important in this statement. It is the reference power level used in computing the distortion products; if the parameter is omitted, a value of 1 mW is used.

Consider, for example, an input signal $V_{in} = V_A \cos \omega_1 t$. The output signal applied on R_L (output load) will be $V_{out} = V_o \cos(\omega_1 t + \varphi)$, so that from Eq. (7-20a)

$$HD_2 = \frac{a_2 V_A}{2 a_1} = \frac{1}{2}\frac{a_2}{a_1^2} V_o = \frac{1}{2}\frac{a_2}{a_1^2} \sqrt{2}\, V_{o,\text{eff}} \qquad (7\text{-}67)$$

Using the relation $P_{out} = V_{o,\text{eff}}^2/R_L$, it can be written

$$HD_2 = \frac{1}{2}\frac{a_2}{a_1^2} \sqrt{2 R_L P_{out}} \qquad (7\text{-}68)$$

Thus, P_{out} represents the REFPWR parameter.

7.2.4 Fourier analysis

For relatively large levels of distortion, the method of fitting a Fourier series of a periodic waveform on the time-domain response yields a good approximation of the distortion components of the waveform.

For large-signal sinusoidal simulations, a Fourier analysis of the output waveform can be specified to obtain the frequency-domain Fourier coefficients: the first nine harmonic components of the specified output are computed by SPICE2. The transient time interval and the Fourier analysis options are specified in the .TRAN and .FOUR control statements. It is very important to note that the sinusoidal waveform is the only periodic waveform allowed.

7.2.5 Brief notes on distortion for SPICE2-based versions

In this section, some brief notes about distortion of the SPICE2-based versions considered in this book are made. HSPICE [3] distortion analysis is based directly on SPICE2 distortion analysis. PSPICE [4] and SPICE3 [18] do not include the SPICE2-based distortion analysis (.DISTO statement); to calculate distortion, PSPICE makes use of spectral analysis techniques (see Ref. [19]). In addition, the .FOUR statement used by SPICE2 to request Fourier analysis of outputs has been replaced by the Fourier command in Nutmeg, the interactive SPICE3 front end.

References

1. P. R. Gray and R. G. Meyer, *Analysis and Design of Analog Integrated Circuits*, Wiley, New York, 1977.
2. L. Nagel, SPICE2: A Computer Program to Simulate Semiconductor Circuits, Electronics Research Laboratory, Rep. No. ERL-M520, University of California, Berkeley, 1975.
3. Meta-Software Inc., *HSPICE User's Manual*, Campbell, Calif., 1991.
4. MicroSim Corporation, *PSPICE User's Manual*, Irvine, Calif., 1989.
5. P. Antognetti, P. Antoniazzi, and E. Meda, Computer-Aided Evaluation of Transient Intermodulation Distortion (TIM) in Monolithic Integrated Amplifiers, *Proc. IEE-EDA*, 1981.
6. W. Sansen, Optimum Design of Integrated Variable-Gain Amplifiers, Electronics Research Laboratory, Rep. No. ERL-M367, University of California, Berkeley, 1972.
7. K. A. Simons, The Decibel Relationships between Amplifiers Distortion Products, *Proc. IEEE*, **58**, 1970.
8. S. Narayanan, Transistor Distortion Analysis Using Volterra Series Representations, *Bell Syst. Tech. J.*, **40**, 1967.
9. S. Chilsholm and L. Nagel, Efficient Computer Simulation of Distortion in Electronic Circuits, *IEEE Trans. Circuit Theory*, 1973.
10. L. Kuo, Distortion Analysis of Bipolar Transistor Circuits, *IEEE Trans. Circuit Theory*, 1972.
11. P. Antognetti, P. Antoniazzi, and E. Meda, TIM Distortion in Monolithic Integrated Circuits: Measurements and Simulation, *AES Convention*, Hamburg, 1981.
12. A. Vladimirescu, K. Zhang, A. R. Newton, D. O. Pederson, and Sangiovanni-Vincentelli, *SPICE Version 2G User's Guide*, Dept. EECS, University of California, Berkeley, 1981.
13. H. C. Poon, Modeling of Bipolar Transistor Using Integral Charge Control Model with Application to Third-Order Distortion Studies, *IEEE Trans. Electron Devices*, ED-19, 1972.
14. S. Narayanan and H. C. Poon, An Analysis of Distortion in Bipolar Transistors Using Integral Charge Control Model and Volterra Series, *IEEE Trans. Circuit Theory*, CT-20, 1973.
15. R. G. Meyer, M. Shensa, and R. Eschenbach, Cross Modulation and Intermodulation in Amplifiers at High Frequencies, *IEEE J. Solid-State Circuits*, SC-7, 1972.
16. H. C. Poon, Implication of Transistor Frequency Dependence on Intermodulation Distortion, *IEEE Trans. Electron Devices*, ED-21, 1974.
17. R. K. Brayton and R. Spence, *Sensitivity and Optimization*, Elsevier, Amsterdam, 1980.
18. T. I. Quarles, SPICE3 Version 3C1 User's Guide, Electronics Research Laboratory, Rep. No. ERL-M46, University of California, Berkeley, 1989.
19. P. W. Tuinenga, *SPICE: A Guide to Circuit Simulation and Analysis Using PSPICE*, Prentice-Hall, Englewood Cliffs, N.J., 1988.

Chapter

8

The SPICE Program

This chapter points out the rudiments needed to modify existing device models or to implement new device models in SPICE2. A complete description of the development and design of SPICE2 can be found in the literature [1–4]; it is beyond the aim of this book. However, we will describe the general capabilities of SPICE2. We refer to SPICE2 version G. The main characteristics of SPICE2-based programs such as SPICE3, HSPICE, and PSPICE are also considered. We have made this choice because SPICE3 represents the "U.C. Berkeley connection and evolution" from the university where it originated in the early 1970s, HSPICE is considered the most representative of the mainframe-based commercially supported versions of SPICE2, and PSPICE is the first SPICE2-based simulator on PCs.

8.1 SPICE2 Capabilities

SPICE2, which stands for *S*imulation *P*rogram with *I*ntegrated *C*ircuit *E*mphasis, is a general-purpose circuit simulation program developed at the University of California at Berkeley for nonlinear dc, nonlinear transient, and linear ac analyses. Circuits can contain resistors, capacitors, inductors, mutual inductors, independent voltage and current sources, four types of dependent sources, transmission lines, and the four most common semiconductor devices: diodes, BJTs, JFETs, and MOSTs.

Note: This chapter has been written with the contribution of Claudio Fasce, who is with SGS-THOMSON Microelectronics, Milan, Italy.

SPICE2 has built-in models for the semiconductor devices, and the user need specify only the pertinent model parameter values. The model for the BJT is based on the integral charge model of Gummel and Poon; however, if the Gummel-Poon parameters are not specified, the model reduces to the simpler Ebers-Moll model. In either case, charge-storage effects, ohmic resistances, and a current-dependent output conductance can be included. The diode model can be used for either junction diodes or Schottky barrier diodes. The JFET model is based on the FET model of Shichman and Hodges. Three MOST models are implemented: MOS1 is described by a square-law I-V characteristic, MOS2 is an analytical model, and MOS3 is a semi-empirical model. Both MOS2 and MOS3 include second-order effects such as channel-length modulation, subthreshold conduction, scattering limited velocity saturation, small-size effects, and charge-controlled capacitances [1].

SPICE2 is node-voltage-oriented, so any node voltage can be requested. Element currents flowing through independent voltage sources can also be requested. Tabular lists and printer plots are available.

8.1.1 Types of analysis

a. DC analysis. The *dc analysis* portion of SPICE2 determines the dc operating point of the circuit with inductors shorted and capacitors opened. A dc analysis is automatically performed prior to a transient analysis to determine the transient initial conditions, and prior to an ac small-signal analysis to determine the linearized, small-signal models for nonlinear devices. If requested, the dc small-signal value of a transfer function (ratio of output variable to input source), input resistance, and output resistance will also be computed as part of the dc solution. The dc analysis can also be used to generate dc transfer curves: A specified independent voltage or current source is stepped over a user-specified range, and the dc output variables are stored for each sequential source value. If requested, SPICE2 also will determine the dc small-signal sensitivities of specified output variables with respect to circuit parameters. The dc analysis options are specified on the .DC, .TF, .OP, and .SENS control statements.

If one desires to see the small-signal models for nonlinear devices in conjunction with a transient analysis operating point, then the .OP statement must be provided. The dc bias conditions will be identical for each case, but the more comprehensive operating-point information is not available for printing when transient initial conditions are computed [1].

b. AC small-signal analysis. The *ac small-signal* portion of SPICE2 computes the ac output variables as a function of frequency. The program first computes the dc operating point of the circuit and determines linearized, small-signal models for all the nonlinear devices in the circuit. The resultant linear circuit is then analyzed over a user-specified range of frequencies. The desired output of an ac small-signal analysis is usually a transfer function (voltage gain, transimpedance, and so forth). If the circuit has only one ac input, it is convenient to set that input to unity and zero phase so that output variables have the same value as the transfer function of the output variable with respect to the input.

The generation of *white noise* by resistors and semiconductor devices can also be simulated with the ac small-signal portion of SPICE2. Equivalent noise-source values are determined automatically from the small-signal operating point of the circuit, and the contribution of each noise source is added at a given summing point. The total output noise level and the equivalent input noise level are determined at each frequency point. The output and input noise levels are normalized with respect to the square root of the noise bandwidth. The output noise and equivalent input noise can be printed or plotted in the same fashion as other output variables. No additional input data are necessary for this analysis.

Flicker-noise sources can be simulated during noise analysis by including values for the parameters KF and AF on the appropriate device model statements (see Chap. 7).

The *distortion* characteristics of a circuit in the small-signal model can be simulated as a part of the ac small-signal analysis. The analysis is performed assuming that one or two signal frequencies are imposed at the input.

The frequency range and the noise and distortion analysis parameters are specified on the .AC, .NOISE, and .DISTO control statements [1].

c. Transient analysis. The *transient analysis* portion of SPICE2 computes the transient output variables as a function of time over a user-specified time interval. The initial conditions are automatically determined by a dc analysis. All sources that are not time-dependent (e.g., power supplies) are set to their dc value. For large-signal sinusoidal simulations, a Fourier analysis of the output waveform can be specified to obtain the frequency-domain Fourier coefficients. The transient time interval and the Fourier analysis options are specified on the .TRAN and .FOUR control statements [1].

d. Analysis at different temperatures. All input data for SPICE2 are assumed to have been measured at 27°C (300 K). The simulation also assumes a nominal temperature of 27°C. The circuit can be simulated at other temperatures by using a .TEMP control statement. Temperature appears explicitly in the exponential terms of the BJT and diode model equations. In addition, saturation currents have a built-in temperature dependence, as do the forward and reverse β and the junction potentials.

Temperature also appears explicitly in the value of surface mobility for the MOST model. The effects of temperature on resistors are also modeled. Temperature effects on the device model parameters are detailed in the chapters where each device is discussed.

8.1.2 Convergence

Both dc and transient solutions are obtained by an iterative process that is terminated when both the following conditions hold:

1. The nonlinear branch currents converge to within a tolerance of 0.1 percent or 10^{-12} A, whichever is larger.

2. The node voltages converge to within a tolerance of 0.1 percent or 10^{-6} V, whichever is larger.

Although the algorithm used in SPICE2 has been found to be very reliable, in some cases it will fail to converge to a solution. When this occurs, the program prints the node voltages at the last iteration and terminates the job. In such cases, the node voltages that are printed are not necessarily correct or even close to the correct solution.

Failure to converge in the dc analysis is usually due to an error in specifying circuit connections, element values, or model parameter values. Regenerative switching circuits or circuits with positive feedback probably will not converge in the dc analysis unless the OFF option is used for some of the devices in the feedback path or the .NODESET statement is used to force the circuit to converge to the desired state (see Sec. 8.3) [1].

8.1.3 Input format

The input format for SPICE2 is of the free-format type. Fields on a statement are separated by one or more blanks, a comma, an equal (=) sign, or a left or right parenthesis; extra spaces are ignored. A statement can be continued by entering a plus sign (+) in column 1 of the subsequent statement; SPICE2 continues reading beginning with column 2.

A name field must begin with a letter (A to Z) and cannot contain any delimiters. Only the first eight characters of the name are used.

A number field can be an integer field (12, −44), a floating-point field (3.14159), either an integer or a floating-point number followed by an integer exponent (1E − 14, 2.65E3), or either an integer or a floating-point number followed by one of the following factors:

$$T = 1E12$$

$$G = 1E9$$

$$MEG = 1E6$$

$$K = 1E3$$

$$MIL = 25.4E - 6$$

$$M = 1E - 3$$

$$U = 1E - 6$$

$$N = 1E - 9$$

$$P = 1E - 12$$

$$F = 1E - 15$$

Letters immediately following a number that are not scale factors are ignored, and letters immediately following a scale factor are ignored. Hence, 10, 10V, 10VOLTS, and 10HZ all represent the same number, and M, MA, MSEC, and MMHOS all represent the same scale factor. Note that 1000, 1000.0, 1000HZ, 1E3, 1.0E3, 1KHZ, and 1K all represent the same number [1].

Control commands begin with a period (.); the other lines are assumed to be circuit elements.

8.1.4 Circuit description

The circuit to be analyzed is described to SPICE2 by a set of element statements which define the circuit topology and element values, and a set of control statements which define the model parameters and the run controls. The first statement in the input file must be a *title* statement, and the last statement must be an .END statement. The order of the remaining statements is *arbitrary* (except, of course, that continuation statements must immediately follow the statement being continued).

Each element in the circuit is specified by an element statement that contains the element name, the circuit nodes to which the element is connected, and the values of the parameters that determine the electrical characteristics of the element. The first letter of the element name specifies the element type.

With respect to branch voltages and currents, SPICE2 uniformly uses the associated reference convention (current flows in the direction of voltage drop).

Nodes *must* be nonnegative integers but need not be numbered sequentially. The datum (ground) node *must* be numbered zero. The circuit *cannot* contain a loop of voltage sources and/or inductors and *cannot* contain a cutset of current sources and/or capacitors. Each node in the circuit *must* have a dc path to ground. Every node *must* have at least two connections, except for transmission-line nodes (to permit unterminated transmission lines), and MOST substrate nodes (which have two internal connections anyway) [1].

Prefixes used for element names are as follows:

R	Resistor
C	Capacitor
L	Inductor
K	Mutual inductor
T	Transmission lines (lossless)
V	Independent voltage source
I	Independent current source
G	(Non)linear voltage-controlled current source
E	(Non)linear voltage-controlled voltage source
F	(Non)linear current-controlled current source
H	(Non)linear current-controlled voltage source
Q	BJT
D	DIODE
J	JFET
M	MOST
X	Subcircuit

The R and T elements must be constants, while the C, L, G, E, F, H elements may be expressed as nonlinear polynomials. The elements Q, D, J, M are defined by model parameters. The subcircuit is a user-defined subcircuit.

The independent sources can be constants or any of the following functions. Any of these functions can be used as values for independent sources:

1. *Constants.* AC or dc constant values. Constants can be suffixed with scaling factors (see Sec. 8.1.3).

2. *Pulse.* Defined with high and low values, rise and fall times, and pulse width and repetition rate.
3. *Sinusoidal.* Exponentially decaying sinusoid.
4. *Exponential.* Sum of two exponential waveforms.
5. *Piecewise linear.* Tabular function of time.
6. *SFFM.* Single-frequency, frequency-modulated waveform.

Figure 8-1 shows a schematic diagram for a single-stage BF preamplifier and the SPICE2 input file that defines the topology of the circuit.

8.1.5 Semiconductor devices

The models for the four semiconductor devices included in SPICE2 require many parameter values. Moreover, many devices in a circuit often are defined by the same set of device model parameters. For these reasons, a set of device model parameters is defined on a separate .MODEL statement and assigned a unique model name. The device element statements in SPICE2 then reference the model name. This scheme alleviates the need to specify all the model parameters on each device element statement.

Each device element statement contains the device name, the nodes to which the device is connected, and the device model name. In addition, other optional parameters can be specified for each device: geometric factors and initial conditions.

The AREA factor used on the DIODE, BJT, and JFET device statements determines the number of equivalent parallel devices of a specified model. Several geometric factors associated with the channel and the drain and source diffusions can be specified on the MOST device statement.

Two different forms of initial conditions can be specified for devices. The first form is included to improve the dc convergence for circuits that contain more than one stable state. If a device is specified OFF, the dc operating point is determined with the terminal voltages for that device set to zero. After convergence is obtained, the program continues to iterate to obtain the exact value of the terminal voltages.

If a circuit has more than one dc stable state, the OFF option can be used to force the solution to correspond to a desired state. If a device is specified OFF when in reality the device is conducting, the program will still obtain the correct solution (assuming the solutions converge), but more iterations will be required since the program must independently converge to two separate solutions. The

(a)

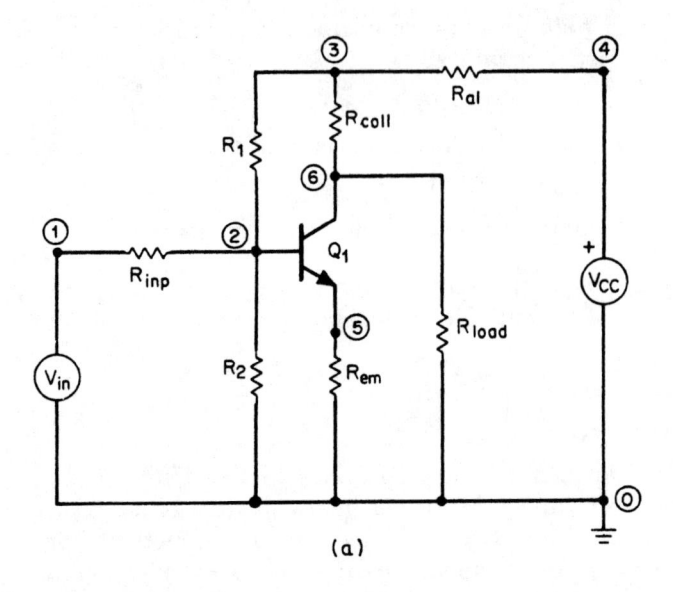

SPICE 2G

BF PREAMPLIFIER

•••• INPUT LISTING   TEMPERATURE = 27.000 °C

••

```
VIN 1 0 SIN (0 0.1 5K 100NS) AC 0.1
VCC 4 0 12
Q1 6 2 5 TRN
RINP 1 2 20K
R1 2 3 10K
R2 2 0 3.3K
RCOLL 3 6 2.2K
REM 5 0 1K
RAL 3 4 100
RLOAD 6 0 100
•MODEL TRN NPN (BF = 80  RB = 100 CCS = 2PF TF = 0.3NS
+TR = 6NS CJE = 3PF CJC = 2PF VA = 50)
•OPT LIST ACCT NODE
•TF V(6) VIN
•DC VIN -0.25 0.25 0.01
•AC DEC 5 1 10GHZ
•TRAN 400US 20MS
•PLOT DC V(6)
•PLOT AC VM(6) VP(6)
•WIDTH OUT=80
•PLOT TRAN V(6)
•END
```

(b)

Figure 8-1 (a) Schematic diagram for a single-stage BF preamplifier and (b) SPICE2 input file.

.NODESET statement serves a purpose similar to that of the OFF option. The .NODESET statement is easier to apply and is the preferred means to aid convergence.

The second form of initial conditions is specified for use with the transient analysis. These are true initial conditions as opposed to the convergence aids specified previously. See the description of the .IC statement and the .TRAN statement for a detailed explanation of initial conditions [1].

8.1.6 Subcircuits

A *subcircuit* that consists of SPICE2 elements can be defined and referenced in a fashion similar to that used for device models. The subcircuit is defined in the input file by a group of element statements; the program then automatically inserts the elements wherever the subcircuit is referenced. There is no limit on the size or complexity of subcircuits, and subcircuits can contain other subcircuits [1].

8.2 SPICE2 Structure

The SPICE2 program (written in FORTRAN language) consists of seven major independent modules together with a main program (see Fig. 8-2). The names of the modules are READIN, ERRCHK, SETUP, DCOP, DCTRAN, ACAN, and OVTPVT.

The complete SPICE2 map is outlined in Fig. 8-3. Individual modules are shown in Figs. 8-4 to 8-11. The ellipses ($\cdots$) in these figures mean that the calling map for the present routine has already been shown elsewhere (i.e., previously). For the reader's convenience, the maps of invididual modules of SPICE2 may be found as follows: SETMEM (Fig. 8-4), READIN (Fig. 8-5), ERRCHK (Fig. 8-6), SETUP (Fig. 8-7), DCTRAN (Fig. 8-8), OVTPVT (Fig. 8-9), DCOP (Fig. 8-10), and ACAN (Fig. 8-11).

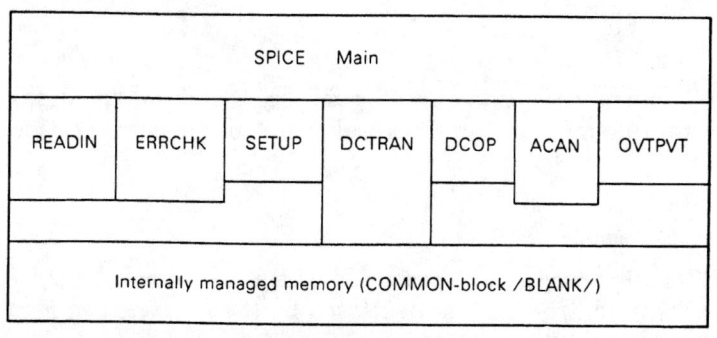

SPICE Main						
READIN	ERRCHK	SETUP	DCTRAN	DCOP	ACAN	OVTPVT
Internally managed memory (COMMON-block /BLANK/)						

Figure 8-2 Memory map for SPICE2 [4].

```
                    ┌── TODALF ────── TIME
                    ├── DATE
                    ├── TIMRB ─────── SECNDS
                    ├── GETCJE ────── SECOND
                    ├── SECOND
                    ├── SETMEM ────── Fig. 8-4
                    ├── ZERO8
                    ├── READIN ────── Fig. 8-5
SPICE ──────────────┼── ERRCHK ────── Fig. 8-6
                    ├── SETUP ─────── Fig. 8-7
                    ├── TMPUPD ────── TITLE
                    ├── DCTRAN ────── Fig. 8-8
                    ├── OVTPVT ────── Fig. 8-9
                    ├── DCOP ──────── Fig. 8-10
                    ├── ACAN ──────── Fig. 8-11
                    ├── TITLE
                    └── TIMRE
```

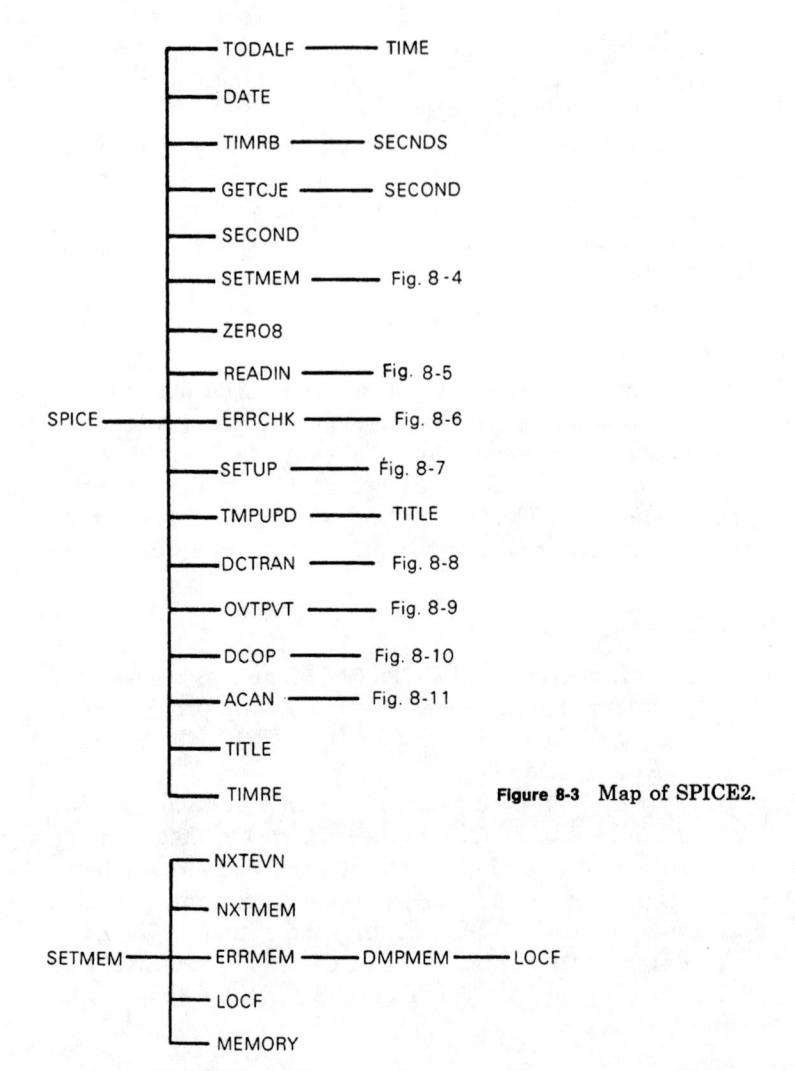

Figure 8-3 Map of SPICE2.

```
                    ┌── NXTEVN
                    ├── NXTMEM
SETMEM ─────────────┼── ERRMEM ────── DMPMEM ────── LOCF
                    ├── LOCF
                    └── MEMORY
```

Figure 8-4 SETMEM map.

In the following sections we will describe the seven modules in addition to the only subroutines relevant to the aim of this book.

8.2.1 The SPICE2 main program

The SPICE2 main program contains the main control loop of the program. This control loop is illustrated by the flowchart in Fig. 8-12.

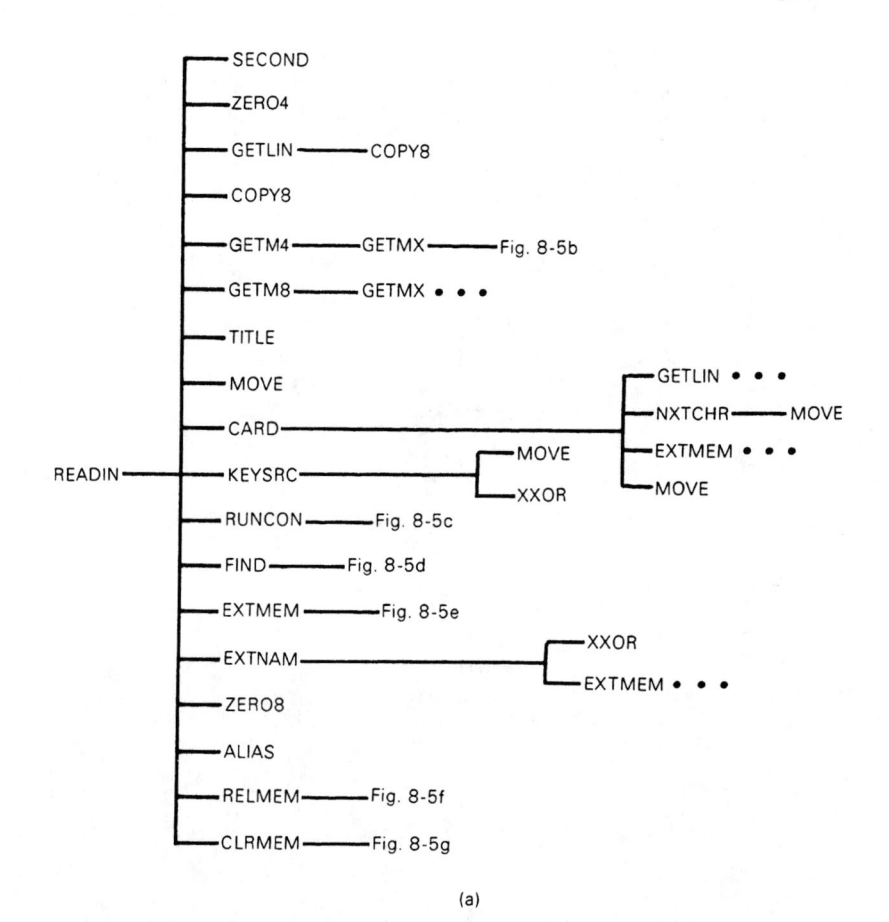

```
                    ┌── SECOND
                    ├── ZERO4
                    ├── GETLIN ──── COPY8
                    ├── COPY8
                    ├── GETM4 ───── GETMX ────── Fig. 8-5b
                    ├── GETM8 ───── GETMX • • •
                    ├── TITLE
                    ├── MOVE                                    ┌── GETLIN • • •
                    ├── CARD ───────────────────────────────────┼── NXTCHR ──── MOVE
                    │                              ┌── MOVE     ├── EXTMEM • • •
 READIN ────────────┼── KEYSRC ────────────────────┤             └── MOVE
                    │                              └── XXOR
                    ├── RUNCON ──── Fig. 8-5c
                    ├── FIND ────── Fig. 8-5d
                    ├── EXTMEM ──── Fig. 8-5e
                    │                              ┌── XXOR
                    ├── EXTNAM ────────────────────┤
                    │                              └── EXTMEM • • •
                    ├── ZERO8
                    ├── ALIAS
                    ├── RELMEM ──── Fig. 8-5f
                    └── CLRMEM ──── Fig. 8-5g
```

(a)

Figure 8-5 READIN map.

The program begins by initializing some program constants and reading the job title statement. If an end-of-file is encountered on the input file, the program terminates. Otherwise, the READIN module is called to read the remainder of the input file. The READIN program stops reading after an .END statement or an input end-of-file is encountered.

As the READIN module reads the input file, the circuit data structure is constructed. This data structure consists of linked lists that define each circuit element, each device model, and each output variable. In addition, common block variables are set to indicate the types of analysis that have been requested as well as the simulation control parameters that have been specified.

After READIN has executed successfully, ERRCHK is called to check the circuit description for common user errors. In addition,

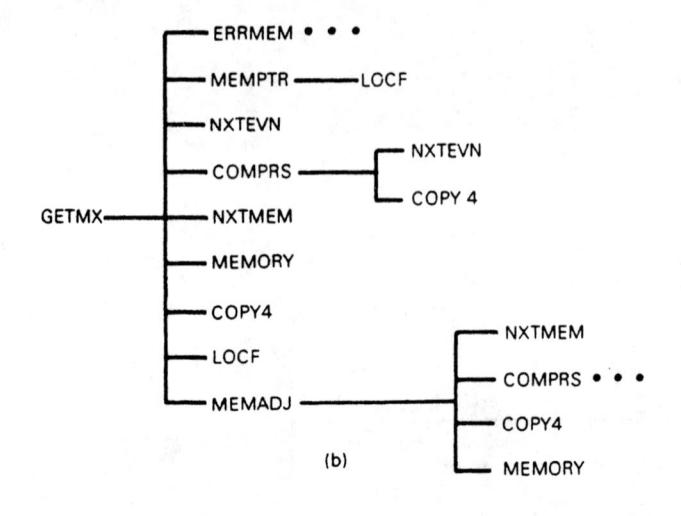

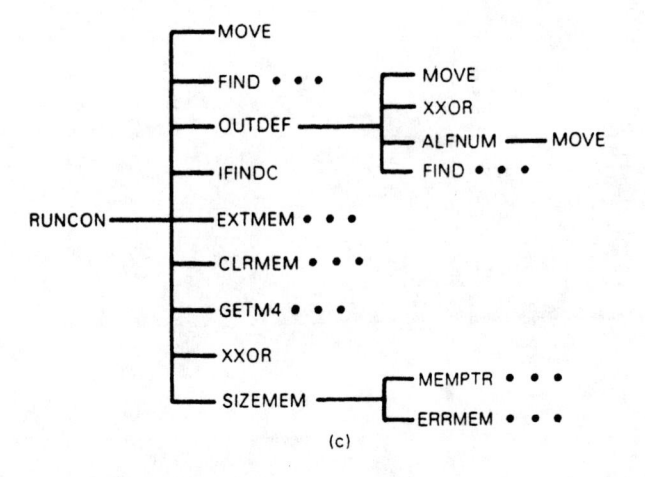

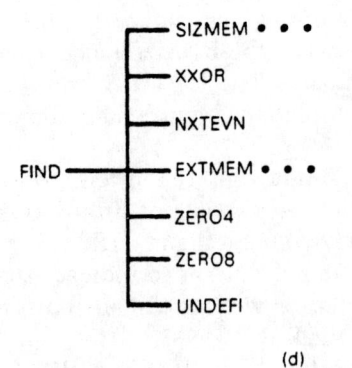

```
            ┌─ ERRMEM • • •
            ├─ MEMPTR ──── LOCF
            ├─ NXTEVN
            ├─ COMPRS ────┬─ NXTEVN
            │             └─ COPY 4
GETMX ──────┼─ NXTMEM
            ├─ MEMORY
            ├─ COPY4
            ├─ LOCF                  ┌─ NXTMEM
            └─ MEMADJ ───────────────┼─ COMPRS • • •
                                     ├─ COPY4
                                     └─ MEMORY
                        (b)


            ┌─ MOVE
            ├─ FIND • • •         ┌─ MOVE
            ├─ OUTDEF ────────────┼─ XXOR
            │                     ├─ ALFNUM ──── MOVE
            ├─ IFINDC             └─ FIND • • •
RUNCON ─────┼─ EXTMEM • • •
            ├─ CLRMEM • • •
            ├─ GETM4 • • •
            ├─ XXOR
            └─ SIZEMEM ───────────┬─ MEMPTR • • •
                                  └─ ERRMEM • • •
                        (c)


            ┌─ SIZMEM • • •
            ├─ XXOR
            ├─ NXTEVN
FIND ───────┼─ EXTMEM • • •
            ├─ ZERO4
            ├─ ZERO8
            └─ UNDEFI
                        (d)
```

Figure 8-5 (*Continued*)

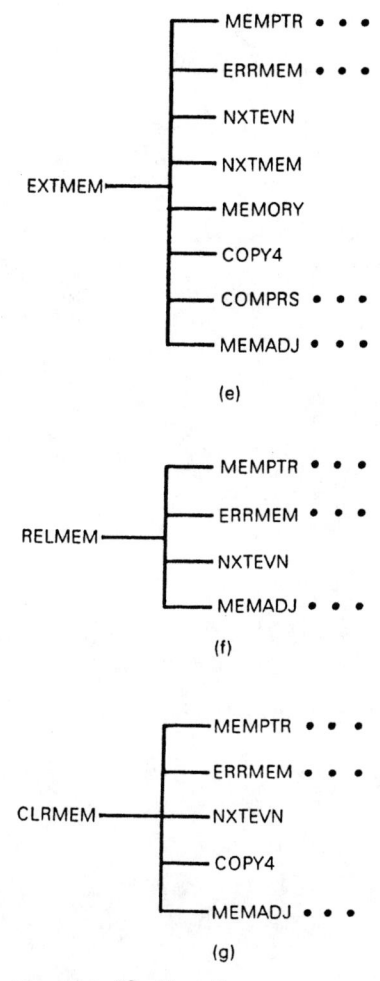

```
                ┌── MEMPTR • • •
                ├── ERRMEM • • •
                ├── NXTEVN
                ├── NXTMEM
EXTMEM ─────────┤── MEMORY
                ├── COPY4
                ├── COMPRS • • •
                └── MEMADJ • • •
                     (e)

                ┌── MEMPTR • • •
                ├── ERRMEM • • •
RELMEM ─────────┤── NXTEVN
                └── MEMADJ • • •
                     (f)

                ┌── MEMPTR • • •
                ├── ERRMEM • • •
CLRMEM ─────────┤── NXTEVN
                ├── COPY4
                └── MEMADJ • • •
                     (g)
```

Figure 8-5 (*Continued*)

ERRCHK also prints the circuit listing, the device model parameter listing, and the node table.

Once READIN and ERRCHK have been executed, the preparation for circuit analysis can begin. SETUP is called to construct the integer pointer structure used by both DCTRAN and ACAN.

After SETUP has executed successfully, the circuit analysis can proceed. As shown in Fig. 8-12, the main analysis loop is repeated for each of the user-specified temperatures. After the main analysis loop is finished for each temperature, the program prints the job statistics and reads the next input [3].

338 Chapter Eight

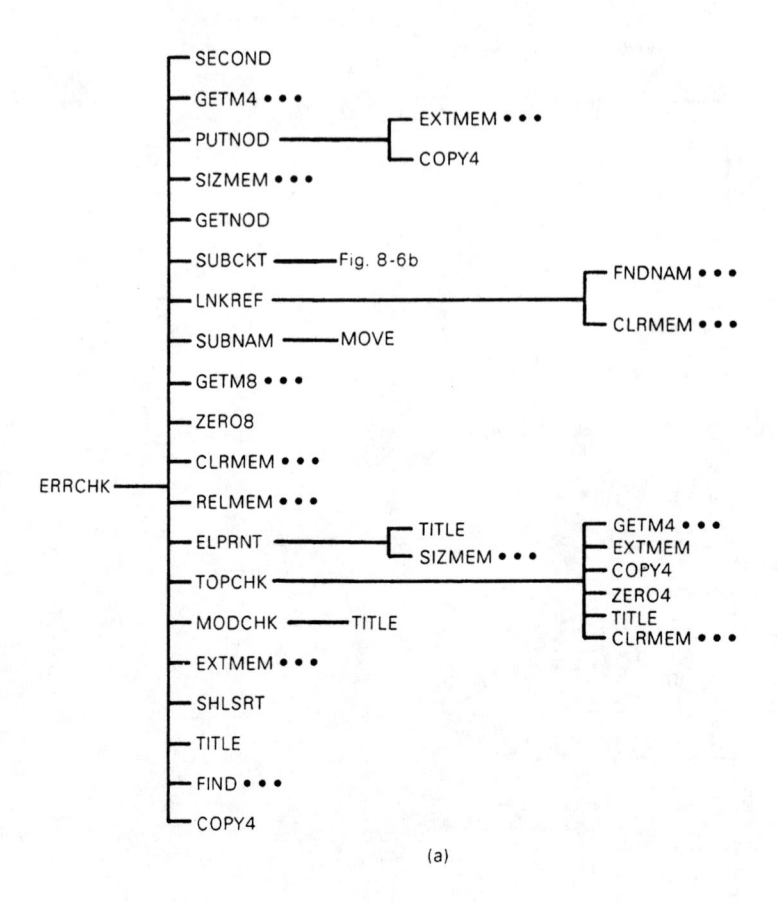

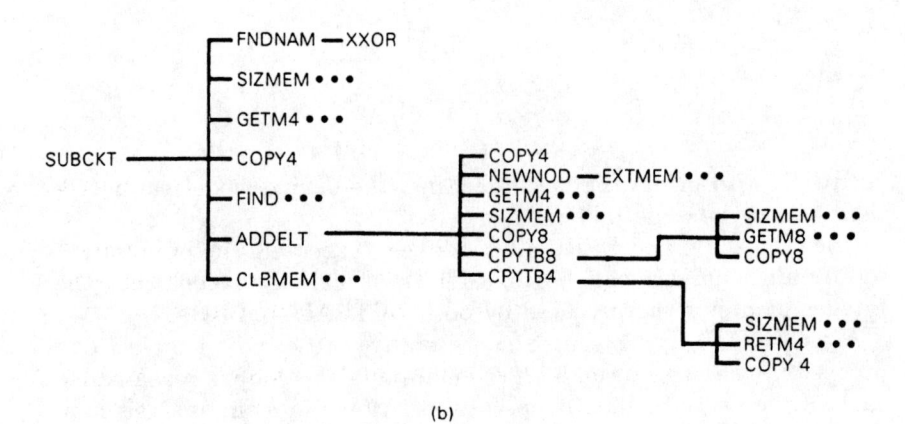

```
           ┌─ SECOND
           ├─ GETM4 •••
           ├─ PUTNOD ──────┬─ EXTMEM •••
           │               └─ COPY4
           ├─ SIZMEM •••
           ├─ GETNOD
           ├─ SUBCKT ───── Fig. 8-6b
           ├─ LNKREF ──────┬─ FNDNAM •••
           │               └─ CLRMEM •••
           ├─ SUBNAM ───── MOVE
           ├─ GETM8 •••
           ├─ ZERO8
           ├─ CLRMEM •••
ERRCHK ────┤
           ├─ RELMEM •••
           │                                  ┌─ GETM4 •••
           ├─ ELPRNT ──────┬─ TITLE           ├─ EXTMEM
           ├─ TOPCHK ──────┴─ SIZMEM •••──────┼─ COPY4
           │                                  ├─ ZERO4
           │                                  ├─ TITLE
           │                                  └─ CLRMEM •••
           ├─ MODCHK ───── TITLE
           ├─ EXTMEM •••
           ├─ SHLSRT
           ├─ TITLE
           ├─ FIND •••
           └─ COPY4
```
(a)

```
            ┌─ FNDNAM ── XXOR
            ├─ SIZMEM •••
            ├─ GETM4 •••
            ├─ COPY4
SUBCKT ─────┤
            ├─ FIND •••
            │              ┌─ COPY4
            │              ├─ NEWNOD ── EXTMEM •••
            │              ├─ GETM4 •••
            ├─ ADDELT ─────┼─ SIZMEM •••           ┌─ SIZMEM •••
            │              ├─ COPY8                ├─ GETM8 •••
            │              ├─ CPYTB8 ──────────────┴─ COPY8
            │              └─ CPYTB4
            │                                      ┌─ SIZMEM •••
            │                                      ├─ RETM4 •••
            └─ CLRMEM •••                          └─ COPY 4
```
(b)

Figure 8-6 ERRCHK map.

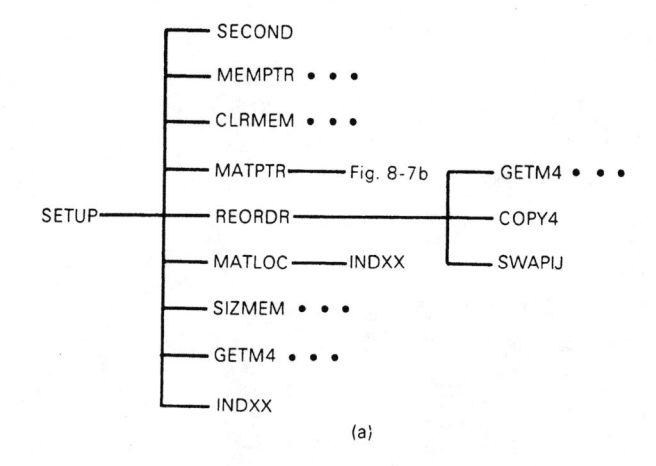

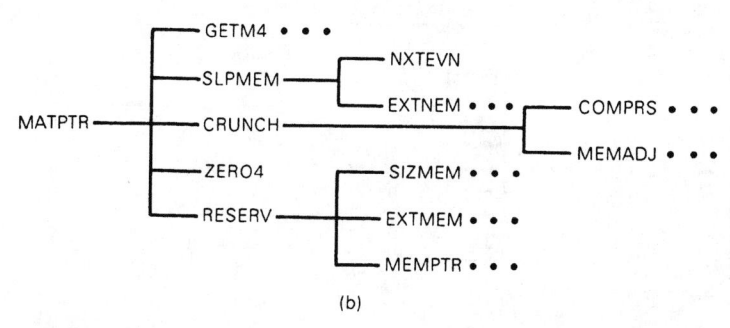

```
                    ┌── SECOND
                    ├── MEMPTR • • •
                    ├── CLRMEM • • •
                    ├── MATPTR ──── Fig. 8-7b ──┬── GETM4 • • •
    SETUP ──────────┼── REORDR ─────────────────┼── COPY4
                    ├── MATLOC ──── INDXX       └── SWAPIJ
                    ├── SIZMEM • • •
                    ├── GETM4 • • •
                    └── INDXX
                                (a)


                    ┌── GETM4 • • •
                    │                   ┌── NXTEVN
                    ├── SLPMEM ─────────┤                     ┌── COMPRS • • •
    MATPTR ─────────┼── CRUNCH ─────────┴── EXTNEM • • • ─────┤
                    │                                         └── MEMADJ • • •
                    ├── ZERO4          ┌── SIZMEM • • •
                    └── RESERV ────────┼── EXTMEM • • •
                                       └── MEMPTR • • •
                                (b)
```

Figure 8-7 SETUP map.

8.2.2 The main analysis loop

The main analysis loop consists of three parts: the *dc transfer-curve loop* (shown in Fig. 8-13), the *dc operating-point* and *ac analysis loop* (shown in Fig. 8-14), and the *transient analysis loop* (shown in Fig. 8-15). These loops are executed in succession.

The dc transfer-curve loop (see Fig. 8-13) first checks to ensure that a dc transfer curve is requested. The flags MODE and MODEDC are set to 1 and 3, respectively, and DCTRAN is called to perform the dc transfer-curve analysis. OVTPVT is then called to construct the tabular listings and line-printer plots requested for the dc transfer-curve analysis.

The dc operating-point and ac analysis are illustrated in Fig. 8-14. First, the loop checks if a dc operating point has been requested. If not, the dc operating-point analysis will be performed only if an ac analysis of a nonlinear circuit is requested, since, for this case, the

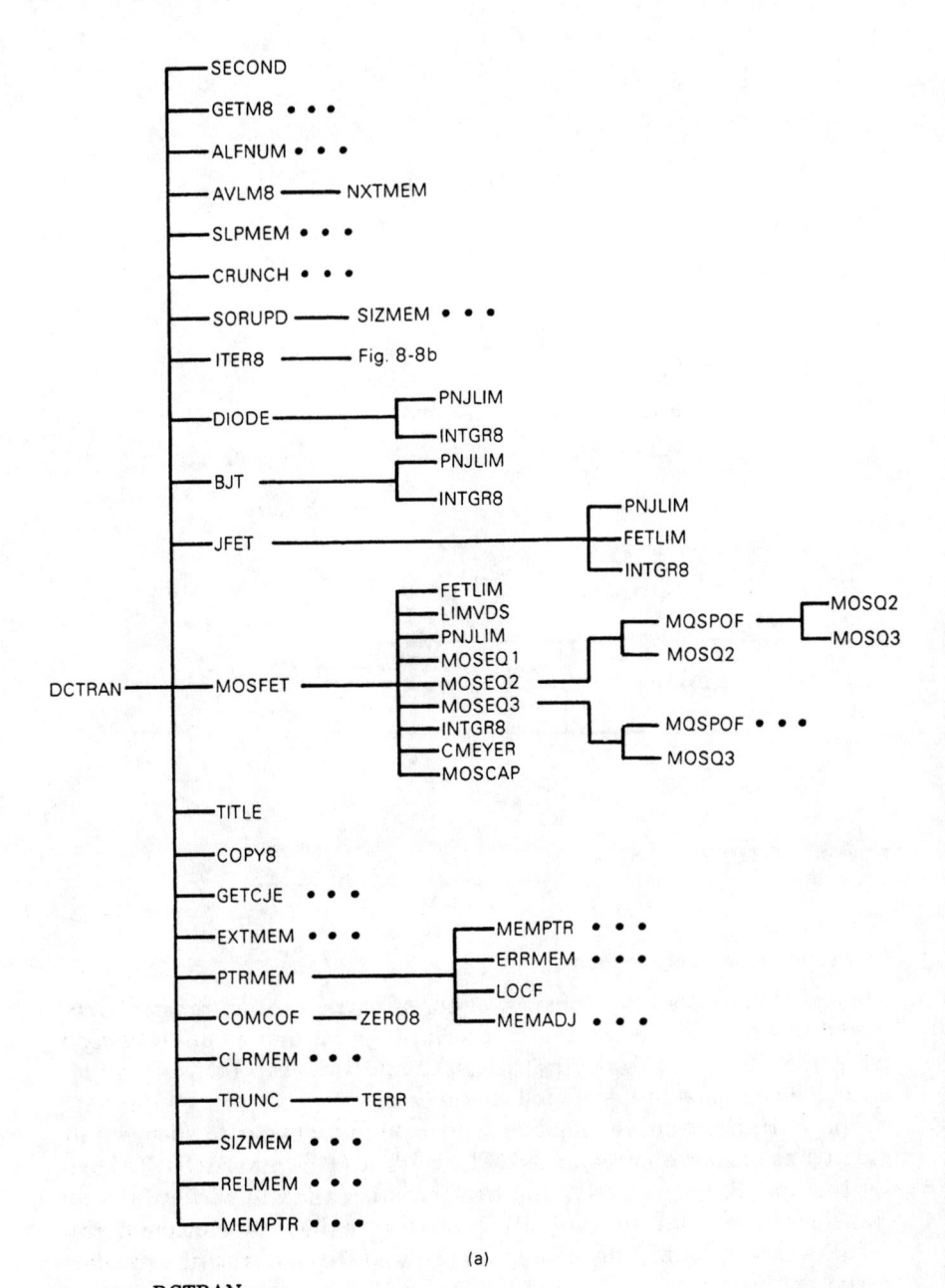

```
DCTRAN ─┬─ SECOND
        ├─ GETM8 • • •
        ├─ ALFNUM • • •
        ├─ AVLM8 ──── NXTMEM
        ├─ SLPMEM • • •
        ├─ CRUNCH • • •
        ├─ SORUPD ──── SIZMEM • • •
        ├─ ITER8 ──── Fig. 8-8b
        ├─ DIODE ──┬─ PNJLIM
        │          └─ INTGR8
        ├─ BJT ────┬─ PNJLIM
        │          └─ INTGR8
        ├─ JFET ───┬─ PNJLIM
        │          ├─ FETLIM
        │          └─ INTGR8
        ├─ MOSFET ─┬─ FETLIM
        │          ├─ LIMVDS
        │          ├─ PNJLIM
        │          ├─ MOSEQ1
        │          ├─ MOSEQ2 ──┬─ MQSPOF ──┬─ MOSQ2
        │          │           │           └─ MOSQ3
        │          │           └─ MOSQ2
        │          ├─ MOSEQ3 ──┬─ MQSPOF • • •
        │          │           └─ MOSQ3
        │          ├─ INTGR8
        │          ├─ CMEYER
        │          └─ MOSCAP
        ├─ TITLE
        ├─ COPY8
        ├─ GETCJE • • •
        ├─ EXTMEM • • •
        ├─ PTRMEM ──┬─ MEMPTR • • •
        │           ├─ ERRMEM • • •
        │           ├─ LOCF
        │           └─ MEMADJ • • •
        ├─ COMCOF ──── ZERO8
        ├─ CLRMEM • • •
        ├─ TRUNC ──── TERR
        ├─ SIZMEM • • •
        ├─ RELMEM • • •
        └─ MEMPTR • • •
```

(a)

Figure 8-8 DCTRAN map.

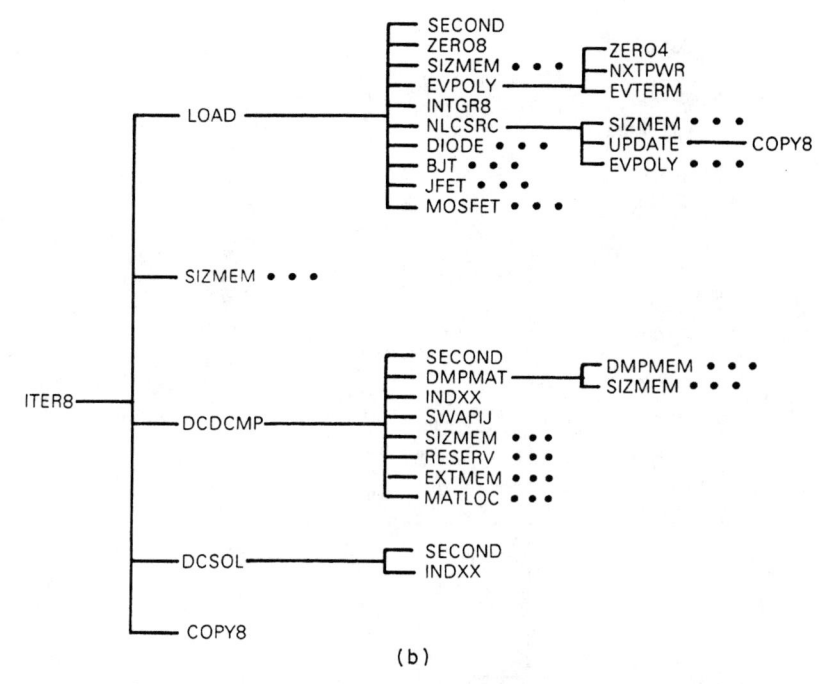

```
                                  ┌─ SECOND
                                  ├─ ZERO8         ┌─ ZERO4
                                  ├─ SIZMEM • • •  ├─ NXTPWR
                                  ├─ EVPOLY ───────┴─ EVTERM
                                  ├─ INTGR8
                    ┌─ LOAD ──────┤                ┌─ SIZMEM • • •
                    │             ├─ NLCSRC ───────┼─ UPDATE ─────── COPY8
                    │             ├─ DIODE • • •   └─ EVPOLY • • •
                    │             ├─ BJT • • •
                    │             ├─ JFET • • •
                    │             └─ MOSFET • • •
                    │
                    ├─ SIZMEM • • •
                    │
                    │             ┌─ SECOND
                    │             ├─ DMPMAT ───────┬─ DMPMEM • • •
                    │             ├─ INDXX         └─ SIZMEM • • •
  ITER8 ────────────┼─ DCDCMP ────┼─ SWAPIJ
                    │             ├─ SIZMEM • • •
                    │             ├─ RESERV • • •
                    │             ├─ EXTMEM • • •
                    │             └─ MATLOC • • •
                    │
                    ├─ DCSOL ─────┬─ SECOND
                    │             └─ INDXX
                    │
                    └─ COPY8
```

(b)

Figure 8-8 (*Continued*)

linearized device model parameters are a necessary prerequisite for the ac analysis.

Once it is established that a dc operating point is necessary, the flags MODE and MODEDC are set to 1 and DCTRAN is called to compute the dc operating point. DCOP is then called to print the linearized device model parameters [3].

If requested, the ac analysis is performed next by setting the flag MODE to 3 and calling ACAN. After the ac analysis has been obtained, OVTPVT is called to generate the tabular listings and line-printer plots that are requested in conjunction with the ac analysis.

The final analysis loop, transient analysis, is shown in Fig. 8-15. First, the loop checks that a transient analysis is requested. Then, the flags MODE and MODEDC are set to 1 and 2, respectively, and DCTRAN is called to determine the transient initial conditions. Next, DCOP is invoked to print the nonlinear device operating points. Finally, the flag MODE is set to 2 and DCTRAN is called again to determine the transient analysis. OVTPVT then is called to generate the tabular listings and line-printer plots that are requested. In addition, OVTPVT will perform, if requested, the Fourier analysis of time-domain waveforms after the transient analysis has been determined [3].

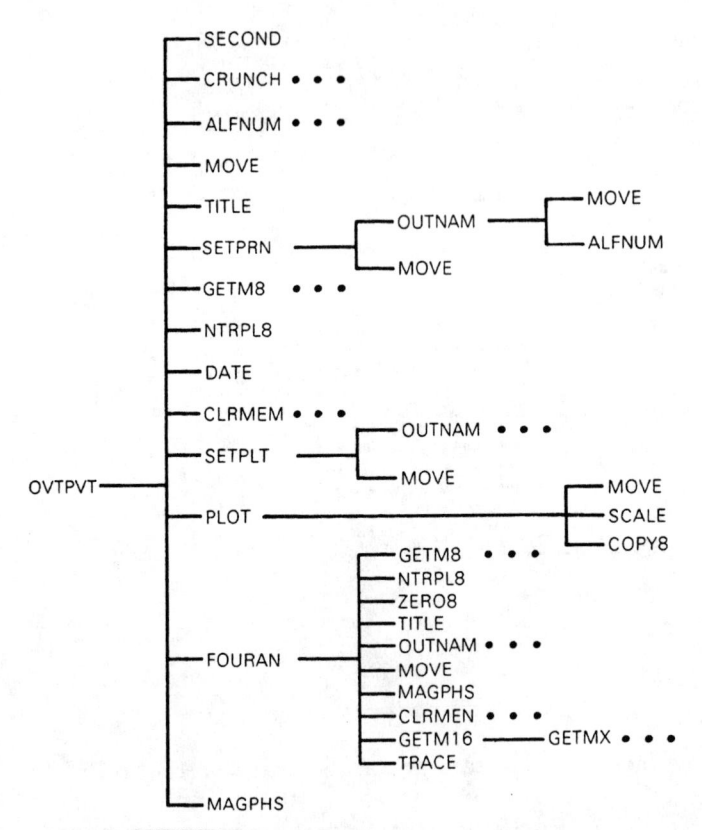

```
                    ┌─ SECOND
                    ├─ CRUNCH • • •
                    ├─ ALFNUM • • •
                    ├─ MOVE
                    ├─ TITLE                         ┌─ MOVE
                    ├─ SETPRN ──┬─ OUTNAM ──┤
                    │           └─ MOVE             └─ ALFNUM
                    ├─ GETM8 • • •
                    ├─ NTRPL8
                    ├─ DATE
                    ├─ CLRMEM • • •
                    ├─ SETPLT ──┬─ OUTNAM • • •
                    │           └─ MOVE
OVTPVT ─────────────┤                                ┌─ MOVE
                    ├─ PLOT ─────────────────────────┼─ SCALE
                    │                                └─ COPY8
                    │             ┌─ GETM8 • • •
                    │             ├─ NTRPL8
                    │             ├─ ZERO8
                    │             ├─ TITLE
                    ├─ FOURAN ────┼─ OUTNAM • • •
                    │             ├─ MOVE
                    │             ├─ MAGPHS
                    │             ├─ CLRMEN • • •
                    │             ├─ GETM16 ─── GETMX • • •
                    │             └─ TRACE
                    └─ MAGPHS
```

Figure 8-9 OVTPVT map.

8.2.3 The READIN module

The READIN module reads the input file, checks each line of input for syntax, and constructs the data structure for the circuit. The data structure that defines the circuit is a set of linked lists.

Each call to subroutine CARD returns the contents of the next line of input, together with the contents of any continuation lines. The input line is broken into fields, with each field containing a decimal number (a number field) or an alphanumeric name (a name field). The first field of the statement determines whether the statement is an element statement or a control statement; every control statement must begin with a period. If the statement is an element statement or a .MODEL statement, then it is processed by the READIN subroutine. If the statement is a control statement, then it is processed by the RUNCON subroutine.

The FIND subroutine handles all matters of the linked-list structure. Whenever a specific element of a linked list (that is, a bead) is

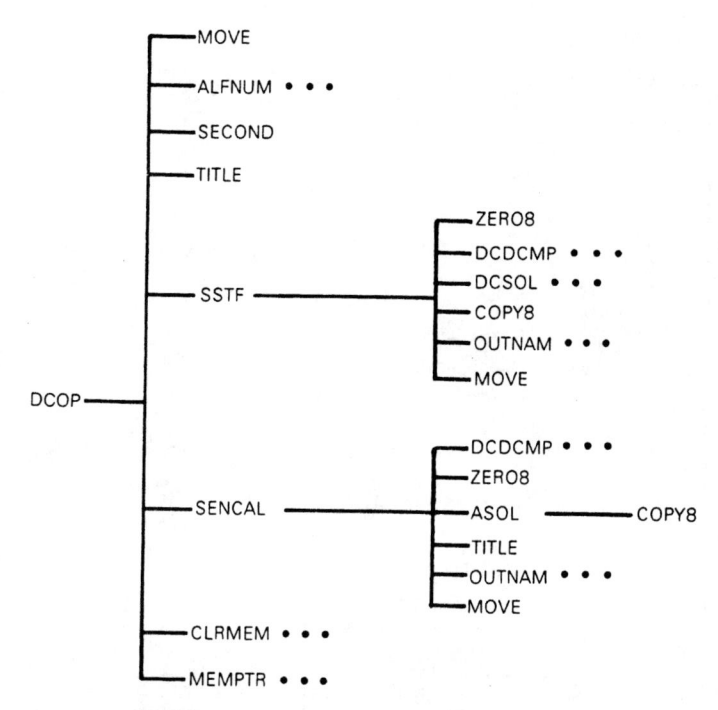

```
                    ┌─ MOVE
                    ├─ ALFNUM • • •
                    ├─ SECOND
                    ├─ TITLE
                    │              ┌─ ZERO8
                    │              ├─ DCDCMP • • •
                    ├─ SSTF ───────┼─ DCSOL • • •
                    │              ├─ COPY8
                    │              ├─ OUTNAM • • •
                    │              └─ MOVE
DCOP ───────────────┤
                    │              ┌─ DCDCMP • • •
                    │              ├─ ZERO8
                    ├─ SENCAL ─────┼─ ASOL ─────── COPY8
                    │              ├─ TITLE
                    │              ├─ OUTNAM • • •
                    │              └─ MOVE
                    ├─ CLRMEM • • •
                    └─ MEMPTR • • •
```

Figure 8-10 DCOP map.

desired, the FIND subroutine is called to locate the bead. If a bead is to be added to the data structure, then the FIND subroutine handles the linking and memory extension [3].

8.2.4 The ERRCHK module

First, ERRCHK checks for undefined elements (as might happen, for example, if a coupled inductor element references an inductor that is not defined). Then, ERRCHK constructs an ordered list of the user node numbers, processes the sources in the circuits, processes the coupled inductors in the circuit, and establishes a list of source breakpoints for later use in the transient analysis.

The ELPRNT subroutine prints the circuit summary. The MODCHK subroutine assigns default values for model parameters that have not been specified, prints the summary of device model parameters, performs some additional model parameter processing (such as inverting series resistances), and, finally, reserves any internal device nodes that are generated by nonzero device ohmic resistances.

The TOPCHK subroutine constructs the node table, prints the node table, and checks that every node in the circuit has at least two

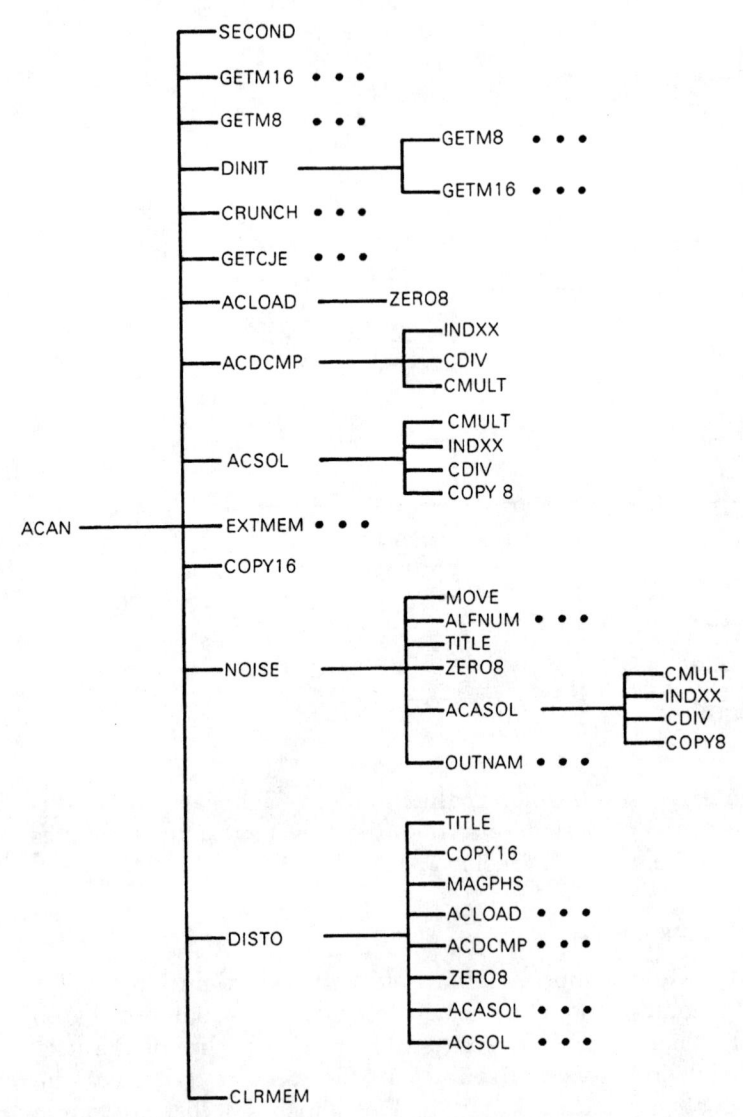

```
ACAN ─┬─ SECOND
      ├─ GETM16 • • •
      ├─ GETM8 • • •
      ├─ DINIT ──┬─ GETM8 • • •
      │         └─ GETM16 • • •
      ├─ CRUNCH • • •
      ├─ GETCJE • • •
      ├─ ACLOAD ── ZERO8
      ├─ ACDCMP ──┬─ INDXX
      │           ├─ CDIV
      │           └─ CMULT
      ├─ ACSOL ──┬─ CMULT
      │          ├─ INDXX
      │          ├─ CDIV
      │          └─ COPY 8
      ├─ EXTMEM • • •
      ├─ COPY16
      ├─ NOISE ──┬─ MOVE
      │          ├─ ALFNUM • • •
      │          ├─ TITLE
      │          ├─ ZERO8
      │          ├─ ACASOL ──┬─ CMULT
      │          │           ├─ INDXX
      │          │           ├─ CDIV
      │          │           └─ COPY8
      │          └─ OUTNAM • • •
      ├─ DISTO ──┬─ TITLE
      │          ├─ COPY16
      │          ├─ MAGPHS
      │          ├─ ACLOAD • • •
      │          ├─ ACDCMP • • •
      │          ├─ ZERO8
      │          ├─ ACASOL • • •
      │          └─ ACSOL • • •
      └─ CLRMEM
```

Figure 8-11 ACAN map.

elements connected to it, that every node in the circuit has a dc path to ground, and that there are no inductor and/or voltage source meshes and no capacitor and/or current source cuts.

8.2.5 The SETUP module

The SETUP routine is the control routine for this module. First, the subroutine MATPTR is called to determine the nodal admittances

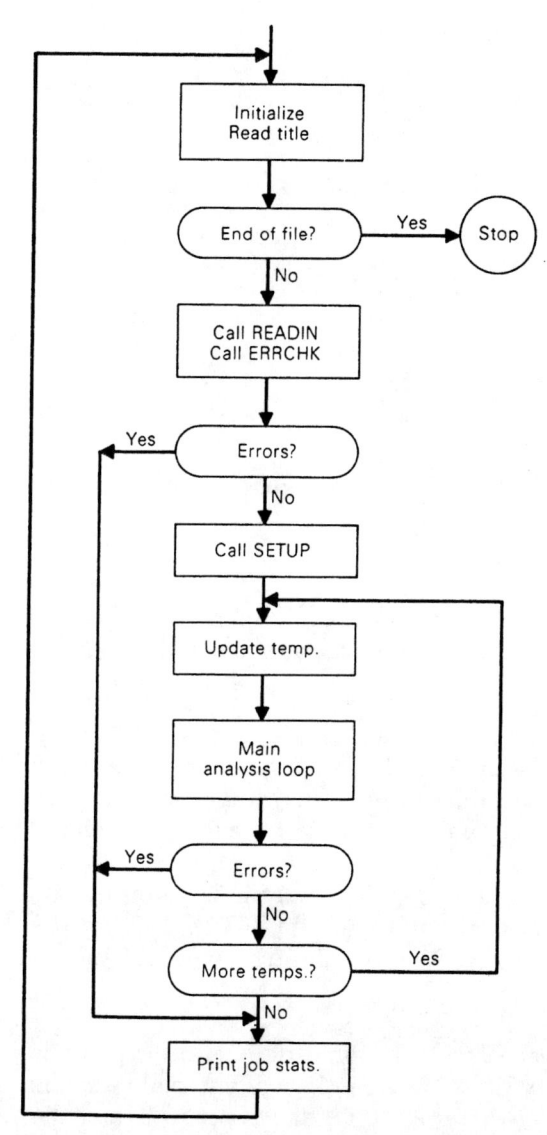

Figure 8-12 Main control flowchart of SPICE2 [3].

and matrix coefficients that are nonzero. MATPTR cycles through the circuit elements and, with the aid of the RESERV subroutine, establishes the initial integer pointer structure of nonzero Y-matrix terms. Next, the pointer structure is reordered to minimize fill-in, and the fill-in that does occur is added to the pointers. The reordering is accomplished by the REORDR subroutine. Finally, the MATLOC subroutine computes and stores the pointer locations that are used to

Figure 8-13 DC transfer-curve flowchart [3].

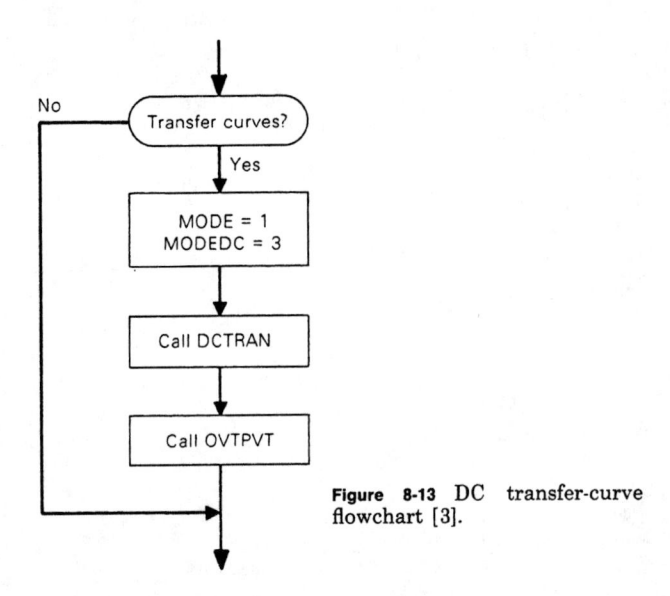

construct the Y matrix. The INDXX routine is used by MATLOC to determine the location of a specific Y-matrix coefficient [3].

8.2.6 The DCTRAN module

DCTRAN is the largest and most complicated module in the SPICE2 program. This subroutine performs the dc operating-point analysis, the transient initial-condition analysis, the dc transfer-curve analysis, and the transient analysis.

The flowchart for the dc operating-point analysis is given in Fig. 8-16, for the transient initial time-point solution in Fig. 8-17, for the dc transfer-curve analysis in Fig. 8-18, for the transient analysis in Fig. 8-19 [3].

a. DC operating-point analysis. The logic for a dc operating point is uncomplicated. As shown in Fig. 8-16, the program first evaluates the source values at time zero. Next, the INITF flag is set to 2, and the subroutine ITER8 is called to determine the dc solution iteratively. If the solution converges, the INITF flag is reset to 4, and the device model routines DIODE, BJT, JFET, and MOSFET are called to compute the linearized, small-signal value of the nonlinear capacitors in the device models. The node voltages for the solution are then printed, and DCTRAN returns to the main program.

b. Transient initial time-point solution. The solution of the initial transient time point is simpler than that of the dc operating point. As

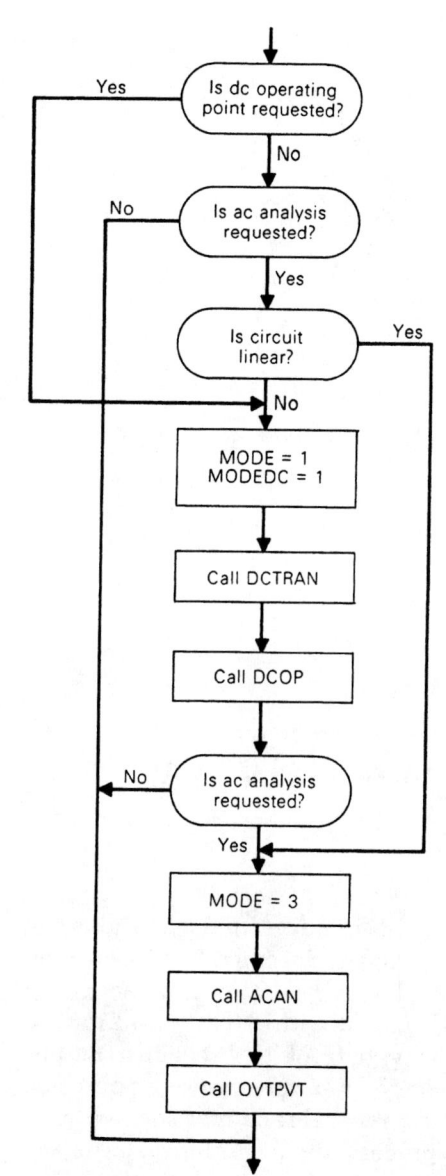

Figure 8-14 DC operating-point and ac analysis flowchart [3].

shown in Fig. 8-17, the difference between these two single-point dc analyses is that, for the case of the transient initial time point, the linearized capacitance values are not computed, since they are of no use in the transient analysis [3].

c. DC transfer-curve analysis. The dc transfer-curve analysis requires multiple-point analysis and the subsequent storage of output vari-

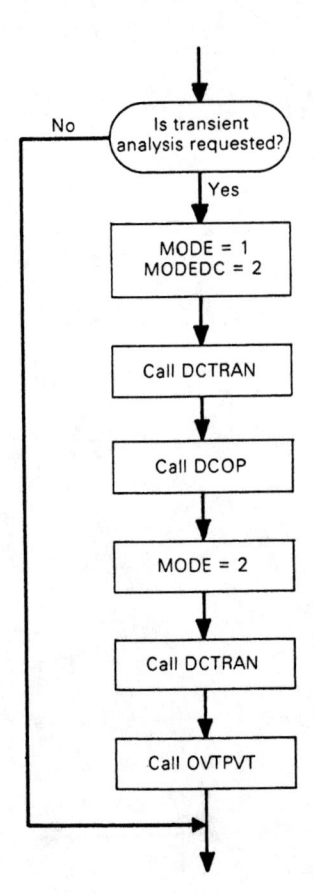

Figure 8-15 Transient analysis flowchart [3].

ables. As shown in Fig. 8-18, the dc transfer-curve analysis begins the same way as the dc operating-point analysis. Specifically, the variable source value is set to the first value, all other source values are set to their time-zero values, and ITER8 is called with INITF equal to 2 to obtain the first-point solution. The values of the specified output variables are stored in central memory after this transfer point has been computed. Then the variable source value is incremented, and the INITF flag is set to 6. This process then continues until the required number of dc transfer-curve points has been computed [3].

d. Transient analysis. The transient analysis loop is shown in Fig. 8-19. As this figure shows, the transient analysis and the dc transfer-curve analysis loops are very similar. The time-zero solution has already been computed before this loop is entered. Therefore, the first step in the analysis is the storage of the time-zero output variables. The first time point is then computed by calling ITER8 with the flag

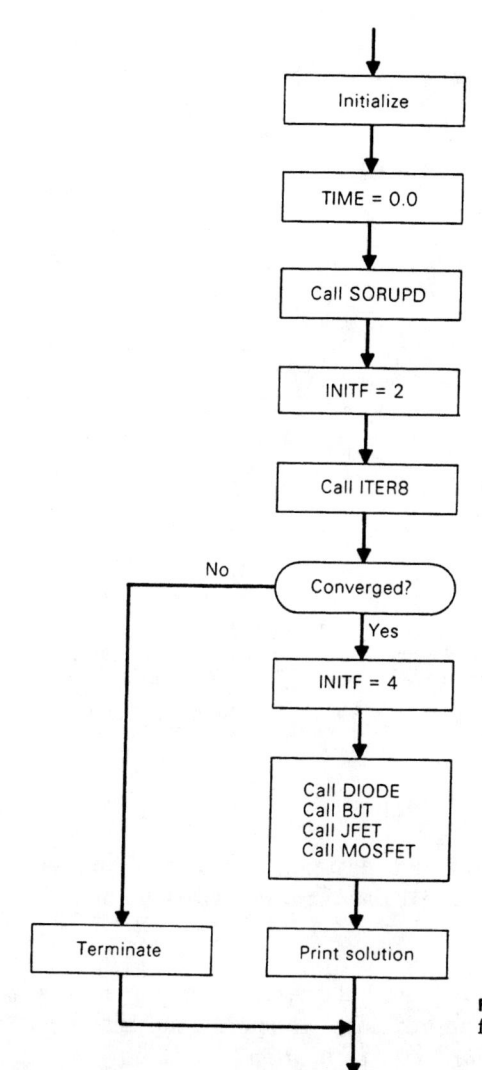

Figure 8-16 DC operating-point flowchart [3].

INITF set to 5. Each subsequent time point is computed with the flag INITF set to 6. If the time point does not converge, the time step DELTA is reduced by a factor of 8 and the time point is reattempted. If the time point does converge, then the subroutine TRUNC is called to determine the new time step that is consistent with the truncation error tolerance that has been specified. As long as the time step is larger than the time step presently in effect, the time point is accepted and the output variables are stored. If the truncation error is too large, as indicated by a time step smaller than the present time

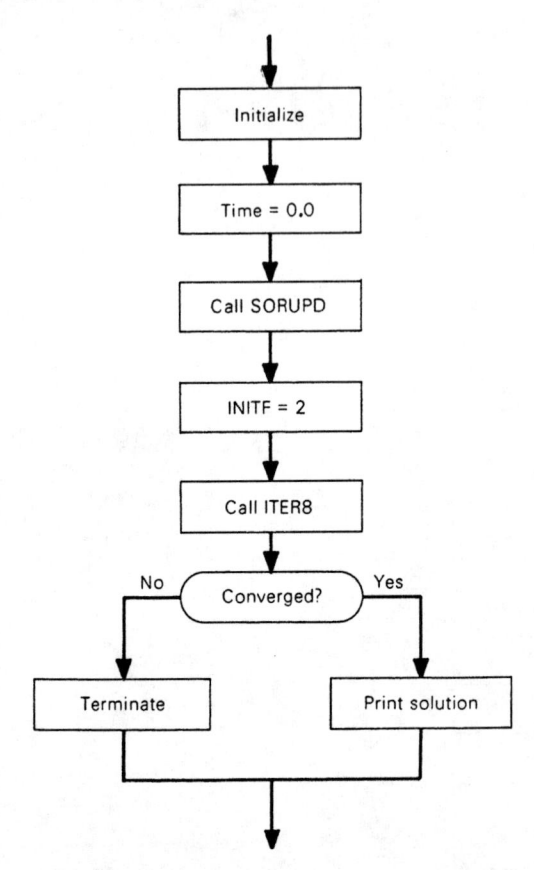

Figure 8-17 Transient initial condition flowchart [3].

step, then the time point is rejected and reattempted with the smaller time step. This process continues until the time interval specified by the user has been covered [3].

e. Automatic time step. There are two methods available to the user to control the computation of the appropriate transient analysis time step: *iteration count* and *truncation error estimation*.

The iteration count method uses the number of Newton-Raphson iterations required to converge at a time point as a measure of the rate of change of the circuit. If the number of iterations is less than ITL3, the time step is doubled; if the number is greater than ITL4, the time step is divided by 8. This method has the advantage of minimal computation overhead; however, it has the disadvantage that it does not take into account the true rate of change of circuit variables.

The second method of time-step determination is based on the estimation of local truncation error. The appropriate function derivative is approximated by a divided difference, hence the *estimation*.

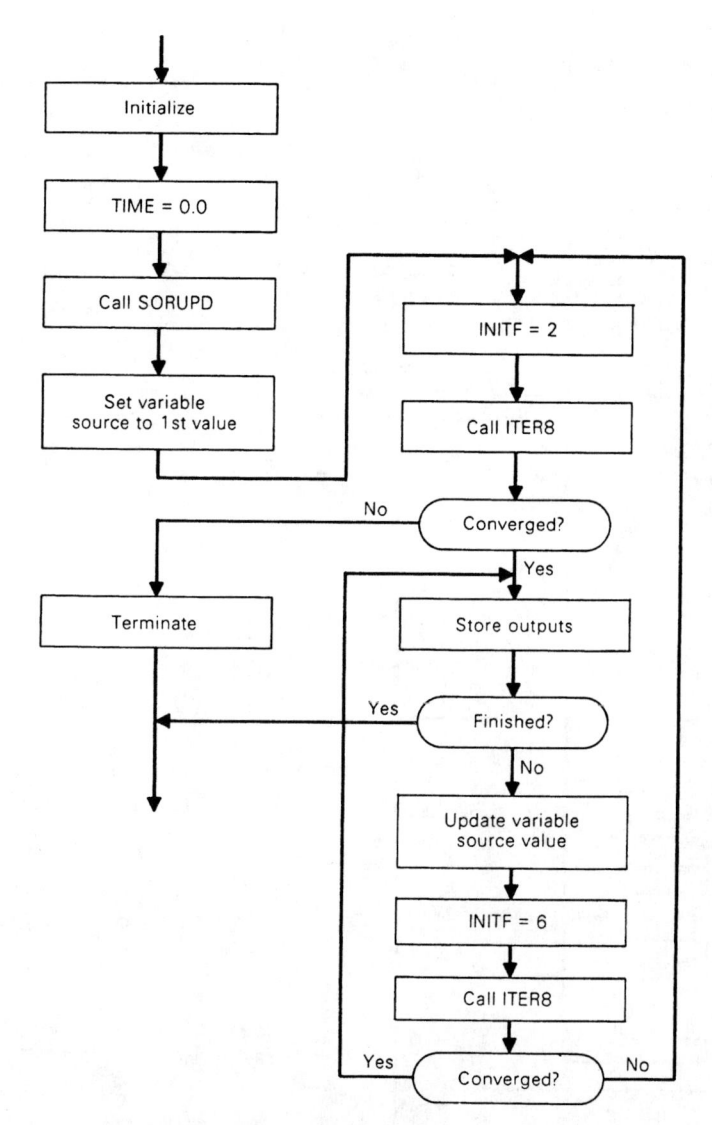

Figure 8-18 DC transfer-curve flowchart [3].

Although requiring somewhat more CPU time than iteration count, the truncation error estimation method allows a meaningful error bound to be set on the computed output values [3,4].

f. Breakpoint table. In transient analysis, SPICE2 always uses a program-calculated time step, regardless of the user-specified print interval. However, the independent source waveforms frequently have

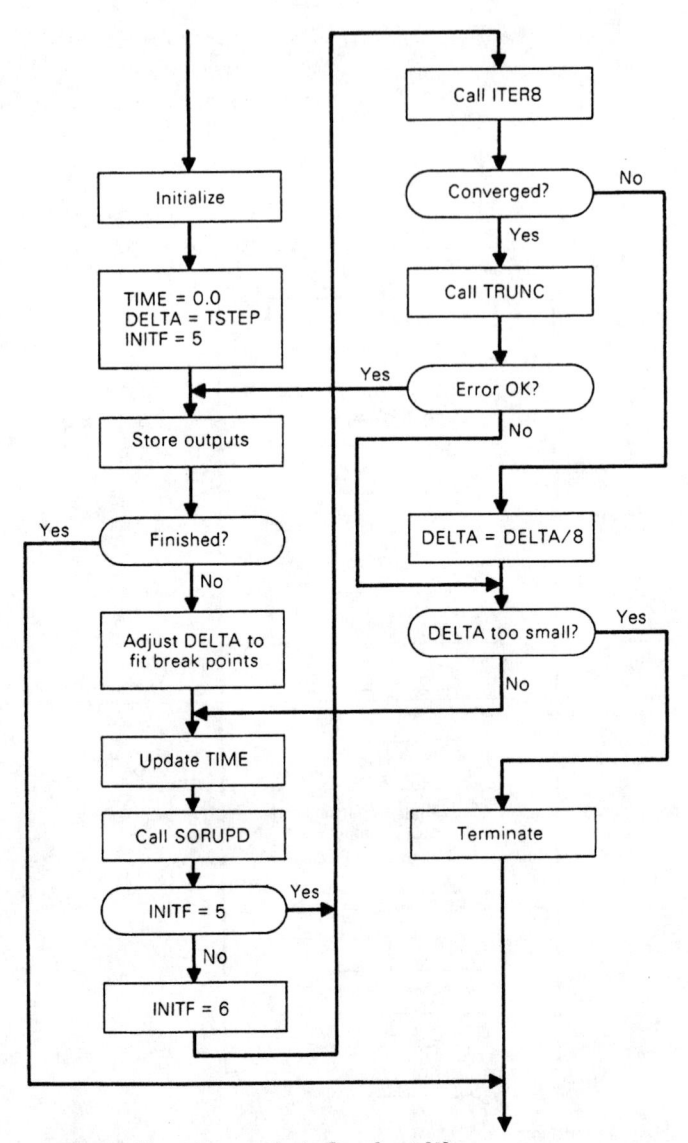

Figure 8-19 Transient analysis flowchart [3].

sharp transitions that could cause an unnecessary reduction in the time step in order to find the exact transition time. To overcome this problem, ERRCHK generates a *breakpoint table*, LSBKPT, which contains a sorted list of all the transition points of the independent sources. During transient analysis, whenever the next time point is sufficiently close to one of the breakpoints, the time step is adjusted so that the program lands exactly on the breakpoint [4].

g. ITER8 subroutine. All four of the analysis procedures implemented in DCTRAN use the subroutine ITER8 to determine the solution of the circuit equations. The flowchart for the ITER8 subroutine is shown in Fig. 8-20. The solution commences as follows. First, the subroutine LOAD is called. This subroutine constructs the linearized system of nodal equations for the specific iteration. The contributions of the Y matrix from nonlinear devices are computed separately in

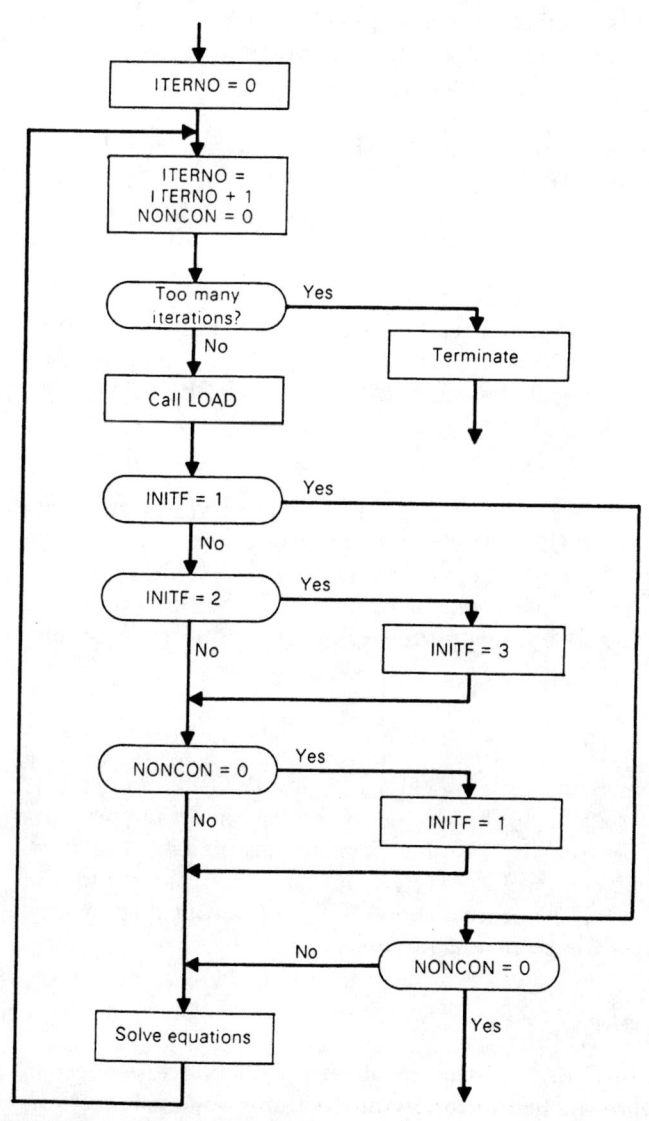

Figure 8-20 ITER8 flowchart [3].

the subroutines DIODE, BJT, JFET, and MOSFET. The NONCON flag is the number of nonconvergent nonlinearities in the circuit. This flag is incremented during the LOAD procedure, since the convergence is checked as an integral part of the linearization procedure. Immediately following the construction of the Y matrix and the forcing vector, a check on NONCON is performed to determine if the solution has converged. If convergence has been obtained, then the ITER8 subroutine is exited. Otherwise, the system of linearized equations is solved for the next iterate value, and the system of circuit equations is linearized again. This process continues until the solutions converge or until the number of iterations exceeds a preset number [3].

The solution of the linearized equations is determined by the routines DCDCMP and DCSOL.

8.2.7 The DCOP module

DCOP performs three functions: it prints out information about the device operating point, it computes the small-signal transfer function (subroutine SSTF), and it computes dc sensitivities of output variables with respect to circuit parameters (subroutine SENCAL).

8.2.8 The ACAN module

The ACAN module in SPICE2 determines the small-signal frequency-domain analysis. The ACLOAD subprogram constructs the complex system of linear equations for each frequency point. This system of equations is then solved with the routines ACDCMP and ACSOL. Subroutines DISTO and NOISE perform distortion and noise analysis, respectively.

8.2.9 The OVTPVT module

The last module in SPICE2 is OVTPVT. The OVTPVT subroutine interpolates the output variables to match the user-specified printing increment and generates the tabular listings specified by the user. The PLOT subroutine is called to generate the line-printer plots of the simulated results. If requested, the FOURAN subroutine performs the Fourier analysis for transient analysis.

8.3 Convergence Problems

It has been seen in Sec. 8.1.2 that problems with convergence can arise during simulation. In the following sections, some suggestions are given to improve and force convergence.

a. DC or transient analysis. First, it is very important to point out that most convergence problems are due to errors in the description of the circuit, both in the value of some parameter (electrical, geometrical, or physical) or in the circuit description itself (wrong connections, wrong model references, floating subcircuit nodes). Consequently, it is very important to examine the input data for possible errors.

To facilitate this error check, use the LIST and NODE options in the .OPTIONS statement. Once this check has been completed, you can decide if you want to modify some of the convergence parameters in the .OPTIONS statement.

Those parameters that may have an influence on convergence in both dc and transient analysis are described in Table 8-1.

The three parameters in Table 8-1 give an estimate of the convergence conditions that must be satisfied. Their modification obtains two main results:

1. Speeding up the convergence in any case, if the suggested values are used, still assuring good solution values (this depends also on the type of circuit analyzed).
2. Forcing the convergence when nothing else seems to work. This can help to isolate the part of the circuit that is creating the problems.

To see the default values used in the actual simulation, the user can see option OPTS in the .OPTIONS statement.

b. DC analysis. Something can be done specifically for dc convergence if this is the main problem.

1. Use the .NODESET statement, giving reasonable initial values to the most critical nodes in the circuit. In some cases it may help to step the dc voltage sources from 0 to their dc value using the intermediate solution as initial conditions (with .NODESET) of the following solution (source stepping).
2. Use the OFF option for those devices in a positive feedback path.
3. Increase the dc iteration limit, ITL1, in the .OPTIONS statement.

TABLE 8-1

Parameter	Default value	Suggested to help or to speed up convergence	Suggested to force convergence
RELTOL	0.001	0.01	0.1
ABSTOL	1 pA	1 nA	1 mA
VNTOL	1 μV	1 mV	10 mV

4. Step the dc voltage sources from 0 to their dc values via a transient analysis. In this way the sources are increased "slowly" and the capacitors are included in the solution. This is actually more similar to the real world and can be very appropriate for those circuits that need a finite setup time. When the dc solution is needed anyway (for an ac analysis), put in the .NODESET statement the solution obtained in this way.

c. Transient analysis. The typical error message when convergence is not achieved in transient analysis is

INTERNAL TIMESTEP TOO SMALL IN TRANSIENT ANALYSIS

This is due to the fact that SPICE2 starts internally to use the time specified in the .TRAN statement. But if convergence is not achieved within a certain criteria (see below), it keeps decreasing the internal time step, dividing it by 8 each time until a certain lower limit (not accessible to the user) is exceeded.

Several things can be done to help convergence:

1. Use the .IC statement (or the IC = *option* in the transistor description) and/or the option UIC in the .TRAN statement. A good method for avoiding initial convergence problems (in the simulated time sense) is to use the option UIC without specifying any IC (which is to say they are all set to zero) and applying the power supply only. In this case, the circuit will reach the steady state (i.e., initial conditions for transient analysis) gradually. After the simulated time necessary for the circuit to reach the steady state (depending on the circuit), the time-dependent input sources can be applied.

2. Act on the parameters LVLTIM, TRTOL, and ITL4 of the .OPTIONS statement. In the latest versions of SPICE2, the truncation error time-step control is the default method for deciding when to double the internal time step if convergence is easy and when to divide the internal time step by 8 if convergence is difficult. If this method is used (LVLTIM = 2, default), then TRTOL is the parameter that must be modified to have some influence on convergence; if TRTOL is increased, less importance is given to the truncation error, and then it is less likely that the internal time step will be divided by 8, which means that the condition described at the beginning is less likely to occur. If LVLTIM = 1, the doubling or the division by 8 of the internal time step depends only on the number of iterations needed to reach convergence, and then TRTOL has no more influence. In this case, ITL4 is the parameter to modify, and, as is evident, increasing ITL4 helps SPICE2 to converge.

3. Change the method of integration from TRAPEZOIDAL (default) to GEAR. The parameter is METHOD in the .OPTIONS statement. If it is changed to GEAR, MAXORD must also be changed. The suggested value to give MAXORD is 4. If GEAR is chosen, the truncation error control is also automatically inserted.

8.4 SPICE2 Operation

8.4.1 Modified nodal analysis

To illustrate the operation of SPICE2 and to show how SPICE2 employs a device model, we will consider some examples. First, consider the simple circuit of Fig. 8-21.

From Kirchhoff's current law at each of the nodes of the circuit, we obtain the following set of equations:

node 1: $\qquad i_1 - i_2 = 0 \qquad$ (8-1)

node 2: $\qquad i_2 - i_3 = 0 \qquad$ (8-2)

which in terms of the voltage drops across each branch, can be written as

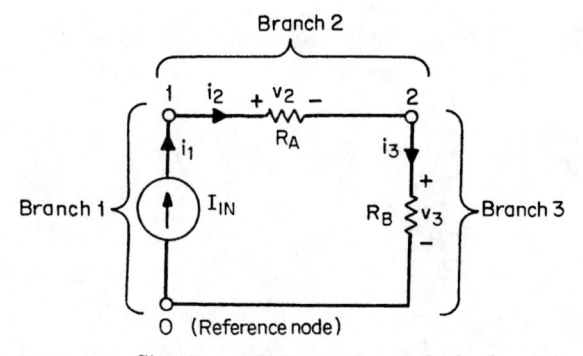

node 1: $\qquad I_{IN} - \dfrac{v_2}{R_A} = 0 \qquad$ (8-3)

node 2: $\qquad \dfrac{v_2}{R_A} - \dfrac{v_3}{R_B} = 0 \qquad$ (8-4)

Figure 8-21 Circuit to illustrate the method of nodal analysis [5]. Reprinted by permission.

Note: The material contained in Sec. 8.4 is taken from David L. Pulfrey and N. Garry Tarr, *Introduction to Microelectronic Devices*, copyright © 1989 by Prentice-Hall, Englewood Cliffs, N.J. Reprinted by permission.

SPICE2 prefers to deal with node voltages rather than the voltage drop across each branch, and so rewrites Eqs. (8-3) and (8-4) as

node 1: $$I_{IN} + \frac{e_2 - e_1}{R_A} = 0 \qquad (8\text{-}5)$$

node 2: $$\frac{e_1 - e_2}{R_A} - \frac{e_2}{R_B} = 0 \qquad (8\text{-}6)$$

The node voltages e_1 and e_2 are measured with respect to a reference or ground node labeled as node 0. Equations (8-5) and (8-6) are referred to as the *nodal analysis* (*NA*) equations. This set of linear equations can be solved to find the unknown node voltages e_1 and e_2.

The circuit of Fig. 8-22, which contains an independent voltage source V_{IN}, presents a slightly more difficult problem, since we cannot relate i_1 to V_{IN} without knowledge of the currents in the other branches. In this case it is necessary to generate another equation by applying Kirchhoff's voltage law to relate V_{IN} to other node voltages, producing the set of *modified nodal analysis* (*MNA*) equations listed below.

node 1: $$i_1 + \frac{e_2 - e_1}{R_A} = 0 \qquad (8\text{-}7)$$

node 2: $$\frac{e_1 - e_2}{R_A} - \frac{e_2}{R_B} = 0 \qquad (8\text{-}8)$$

$$e_1 = V_{IN} \qquad (8\text{-}9)$$

Linear algebra provides a variety of techniques that can be used to solve this system of linear equations for the unknowns e_1, e_2 and i_1.

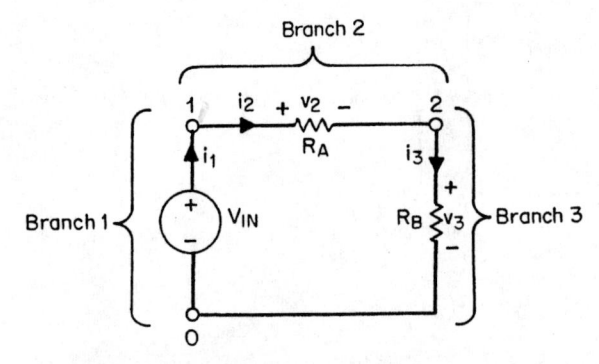

Figure 8-22 Circuit to illustrate the method of modified nodal analysis [5]. Reprinted by permission.

8.4.2 Nonlinear circuit elements

Consider the circuit shown in Fig. 8-23, in which one of the resistors of Fig. 8-22 has been replaced by a diode. Assuming that the polarity of the voltage source V_{IN} is such as to place the diode into forward-bias operation, the steady-state current-voltage relation for the diode is

$$I_D = I_S e^{qV_D/kT} \qquad (8\text{-}10)$$

where I_S is the diode saturation current. The MNA equations for Fig. 8-23 are therefore Eqs. (8-7), (8-9), and

$$\frac{e_1 - e_2}{R_A} - I_S e^{qe_2/kT} = 0 \qquad (8\text{-}11)$$

The MNA equations are now nonlinear, so they cannot be solved by conventional linear algebra techniques. SPICE2 solves these equations using an iterative Newton-Raphson linearization procedure. Specifically, suppose that estimates e_1^k and e_2^k for the node voltages are known (here the superscript k will come to denote an iteration number). Applying a Taylor series expansion of the diode current I_D about some bias point V_0 and retaining only first-order terms gives

$$\begin{aligned}I_D &\simeq I_S e^{qV_0/kT} + \left.\frac{dI_D}{dV_D}\right|_{V_D = V_0} (V_D - V_0) \\ &\simeq I_S e^{qV_0/kT} + \frac{q(V_D - V_0)}{kT} I_S e^{qV_0/kT}\end{aligned} \qquad (8\text{-}12)$$

With this linearization the MNA equations for Fig. 8-23 become

node 1: $\quad i_1^{k+1} + \dfrac{e_2^{k+1} - e_1^{k+1}}{R_A} = 0 \qquad (8\text{-}13)$

node 2: $\quad \dfrac{e_1^{k+1} - e_2^{k+1}}{R_A} - I_S e^{qe_2^k/kT} - \dfrac{q(e_2^{k+1} - e_2^k)}{kT} I_S e^{qe_2^k/kT} = 0 \qquad (8\text{-}14)$

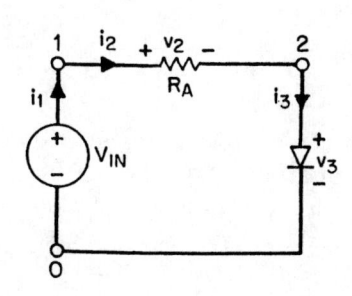

$$e_1^{k+1} = V_{IN} \qquad (8\text{-}15)$$

Figure 8-23 Circuit to illustrate the method of modified nodal analysis when nonlinear elements are present [5]. Reprinted by permission.

The linearized MNA Eqs. (8-13) to (8-15) can be solved using standard matrix techniques to determine new approximations e_1^{k+1} and e_2^{k+1} to the node voltages. This procedure is repeated until the change in the "best approximation" values for these voltages between iterations is acceptably small.

8.4.3 Transient analysis

In this section, how SPICE2 carries out transient analysis is considered. This will be done in reference to the circuit of Fig. 8-23, but we will now suppose that voltage source V_{IN} changes with time.

In a transient situation, it is convenient to divide the current at each terminal of a device into two parts: a *transport current*, which is simply the steady-state current that would exist under the given bias conditions, and a *charging current*, which supplies the change in stored charge associated with the terminal in question. SPICE2 uses a *quasi-static* approach to computation of stored charge. Specifically, it is assumed that the stored charge is a function only of the present terminal voltages and does not depend on the previous biasing history of the device. When using quasi-static analysis, the transient response of a device can be determined completely by deriving expressions for the steady-state transport current and the steady-state stored charge as functions of applied bias. It is also convenient if analytic expressions for the derivatives of the stored charge with respect to the terminal voltages are available. In an equivalent-circuit representation, these derivatives are viewed as small-signal capacitances associated with a device.

At moderate forward bias, the main component of stored charge in the diode is associated with the injected electrons. The stored charge Q_D is related to the applied bias V_D by

$$Q_D = Q_0 \, e^{qV_D/kT} \tag{8-16}$$

where Q_0 is a constant. Adding this charging current to branch 3 of Fig. 8-23 results in the modification of MNA Eq. (8-11) to the following:

$$\frac{e_1 - e_2}{R_A} - I_S e^{qe_2/kT} - \frac{dQ_D}{dt} = 0 \tag{8-17}$$

To determine the node voltages, it is therefore now necessary to solve a set of coupled ordinary differential equations [Eqs. (8-7), (8-9), and (8-17)]. A variety of numerical methods is available for this purpose. SPICE2 proceeds as follows. Suppose that the node voltages $e_2(t_n)$ and $e_1(t_n)$ are known at some time t_n (here the subscript n will serve as a time-step index). The values of $e_2(t_{n+1})$ and $e_1(t_{n+1})$ at some time Δt later are then specified implicitly by

$$Q_D(e_2(t_{n+1})) - Q_D(e_2(t_n)) = \frac{\Delta t}{2}\left[\left(\frac{e_1(t_{n+1}) - e_2(t_{n+1})}{R_A} - I_S e^{qe_2(t_{n+1})/kT}\right)\right.$$
$$\left. + \left(\frac{e_1(t_n) - e_2(t_n)}{R_A} - I_S e^{qe_2(t_n)/kT}\right)\right] \quad (8\text{-}18)$$

and Eqs. (8-7) and (8-9). The right-hand side of Eq. (8-18) represents the average of the charging currents at time t_{n+1} and at time t_n. This set of equations is then linearized and solved iteratively, following the approach discussed above for the steady-state analysis of circuits containing nonlinear elements. The linearization to the steady-state transport current I_D in the diode has already been discussed. The linearization of the expression for the stored charge Q_D is carried out in an analogous fashion. Let $e_2^k(t_{n+1})$ represent the "best guess" at iteration k for node voltage e_2 at time t_{n+1}; then

$$Q_D(e_2^{k+1}(t_{n+1})) = Q_D(e_2^k(t_{n+1})) + \left.\frac{\partial Q_D}{\partial V_D}\right|_{V_D = e_2^k(t_{n+1})} [e_2^{k+1}(t_{n+1}) - e_2^k(t_{n+1})]$$
$$= Q_D(e_2^k(t_{n+1})) + C_D(e_2^{k+1}(t_{n+1}) - e_2^k(t_{n+1})) \quad (8\text{-}19)$$

where C_D is the diode capacitance.

To obtain an accurate estimate for $e_2(t_{n+1})$ and $e_1(t_{n+1})$ at the end of this iterative procedure, it is necessary that the time step Δt be suitably small. Sophisticated procedures are followed by SPICE2 to adjust the size of the time-step interval to preserve accuracy as the transient analysis proceeds.

Although this discussion of the circuit analysis procedure followed by SPICE2 has been presented for simple examples, the same basic principles are applied to more complex circuits as described in Ref. [5].

8.5 Linked-List Specification

After reading the input data, SPICE2 builds a list of vectors that contain the basic information for all the elements and for all the analysis and output calls.

This section details an analysis of the vector structure limited to the most important ones, the semiconductor devices; for all the other elements, which are beyond the aim of this book, read Ref. [4].

The vector LOCATE (50) contains the pointers to each list head element; element types are referenced through the ID identifier. Inside this vector all integer data are referenced using the LOC pointer, while all real data are accessed using the LOCV pointer. In the following section, the expression enclosed in parentheses specifies in which subroutine the data are defined.

8.5.1 Semiconductor devices

a. DIODE (identifier ID = 11)

	−1	Subcircuit information	(FIND)
LOC	+0	Next pointer	(FIND)
	+1	LOCV pointer	(FIND)
	+2	Positive node np	(READIN)
	+3	Negative node nn	(READIN)
	+4	Model added node np'	(MODCHK)
	+5	Pointer to the device model	(FNDNAM)
	+6	OFF condition indicator	(READIN)
LOCV	+0	Element name	(FIND)
	+1	Area factor (default = 1)	(READIN)
	+2	Initial condition specification V_D	(READIN)

b. BJT (ID = 12)

	−1	Subcircuit information	(FIND)
LOC	+0	Next pointer	(FIND)
	+1	LOCV pointer	(FIND)
	+2	Collector node nc	(READIN)
	+3	Base node nb	(READIN)
	+4	Emitter node ne	(READIN)
	+5	Model added node nc'	(MODCHK)
	+6	Model added node nb'	(MODCHK)
	+7	Model added node ne'	(MODCHK)
	+8	Pointer to the device model	(FNDNAM)
	+9	OFF condition indicator	(READIN)
LOCV	+0	Element name	(FIND)
	+1	Area factor (default = 1)	(READIN)
	+2	Initial condition specification V_{BE}	(READIN)
	+3	Initial condition specification V_{CE}	(READIN)

c. JFET (ID = 13)

	−1	Subscript information	(FIND)
LOC	+0	Next pointer	(FIND)

	+1	LOCV pointer	(FIND)
	+2	Drain node nd	(READIN)
	+3	Gate node ng	(READIN)
	+4	Source node ns	(READIN)
	+5	Model added node nd'	(MODCHK)
	+6	Model added node ns'	(MODCHK)
	+7	Pointer to the device model	(FNDNAM)
	+8	OFF condition indicator	(READIN)
LOCV	+0	Element name	(FIND)
	+1	Area factor (default = 1)	(READIN)
	+2	Initial condition specification V_{DS}	(READIN)
	+3	Initial condition specification V_{GS}	(READIN)

d. MOST (ID = 14)

	−1	Subcircuit information	(FIND)
LOC	+0	Next pointer	(FIND)
	+1	LOCV pointer	(FIND)
	+2	Drain node nd	(READIN)
	+3	Gate node ng	(READIN)
	+4	Source node ns	(READIN)
	+5	Bulk node nb	(READIN)
	+6	Model added node nd'	(MODCHK)
	+7	Model added node ns'	(MODCHK)
	+8	Pointer to the device mode	(FNDNAM)
	+9	OFF condition indicator	(READIN)
LOCV	+0	Element name	(FIND)
	+1	Channel length L	(READIN)
	+2	Channel width W	(READIN)
	+3	Drain diffusion area A_D	(READIN)
	+4	Source diffusion area A_S	(READIN)
	+5	Initial condition specification V_{DS}	(READIN)
	+6	Initial condition specification V_{GS}	(READIN)
	+7	Initial condition specification V_{BS}	(READIN)
	+8	Device mode (normal = +1; inverse = −1)	(MOSFET)
	+9	Threshold voltage	(MOSFET)
	+10	Saturation voltage	(MOSFET)
	+11	Drain diffusion perimeter P_D	(READIN)
	+12	Source diffusion perimeter P_S	(READIN)

	+13	Drain diffusion equivalent squares NR_D	(READIN)
	+14	Source diffusion equivalent squares NR_S	(READIN)
	+15	Channel charge parameter XQC	(READIN)

8.5.2 Semiconductor device models

a. DIODE model (ID = 21)

	− 1	Subcircuit information	(FIND)
LOC	+ 0	Next pointer	(FIND)
	+ 1	LOCV pointer	(FIND)
	+ 2	Unused (zero)	(FIND)
	+ 3	Unused (zero)	(FIND)
	+ 4	Unused (zero)	(FIND)
LOCV	+ 0	Model name	(FIND)
	+ 1	IS	(READIN)
	+ 2	RS (1/RS in MODCHK)	(READIN)
	+ 3	N	(READIN)
	+ 4	TT	(READIN)
	+ 5	CJO	(READIN)
	+ 6	VJ	(READIN)
	+ 7	M	(READIN)
	+ 8	EG	(READIN)
	+ 9	XTI	(READIN)
	+10	KF	(READIN)
	+11	AF	(READIN)
	+12	FC (FC ∗ VJ in MODCHK)	(READIN)
	+13	BV	(READIN)
	+14	IBV	(READIN)
	+15	$VJ * (1 - \exp((1 - M) * \ln(1 - FC)))/(1 - M)$	(MODCHK)
	+16	$\exp((1 + M) * \ln(1 - FC))$	(MODCHK)
	+17	$1 - FC * (1 + M)$	(MODCHK)
	+18	$N * kT/q * \ln(N * (kT/q)/(\sqrt{2} * IS))$	(MODCHK)

b. BJT model (ID = 22)

	− 1	Subcircuit information	(FIND)
LOC	+ 0	Next pointer	(FIND)
	+ 1	LOCV pointer	(FIND)
	+ 2	Model type (NPN = 1; PNP = −1)	(READIN)

	+ 3	Unused (zero)	(FIND)
	+ 4	Unused (zero)	(FIND)
LOCV	+ 0	Model name	(FIND)
	+ 1	IS	(READIN)
	+ 2	BF	(READIN)
	+ 3	NF	(READIN)
	+ 4	VAF (1/VAF in MODCHK)	(READIN)
	+ 5	IKF (1/IKF in MODCHK)	(READIN)
	+ 6	ISE	(READIN)
	+ 7	NE	(READIN)
	+ 8	BR	(READIN)
	+ 9	NR	(READIN)
	+10	VAR (1/VAR in MODCHK)	(READIN)
	+11	IKR (1/IKR in MODCHK)	(READIN)
	+12	ISC	(READIN)
	+13	NC	(READIN)
	+14	0	(READIN)
	+15	0	(READIN)
	+16	RB	(READIN)
	+17	IRB	(READIN)
	+18	RBM	(READIN)
	+19	RE (1/RE in MODCHK)	(READIN)
	+20	RC (1/RC in MODCHK)	(READIN)
	+21	CJE	(READIN)
	+22	VJE	(READIN)
	+23	MJE	(READIN)
	+24	TF	(READIN)
	+25	XTF	(READIN)
	+26	VTF (1.44/VTF in MODCHK)	(READIN)
	+27	ITF	(READIN)
	+28	PTF [PTF/($2\pi *$ TF) in MODCHK]	(READIN)
	+29	CJC	(READIN)
	+30	VJC	(READIN)
	+31	MJC	(READIN)
	+32	XCJC	(READIN)
	+33	TR	(READIN)
	+34	0	(READIN)
	+35	0	(READIN)

	+36	0	(READIN)
	+37	0	(READIN)
	+38	CJS	(READIN)
	+39	VJS	(READIN)
	+40	MJS	(READIN)
	+41	XTB	(READIN)
	+42	EG	(READIN)
	+43	XTI	(READIN)
	+44	KF	(READIN)
	+45	AF	(READIN)
	+46	FC (FC * VJE in MODCHK)	(READIN)
	+47	$VJE * (1 - \exp((1 - MJE) * \ln(1 - FC)))/(1 - MJE)$	(MODCHK)
	+48	$\exp((1 + MJE) * \ln(1 - FC))$	(MODCHK)
	+49	$1 - FC * (1 + MJE)$	(MODCHK)
	+50	FC * VJC	(MODCHK)
	+51	$VJC * (1 - \exp((1 - MJC) * \ln(1 - FC)))/(1 - MJC)$	(MODCHK)
	+52	$\exp((1 + MJC) * \ln(1 - FC))$	(MODCHK)
	+53	$1 - FC * (1 + MJC)$	(MODCHK)
	+54	$kT/q * \ln((kT/q)/(\sqrt{2} * IS))$	(MODCHK)

c. **JFET model (ID = 23)**

	−1	Subcircuit information	(FIND)
LOC	+0	Next pointer	(FIND)
	+1	LOCV pointer	(FIND)
	+2	Model type (NJF = 1; PJF = −1)	(READIN)
	+3	Unused (zero)	(FIND)
	+4	Unused (zero)	(FIND)
LOCV	+0	Model name	(FIND)
	+1	VTO	(READIN)
	+2	BETA	(READIN)
	+3	LAMBDA	(READIN)
	+4	RD (1/RD in MODCHK)	(READIN)
	+5	RS (1/RS in MODCHK)	(READIN)
	+6	CGS	(READIN)
	+7	CGD	(READIN)
	+8	PB	(READIN)

+ 9	IS	(READIN)
+10	KF	(READIN)
+11	AF	(READIN)
+12	FC (FC * PB in MODCHK)	(READIN)
+13	PB * (1 − exp(0.5 * ln(1 − FC)))/0.5	(MODCHK)
+14	exp(1.5 * ln(1 − FC))	(MODCHK)
+15	1 − FC * 1.5	(MODCHK)
+16	$kT/q * \ln((kT/q)/(\sqrt{2} * IS))$	(MODCHK)

d. MOST model (ID = 24)

	− 1	Subcircuit information	(FIND)
LOC	+ 0	Next pointer	(FIND)
	+ 1	LOCV pointer	(FIND)
	+ 2	Model type (NMOST = 1; PMOST = −1)	(READIN)
	+ 3	Unused (zero)	(FIND)
	+ 4	Unused (zero)	(FIND)
LOCV	+ 0	Model name	(FIND)
	+ 1	LEVEL	(READIN)
	+ 2	VTO	(READIN)
	+ 3	KP	(READIN)
	+ 4	GAMMA	(READIN)
	+ 5	PHI	(READIN)
	+ 6	LAMBDA [$2 * K_{Si} * \varepsilon_0/q * $ NSUB (LEVEL = 3) in MODCHK]	(READIN)
	+ 7	RD (1/RD in MODCHK)	(READIN)
	+ 8	RS (1/RS in MODCHK)	(READIN)
	+ 9	CBD	(READIN)
	+10	CBS	(READIN)
	+11	IS	(READIN)
	+12	PB	(READIN)
	+13	CGSO	(READIN)
	+14	CGDO	(READIN)
	+15	CGBO	(READIN)
	+16	RSH (1/RSH in MODCHK)	(READIN)
	+17	CJ	(READIN)
	+18	MJ	(READIN)
	+19	CJSW	(READIN)
	+20	MJSW	(READIN)

+21	JS	(READIN)
+22	TOX (COX = $K_{SiO_2} * \varepsilon_0$/TOX in MODCHK)	(READIN)
+23	NSUB (NSUB * 10^6 in MODCHK)	(READIN)
+24	NSS (NSS * 10^4 in MODCHK)	(READIN)
+25	NFS (NFS * 10^4 in MODCHK)	(READIN)
+26	TPG	(READIN)
+27	XJ	(READIN)
+28	LD	(READIN)
+29	UO (UO * 10^{-4} in MODCHK)	(READIN)
+30	UCRIT [UCRIT * 10^2 * K_{Si} * ε_0/COX (LEVEL = 2) in MODCHK] [THETA (LEVEL = 3) in MODCHK]	(READIN)
+31	UEXP [ETA * 8.15 * 10^{-22}/COX (LEVEL = 3) in MODCHK]	(READIN)
+32	UTRA [KAPPA (LEVEL = 3) in MODCHK]	(READIN)
+33	VMAX	(READIN)
+34	NEFF	(READIN)
+35	XQC [1 (LEVEL = 3) in MODCHK]	(READIN)
+36	KF	(READIN)
+37	AF	(READIN)
+38	FC (FC * PB in MODCHK)	(READIN)
+39	DELTA [DELTA * 0.25 * 2π * K_{Si} * ε_0/COX (LEVEL = 3) in MODCHK]	(READIN)
+40	THETA [PB * (1 − exp(0.5 * ln(1 − FC)))/0.5 in MODCHK]	(READIN)
+41	ETA [exp(1.5 * ln(1 − FC)) in MODCHK]	(READIN)
+42	KAPPA (1 − FC * 1.5 in MODCHK)	(READIN)
+43	−1	(MODCHK)
+44	VTO − GAMMA * $\sqrt{PHI}$ + PHI	(MODCHK)
+45	$\sqrt{2 * K_{Si} * \varepsilon_0/q * NSUB}$	(MODCHK)

8.5.3 Useful flags

In analyzing the device model subroutines, the user must know the variables used by SPICE2 to keep track of the state of the analysis. The values of these flags (and the corresponding meanings) are as shown in Table 8-2.

Other useful flags the user can specify in the input are included in the .OPTIONS statement [1]. This statement allows the user to reset

TABLE 8-2

Flag	Value	Meaning
MODE	1	DC analysis (subtype defined by MODEDC)
	2	Transient analysis
	3	AC analysis (small-signal)
MODEDC*	1	DC operating point
	2	Initial operating point for transient analysis
	3	DC transfer-curve computation
INITF	1	Converge with OFF devices allowed to float
	2	Initialize junction voltages
	3	Converge with OFF devices held "off"
	4	Store small-signal parameters
	5	First time point in transient analysis
	6	Prediction step
NONCON		Number of nonconvergent branches
ITERNO		Iteration number for current sweep point
ITEMNO		Temperature number
NOSOLV	1	Use initial conditions (UIC) specified for transient analysis

*MODEDC is significant only if MODE = 1.

program control and options for specific simulation purposes. Any combination of the following options can be included in any order, where x represents some positive number.

Flag	Meaning
ACCT	Causes accounting and run-time statistics to be printed.
LIST	Causes the summary listing of the input data to be printed.
NOMOD	Suppresses printout of the model parameters.
NOPAGE	Suppresses page ejects.
NODE	Causes printing of the node table.
OPTS	Causes option values to be printed.
GMIN = x	Resets the value of GMIN, the minimum conductance allowed by the program; the default value is 10^{-12} mho.
RELTOL = x	Resets the relative error tolerance of the program; the default value is 0.001 (0.1 percent).
ABSTOL = x	Resets the absolute current error tolerance of the program; the default value is 10^{-12} A.
VNTOL = x	Resets the absolute voltage error tolerance of the program; the default value is 10^{-6} V.

TRTOL = x	Resets the transient error tolerance; the default value is 7.0. This parameter is an estimate of the factor by which SPICE2 overestimates the actual truncation error.
CHGTOL = x	Resets the charge tolerance of the program; the default value is 10^{-14}.
NUMDGT = x	Resets the number of significant digits printed for output variable values; x must satisfy the relation $0 < x < 8$; the default value is 4. This option is independent of the error tolerance used by SPICE2.
TNOM = x	Resets the nominal temperature; the default value is 27°C (300 K).
PIVTOL = x	Resets the absolute minimum value for a matrix entry to be accepted as a pivot; the default value is 10^{-13}.
PIVREL = x	Resets the relative ratio between the largest column entry and an acceptable pivot value; the default value is 10^{-3}. In the numerical pivoting algorithm, the allowed minimum pivot value is determined by EPSREL = AMAX1 (PIVREL * MAXVAL, PIVTOL), where MAXVAL is the maximum element in the column where a pivot is sought.
ITL1 = x	Resets the dc iteration limit; the default is 100.
ITL2 = x	Resets the dc transfer-curve iteration limit; the default is 50.
ITL3 = x	Resets the lower transient analysis iteration limit; the default value is 4.
ITL4 = x	Resets the transient analysis time-point iteration limit; the default value is 10.
ITL5 = x	Resets the transient analysis total iteration limit; the default is 5000. Set ITL5 = 0 to omit this test.
LIMTIM = x	Resets the amount of CPU time reserved by SPICE2 for generating plots should a limit on CPU time cause job termination; the default value is 2 s.
LIMPTS = x	Resets the total number of points that can be printed or plotted in dc, ac, or transient analysis; the default value is 201.
LVLTIM = x	If x is 1, the iteration time-step control is used. If x is 2, the truncation error time step is used. The default value is 2. If METHOD = GEAR and MAXORD > 2, SPICE2 sets LVLTIM to 2.
METHOD = name	Sets the numerical integration method used by SPICE2. Possible names are GEAR or TRAPEZOIDAL. The default is TRAPEZOIDAL.
MAXORD = x	Sets the maximum order for the integration method if GEAR's variable-order method is used; x must be between 2 and 6. The default value is 2.

DEFL = x   Resets the value for MOST channel length; the default is 10^{-4} m.
DEFW = x   Resets the value for MOST channel width; the default is 10^{-4} m.
DEFAD = x   Resets the value for MOST drain diffusion area; the default is 0.
DEFAS = x   Resets the value for MOST source diffusion area; the default is 0.

8.6 SPICE3 Capabilities versus SPICE2

The *SPICE3* [6,7] (written in C language) is directly based on SPICE2 Version G.6. While SPICE3 is being developed to include new features, it will continue to support those capabilities and models which remain in extensive use in the SPICE2 program.

In addition to the pole-zero analysis for studying the behavior of a linear time-invariant network and to the switch element, the main features concern the semiconductor devices: i.e., the introduction of the new BSIM (Berkeley Short-Channel IGFET Model) model for the MOST, and the implementation of the GaAs MESFET.

8.7 HSPICE Capabilities versus SPICE2

HSPICE [8] (written in C language) incorporates features found in University of California at Berkeley SPICE2 and other circuit analysis programs, as well as enhancements and features developed in-house.

HSPICE is compatible with most SPICE2 versions and adds the following features [8]. (1) Improvement of the convergence capability: a well-known problem of SPICE2, especially when the circuit being simulated contains square-law devices. (2) IC design for single circuit (a circuit located on a single semiconductor sharing the same manufacturing) and for multiple circuits (interaction between different circuits—different simulation temperature, different way of variation of transistor model parameters, etc.—that were manufactured independently). (3) Accurate modeling with many foundry models included. (4) Scaling, which allows selected model and element parameters to be scaled. (5) Input, output, and behavioral algebraics for parametrizable cells: any parameter defined in the netlist can be replaced by an algebraic expression. (6) Spec sheet generator for cell characterization. (7) Optimization, which allows the automatic generation of model parameters and component values from a given set of electrical specifications or measured data. The most powerful feature of the HSPICE approach is its incremental optimization technique (in addition to the simultaneous one). Incremental optimization allows the dc parameters to be solved first, then the ac parameters, and finally the transient parameters. (8) Geometric lossy coupled transmission lines for PCB (Printed Circuit Board), multichip, package, and IC technologies. (9) Discrete component, pin, package, and vendor IC libraries. (10) Graphical utility for interactive waveform graphing from multiple simulations. (11) Pole-zero analysis: an alternative method for studying the behavior of a linear time-invariant network, described by its network function. (12) Monte Carlo analysis, which allows the user to do statistical, sensitivity, and worst-case analyses on circuits. (13) New semiconductor device models. DIODE model types include the junction diode model (with two variations: the geometric and non-geometric) used to simulate Zener, silicon diffused, and Schottky diodes and the Fowler-Nordheim model (used for the modeling of diode effects in nonvolatile EEPROM memory). The BJT model is an adaptation of the integral charge control model of Gummel-Poon to include several effects at high bias levels. Vertical and lateral BJT geometries are also modeled, in addition to a BJT quasi-saturation model used to simulate BJTs that exhibit quasi-saturation or base pushout effects (power BJTs). JFET model types include two models for the silicon-based JFET and a third one (with different submodels) for the MESFET. MOST models consist of client private and public models selected by a parameter specified by the user. New models (more than 30 models are now available) are constantly being added. For the semiconductor devices, to allow the use of model parameters from earlier versions of SPICE2, many of the model parameters have aliases or other names which are included in the .MODEL parameter list. Different sets of temperature equations for the semiconductor device parameters which may be selected by setting the temperature equation selectors in the .MODEL statement are also available.

8.8 PSPICE Capabilities versus SPICE2

PSPICE [9,10] (written in C language) is the first SPICE2-based simulator on the PC. PSPICE uses the same numeric algorithms as SPICE2 and also conforms to the SPICE2 format for input and output files. However, there are a number of features (utilities) in PSPICE that are not available in SPICE2. They are [9]: (1) A utility (*Probe*) for graphical output of analysis results; it can be defined as a "software oscilloscope." (2) A utility (*Parts*) that aids the user in modeling components using information from the data sheet. The result is a file representing a "real" device which can be added to the user library of devices. (3) A utility (*Analog Behavioral Modeling*) which allows the user a more flexible description of electronic components in terms of a transfer function. (4) A utility (*Monte Carlo*) which allows the user to do statistical, sensitivity, and worst-case analyses on circuits. (5) A utility (*Device Equations*) which allows the user to change the built-in model equations for the semiconductor devices. (6) A utility (*Digital Simulation*) which allows the user to simulate digital components. (7) A utility (*Digital Files*) which gives the user a flexibility in modeling the analog-to-digital and digital-to-analog interface. (8) Analog and digital device libraries. (9) Improvement in some analysis types (e.g., distortion, temperature, parametric, sweep, etc.) and extended syntax for input/output variables. (10) Additional nonlinear magnetics model for power inductors and transformers. (11) Ideal switch model. (12) Additional semiconductor devices: the GaAs MESFET and the BSIM model for the MOST. Some improvements in the other device models also have been made.

All the above main features that PSPICE exhibits are supported by a successful improvement of the convergence capability.

References

1. A. Vladimirescu, K. Zhang, A. R. Newton, D. O. Pederson, and A. Sangiovanni-Vincentelli, *SPICE Version 2G User's Guide*, Dept. EECS, University of California, Berkeley, 1981.
2. L. W. Nagel and D. O. Pederson, Simulation Program with Integrated Circuit Emphasis (SPICE), Electronics Research Laboratory, Rep. No. ERL-M382, University of California, Berkeley, 1973.
3. L. W. Nagel, SPICE2: A Computer Program to Simulate Semiconductor Circuits, Electronics Research Laboratory, Rep. No. ERL-M520, University of California, Berkeley, 1975.
4. E. Cohen, Program Reference for SPICE2, Electronics Research Laboratory, Rep. No. ERL-M592, University of California, Berkeley, 1976.
5. D. L. Pulfrey and N. G. Tarr, *Introduction to Microelectronic Devices*, Prentice-Hall, Englewood Cliffs, N.J., 1989.
6. T. I. Quarles, The SPICE3 Implementation Guide, Electronics Research Laboratory, Rep. No. ERL-M44, University of California, Berkeley, 1989.
7. T. I. Quarles, *SPICE3 Version 3C1 User's Guide*, Electronics Research Laboratory, Rep. No. ERL-M46, University of California, Berkeley, 1989.
8. Meta-Software Inc., *HSPICE User's Manual*, Campbell, Calif., 1991.
9. MicroSim Corporation, *PSPICE User's Manual*, Irvine, Calif., 1989.
10. P. W. Tuinenga, *SPICE: A Guide to Circuit Simulation and Analysis Using PSPICE*, Prentice-Hall, Englewood Cliffs, N.J., 1988.

Chapter

9

Metal-Semiconductor Field-Effect Transistor, Ion-Sensitive Field-Effect Transistor, and Semiconductor-Controlled Rectifier

Not all the semiconductor devices that one would like to simulate are built into the original University of California at Berkeley SPICE2. Thus many companies have an in-house version of SPICE2 that has modifications to suit particular needs. On the other hand, SPICE3 and some of the popular commercially supported versions of SPICE2 (HSPICE and PSPICE, considered in this book) present improvements in the semiconductor device models, but, as far as new semiconductor devices are concerned, these programs implement only the MESFET.

This chapter deals with three semiconductor devices not included in the original University of California at Berkeley SPICE2: MESFET, ISFET, and THYRISTOR. The MESFET has been chosen because now it is a well-established device implemented, as stated before, in SPICE3, HSPICE, and PSPICE; the ISFET has been chosen to point out the flexibility of the program to allow the user to implement also "hybrid" structures (i.e., in this case, solid-state and chemical technology interaction); the THYRISTOR has been chosen to point out that complex devices can be built out of components (in this case BJTs) which are included in the program.

Brief descriptions of the principles of operation of these devices, and of their electrical performances to obtain models suitable to be implemented in SPICE2 and/or SPICE2-based versions, are presented.

9.1 MESFET

9.1.1 Principles of operation

The *M*etal-*S*emiconductor *F*ield-*E*ffect *T*ransistor (*MESFET*), whose structure is shown in Fig. 9-1, is a Schottky-barrier gate FET made of silicon (Si) or of gallium arsenide (GaAs).

The MESFET basic operation is like the JFET (see Chap. 3) from which it differs in that a Schottky barrier junction (see Appendix C), rather than a *pn* junction (see Appendix A), is used as the gate element to modulate the depletion region width of the MS junction and therefore the resistance of the channel (between source and drain) through which the majority carriers flow. Thus one might expect that both types of devices could be modeled by the same equations. This is quite true for Si MESFETs, but this cannot be done for GaAs MESFETs. The physical reason lies in the fact that in GaAs the electron velocity saturates near the rather low electric field of 3×10^3 V/cm (see Fig. 9-2), whereas Si shows ohmic behavior over a range about 10 times larger (see Fig. A-1). Thus in GaAs the saturation of drain current with increasing drain-source voltage is caused by carrier-velocity saturation, whereas in Si it is channel pinchoff that causes the drain current to saturate.

Although MESFETs can be made of Si, such devices are not so common as GaAs MESFETs. In fact, Si MOSTs and Si JFETs are so

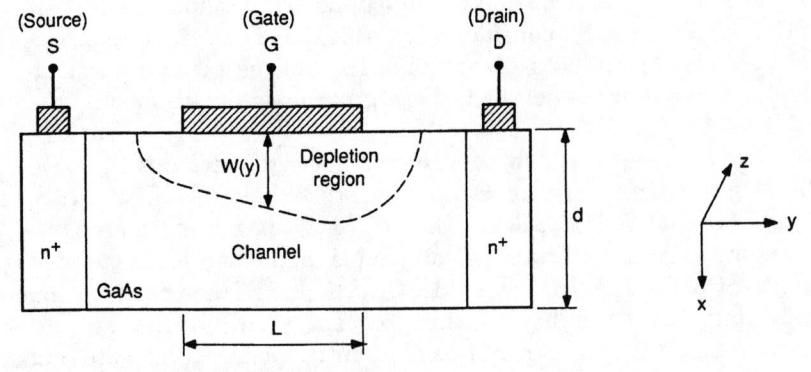

Figure 9-1 GaAs MESFET structure.

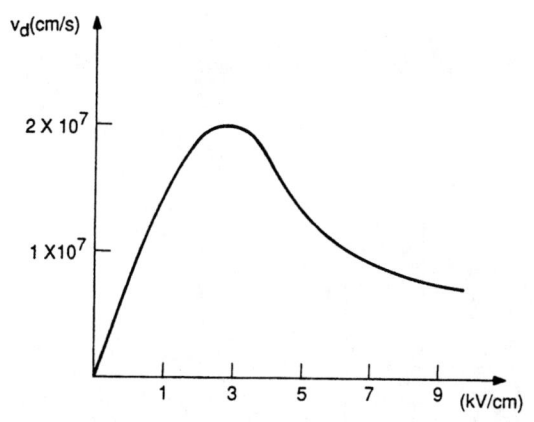

Figure 9-2 Effect of an electric field on the magnitude of the drift velocity of electrons in GaAs.

well-established structures and able of such high performances that there is no need to develop another Si FET structure, such as a Si MESFET. On the other hand, the GaAs technology gives the possibility of realizing devices which take advantage of the electron high mobility that this material exhibits. This characteristic, associated with the possibility to realize devices with very short channel lengths, make GaAs MESFETs of use in very high frequency and fast switching applications. This reason, and the fact that the Si MESFET is operationally identical (same equations and models) to the Si JFET which is treated in detail in Chap. 3, lead us to consider here only the GaAs MESFET. In addition, as for GaAs the mobility of electrons is much greater than that of holes, only n-type MESFETs are considered.

Enhancement- and depletion-type MESFETs are also possible, and are characterized by a channel that is closed or open, respectively, at zero gate-source bias V_{GS}. For values of V_{GS} above a particular voltage, the threshold voltage V_{T0}, the channel conducts. For low values of drain-source voltage V_{DS}, the channel behaves as a resistor. When V_{DS} exceeds $(V_{GS} - V_{T0})$, the drain current I_D saturates. Saturation can occur due to either pinchoff of the channel at the drain end (Si MESFETs), or to velocity saturation of the mobile carriers (electrons) in the channel in short channel GaAs MESFETs.

There is no need to consider enhancement and depletion MESFETs separately. Their principles of operation are the same and the two types of device are characterized by their threshold voltage V_{T0} (positive in an enhancement device, and negative in a depletion device).

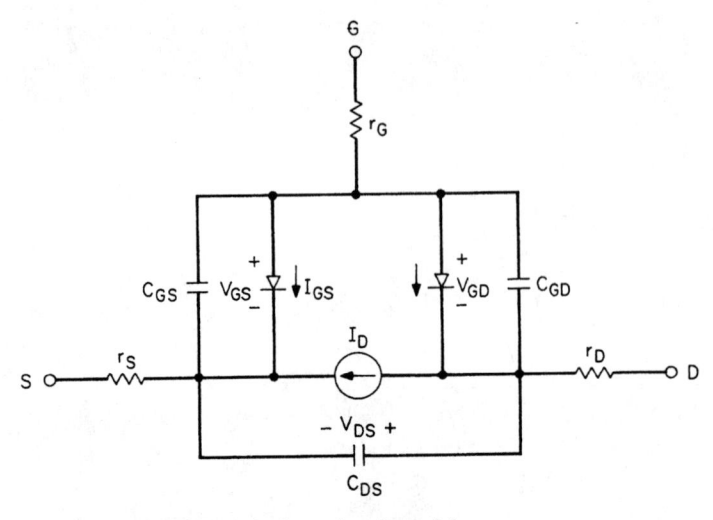

Figure 9-3 MESFET equivalent circuit model.

As the development of the MESFET basic theory follows directly from the JFET theory (described in Chap. 3) with the only restriction of keeping in mind the carrier-velocity saturation effect [1], in the following it is presented directly the basic-starting GaAs MESFET models proposed in the literature for use in circuit simulation programs such as SPICE3, HSPICE, and PSPICE.

For all the GaAs MESFET models, the complete equivalent circuit model follows directly from the one of the JFET and repeated for convenience in Fig. 9-3.

The resistors account for contact resistances and, in the case of r_S and r_D, the resistances of the n-depleted regions between the ends of the channel and the source and drain, respectively. The diodes represent the gate-source and the gate-drain Schottky barrier junctions. The currents I_{GS} and I_{GD} through the diodes follow the diode law indicated in Eqs. (C-2) and (C-3). The nonlinear drain current source I_D assumes different expressions according to linear or saturation regions. The capacitances C_{GS} and C_{GD} account for the charge-storage effects. The current I_D and the capacitances were investigated carefully in the literature as their formulation characterizes the GaAs MESFET respective to the JFET.

9.1.2 Curtice model

When SPICE2 has been used in the past to simulate GaAs MESFETs, the standard JFET model based on Shichman-Hodges relations (see Chap. 3) was often employed instead of a real GaAs MESFET model.

However, this approach has been shown to yield considerable errors with regard to I_D-V_{DS} characteristics below current saturation. Furthermore, electron transit-time effects under the gate are omitted.

a. Static model. As stated before, the current saturation in GaAs MESFETs occurs at lower voltages than in Si devices because of the much larger low-field mobility. The use of the hyperbolic tangent function applied to Shichman-Hodges equations below saturation has been proposed by Curtice [2] to provide a good analytical expression for the description of the drain current source I_D in GaAs MESFETs, both in linear and in saturation regions, suitable for use with circuit simulation programs. The basic static relationship for the proposed model is then

$$I_D = \beta(V_{GS} - V_{T0})^2(1 + \lambda V_{DS}) \tanh(\alpha V_{DS}) \qquad (9\text{-}1)$$

where β is the transconductance parameter, λ is the channel-length modulation parameter (see Sec. 3.1.2b); V_{T0} is the threshold voltage, and α is the hyperbolic tangent function parameter which determines the voltage at which the drain current characteristics saturate.

Equation (9-1) was used by Curtice [2] to simulate a set of experimental data, obtaining an accurate matching. For comparison, the SPICE2 JFET model was also used, showing a significant error below current saturation due to the lack of a parameter (i.e., α) to adjust the saturation point.

The *Curtice static model* equivalent circuit for the GaAs MESFET is shown in Fig. 9-3 when the capacitances are not taken into account, and the current source I_D is given by Eq. (9-1).

b. Transit-time effects. In transient operation, a change in $V_{GS}(t)$ does not cause an instantaneous change in I_D. In fact it takes some time for the depletion region charge to respond to the gate voltage change. The most important result of this effect is a time delay between V_{GS} and I_D. Thus Eq. (9-1) should be modified as [2]

$$I_D[V_{GS}(t - \tau), V_{DS}] \qquad (9\text{-}2)$$

where $\tau = L/v_{\text{drift}}$ is the transit time under the gate, L being the channel length, and v_{drift} the scatter-limited velocity for the electrons. Curtice [2] proposed a simple way to implement the time delay effect as a model parameter for circuit simulation programs.

The current source is assumed to be of the form [2]

$$I_D(V) - \tau \frac{dI_D(V)}{dt} \tag{9-3}$$

The second term of Eq. (9-3) can be considered as the first term of the expansion of $I_D(t - \tau)$ in time.

c. Large-signal model. As the MESFET is a majority-carrier device, the minority-carrier charge-storage effects (diffusion charge component) is not so relevant as in the BJT or in the inversion layer of the MOST. Consequently the only charge involved in the charge-storage effects is the ionized donors present in the depletion region under the gate (junction charge component). Under these assumptions, the *Curtice large-signal model* (Fig. 9-3) uses a SBD-like capacitance between source and gate (C_{GS}), where the depletion region thickness and thus the capacitance are determined by the gate-source voltage V_{GS}. A similar diode model is used to describe the capacitance between gate and drain (C_{GD}) with voltage V_{GD}-dependent capacitance. The analytical expressions for these capacitances are then

$$C_{GS} = C_{GS}(0)\left(1 - \frac{V_{GS}}{\phi_{s0}}\right)^{-1/2} \tag{9-4}$$

and

$$C_{GD} = C_{GD}(0)\left(1 - \frac{V_{GD}}{\phi_{s0}}\right)^{-1/2} \tag{9-5}$$

where ϕ_{s0} is the built-in voltage of the Schottky barrier junction and $C_{GS}(0)$ and $C_{GD}(0)$ are the zero-bias gate-source and gate-drain capacitances, respectively. It must be noted that for a negative gate-source voltage and for an increasing drain-source voltage beyond the point of current saturation, C_{GD} becomes much more heavily back-biased than C_{GS}. Thus C_{GD} can be neglected in the equivalent circuit of Fig. 9-3.

d. Small-signal model. The GaAs MESFET *small-signal model* is shown in Fig. 9-4. The conductances g_{GD} and g_{GS} are the two SBD small-signal conductances, defined as

$$g_{GS} = \left.\frac{dI_{GS}}{dV_{GS}}\right|_{op} \tag{9-6}$$

$$g_{GD} = \left.\frac{dI_{GD}}{dV_{GD}}\right|_{op} \tag{9-7}$$

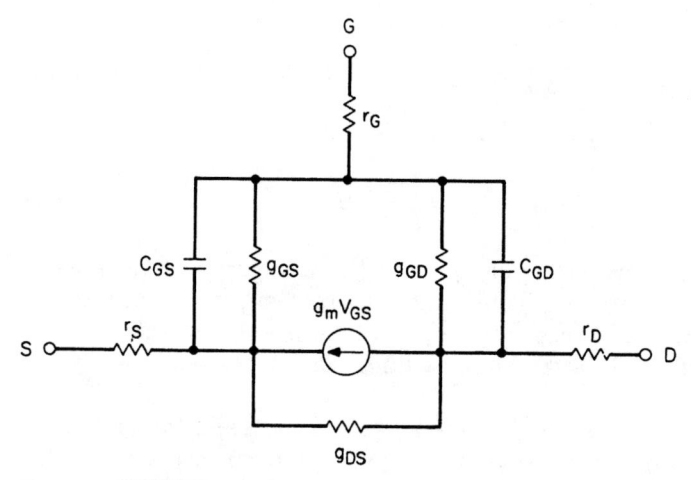

Figure 9-4 MESFET small-signal model.

and the transconductance g_m and the channel or output conductance g_{DS} are defined as

$$g_m = \frac{dI_D}{dV_{GS}}\bigg|_{op} \quad (9\text{-}8)$$

$$g_{DS} = \frac{dI_D}{dV_{DS}}\bigg|_{op} \quad (9\text{-}9)$$

where op indicates an operating point determined by the dc voltages V_{GS} and V_{GD}.

The model for the n-channel GaAs MESFET developed by Curtice was implemented into SPICE2 by Sussman-Fort et al. [3].

9.1.3 Statz model

The model proposed by Statz [4] has two improvements over the Curtice model [2]: (1) an enhanced drain current formulation and (2) a new capacitance model.

a. Static model. The Curtice static model [2] defined by Eq. (9-1) is a good representation of the drain current I_D versus the drain-source voltage V_{DS} for a given V_{GS}, but the behavior of I_D as a function of V_{GS} is not well represented, especially if the pinchoff voltage of the MESFET is large. Using observed operation, the quadratic term of the Curtice expression, defining the saturated drain current $I_{D,\text{sat}}$,

$$I_{D,\text{sat}} = \beta(V_{GS} - V_{T0})^2 \tag{9-10}$$

was changed to the empirical expression [4]

$$I_{D,\text{sat}} = \frac{\beta(V_{GS} - V_{T0})^2}{1 + b(V_{GS} - V_{T0})} \tag{9-11}$$

where V_{T0} is the threshold voltage, and b is a measure of the doping profile extending into the insulating substrate and thus depends on fabrication process. For small values of $(V_{GS} - V_{T0})$, Eq. (9-11) is indeed quadratic, while for larger values, Eq. (9-11) becomes almost linear in $(V_{GS} - V_{T0})$.

To increase computational efficiency, the hyperbolic tangent function was replaced with a polynomial approximation of the form [4]

$$\tanh(\alpha V_{DS}) \simeq 1 - \left(1 - \frac{\alpha V_{DS}}{n}\right)^n \tag{9-12}$$

with $n = 2$ or 3. In the saturated region ($V_{DS} > n/\alpha$), the tanh function is replaced by unity. Moreover, it has been shown [4] that the polynomial [Eq. (9-12)] with $n = 3$ gives the best fit.

In summary, Statz [4] developed a *GaAs MESFET static model* suitable for circuit simulation programs such as SPICE by the following equations:

$$I_D = \begin{cases} \dfrac{\beta(V_{GS} - V_{T0})^2}{1 + b(V_{GS} - V_{T0})} \left[1 - \left(1 - \dfrac{\alpha V_{DS}}{3}\right)^3\right](1 + \lambda V_{DS}) & \text{for } 0 < V_{DS} < \dfrac{3}{\alpha} \\ \dfrac{\beta(V_{GS} - V_{T0})^2}{1 + b(V_{GS} - V_{T0})}(1 + \lambda V_{DS}) & \text{for } V_{DS} \geq \dfrac{3}{\alpha} \end{cases} \tag{9-13}$$

The *Statz static model* equivalent circuit of the GaAs MESFET is still represented by Fig. 9-3 when the capacitances are not taken into account, and the current source I_D is given by Eqs. (9-13).

b. Large-signal model. The Curtice large-signal model [2] used a diode-like capacitance to describe the gate-source and gate-drain capacitances. This model introduced inaccuracies in the results for low source-drain voltages or when the MESFET is in reverse-biased operation. While the above remarks apply to both Si and GaAs MESFETs, the behavior of GaAs is further complicated by the early onset of carrier-velocity saturation. The capacitance model proposed by Statz [4] and developed with circuit simulation programs in mind aims to solve the above problems.

The analytical expressions for the gate-source capacitance C_{GS} and the gate-drain capacitance C_{GD} are [4]

$$C_{GS,GD} = \frac{C_{GS}(0)}{\sqrt{1 - \frac{V_n}{\phi_{s0}}}} \frac{1}{2}\left[1 + \frac{V_e - V_{TO}}{\sqrt{(V_e - V_{TO})^2 + \delta^2}}\right]$$

$$\times \frac{1}{2}\left[1 \pm \frac{V_{GS} - V_{GD}}{\sqrt{(V_{GS} - V_{GD})^2 + \left(\frac{1}{\alpha}\right)^2}}\right]$$

$$+ C_{GD}(0) \frac{1}{2}\left[1 \mp \frac{V_{GS} - V_{GD}}{\sqrt{(V_{GS} - V_{GD})^2 + \left(\frac{1}{\alpha}\right)^2}}\right] \quad (9\text{-}14)$$

The upper sign in the square brackets holds for C_{GS} and the lower sign holds for C_{GD}. In Eq. (9-14),

$$V_e = \frac{1}{2}\left[V_{GS} + V_{GD} + \sqrt{(V_{GS} - V_{GD})^2 + \left(\frac{1}{\alpha}\right)^2}\right] \quad (9\text{-}15)$$

$$V_n = \frac{1}{2}[V_e + V_{TO} + \sqrt{(V_e - V_{TO})^2 + \delta^2}] \quad \text{for } V_n \leq V_{max} \quad (9\text{-}16)$$

$$V_n = V_{max} \quad \text{for } V_n > V_{max} \quad (9\text{-}17)$$

$C_{GS}(0)$ and $C_{GD}(0)$ are the zero-bias gate-source and gate-drain junction capacitances, respectively; V_{TO} is the threshold voltage; ϕ_{s0} is the built-in junction potential; $1/\alpha$ is a parameter that determines the V_{DS} voltage at which the onset of drain current saturation occurs; V_e yields a smooth interpolation of C_{GS} and C_{GD} through the point of discontinuity at $V_{DS} = 0$; V_n is essentially equal to V_e before pinchoff and to V_{TO} beyond pinchoff, or in other words, V_n is to select the smaller of the two values of $-V_{TO}$ and $-V_e$; δ represents the voltage range over which the transition between these two values is accomplished (a suggested value is $\delta = 0.2$ V); V_{max} (whose proposed value is 0.5 V) limits the value of V_n in order to avoid imaginary numbers or dividing by zero in Eq. (9-14) when V_n becomes positive and equal or larger than ϕ_{s0}. This procedure limits the value (V_{max}) of the junction capacitance; for voltages beyond V_{max} the Statz model takes the capacitances to remain constant.

Analyzing Eq. (9-14), some considerations can be discussed. For $V_{DS} \gg 0$ (normal biasing conditions), C_{GS} follows a diode-like capacitance model as a function of V_{GS}. However, when V_{GS} approaches the pinchoff voltage V_{TO}, C_{GS} falls rapidly to zero within a voltage range

δ. For $V_{DS} < 0$, C_{GS} is really a gate-drain capacitance because the reverse bias interchanges the roles of source and drain. The capacitance in this range becomes small and independent of V_{GS}. Because of smooth transition from positive to negative drain-source voltages, the situation for $V_{DS} = 0$ is intermediate between the two cases described above [4].

The Statz large-signal model equivalent circuit of the GaAs MESFET is still represented by Fig. 9-3.

c. Small-signal model. The *GaAs MESFET small-signal model* is shown in Fig. 9-4, and it is defined by Eqs. (9-6) to (9-9).

9.1.4 SPICE3 models

The *SPICE3* [5] GaAs MESFET model is derived from the GaAs FET model of Statz described in Sec. 9.1.3. Thus the *static characteristics* are modeled by Eqs. (9-13) as regards I_D, and by Eqs. (3-25) and (3-26) as regards the currents I_{GD} and I_{GS}. The static behavior of the GaAs MESFET can then be simulated by the user of SPICE3 specifying the following parameters: in the .MODEL statement VTO (V_{T0}), the threshold voltage (default = -2 V); B (b), the doping tail extending parameter (default = 0.3 V^{-1}); BETA (β), the transconductance parameter (default = 10^{-4} A · V^{-2}); ALPHA (α), the saturation voltage parameter (default = 2 V^{-1}); LAMBDA (λ), the channel-length modulation parameter (default = 0 V^{-1}); IS (I_S), the gate junction saturation current (default = 10^{-14} A); RS (r_S), and RD (r_D), the source and drain resistances, respectively (default = 0 Ω).

The *charge-storage effects* are modeled by Eqs. (9-14) to (9-17) specifying in the .MODEL statement the following parameters: CGS ($C_{GS}(0)$), and CGD ($C_{GD}(0)$), the zero-bias gate-source and gate-drain junction capacitances, respectively (default = 5 pF and 1 pF, respectively); PB (ϕ_{s0}), the gate junction potential (default = 1 V). The parameters δ and $V_{\max}$ are not .MODEL parameters, but program constants set to 0.2 and 0.5, respectively. The keyname of the parameters used in the text is indicated in parentheses.

The SPICE3 GaAs MESFET *small-signal model* is automatically computed inside the program by the linearization of the static and large-signal model equations. The user of SPICE3 has to specify no new parameters in the .MODEL statement.

The GaAs MESFET *noise model* is defined by the same relations encountered in the analysis of the JFET noise model (see Chap. 7), and thus the .MODEL statement parameters involved in describing this phenomenon are KF (default = 0), the flicker noise coefficient and AF (default = 1), the flicker noise exponent.

The parameters common to the GaAs MESFET models and JFET models that depend on the AREA factor maintain the same dependence; in addition, the dependence on the AREA factor of B and ALPHA parameters must be taken into account.

Figure 9-3 shows the SPICE3 GaAs MESFET complete static and large-signal model, while Fig. 9-4 shows the small-signal model. For SPICE3 models, the components r_G and C_{DS} must not be considered.

9.1.5 HSPICE models

The *HSPICE* [6] GaAs MESFET models are included in the HSPICE JFET models and are selected by using the .MODEL statement parameter LEVEL = 3. Different *static submodels* for the GaAs MESFET LEVEL = 3 are available and can be selected by the .MODEL statement parameter SAT. In particular, setting SAT = 0, the user selects the Curtice model [2]; setting SAT = 1, the user selects the Curtice model with hyperbolic tangent coefficient and a user-defined exponent; setting SAT = 2, the user selects a cubic approximation of the Curtice model with gate field degradation (Statz model [4]); setting SAT = 3, the user selects the Meta-Software variable saturation model [6].

As far as the large-signal model is concerned, the .MODEL statement parameter CAPOP can be used to select the type of capacitor model; CAPOP = 0 selects the SPICE2 depletion capacitor model (a diode-like capacitance between the gate and the source and drain regions); CAPOP = 1 selects the Statz charge-conserving, symmetric capacitor model.

Scaling and area effects on the model parameters are common to both the GaAs MESFET and JFET (see Sec. 3.4.2).

a. Static model. The HSPICE GaAs MESFET *static model* is shown in Fig. 9-3 when the capacitances are not considered. The basic relationship for the model is represented by the nonlinear hyperbolic tangent current source I_D whose value is determined by different equations according to the selected submodel.

To analyze the different expressions for I_D, we firstly make the following assumptions:

$$V_{GST} = V_{GS} - [V_{T0} - \gamma_{DS} V_{DS} + K_1(V_{BS})] \equiv V_{GS} - V_{TH} \quad (9\text{-}18)$$

$$\beta^* = \frac{\beta}{1 + E_c V_{GST}} \quad (9\text{-}19)$$

where V_{T0} (VTO) is the threshold voltage (default = -2 V) (a negative value considers a depletion GaAs MESFET, while a positive value considers an enhancement GaAs MESFET regardless of the channel type); γ_{DS} (GAMDS) is the drain voltage, induced threshold voltage shift coefficient; K_1 (K1) is the threshold voltage sensitivity to bulk node (default $= 0$ V$^{1/2}$); β (BETA) is the transconductance parameter, gain (default $= 10^{-4}$ A · V^{-2}); E_c (UCRIT) is the critical field for mobility degradation in units V · cm^{-1}. The .MODEL statement keywords are indicated in parentheses. If the .MODEL statement parameter VTO is not specified, it is calculated as follows:

$$V_{TO} = V_p + \phi_{s0} \tag{9-20}$$

where V_p, if not specified in the .MODEL statement, is calculated as

$$V_p = \frac{qN_{\text{chan}}}{2\varepsilon_0 \varepsilon_s} \tag{9-21}$$

In Eqs. (9-20) and (9-21), V_p (VP) is the pinchoff voltage in units V; ϕ_{s0} (VBI) is the gate diode built-in voltage (default $= 1$ V); N_{chan} (NCHAN) is the effective dopant concentration in the channel (default 1.55×10^{16} atoms/cm^3); ε_s (D) is the semiconductor dielectric constant (Si $= 11.8$, GaAs $= 10.9$). The .MODEL statement keyword is indicated in parentheses.

Taking into account the above relations, we have

1. For $V_{GST} < 0$ (*channel pinched off*)

$$I_D = I_{DSUBTH}(N_G, N_D, V_{DS}, V_{GS}, \beta, V_{TH}) \tag{9-22}$$

where I_{DSUBTH} is a special function that calculates the subthreshold currents given the .MODEL statement parameters N_G (NG), the gate subthreshold factor (default $= 0$ V^{-1}), and N_D (ND), the drain subthreshold factor (default $= 0$ V^{-1}).

2. For $V_{GST} > 0$ (*on region*)

$$I_D = \beta^* V_{GST}^{VGEXP}(1 + \lambda V_{DS}) \tanh(\alpha V_{DS}) + I_{DSUBTH} \quad \text{for SAT} = 0 \tag{9-23}$$

$$I_D = \beta^* V_{GST}^{VGEXP}(1 + \lambda V_{DS}) \tanh\left(\alpha \frac{V_{DS}}{V_{GST}}\right) + I_{DSUBTH} \quad \text{for SAT} = 1 \tag{9-24}$$

$$I_D = \beta^* V_{GST}^2 (1 + \lambda V_{DS})\left[1 - \left(1 - \alpha \frac{V_{DS}}{3}\right)^3\right]$$

$$+ I_{DSUBTH} \quad \text{for} \begin{cases} \text{SAT} = 2 \\ 0 < V_{DS} < \dfrac{3}{\alpha} \end{cases} \tag{9-25}$$

$$I_D = \beta^* V_{GST}^2 (1 + \lambda V_{DS}) + I_{DSUBTH} \quad \text{for} \quad \begin{cases} \text{SAT} = 2 \\ V_{DS} > \dfrac{3}{\alpha} \end{cases} \quad (9\text{-}26)$$

$$I_D = \beta^* V_{GST}^{V_{GEXP}} (1 + \lambda V_{DS}) \left[1 - \left(1 - \alpha \dfrac{V_{DS}}{S_{\text{EXP}}}\right)^{S_{\text{EXP}}} \right]$$

$$+ I_{DSUBTH} \quad \text{for} \quad \begin{cases} \text{SAT} = 3 \\ 0 < V_{DS} < \dfrac{S_{\text{EXP}}}{\alpha} \end{cases} \quad (9\text{-}27)$$

$$I_D = \beta^* V_{GST}^{V_{GEXP}} (1 + \lambda V_{DS}) + I_{DSUBTH} \quad \text{for} \quad \begin{cases} \text{SAT} = 3 \\ V_{DS} > \dfrac{S_{\text{EXP}}}{\alpha} \end{cases} \quad (9\text{-}28)$$

In Eqs. (9-23) to (9-28), SAT is the saturation factor (default = 0); V_{GEXP} (VGEXP) is the gate voltage exponent (default = 2); λ (LAMBDA) is the channel-length modulation parameter (default = 0 V^{-1}); α (ALPHA) is the hyperbolic tangent drain voltage multiplier which determines the voltage at which the drain current saturates due to velocity saturation (default = 2 V^{-1}); S_{EXP} (SATEXP) is the drain voltage exponent (default = 3).

The currents I_{GD} and I_{GS} through the diodes follow Eqs. (3-25) and (3-26). They are determined by the parameters I_S (IS), the gate junction saturation current (default = 10^{-14} A) and n(N), the emission coefficient for gate-drain and gate-source diodes (default = 1), to be specified in the .MODEL statement.

In addition to the above parameters, the HSPICE static model parameter set includes the resistive elements r_G (RG), r_D (RD), and r_S (RS), whose default values are 0 Ω. The .MODEL statement keyword is indicated in parentheses.

b. Large-signal model. The HSPICE GaAs MESFET *large-signal model* is shown in Fig. 9-3, neglecting C_{DS} capacitance). The relationships for the capacitances are selected between the Curtice and Statz models using the .MODEL statement parameter CAPOP (default = 0). Thus if CAPOP = 0 (Curtice-like model, or diode-like capacitance model), the capacitances C_{GS} and C_{GD} are described by Eqs. (3-29) to (3-31) when ϕ_0 is replaced by ϕ_{s0}. The .MODEL statement parameters that the user of HSPICE has to specify to simulate the charge-storage effects are: CGD ($C_{GD}(0)$), the zero-bias gate-drain junction capacitance (default = 0 F); CGS ($C_{GS}(0)$), the zero-bias gate-source junction capacitance (default = 0 F); PB (ϕ_{s0}), the gate junction potential (default = 0.8 V); FC (FC), the coefficient for forward-bias depletion capacitance formula (default = 0.5); M (m), the grading coefficient for gate-drain and gate-source diodes (default = 0.5); TT (τ), the transit time (default = 0 s). The last three parameters apply only for CAPOP = 0. The parameter keyname used in the text is indicated in parentheses.

If CAPOP = 1 (Statz model), the capacitances C_{GS} and C_{GD} are described by Eqs. (9-14) to (9-17). In this case, the .MODEL statement parameters are VTO, CGS, CGD, PB, and ALPHA.

c. Small-signal model. The HSPICE GaAs MESFET *small-signal model* is automatically computed inside the program by the linearization of the static and large-signal model equations. The user of HSPICE has to specify no new parameters in the .MODEL statement. The HSPICE small-signal model is shown in Fig. 9-4.

d. Noise, temperature, and area effects on the model parameters. The *noise model* of the GaAs MESFET is described in Chap. 7 in conjunction with the JFET noise model. As far as *temperature* and *area* effects on the model parameters are concerned, it can be stated that the dependence of the GaAs MESFET parameters on temperature and the AREA factor follows the one of the analogous parameters of the JFET (see Chap. 3).

9.1.6 PSPICE models

PSPICE [7,8] has two built-in models for the GaAs MESFET: Curtice model (LEVEL = 1) and Statz model (LEVEL = 2), LEVEL being the .MODEL statement parameter that allows the user to switch (choose) between the two models.

a. Static model. The PSPICE GaAs MESFET *static model* is shown in Fig. 9-3 when the capacitances are not considered.

The current sources I_{GD} and I_{GS} are defined by Eqs. (3-25) and (3-26), respectively, and are determined by the parameters IS, the gate junction saturation current (default = 10^{-14} A), and N, the emission coefficient for gate-drain and gate-source diodes (default = 1), to be specified in the .MODEL statement. Equations (3-25) and (3-26) are valid for both LEVEL = 1 and LEVEL = 2.

The current source I_D assumes different expressions according to the selected model (LEVEL parameter). In particular, for LEVEL = 1 (Curtice model [2]), it can be written

$$I_D = \begin{cases} 0 & \text{for } V_{GS} - V_{T0} < 0 \\ \beta(V_{GS} - V_{T0})^2(1 + \lambda V_{DS}) \tanh(\alpha V_{DS}) & \text{for } V_{GS} - V_{T0} \geq 0 \end{cases} \quad (9\text{-}29)$$

Equation (9-29) is valid when the GaAs MESFET is in normal mode operation ($V_{DS} > 0$); in inverted mode ($V_{DS} < 0$), the source and drain must be switched in the above equation.

The .MODEL statement parameters that the user of PSPICE has to specify are: ALPHA (α), the saturation voltage parameter (default = $2\,V^{-1}$); BETA (β), the transconductance coefficient (default = $0.1\,A \cdot V^{-2}$); VTO (V_{TO}), the pinchoff voltage (default = $-2.5\,V$); LAMBDA (λ), the channel-length parameter (default = $0\,V^{-1}$). The parameter keyname used in the text is indicated in parentheses.

When LEVEL = 2 is specified, the Statz model [4] is selected, and I_D assumes the following expression [see Eqs. (9-13)]:

$$I_D = \begin{cases} 0 & \text{for } V_{GS} - V_{TO} < 0 \\[2mm] \dfrac{\beta(V_{GS} - V_{TO})^2}{1 + b(V_{GS} - V_{TO})}\left[1 - \left(\dfrac{\alpha V_{DS}}{3}\right)^3\right](1 + \lambda V_{DS}) & \\ \quad \text{for } \begin{cases} V_{GS} - V_{TO} \geq 0 \\ 0 < V_{DS} < \dfrac{3}{\alpha} \end{cases} \text{(linear)} & \\[2mm] \dfrac{\beta(V_{GS} - V_{TO})^2}{1 + b(V_{GS} - V_{TO})}(1 + \lambda V_{DS}) & \\ \quad \text{for } \begin{cases} V_{GS} - V_{TO} \geq 0 \\ V_{DS} \geq \dfrac{3}{\alpha} \end{cases} \text{(saturation)} \end{cases} \quad (9\text{-}30)$$

Equation (9-30) is valid when the GaAs MESFET is in normal mode operation ($V_{DS} > 0$); in inverted mode ($V_{DS} < 0$), the source and drain must be switched in the above equation.

The .MODEL statement new parameter that the user of PSPICE has to specify is B (b), the doping tail extending parameter (default = $0.3\,V^{-1}$). In addition, the PSPICE static model parameter set includes the resistive elements r_G (RG), r_S (RS), and r_D (RD), whose default values are $0\,\Omega$. The .MODEL statement keyword is indicated in parentheses.

b. Large-signal model. The PSPICE GaAs MESFET *large-signal model* is shown in Fig. 9-3. The relationships for the capacitances are selected between Curtice and Statz models using the .MODEL statement parameter LEVEL.

Thus if we are dealing with LEVEL = 1 (Curtice model), the capacitances C_{GS} and C_{GD} are described by Eqs. (3-29) to (3-31) when ϕ_0 is replaced by ϕ_{s0}. The .MODEL statement parameters that the user of PSPICE has to specify to simulate the charge storage effects, according to Curtice, are: CGD ($C_{GD}(0)$), the zero-bias gate-drain junction capacitance (default = 0 F); CGS ($C_{GS}(0)$), the zero-bias gate-source junction capacitance (default = 0 F); VBI (ϕ_{s0}), the gate junction potential (default = 1 V); FC (FC), the forward-bias depletion capacitance coefficient (default = 0.5); M (m), the gate junction grading coefficient (default = 0.5); and TAU (τ), the conduction current delay time (default = 0 s). The CDS (C_{DS}) parameter represents the drain-source capacitance and it is considered constant (default = 0 F). The parameter keyname used in the text is indicated in parentheses.

If we are dealing with LEVEL = 2 (Statz model), the capacitances C_{GS} and C_{GD} are described by Eqs. (9-14) to (9-17). In this case, the .MODEL statement parameters are VTO, CGS, CGD, VBI, and ALPHA.

c. Small-signal model. The PSPICE GaAs MESFET *small-signal model* is automatically computed inside the program by the linearization of the static and large-signal model equations. The user of PSPICE has to specify no new parameters in the .MODEL statement.

The PSPICE small-signal model is shown in Fig. 9-4.

d. Noise, temperature, and area effects on the model parameters. The *noise model* of the GaAs MESFET is described in Chap. 7 in conjunction with the JFET noise model. As regards the *temperature* and *area* effects on the model parameters, it can be stated that the dependence of the GaAs MESFET parameters on temperature and AREA factor follows the one of the analogous parameters of the JFET (see Chap. 3).

9.2 ISFET*

9.2.1 Principles of operation

The structure of the *I*on-*S*ensitive *F*ield-*E*ffect *T*ransistor (*ISFET*) [9] is shown in Fig. 9-5. The ISFET is essentially a MOST, made ion-sensitive by eliminating its metal gate electrode in order to expose the gate insulator to the electrolyte solution. By choosing appropriate inorganic (or organic) insulators, the device can be made sensitive to different ions or other chemical species, generating an interface potential on the gate: the corresponding drain-source current change in the semiconductor channel is observed. In the case

---

*Reprinted, with permission, from M. Grattarola, G. Massobrio, and S. Martinoia, Modeling H^+-Sensitive FET's with SPICE, *IEEE Trans. Electron Devices*, **ED-39**, 4, 1992.

MESFET, ISFET, and SCR-THYRISTOR 391

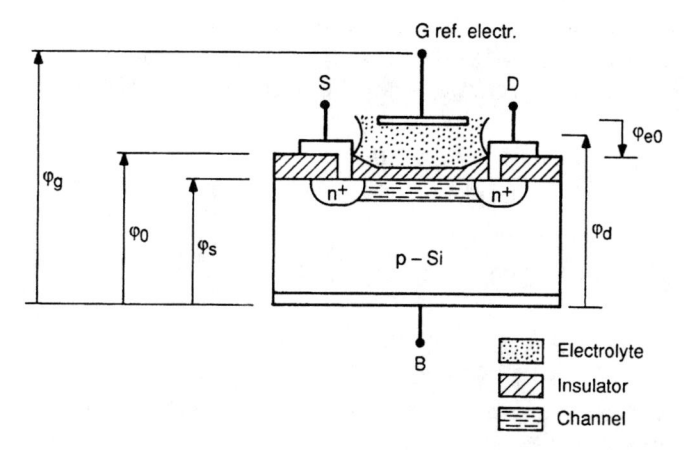

Figure 9-5 n-channel ISFET structure.

considered here, the gate insulator surface interacts specifically with the protons present in the solution. As a result, changes in H$^+$ concentration (i.e., changes in pH) affect the threshold voltage of the ISFET, by modifying the electrolyte-insulator potential.

The response of these devices to pH is commonly explained by using the so-called site-binding model [10], which states that H$^+$ specific binding sites at the surface of the insulator exposed to the electrolyte are responsible for the pH dependence of the charge distribution of the ISFET insulating layer. In the following, this theory (generalized to two kinds of binding sites), together with the *Gouy-Chapman-Stern* description of the potential profile in the electrolyte, is coupled to the MOS physics, in order to give a complete description of the ISFET behavior. As a result, a set of equations that allows one to define a relationship between pH and the electrolyte-insulator voltage φ_{e0}, for inorganic insulators, including those with more than one kind of site (i.e., Si$_3$N$_4$), is derived and introduced into a modified version of the electronic circuit simulation program SPICE2 [11], defining a SPICE2 built-in ISFET model [12–21].

9.2.2 EIS Structure

In this section, a set of equations is derived to fully characterize the basic *E*lectrolyte-*I*nsulator-*S*emiconductor (*EIS*) structure of the ISFET (Fig. 9-5).

In the next section, these equations will then be adapted to describe the ISFET model. The meaning of the symbols of the model is given in Fig. 9-5.

At the surface of the Si$_3$N$_4$ insulator, silanol sites, tertiary amine sites, secondary amine sites, and primary amine sites are present. We assume that only silanol sites and basic primary amine sites are present on the surface after oxidation [22]. Thus, the dissociation constants for the chemical reactions at the insulator interface are [23]

$$K_+ = \frac{[\text{SiOH}][\text{H}^+]_s}{[\text{SiOH}_2^+]} = \frac{\theta_0}{\theta_+}[\text{H}^+]_s \qquad (9\text{-}31)$$

$$K_- = \frac{[\text{SiO}^-][\text{H}^+]_s}{[\text{SiOH}]} = \frac{\theta_-}{\theta_0}[\text{H}^+]_s \qquad (9\text{-}32)$$

$$K_{N+} = \frac{[\text{SiNH}_2][\text{H}^+]_s}{[\text{SiNH}_3^+]} = \frac{\theta_{N0}}{\theta_{N+}}[\text{H}^+]_s \qquad (9\text{-}33)$$

where $[\text{H}^+]_s$ is the concentration of protons on the surface of the insulator. The normalization condition yields

$$\theta_+ + \theta_- + \theta_0 = \frac{N_{sil}}{N_s} \qquad (9\text{-}34)$$

$$\theta_{N+} + \theta_{N0} = \frac{N_{nit}}{N_s} \qquad (9\text{-}35)$$

$$\theta_0 + \theta_+ + \theta_- + \theta_{N+} + \theta_{N0} = 1 \qquad (9\text{-}36)$$

where N_{sil} and N_{nit} are the numbers of silanol sites and of primary amine sites per unit area, respectively, and N_s is the total number of available binding sites.

The charge density of the surface sites σ_0 on the insulator is given by

$$\theta_+ + \theta_{N+} - \theta_- = \frac{\sigma_0}{qN_s} \qquad (9\text{-}37)$$

By combining Eqs. (9-31) through (9-37) we get

$$\frac{\sigma_0}{qN_s} = \left(\frac{[\text{H}^+]_s^2 - K_+ K_-}{[\text{H}^+]_s^2 + K_+[\text{H}^+]_s + K_+ K_-}\right)\frac{N_{sil}}{N_s} + \left(\frac{[\text{H}^+]_s}{[\text{H}^+]_s + K_{N+}}\right)\frac{N_{nit}}{N_s} \qquad (9\text{-}38)$$

To obtain the relationship $\text{pH} = f(\varphi_{e0})$ we are looking for, the above equation must be completed with the model for the electrical double layer and with the condition of charge neutrality of the system in Fig. 9-5, when the relation between the concentration of protons $[\text{H}^+]_b$ in the solution and at the surface $[\text{H}^+]_s$ of the insulator is given by the Boltzmann equation

$$[\text{H}^+]_s = [\text{H}^+]_b \exp\left(-\frac{q}{kT}\varphi_{e0}\right) \qquad (9\text{-}39)$$

where $\varphi_{e0} = (\varphi_0 - \varphi_g)$ is the potential of the electrolyte-insulator interface.

Thus, by the Gouy-Chapman-Stern theory, the potential φ_{e0} is expressed as a function of the charge density σ_d in the diffuse layer and of the ion concentration c_0 in the solution [24]

$$\varphi_{e0} = -\frac{2kT}{q} \sinh^{-1}\left[\frac{\sigma_d}{\sqrt{8\varepsilon_w kTc_0}}\right] - \frac{\sigma_d}{C_h} \quad (9\text{-}40)$$

where ε_w is the permittivity of the solution, and C_h is the capacitance of the Helmholtz layer.

This well-known result can be expressed in a form more suitable for SPICE2 by considering the potentials of the system in Fig. 9-5.

By inspection, the following relationships can be written

(Helmholtz layer)

$$\varphi_d - \varphi_0 = \frac{\sigma_d}{C_h} \quad (9\text{-}41)$$

(Gouy-Chapman layer)

$$\sigma_d = \sqrt{8\varepsilon_w kTc_0} \sinh\left[\frac{q(\varphi_g - \varphi_d)}{2kT}\right] \quad (9\text{-}42)$$

By combining Eqs. (9-41) and (9-40), we obtain

$$\varphi_g = \varphi_d + \frac{2kT}{q} \sinh^{-1}\left[\frac{C_h(\varphi_d - \varphi_0)}{\sqrt{8\varepsilon_w kTc_0}}\right] \quad (9\text{-}43)$$

Equation (9-43) is identical to Eq. (9-40) and provides a relation between the charge density σ_d in the diffuse layer and the electrolyte-insulator potential φ_{e0} through the potential differences at the interfaces in the system in Fig. 9-5.

In addition, the condition of charge neutrality for the system in Fig. 9-5 gives

$$\sigma_d + \sigma_0 + \sigma_{mos} = 0 \quad (9\text{-}44)$$

where σ_{mos} is the charge density in the semiconductor.

By taking into account Eq. (9-41), Eq. (9-44) can be rewritten as

$$\sigma_0 = C_h(\varphi_0 - \varphi_d) - \sigma_{mos} \quad (9\text{-}45)$$

where σ_{mos} is given by Eq. (B-25), and here rewritten for convenience in the form

$$\sigma_{mos} = \pm\sqrt{2\varepsilon_s kTp_{p0}} \left[\left(\frac{q\varphi_s}{kT} - 1 + \exp\left(-\frac{q\varphi_s}{kT}\right)\right)\right.$$
$$\left. + \frac{n_{p0}}{p_{p0}}\left(-\frac{q\varphi_s}{kT} - 1 + \exp\left(\frac{q\varphi_s}{kT}\right)\right)\right]^{1/2} \quad (9\text{-}46)$$

where n_{p0} and p_{p0} are the equilibrium concentrations of electrons and holes, respectively. The charge density σ_{mos} within the semiconductor is taken positive for $\varphi_s < 0$, and negative for $\varphi_s > 0$. The charge density in the semiconductor σ_{mos} can also be written as

$$\sigma_{mos} = C_{ox}(\varphi_s - \varphi_0) \tag{9-47}$$

where C_{ox} is the insulator capacitance.

The equations obtained by substituting Eq. (9-46) into Eq. (9-47) and Eq. (9-45) into Eq. (9-38) form, together with Eqs. (9-39) and (9-43), a system that describes the EIS structure of the ISFET with two kinds of binding sites (i.e., Si_3N_4). The system of equations is solved for the four variables φ_d, φ_0, φ_s and $[H^+]_s$ while φ_g is an externally defined parameter. This set of equations, from now on referred to as the "EIS system," will be used to calculate the electrolyte-insulator potential φ_{e0}.

9.2.3 ISFET model

The ISFET behavior can be deduced by combining the aforesaid EIS structure with the physics of the MOST that will be summarized in this section.

a. Static model. The equations describing the behavior of the ISFET can be derived from the analogous equations of the MOST, by taking into account the potential differences among the new elements of the system in Fig. 9-5. The result is a modification of the MOST threshold voltage

$$V_{TH}(\text{MOST}) = \varphi_{MS} + 2\varphi_p - \frac{Q_{ss} + Q_{sc}}{C_{ox}} \tag{9-48}$$

where φ_{MS} is the metal-semiconductor work function difference, φ_p is the Fermi potential of the semiconductor (p-type for the n-channel MOST considered here), Q_{ss} is the fixed surface-state charge density per unit area at the insulator-semiconductor interface, and Q_{sc} is the semiconductor surface depletion-region charge per unit area.

By including all the potential terms of the ISFET configuration in the definition of the ISFET threshold voltage V_{TH}^*, it can be written

$$V_{TH}^*(\text{ISFET}) = (E_{ref} + \varphi_{lj}) - (\varphi_{e0} - \chi_{e0}) - \left[\frac{Q_{ss} + Q_{sc}}{C_{ox}} - 2\varphi_p + \frac{\varphi_{sc}}{q}\right] \tag{9-49}$$

Terms common to Eqs. (9-48) and (9-49), such as Q_{ss}, Q_{sc}, φ_p and C_{ox}, retain their conventional definition according to the standard MOST theory, while E_{ref} is the potential of the reference electrode, φ_{lj} is the liquid-junction potential difference between reference solution and the electrolyte, φ_{e0} is the potential of electrolyte-insulator interface, χ_{e0} is the electrolyte-insulator surface dipole potential, and φ_{sc} is the semiconductor work function. The potential φ_{e0} comes from the solution of the EIS system. For the n-channel ISFET considered, we have utilized an Ag|AgCl reference electrode, characterized by

$$E_{ref} = E_{abs}(\text{H}^+|\text{H}_2) + E_{rel}(\text{Ag}|\text{AgCl}) \qquad (9\text{-}50)$$

where $E_{abs}(\text{H}^+|\text{H}_2)$ is the potential of the standard hydrogen electrode and $E_{rel}(\text{Ag}|\text{AgCl})$ is the potential of the Ag|AgCl reference electrode (relative to the hydrogen electrode).

By comparing Eqs. (9-48) and (9-49), it follows that

$$V_{TH}^*(\text{ISFET}) = V_{TH}(\text{MOST}) + E_{ref} + \varphi_{lj} + \chi_{e0} - \varphi_{e0} - \frac{\varphi_M}{q} \qquad (9\text{-}51)$$

where φ_M is the work function of the metal gate (reference electrode) relative to vacuum. We obtain the ISFET static model by replacing the MOST threshold voltage V_{TH} with V_{TH}^* in the I-V equations of the MOST (see Chap. 4).

b. Large-signal model. The ISFET large-signal model is based on the MOST large-signal model, characterized by two pn-junction depletion region capacitances and by the gate-source, gate-drain, gate-bulk capacitances which in turn are functions of the insulator capacitance C_{ox}. For the ISFET, the insulator capacitance is sensitive to the ionic activity of the electrolyte, and consists of two elements connected in series: the MOST insulator capacitance C_{ox}, and the Helmholtz layer capacitance C_h, which has to be considered to model the large-signal behavior of the ISFET. Thus for the ISFET it is necessary to consider a new capacitance C_{ox}^* given by

$$C_{ox}^* = \frac{C_{ox} C_h}{(C_{ox} + C_h)} \qquad (9\text{-}52)$$

c. Thermal model. The close similarity between the MOST and the ISFET can be exploited in the analysis of the temperature dependence of the ISFET. In particular, the most significant ISFET terms in regard to their temperature dependence, neglecting the semiconductor terms that are described in Chap. 4, are the dissociation constant K's and the reference electrode potential E_{ref}.

Following Barabash et al. [25], it can be written

$$K(T) = K(300)^{300/T} \qquad (9\text{-}53)$$

Assuming an Ag|AgCl electrode with filling solution of 3.5M KCl, saturated with AgCl, the temperature dependence of the potential E_{ref} can be written as [25]

$$E_{ref}(T) = E_{abs}(\text{H}^+|\text{H}_2) + E_{rel}(\text{Ag}|\text{AgCl}) + \left(\frac{dE_{ref}}{dT}\right)(T - 298.16)$$

$$= 4.7 + 0.205 + 1.4 \times 10^{-4}(T - 298.16) \quad (9\text{-}54)$$

The liquid-junction potential φ_{lj} is typically quite small (a few millivolts), thus its dependence on temperature is insignificant in comparison to that of any other temperature-dependent term and then it is not considered.

d. Slow response effect model: hysteresis. The theory developed above was only concerned with the fast part of the response after a pH step. There is also an additional slow response. In fact, apart from a continuous drift of the threshold voltage, which does not depend on pH, it was observed a hysteresis in the pH response in the direction which corresponds to a memory effect. It occurs with a delay of the order of minutes to hours after the pH variation and it is due mainly to the presence of OH$^-$ sites buried beneath the surface [26]. (Note that the fast part of the response after a pH step usually is of the order of 1 ms or faster.) The hysteresis phenomenon can be modeled by assuming that a small fraction of binding sites is buried a few nm inside the surface, so that OH$^-$ and H$^+$ groups can react with them at the end of a diffusion process.

The diffusion of groups inside the insulator is described by the following relationship:

$$[\text{H}^+]_{bur} = [\text{H}^+]_s \left[1 + erf\left(\frac{x}{2\sqrt{D_{eff}t}}\right)\right] \quad (9\text{-}55)$$

where $[\text{H}^+]_{bur}$ is the proton concentration in the buried layer, and D_{eff} is the effective diffusion constant, given by [26]

$$D_{eff} = \frac{D}{K_{bur}(N_{sil} + N_{nit})\eta} \quad (9\text{-}56)$$

where K_{bur} is the reaction equilibrium constant, and D is the diffusion constant of H$^+$ and OH$^-$ ions.

The addition of the buried layer modifies Eq. (9-44), and introduces a new equation into the model:

$$\sigma_d + \sigma_0 + \sigma_{mos} + \sigma_b(t) = 0 \quad (9\text{-}57)$$

where σ_b is the charge density of the buried layer

$$\frac{\sigma_b}{qN_s} = \left(\frac{[H^+]_{bur}^2 - K_+ K_-}{[H^+]_{bur}^2 + K_+[H^+]_{bur} + K_+ K_-}\right)\eta\frac{N_{sil}}{N_s}$$
$$+ \left(\frac{[H^+]_{bur}}{[H^+]_{bur} + K_{N+}}\right)\eta\frac{N_{nit}}{N_s} \quad (9\text{-}58)$$

where $N_s = (1 + \eta)(N_{sil} + N_{nit})$.

e. Slow response effect model: REFET. The REFET is an ISFET whose gate surface has, by suitable chemical surface modification, been made as insensitive as possible to all kinds of ionic species.

The differential signal of a pair of electronically identical FETs, one ISFET (ion-sensitive), and one REFET (ion-insensitive) is then automatically free of all kinds of slight perturbations of the potential of the electrolyte. For this reason, the REFET can be used as reference electrode. Examples of materials used for the REFET gate and thus partially pH-insensitive are Teflon or Parylene. These materials can be modeled by considering the chemical equilibrium equation between the surface sites —COOH and —COH of the organic membrane and protons [27]. This equation can be obtained for the model in a straightforward way by modifying Eq. (9-38) into

$$\frac{\sigma_0}{qN_s} = \left(\frac{K(COOH)}{[H^+]_s + K(COOH)}\right)\frac{N(COOH)}{N_s} + \left(\frac{K(COH)}{[H^+]_s + K(COH)}\right)\frac{N(COH)}{N_s} \quad (9\text{-}59)$$

where K(COOH) and K(COH) are the dissociation constants for the chemical reaction, and N(COOH) and N(COH) are the numbers of —COOH sites and —COH sites per unit area, respectively.

9.2.4 SPICE2 ISFET model

a. Static model. The equations describing the MOST static model (see Chap. 4) in the MOSFET subroutine of SPICE2 have been kept for the ISFET static model, by replacing the expression for V_{TH} of the MOST with Eq. (9-51) and using the φ_{e0} value resulting from the solution of the EIS system. For this purpose, three new subroutines have been added into the source code of SPICE2. Thus, in addition to the MOST static model parameters of SPICE2, we have introduced eight more parameters that the user of SPICE2 can specify in the .MODEL statement to model the static behavior of the ISFET. These parameters are: ZSCON(c_0), the ion concentration in the electrolyte (default = 0.1 mol/l); ZNKAP(K_{N+}), the primary amine dissociation constant (default = 0 mol/l); ZAKAP(K_+), dissociation constant for positive silanol groups (default = 2×10^{-6} mol/l); ZBKAP(K_-), the dissociation constant for negative silanol groups (default = 3×10^{-12} mol/l); ZCHL(C_h), the Helmholtz layer capacitance (default = $20\ \mu\text{F/cm}^2$); ZNNIT(N_{nit}), the number of primary amine sites (default = 0 sites/cm^2); ZPH(PH), pH operating point (default = 7); ZNOX(N_{sil}), the number of silanol sites (default = 8×10^4 sites/cm^2). The other parameters are set internally to SPICE.

Finally, in order to simulate ISFETs with more than one insulator layer (in fact SPICE2 has built-in parameters only for SiO$_2$), two new parameters for the .MODEL statement have been introduced: ZTOXN(t_{ox1}), the thickness of the added insulator (default = 0 m); ZEPSOX(ε_{ox1}), the dielectric constant for the added insulator (default = 0 F/m). The keyname of the parameters used in the text is given in parentheses.

b. Large-signal model. The equations describing the MOST large-signal model in the CMEYER subroutine of SPICE2 are then modified by introducing Eq. (9-52). Thus to simulate the large-signal behavior of the ISFET, the user of SPICE2 has to introduce the C_h(ZCHL) parameter in the .MODEL statement.

c. Thermal model. As far as the temperature dependence is concerned, the parameters common to the MOST and the ISFET retain their definition in the TMPUPD subroutine of SPICE2. Therefore only Eqs. (9-53) and (9-54) has been implemented in SPICE2 to characterize the ISFET thermal model.

d. Hysteresis model. The implementation of the hysteresis model has required that a new subroutine be added to the source code of SPICE2. To simulate this phenomenon, the user of SPICE2 has to specify in the .MODEL statement the following parameters: ZEPSIL(η), the fraction of buried binding sites (default = 0 sites/cm^2); ZDEFF($D_{\textit{eff}}$), the effective diffusion coefficient (default = 1 cm^2/s). The keyname of the parameter used in the text is indicated in parentheses.

e. REFET model. The REFET simulation is automatically performed if ad hoc equilibrium constants K values and total number of available sites $N_s = N_{sil} + N_{nit}$ are chosen.

f. Independent source. A time-varying pH source has also been implemented in the SORUPD subroutine of SPICE2 in order to simulate the pH behavior in different experimental situations. The pH source has been introduced as a dummy voltage source. In this way the voltage source can be introduced without disturbing the electrical behavior of the circuit and the user can easily plot the pH waveform, while the running pH value (time dependent) is used to set the value of [H^+] in the Boltzmann equation.

g. ISFET model simulation results. The input characteristics of "ideal" SiO_2-, Al_2O_3-, Si_3N_4-gate ISFETs have been simulated by introducing appropriate values for the equilibrium constants K's (Table 9-1) into the EIS system, via the .MODEL statement of SPICE2, and by utilizing the other physicochemical parameters.

Figure 9-6 shows, as an example, the resulting I_D vs. V_{GS} curves (with source and bulk grounded) for different values of pH and for a given V_{DS} in the case of Si_3N_4. Other results can be found in Refs. [16–19].

TABLE 9-1 Insulator Electrochemical Parameters (25°C)

Insulator	Surface ionization constants*			Surface site density†		pH_{pzc}‡
	K_+	K_-	K_{N+}	$N_{sil/al}$	N_{nit}	
SiO_2	15.8	63.1×10^{-9}	—	5.0×10^{14}	—	3.0
Si_3N_4	15.8	63.1×10^{-9}	1×10^{-10}	3.0×10^{14}	2.0×10^{14}	6.8
Al_2O_3	12.6×10^{-9}	79.0×10^{-10}	—	8.0×10^{14}	—	8.5

*Moles per liter.
†Number per cm².
‡pH_{pzc} = pH value at the point of zero charge (σ_0).

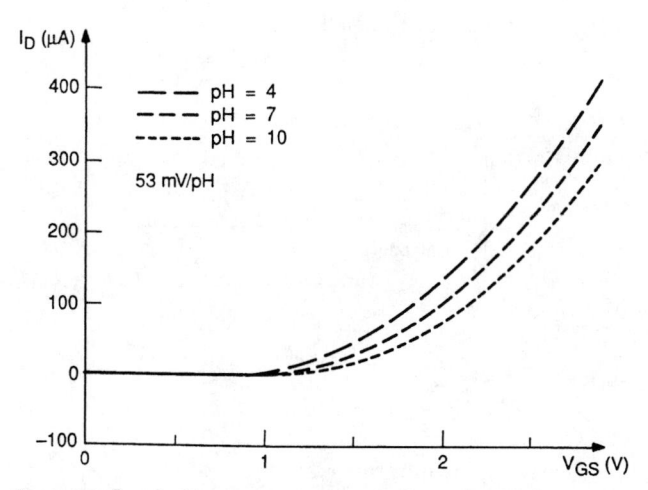

Figure 9-6 Input characteristics of a Si_3N_4-gate ISFET.

9.3 THYRISTOR

The *THYRISTOR* is a semiconductor device composed of four regions of alternate conductivity-type semiconductor material (*pnpn*) that exhibits intrinsic regenerative action in its operation. The device operates as a switch and can handle large amounts of power due to its capability to hold off high voltages at low reverse currents when in its blocking state, as well as to conduct high currents with low voltage drops in its forward conduction state.

9.3.1 Principles of operation

The basic structure of a *pnpn THYRISTOR* is shown in Fig. 9-7 and its current-voltage characteristic is shown in Fig. 9-8.

The structure of Fig. 9-7 can be considered as composed by three *pn* junctions J_1, J_2, and J_3 in series. The contact electrode to the outer *p* region (*P*1) is called the *anode* (*A*) and that to the outer *n* region (*N*2) is called the *cathode* (*K*). A contact electrode is also made to the other *p* region (*P*2) of Fig. 9-7 to form the *gate* (*G*).

Under reverse-bias conditions (anode negative with respect to cathode), the junctions J_1 and J_3 are reverse biased while J_2 is forward biased. In this condition the I_A-V_{AK} characteristic of the device is very similar to that of two *pn* junctions in series. The device is said to be in its *reverse blocking state* [region (0)–(4) in Fig. 9-8]. *Breakdown* may also occur at a given reverse voltage [region (4)–(5) in Fig. 9-8].

Under forward-bias conditions (anode positive with respect to cathode), the junctions J_1 and J_3 are forward biased, while J_2 is reverse biased, so the initial current flowing through the device is very small. This condition corresponds to the *off* or *forward blocking state* of the device [region (0)–(1) in Fig. 9-8].

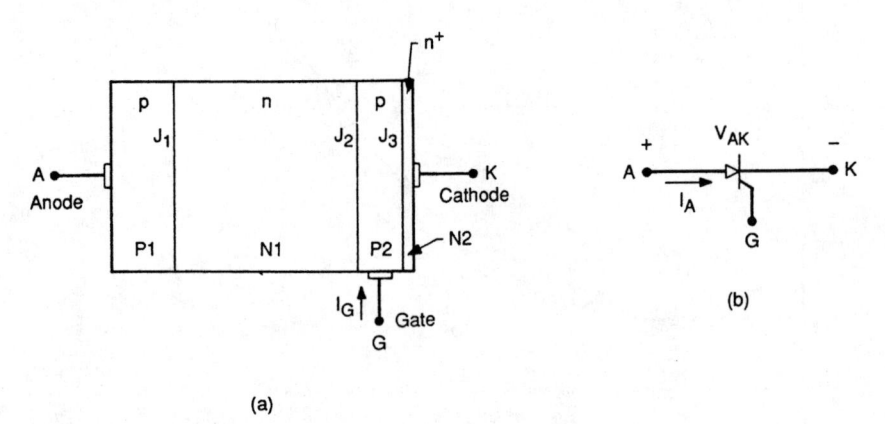

Figure 9-7 (*a*) THYRISTOR structure, (*b*) circuit symbol.

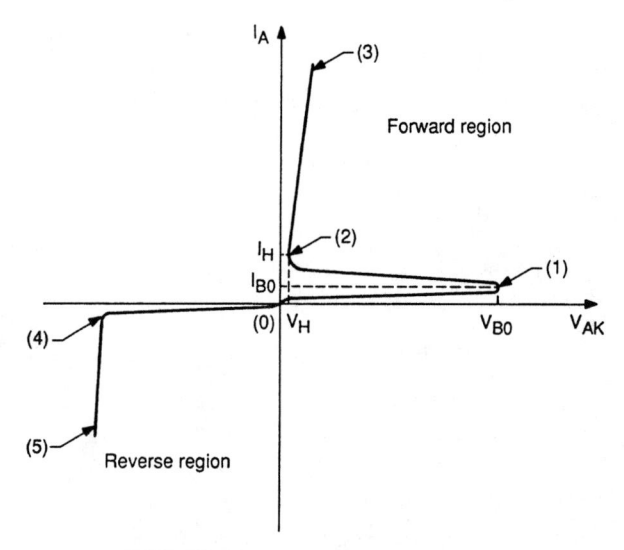

Figure 9-8 THYRISTOR static characteristic.

Thus at low currents the components $N2P2N1$ and $P1N1P2$ of the device behave like two BJTs with $N2$ and $P1$ as their emitters, respectively.

A gate current I_G injected at the $P2$ region increases the forward current between $N1$ and $N2$, by transistor action. This, increases the base current of the $P1N1P2$ BJT, which in turn begins to conduct providing additional base drive for the $N2P2N1$ BJT, causing a positive feedback process. If the gain of the feedback loop is sufficient, the current will increase until the THYRISTOR shows a very low voltage drop in its forward bias, even in the absence of any further gate drive at $P2$. The THYRISTOR is now said to be in its *on* or *forward conducting state* [region (2)–(3) in Fig. 9-8]. The on state holds as long as sufficient current I_A flows in the anode-cathode circuit. If this current is reduced to a value where the gains of each BJT drop so that the regenerative process cannot be supported, the THYRISTOR will be driven to its *nonconducting state* [28].

9.3.2 THYRISTOR model

The effects of the three junctions on one another can be explained considering the equivalent structure of Fig. 9-9a and its circuit representation of Fig. 9-9b. The above model is the well-known *two-transistor analogy*, and an analysis of the THYRISTOR can be made in terms of the parameters of the two BJTs.

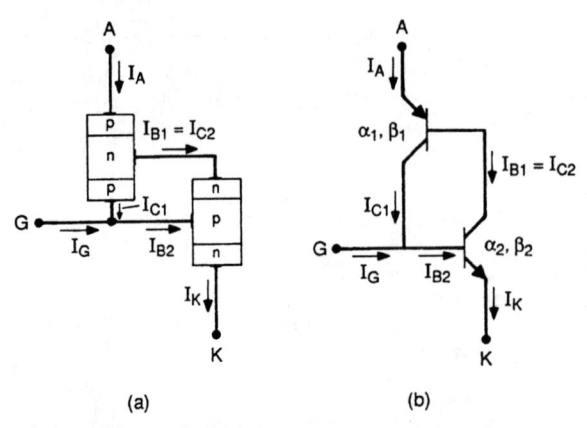

Figure 9-9 THYRISTOR model. (a) Structure partitioned, and (b) two-BJT analogy.

From Fig. 9-9b, the collector of the *npn* BJT provides the base drive for the *pnp* BJT. In addition, the collector of the *pnp* BJT with the gate current I_G supplies the base drive for the *npn* BJT. Thus a regenerative action occurs when the total loop gain exceeds unity. Let us now examine the mathematical condition for turning on the THYRISTOR [29].

From BJT theory, it is known that the base current I_{B1} of the *pnp* BJT, which is supplied by the collector of the *npn* BJT, is

$$I_{B1} = (1 - \alpha_1)I_A - I_{CS1} \tag{9-60}$$

where α is the dc common-base current gain of the BJT and I_{CS} is the collector-base reverse saturation current of the BJT.

The collector current I_{C2} of the *npn* BJT is given by

$$I_{C2} = \alpha_2 I_K + I_{CS2} \tag{9-61}$$

By equating I_{B1} and I_{C2} as is shown in Fig. 9-9, it can be written

$$(1 - \alpha_1)I_A - I_{CS1} = \alpha_2 I_K + I_{CS2} \tag{9-62}$$

As $I_K = I_A + I_G$, from Eq. (9-61), and rearranging, it follows,

$$I_A = \frac{\alpha_2 I_G + (I_{CS1} + I_{CS2})}{1 - (\alpha_1 + \alpha_2)} \tag{9-63}$$

where α_1 varies with I_A and α_2 varies with $(I_A + I_G)$ [29], as stated by the transistor theory, which states that α increases rapidly with emitter current.

If the base drive current $I_G + I_{C1}$ is equated to the base current of the *npn* BJT, exactly the same solution results.

Equation (9-63) characterizes the behavior of the THYRISTOR. From this equation, supposing to neglect I_G, it follows that I_A is small unless $(\alpha_1 + \alpha_2)$ approaches unity: in this situation the denominator of the equation approaches zero and switching occurs.

Using the more familiar common-emitter current gain β, the base-current-to-collector-current relationship for the *pnp* BJT is

$$I_{C1} = \beta_1 I_{B1} + (1 + \beta_1) I_{CS1} \tag{9-64}$$

and for the *npn* BJT is

$$I_{C2} = \beta_2 I_{B2} + (1 + \beta_2) I_{CS2} \tag{9-65}$$

As $I_{B2} = I_G + I_{C1}$, then Eq. (9-65) becomes

$$I_{C2} = \beta_2 (I_G + I_{C1}) + (1 + \beta_2) I_{CS2} \tag{9-66}$$

Substituting Eq. (9-64) into Eq. (9-66), and taking into account that $I_{C2} = I_{B1}$, gives

$$I_{B1} = \beta_2 I_G + \beta_2 \beta_1 I_{B1} + \beta_2 (1 + \beta_1) I_{CS1} + (1 + \beta_2) I_{CS2} \tag{9-67}$$

Solving Eq. (9-67) for I_{B1}, gives

$$I_{B1} = \frac{\beta_2 I_G + \beta_2 (1 + \beta_1) I_{CS1} + (1 + \beta_2) I_{CS2}}{1 - \beta_1 \beta_2} \tag{9-68}$$

By using the relationship between base current and emitter current, it can be written

$$I_A = (1 + \beta_1)(I_{B1} + I_{CS1}) \tag{9-69}$$

By substituting Eq. (9-68) into Eq. (9-69), and rearranging yields

$$I_A = \frac{(1 + \beta_1)(1 + \beta_2)(I_{CS1} + I_{CS2}) + \beta_2 (1 + \beta_1) I_G}{1 - \beta_1 \beta_2} \tag{9-70}$$

From Eq. (9-70) it can be seen that the anode current tends to increase (switch) when $\beta_1 \beta_2$ approaches unity.

If the polarities of the anode and cathode are reversed, there is no switching action, since only the center junction behaves as an emitter, and no regenerative process can occur.

In conclusion, to switch the THYRISTOR on, α_1 and α_2 must be increased. We now briefly consider methods of increasing α_1 and α_2: *voltage, gate, temperature,* and *radiation triggering*.

a. **Voltage triggering.** A method of increasing α is to increase the voltage between the collector and the emitter. If this voltage is increased to a large enough value, J_2 will break down and allow a reverse current to flow (*avalanche breakdown*). If J_2 exhibits avalanche breakdown, then the current through J_2 increases, which in turn increases α_1 and α_2 causing the device to go into its on state. In a THYRISTOR, the anode-cathode voltage which causes this to occur is called the *breakover voltage*, V_{B0} [point (1) in Fig. 9-8]. After the THYRISTOR has gone into its on state, it will remain on as long as the current through J_2 is sufficient to cause $(\alpha_1 + \alpha_2)$ to be equal to unity. The minimum current through J_2, and then through the external circuit, that will cause the device to remain in the on state is called the *holding current* I_H [point (2) in Fig. 9-8].

This voltage triggering is considered to be obtained by slowly increasing the forward voltage until the breakover voltage is reached. The same result can be obtained by applying the anode voltage rapidly (*dV/dt triggering*) which gives rise to a displacement current $I_d = d(CV)/dt$, C being the capacitance of the collector junction J_2. This current in turn can cause the alpha sum to become unity; then a "premature" switching occurs.

b. **Gate triggering.** If a gate current is injected at the $P2$ region, a current is caused to flow across J_3: thus the current through the THYRISTOR is increased. This increase in current is due to an increase in α_1 and α_2. If the current going into the p region between J_3 and J_2 is made greater than zero, the breakover voltage V_{B0} is lowered. As the gate current gets larger, V_{B0} becomes smaller, as seen in Fig. 9-10 and becomes equal to the forward-voltage drop of a *pn* junction at some value of gate current.

From the above discussion, it follows that if a THYRISTOR is forward biased without the avalanche voltage being reached, it will remain in the off state. Upon application of sufficient gate current, V_{B0} will decrease, and the THYRISTOR will turn on. Once the THYRISTOR is on, the gate no longer has any control over the current through the device. To return the THYRISTOR to its off state, the anode current must be reduced below the holding current I_H. This may be accomplished either by removing the anode voltage or by decreasing the anode current to a value less than the holding current. As I_G flows into the gate (base) of the *npn* transistor, it is amplified and appears at the collector as $\beta_2 I_G$. This current flows into the base of the *pnp* transistor and appears at the collector as $\beta_1 \beta_2 I_G$. If, however, α_1 is large, the current $\beta_1 \beta_2 I_G$ also appears at the base of the *npn* transistor, creating a regenerative process. Thus a small gate current will fire the THYRISTOR.

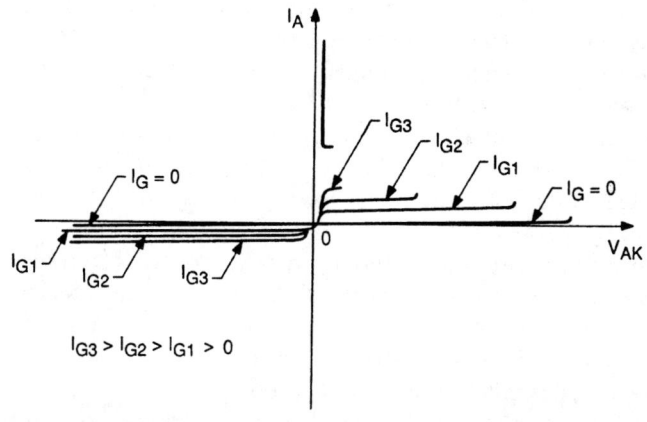

Figure 9-10 Effect of the gate current I_G on the THYRISTOR characteristics.

c. Temperature triggering. Temperature triggering is based on the fact that as temperature increases, more hole-electron pairs are generated and collected by the blocking junction J_2. The sum of α's then rapidly approaches unity and the switching of the THYRISTOR occurs. As for the case of I_G triggering, also in this case, the breakover voltage decreases with increasing temperature.

d. Radiation triggering. Electron-hole pairs may also be produced when a semiconductor is exposed to light or radiation. Thus the basic switching mechanism and I_A-V_{AK} characteristics due to these triggering methods are essentially the same as temperature triggering.

9.3.3 SPICE2 THYRISTOR model

The model [30,31] proposed here as a built-in model for SPICE2 [11] makes use of the basic two-transistor analogy of the THYRISTOR (Fig. 9-9b) with the addition of a gate-cathode resistance r_G to control the α of the *npn* BJT [28].

The model therefore relies on the well-established BJT model derived from the Ebers-Moll and Gummel-Poon models implemented in SPICE2. The number of variables in the model is kept to a minimum by varying the parameter values simultaneously on the two BJTs, and retaining only those transistor model parameters having an appreciable effect on the THYRISTOR electrical parameter.

As we are dealing with a "macromodel" built out of components (BJTs) that are already implemented in the program, the proposed model results suitable for direct applications as built-in model also for SPICE3, HSPICE, and PSPICE.

The static S-shaped I_A-V_{AK} characteristic of the THYRISTOR can be modeled by varying the static model parameters of the two BJTs and the value of r_G. The most important parameters are the forward common emitter current gain β_F (BF), and the saturation current I_S (IS), which affect the shape of the forward blocking region, the values of the breakover voltage V_{B0}, and those of the holding current I_H. The BJT .MODEL statement parameter is indicated in parentheses.

The above information can be matched into families of curves which should help the designer to "synthetize" the static characteristics of a given THYRISTOR, i.e., the parameters V_{B0} and I_H. A third model parameter involved in the static simulation of the THYRISTOR is the BJT base resistance r_B (RB), which only affects the slope of the on-state region of the I_A-V_{AK} curve. As an example of the above-mentioned families of curves, in Fig. 9-11 the breakover voltage V_{B0} vs. β_F for different I_S and r_G values is shown. Other curves can be found in Refs. [30,31].

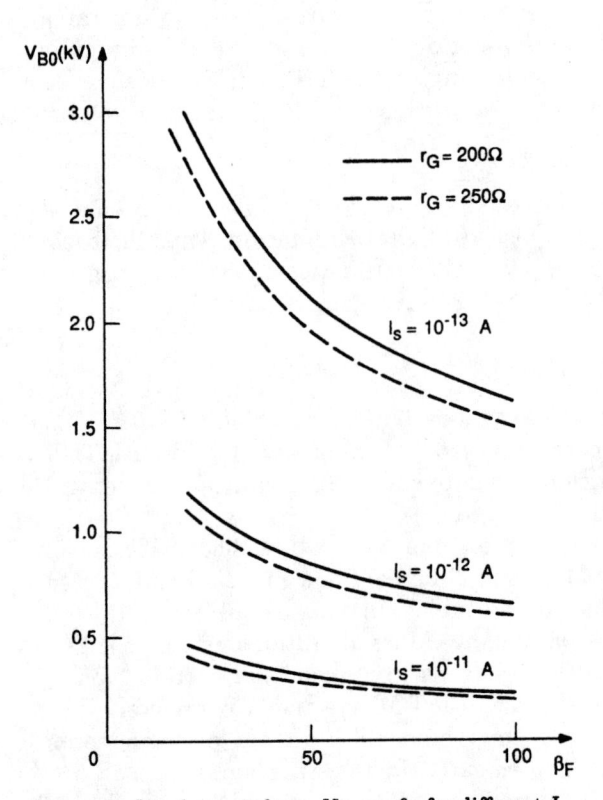

Figure 9-11 Breakover voltage V_{B0} vs. β_F for different I_S and r_G values.

Triggering mechanisms can also be simulated by the model: in particular, the gate current I_G and the temperature T. For large values of I_G, as well as of T, the I_A-V_{AK} curve loses its forward blocking capability, until it becomes a diode-like curve as stated from theory.

Finally, as expected from theory, the value of r_G has a strong influence both on the breakover voltage V_{B0} and on the holding current I_H: when the r_G value is lowered, the breakover voltage V_{B0} increases. A possible limitation is shown by the model in the value of the holding current I_H for a given r_G. However, such a limitation can be overcome by increasing the area of the THYRISTOR (i.e., the AREA factor of the two BJTs constituting the model).

The two-transistor model analogy of the THYRISTOR can also be used to simulate the different dynamic operating conditions such as dV/dt *triggering*, *turn-on*, *turn-off*, and *recovery processes*. The results, expressed as graphs [30,31] that directly relate the THYRISTOR electrical parameters to the BJT model parameters, show which parameters of the two BJTs defining the THYRISTOR model the user has to specify in the .MODEL statement to simulate each process.

Considering only those parameters that have an appreciable effect on each process, and then choosing the parameters in order to separately simulate each process, it follows from Refs. [30,31] that dV/dt triggering is affected by the forward transit time τ_F (TF) and the gate-cathode resistance r_G; the turn-off process is affected by the reverse transit-time τ_R (TR); the recovery process is affected by the

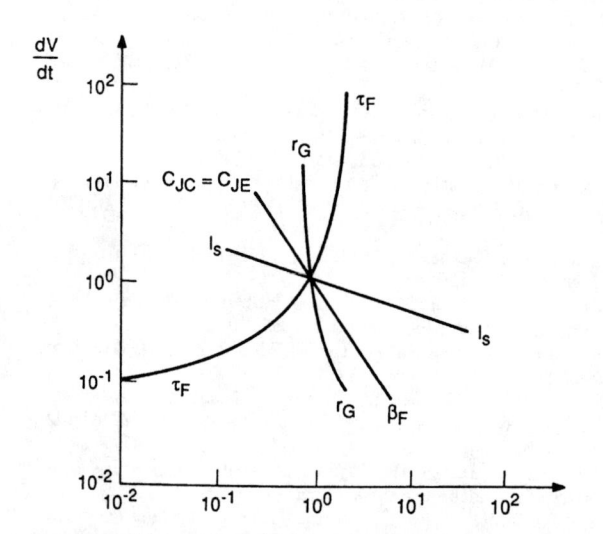

Figure 9-12 Normalized behavior of the dV/dt with respect to the BJT most important parameters.

reverse common emitter current gain β_R (BR) and τ_R (TR); the turn-on process is affected by τ_F (TF). As an example, Fig. 9-12 shows the normalized behavior of the dV/dt with respect to the most important parameters of the BJT.

References

1. D. L. Pulfrey and N. Garry Tarr, *Introduction to Microelectronics Devices*, Prentice-Hall, Englewood Cliffs, N.J., 1989.
2. W. R. Curtice, A MESFET Model for Use in the Design of GaAs Integrated Circuits, *IEEE Trans. Microwave Theory and Techniques*, **MTT-28**, 1980.
3. S. E. Sussman-Fort, S. Narasimhan, and K. Mayaram, A Complete GaAs MESFET Computer Model for SPICE, *IEEE Trans. Microwave Theory and Techniques*, **MTT-32**, 1984.
4. H. Statz, P. Newman, I. W. Smith, R. A. Pucel, and H. A. Hans, GaAs FET Device and Circuit Simulation in SPICE, *IEEE Trans. Electron Devices*, **ED-34**, 1987.
5. T. I. Quarles, SPICE3 Version 3C1 User's Guide, Electronics Research Laboratory, Rep. No. ERL-M46, University of California, Berkeley, 1989.
6. Meta-Software Inc., *HSPICE User's Manual*, Campbell, Calif., 1991.
7. MicroSim Corporation, *PSPICE User's Manual*, Irvine, Calif., 1989.
8. P. W. Tuinenga, *SPICE: A Guide to Circuit Simulation and Analysis Using PSPICE*, Prentice-Hall, Englewood Cliffs, N.J., 1988.
9. P. Bergveld, Development of an Ion-Sensitive Solid State Device for Neurophysiological Measurements, *IEEE Trans. Biomed. Eng.*, **BME-17**, 1970.
10. D. E. Yates, S. Levine, and T. W. Healy, Site-Binding Model of the Electrical Double Layer at the Oxide/Water Interface, *J. Chem. Soc. Faraday*, **70**, 1974.
11. L. W. Nagel, SPICE2: A Computer Program to Simulate Semiconductor Circuits, Electronics Research Laboratory, Rep. No. ERL-M520, University of California, Berkeley, 1975.
12. D. Caviglia, G. Massobrio, and G. Mattioli, Sviluppo di un Modello dell'ISFET per SPICE: una Nuova Via nello Studio dei Biosensori, *Pixel*, 7/8, 1987.
13. G. Massobrio, M. Grattarola, G. Mattioli, and F. Mattioli, Jr., ISFET-Based Biosensors Modeling with SPICE, *Sensors and Actuators*, **B1**, 1990.
14. A. Cambiaso, M. Grattarola, G. Arnaldi, S. Martinoia, and G. Massobrio, Detection of Cell Activity via ISFET Devices: Modelling and Computer Simulations. *Sensors and Actuators*, **B1**, 1990.
15. M. Grattarola, S. Martinoia, G. Massobrio, A. Cambiaso, R. Rosichini, and M. Tetti, Computer Simulations of the Responses of Passive and Active Integrated Microbiosensors to Cell Activity, *Sensors and Actuators*, **B4**, 1991.
16. G. Massobrio, S. Martinoia, and M. Grattarola, Ion Sensitive Field-Effect Transistor (ISFET) Model Implemented in SPICE, *Proc. of 4th SSDP Conf.*, Vol. 4, Zurich, Switzerland, September 1991.
17. M. Grattarola, G. Massobrio, and S. Martinoia, Modeling H^+-Sensitive FET's with SPICE, *IEEE Trans. Electron Devices*, **ED-39**, 4, 1992.
18. S. Martinoia, M. Grattarola, and G. Massobrio, Modelling Non-Ideal Behaviours of H^+-Sensitive FET's with SPICE, *Sensors and Actuators*, **B7**, 1992.
19. S. Martinoia, G. Massobrio, and M. Grattarola, An ISFET Model for CAD Applications, *Sensors and Actuators*, **B8**, 1992.
20. M. Grattarola, G. Arnaldi, A. Cambiaso, S. Martinoia, and G. Massobrio, Interfacing Excitable Cells with Integrated Devices, *Proc. IEEE Engineering in Medicine and Biology Society 11th Conf.*, Seattle, 1989.
21. G. Massobrio, S. Martinoia, and M. Grattarola, Light-Addressable Chemical Sensors: Modelling and Computer Simulations, *Sensors and Actuators*, **B7**, 1992.
22. D. L. Harame, L. J. Bousse, J. D. Shott, and J. D. Meindl, Ion-Sensing Devices with Silicon Nitride and Borosilicate Glass Insulators, *IEEE Trans. Electron Devices*, **ED-34**, 1987.

23. C. D. Fung, P. W. Cheung, and W. H. Ko, A Generalized Theory of an Electrolyte-Insulator-Semiconductor Field-Effect Transistor, *IEEE Trans. Electron Devices*, **ED-33**, 1986.
24. L. Bousse, N. F. de Rooij, and P. Bergveld, Operation of Chemically Sensitive Field-Effect Sensors as a Function of the Insulator-Electrolyte Interface, *IEEE Trans. Electron Devices*, **ED-30**, 1983.
25. P. R. Barabash, R. S. C. Cobbold, and W. B. Wlodarski, Analysis of the Threshold Voltage and its Temperature Dependence in Electrolyte-Insulator-Semiconductor Field-Effect Transistor (EISFET's), *IEEE Trans. Electron Devices*, **ED-34**, 1987.
26. L. Bousse and P. Bergveld, The Role of Buried OH Sites in the Response Mechanism of Inorganic-Gate pH-Sensitive ISFET's, *Sensors and Actuators*, **6**, 1984.
27. W. Leimbrock, U. Landgraf, and G. Kampfrath, An Extended Site-Binding Model and Experimental Results of Organic Membranes for Reference ISFET's, *Sensors and Actuators*, **B2**, 1990.
28. S. K. Ghandhi, *Semiconductor Power Devices*, Wiley, New York, 1977.
29. F. E. Gentry, F. W. Gutzwiller, N. Holonyak, and E. E. von Zastrow, *Semiconductor Controlled Rectifiers: Principles and Applications of pnpn Devices*, Prentice-Hall, Englewood Cliffs, New Jersey, 1964.
30. P. Antognetti and G. Massobrio, Thyristor Model for CAD Applications, *Proc. Journées d'Electronique*, Lausanne, Switzerland, October 1977.
31. P. Antognetti, G. Massobrio, M. La Regina, and S. Parodi, Computer Aided Analysis of Power Electronic Circuits Containing Thyristors, *Proc. Computer Aided Design and Manufacture of Electronic Components, Circuits and Systems Conf.*, London, July 1979.

Appendix A

PN Junction

A.1 Elements of Semiconductor Physics

A brief review of semiconductor physics is presented. Only those results concerning the aim of the book will be given. Not all the results, however, will be proved here; such proofs can be found in most classic books on semiconductor devices [1–5].

A.1.1 Carrier concentrations for semiconductors in equilibrium

Semiconductors derive their name from the fact that they can conduct current better than insulators, but not as well as conductors.

Semiconductors are elements of group IV of the periodic table; the most widely used semiconductor material currently is silicon (Si). The discussion throughout this book is focused on this material, but the arguments treated are valid for other semiconductors as well. Throughout this book, we assume (unless indicated otherwise) that no illumination, no radiation, no mechanical stress, and no magnetic fields are present, that all points of the semiconductor are at the same temperature (room temperature), and that no external voltage and current are applied to it. The semiconductor is then said to be in *equilibrium*.

Four classes of charged particles can be observed in a semiconductor:

Particles that have positive charge $\begin{cases} 1.\ \text{Mobile holes } p \\ 2.\ \text{Immobile donor ions } N_D \end{cases}$

Note: The material preceding Sec. A.1.1a was taken from P. E. Gray and C. L. Searle, *Electronic Principles: Physics, Models, and Circuits,* copyright © 1969 by John Wiley & Sons, Inc. Reprinted by permission.

Particles that have negative charge $\begin{cases} \text{1. Mobile electrons } n \\ \text{2. Immobile acceptor ions } N_A \end{cases}$

In each case the symbol used above represents the *volume concentration* of the corresponding charged particle. Each of these types of charged particles carries a charge q ($q = 1.60 \times 10^{-19}$ C), the electronic charge. Therefore, the local charge density ρ can be written as

$$\rho = q(p + N_D - n - N_A) \qquad \text{(A-1)}$$

In a homogeneous (uniformly doped) semiconductor, the space-charge density must vanish at every point; local neutrality is required as a consequence of Gauss's law. Therefore, the concentrations of the several charged particles must satisfy this condition, which is obtained by setting $\rho = 0$ in Eq. (A-1). Then

$$n - p = N_D - N_A \qquad \text{(A-2)}$$

The impurity concentrations N_D and N_A are determined solely by the fabrication and subsequent processing of the semiconductor. Therefore, Eq. (A-2) can be regarded as one constraint on the concentrations p and n of the mobile carriers.

A second constraint can be obtained from considerations, on a quantum-statistical basis, of the hole and electron concentrations in a semiconductor. The result, which is introduced here as a postulate, is: The *equilibrium* hole and electron concentrations are coupled in such a way that for any particular semiconductor in equilibrium, the *product* of the hole and electron concentrations is a function of temperature alone and is independent of the concentrations of donor and acceptor impurities. That is,

$$np = f(T) \qquad \text{(A-3)}$$

where T denotes the absolute temperature and $f(T)$ is a function independent of the concentrations of the impurities. By common convention, the equilibrium np product is denoted by $n_i^2(T)$. That is,

$$np = n_i^2(T) \qquad \text{(A-4)}$$

where n_i is called the *intrinsic carrier concentration*; for Si it has the value of 1.45×10^{10} cm^{-3} at 300 K.

The equilibrium carrier concentrations in a semiconductor are then governed by Eqs. (A-2) and (A-4). These equations can be solved for n and p explicitly. However, in many practical circumstances the impurity concentrations are large enough compared with the intrinsic concentration n_i to make simple approximate solutions quite accurate [1].

a. **Intrinsic semiconductor.** No dopant (impurity) atoms are present in *intrinsic* (pure) semiconductors, and the ionized atoms coming from thermal excitation contribute electron-hole *pairs*; from this fact and from Eq. (A-4), it follows that

$$n = p = n_i \qquad (A\text{-}5)$$

b. ***n*-Type semiconductor.** Consider a semiconductor material uniformly doped with an impurity that can provide free electrons to the semiconductor; such an impurity is called a *donor* impurity (it comes from group V of the periodic table).

The donor atoms donate (becoming immobile positive ions) electrons to the semiconductor and increase n at the expense of p (as compared to the case of intrinsic semiconductor), while the product np is still given by Eq. (A-4). These electrons become the *majority carriers*, and holes become the *minority carriers*.

Since the majority carriers carry a negative charge, a semiconductor doped with donor impurities is said to be *n-type*. At room temperature practically all the donor atoms are ionized, each contributing one free electron; if $N_D \gg n_i$, practically all the free electrons originate from the donor atoms, and it can be written that

$$n \simeq N_D \qquad (A\text{-}6)$$

Combining Eq. (A-4) with Eq. (A-6) yields

$$p \simeq \frac{n_i^2}{N_D} \qquad (A\text{-}7)$$

If the doping concentration is very high ($>10^{18}$ cm^{-3}), the preceding two relations will not hold, since the assumptions used in deriving them are not valid; semiconductors with very high N_D are called *degenerate*. Also, the two relations will not hold at extremely low temperatures, where the dopant atoms will not all be ionized, or at very high temperatures, where n_i rises to the point that the assumption $N_D \gg n_i$ is not valid. Whenever the preceding relations are used, it is implied that none of these extreme situations is in effect.

c. ***p*-Type semiconductor.** Consider now a semiconductor material doped with an impurity that can capture free electrons from the semiconductor, which is equivalent to providing free holes to it; such an impurity is called an *acceptor* impurity (it comes from group III of the periodic table).

The acceptor atoms donate (becoming immobile negative ions) holes to the semiconductor, thus increasing p at the expense of n;

the majority carriers are now holes, and the minority carriers are electrons.

Since the majority carriers carry a positive charge, a semiconductor doped with acceptor impurities is said to be *p-type*. It is assumed again that practically all the dopant atoms are ionized; each contributes one hole to the semiconductor.

If $N_A \gg n_i$, practically all the holes originate from the acceptor atoms, and it can be written that

$$p \simeq N_A \qquad (A\text{-}8)$$

Combining Eq. (A-4) with Eq. (A-8) yields

$$n \simeq \frac{n_i^2}{N_A} \qquad (A\text{-}9)$$

The preceding two approximations will fail at extreme temperatures or if N_A is too high, as explained for the case of n-type semiconductors.

d. Boltzmann and Poisson equations. The preceding discussion has been limited to the case where no electric fields exist within the semiconductor material. Consider now the case where such fields do exist.

First, note that the existence of electric fields does not necessarily imply current flow; this statement is true for a capacitor, for example. It is, however, valid for semiconductor materials as well. In such materials, current flow is the result of two effects: the *diffusion* of carriers, which is caused by carrier concentration gradients within the semiconductor, and the *drift* of carriers, which is caused by electric fields in the semiconductor.

Knowledge of the electric field intensity by itself is *not* enough to predict current flow; one must take into account diffusion, too. In many cases, an electric field can exist *without* any associated current flow, since its effect is balanced by that of the carrier concentration gradient (this is sometimes described by saying *the diffusion current cancels the drift current*). Such a situation occurs, for example, in the depletion region of a *pn* junction in the absence of external bias.

If an electric field within the semiconductor is not zero, the relations presented in the previous sections will not be valid in general. The values of n and p will be different from the values found in absence of electric field. Thus, a region of semiconductor material (intrinsic, n-type, or p-type) in equilibrium is considered and an electrostatic potential difference ϕ_{sb} between two points s and b is assumed. Then it can be shown that the electron concentration n_s and

n_b at points s and b, respectively are related to ϕ_{sb} by

$$n_s = n_b \, e^{q\phi_{sb}/kT} \tag{A-10a}$$

with k ($k = 1.38 \times 10^{-23}$ J/K) the Boltzmann's constant, q ($q = 1.60 \times 10^{-19}$ C) the magnitude of the electron charge, and T the absolute temperature. From the above values, the value of the thermal voltage kT/q at room temperature (300 K) is 25.86×10^{-3} V.

The relation corresponding to Eq. (A-10a) for holes is (again assuming equilibrium)

$$p_s = p_b \, e^{-q\phi_{sb}/kT} \tag{A-10b}$$

Equations (A-10) are called *Boltzmann's relations*.

In the presence of an electric field, the charge density ρ can vary from point to point. Assuming equilibrium, n and p in Eq. (A-1) must be such that for any two points, they are related to the electrostatic potential by Eqs. (A-10). In addition to these relations, the total charge density ρ must satisfy *Poisson's equation*, which is a general relation in electrostatics, and it is not restricted to semiconductors. Considering the one-dimensional case (x is the dimension involved), the Poisson equation is

$$\frac{d^2\phi}{dx^2} = -\frac{\rho(x)}{\varepsilon_s} \tag{A-11}$$

where ϕ is the electrostatic potential, ρ is the total charge density [Eq. (A-1)], and ε_s is the permittivity of the semiconductor given by

$$\varepsilon_s = \varepsilon_r \varepsilon_0 \tag{A-12}$$

with ε_0 the permittivity of free space ($\varepsilon_0 = 8.85 \times 10^{-14}$ F/cm) and ε_r the dielectric constant of the material ($\varepsilon_r = 11.8$ for Si).

A.1.2 Transport of electric current

The mobile charge carriers in a semiconductor are in constant motion, even under thermal equilibrium conditions. This motion is a manifestation of the random thermal energy of atoms and electrons that form the semiconductor. The motion of an individual hole or electron is irregular; the charge carrier has frequent collisions with the atoms, including impurities, of the semiconductor. The direction

---

Note: The material in Sec. A.1.2a and b, is derived in part from P. E. Gray and C. L. Searle, *Electronic Principles: Physics, Models, and Circuits*, copyright © 1969 by John Wiley & Sons, Inc. Reprinted by permission.

of motion of the charge carrier usually changes as a consequence of these collisions.

Under equilibrium conditions, no average electric current results from this random thermal motion. The equilibrium situation can be disturbed in two ways:

1. An electric field can be applied
2. The carrier distributions can be made nonuniform

In both of these cases, a motion of the charge carriers results, thus producing electric currents [1].

We shall treat both of these cases as one-dimensional, of unit cross-sectional area; x is the dimension involved.

The subscripts n or p on the symbols indicate that the parameters refer to the n-type or p-type semiconductor respectively; the use of the subscript 0 denotes the thermal equilibrium value of the corresponding parameter. This notation is used throughout the book.

a. Drift current. An electric field affects the random thermal motion of charge carriers in the intervals between collisions by giving a small but uniform acceleration to all the carriers exposed to the field. Although, for any carrier, the velocity increment produced by this acceleration is wiped out by the collisions that the carrier has with its environment, the electric field has a net effect simply because the velocity increments that it gives to carriers are all directed along the field. This net effect is called *drift*.

The net effect of the collisions and the intervals of acceleration between collisions can be described by assigning to a group of carriers a *drift velocity* v_p, proportional to the electric field [1].

Thus the drift velocity of the holes is

$$v_p = \mu_p E \qquad (A\text{-}13a)$$

where the parameter μ_p is called the *hole mobility*, and its value is about 480 cm^2/(V · s).

Similarly, the drift velocity of the electrons is

$$v_n = -\mu_n E \qquad (A\text{-}13b)$$

where μ_n is the *electron mobility* and its value is about 1350 cm^2/(V · s). The minus sign appears here because the electrons, having negative charge, are accelerated in a direction opposite to the field.

This treatment assumes that the mobility μ is independent of the applied electric field. This is a reasonable assumption only so long as the drift velocity is small in comparison to the thermal velocity of carriers, which is about 10^7 cm/s for Si at room temperature. As the drift velocity becomes comparable to the thermal velocity, its dependence on the electric field will begin to depart from the simple relationships given above. This is illustrated by experimental measurements of the drift velocity of electrons and holes in Si as a function of the electric field, as shown in Fig. A-1. Evidently an initial straight-line dependence is followed by a less rapid increase as the electric field is increased. At large enough fields, a maximum drift velocity seems to be approached [2,3].

Thus the *hole current density* can be written as

$$J_{p,\text{drift}} = qp\mu_p E \tag{A-14a}$$

Similarly, the *electron current density* can be written as

$$J_{n,\text{drift}} = qn\mu_n E \tag{A-14b}$$

The electron current density has the same direction as the electric field because, while the electrons move opposite to the field, they carry negative charge: the two minus signs cancel.

The *total electric current* produced by an electric field in a semiconductor is the sum of the currents carried by the holes and electrons. Thus

$$J_{\text{drift}} = q(\mu_p p + \mu_n n)E \tag{A-15}$$

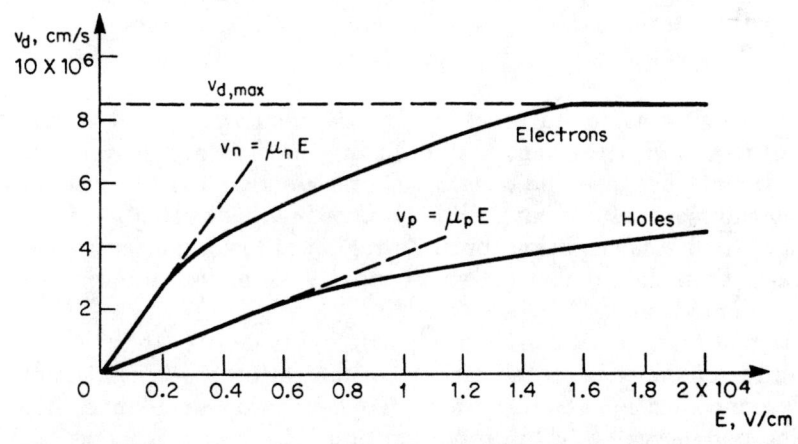

Figure A-1 Effect of an electric field on the magnitude of the drift velocity of carriers in Si. (*From A. S. Grove, Physics and Technology of Semiconductor Devices, copyright © 1967 by John Wiley & Sons, Inc. Used by permission.*)

The coefficient of E in this result is simply the electrical *conductivity* σ of the semiconductor:

$$\sigma = q(\mu_p p + \mu_n n) \tag{A-16}$$

b. Diffusion current. Diffusion is a manifestation of the random thermal motion of particles; it shows up as a current that appears whenever mobile particles are nonuniformly distributed in a system.

The diffusion process cancels the carrier concentration imbalances in the semiconductor.

It can be demonstrated [1] that the *hole current density* associated with diffusion is

$$J_{p,\text{diff}} = -qD_p \frac{dp}{dx} \tag{A-17a}$$

and the corresponding *electron current density* is

$$J_{n,\text{diff}} = qD_n \frac{dn}{dx} \tag{A-17b}$$

where D_p and D_n are the *diffusion coefficients* for holes and electrons, respectively, here considered as constants and independent of x.

The minus sign in Eq. (A-17a) appears because holes (positive) flow down the concentration slope from regions of high concentration to regions of lower concentration and then with a negative derivative (for electrons that carry a negative charge, the two minus signs cancel). The value for D_p is about 13 cm²/s and for D_n is about 35 cm²/s.

Diffusive flow is in no sense a cooperative process, and it has nothing to do with the fact that diffusing carriers are charged. It occurs simply because the *number of carriers* that (as a result of random thermal motion) have velocity components directed from the region of high concentration toward a region of lower concentration is greater than the number of carriers that have oppositely directed velocity components.

It is important to notice that the particle flux density that results from diffusion depends on the carrier concentration *gradient* and *not* on the concentration itself; it is the concentration imbalance that matters, not the value of the concentration [1].

The *total electric current* produced by the diffusion process in a semiconductor is the sum of the currents carried by the holes and electrons. Thus,

$$J_{\text{diff}} = q\left(-D_p \frac{dp}{dx} + D_n \frac{dn}{dx}\right) \quad \text{(A-18)}$$

c. Simultaneous presence of drift and diffusion currents. In many situations, an electric field and carrier concentration gradients are simultaneously present in a semiconductor. For small deviations from equilibrium, it is reasonable to regard the total hole or electron current density as a linear combination of the drift and diffusion current densities. Thus the *net hole current density* can be written, using Eqs. (A-14) and (A-17), as

$$J_p = q\left(p\mu_p E - D_p \frac{dp}{dx}\right) \quad \text{(A-19a)}$$

and the *net electron current density* can be written as

$$J_n = q\left(n\mu_n E + D_n \frac{dn}{dx}\right) \quad \text{(A-19b)}$$

This description of hole and electron motion in nonequilibrium situations in terms of drift and diffusion can be justified in detail in terms of statistical-mechanical concepts and techniques [1]. Here these descriptions will be adopted as postulates.

d. Einstein relations. *Drift* and *diffusion* are both manifestations of the random thermal motion of the carriers. Consequently, the mobility μ and the diffusion coefficient D are not independent. More precisely, they are related as follows:

$$\frac{D_p}{\mu_p} = \frac{kT}{q}$$
$$\frac{D_n}{\mu_n} = \frac{kT}{q} \quad \text{(A-20)}$$

These equations are known as *Einstein relations*.

The Einstein relations can be fully justified by considering the statistical-mechanical implications of the equilibrium situation in a semiconductor that is not uniformly doped. In accordance with our previous position, we will adopt these relations as postulates.

A.1.3 Carrier concentrations for semiconductors in nonequilibrium

Most semiconductor devices operate under nonequilibrium conditions, i.e., under conditions in which the carrier concentration product np differs from its equilibrium value n_i^2. The performance of many semiconductor devices is determined by their tendency to return to equilibrium.

Consider, then, nonequilibrium situations in which Eq. (A-4) is violated. Accordingly, we can distinguish between two types of deviation from equilibrium:

$$np > n_i^2 \quad \textit{injection} \text{ of excess carriers}$$

$$np < n_i^2 \quad \textit{extraction} \text{ of carriers}$$

a. Injection-level concept. Consider now the case when somehow excess carriers of both types are introduced into the semiconductor in equal concentrations in order to preserve space-charge neutrality. Two situations are relevant:

1. The excess minority-carrier concentration is much less than the equilibrium majority-carrier concentration. This condition is referred to as *low-level injection*.

2. The excess minority-carrier concentration approaches the equilibrium majority-carrier concentration. This condition is referred to as *high-level injection*.

b. Return to equilibrium. Whenever the carrier concentrations are disturbed from their equilibrium values, they attempt to return to equilibrium.

In the case of *injection* of excess carriers, return to equilibrium is through *recombination* of the injected minority carriers with the majority carriers. In the case of *extraction* of carriers, return to equilibrium is through the process of *generation* of electron-hole pairs.

Analysis of the process by which excess carriers recombine shows that the average time-space rate of recombination is in many practical cases approximately proportional to the excess minority-carrier concentration; this is true in the assumption of low-level injection. That is, in an n-type semiconductor, for example, the *net rate of recombination*, R_p, of excess holes with excess electrons can be approximated by

$$R_p = \frac{p_n - p_{n0}}{\tau_p} = \frac{p_n'}{\tau_p} \qquad \text{(A-21a)}$$

where τ_p is the *lifetime* of the excess holes and $p_n - p_{n0}$ denotes the *excess* hole concentration, i.e., the amount by which the actual minority-carrier hole concentration p_n exceeds the equilibrium hole concentration p_{n0}.

Similarly, in a *p*-type material, the recombination rate would be described by

$$R_n = \frac{n_p - n_{p0}}{\tau_n} = \frac{n'_p}{\tau_n} \tag{A-21b}$$

A more accurate expression for R, following the Shockley-Hall-Read generation-recombination theory, can be found in Ref. [2].

A.2 Physical Operation of the *PN* Junction

To emphasize the salient features of the *pn*-junction behavior without introducing unessential and complicated details not within the aim of this book, we will consider a *step*, or *abrupt, pn junction*. This is a junction in which the transition from a *p*-type to an *n*-type semiconductor occurs over a region of negligible thickness, *abruptly*. It is assumed that the *metallurgical junction* or boundary between the *n*-type and *p*-type region is a plane, that all carrier distributions and currents are uniform on planes parallel to the boundary plane, and that all currents are directed perpendicular to the boundary plane. This *one-dimensional* model is reasonable for the representative structure shown in Fig. A-2 and is justifiable for most *pn* junctions.

Deviations and limitations of this idealized *pn*-junction model are not discussed here because they are developed step by step in Chap. 1, which related to *building* the real model. Here, only the basic physical operation of the ideal *pn* junction is discussed.

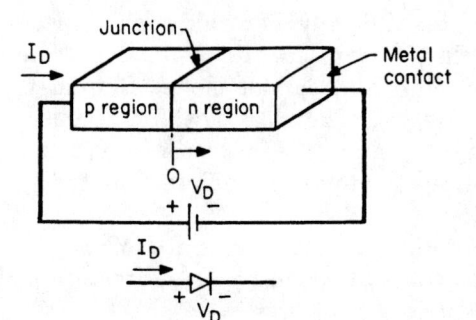

Figure A-2 Physical structure of a *pn* junction.

A.2.1 *PN* junction in equilibrium

A *pn* junction is said to be in *equilibrium* when it is at a uniform temperature and when no external disturbances, such as light or a bias voltage, are acting on it. Under equilibrium conditions the hole current and the electron current must *each* vanish at every point in the semiconductor.

It can be imagined, now, that a *barrier* exists at the boundary plane at $x = 0$ in the structure of Fig. A-2 so that no carrier (electrons or holes) flows between the *p* region and *n* region (see Fig. A-4a). The two regions, then, behave like two homogeneous and separated semiconductor materials whose electron and hole concentrations are uniform and are determined solely by the acceptor or donor concentration and the temperature.

Thus, from Eqs. (A-6) to (A-9), it can be written:

For $x > 0$ (*n* region):

$$n_{n0} \simeq N_D$$
$$p_{n0} \simeq \frac{n_i^2(T)}{N_D} \quad \text{(A-22)}$$

For $x < 0$ (*p* region):

$$n_{p0} \simeq \frac{n_i^2(T)}{N_A}$$
$$p_{p0} \simeq N_A \quad \text{(A-23)}$$

Clearly, concentration gradients of both holes and electrons must exist at the junction, as indicated in Fig. A-3.

Now, if the *barrier* (in reality it does not exist) at $x = 0$ is removed, the *pn* junction is formed and diffusion takes place in its vicinity. Electrons diffuse from the region where they are in high concentration (*n* side) to the region where their concentration is low (*p* side). Each such electron leaves behind a donor atom with a net positive charge, since the donor atoms are initially neutral; such ionized donor atoms are indicated by the $\oplus$ signs in Fig. A-4b, the circle suggesting a bound charge (i.e., one that cannot move). Similarly, holes diffuse from the *p* to the *n* region. They leave behind acceptor atoms with a net negative charge, since the acceptor atoms are initially neutral; such ionized acceptor atoms are indicated by the $\ominus$ signs in Fig. A-4b.

In the region near the junction, then, ionized atoms are left *uncovered*. The electric field created by these atoms has such a direction as to inhibit the diffusion of free carriers.

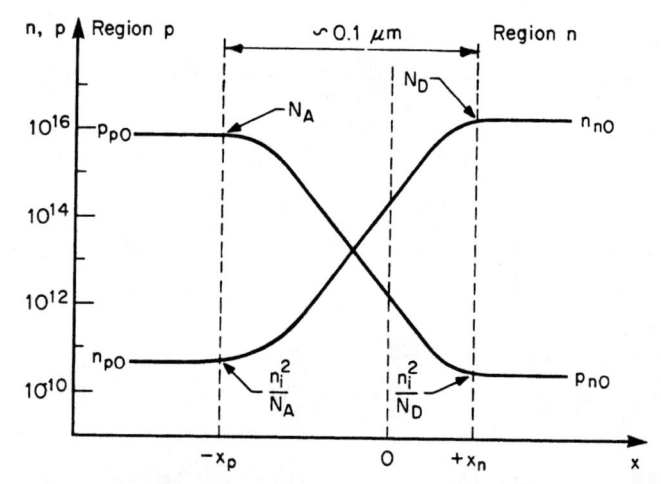

Figure A-3 Approximate distributions of carriers near the boundary plane of a *pn* junction in equilibrium. (*From Semiconductor Electronics Education Committee [4]. Copyright © 1964, Education Development Center, Inc., Newton, Mass. Used by permission.*)

In other words, any flow of charge across the junction is a self-limiting process because the electric field at the junction, which is a direct consequence of the charge transport, increases to exactly the value required to counterbalance the diffusive tendencies of holes and electrons. The field lines extend from the donor ions on the n-type side of the junction to the acceptor ions on the p-type side. The presence of this field causes a potential barrier ϕ_0 between the two types of semiconductor; this potential is frequently referred to as the *contact potential* or the *built-in potential*.

Therefore, a *space-charge region* where, by diffusion, the neutrality condition is violated would be established at the junction. The space-charge region is more often referred to as *depletion region* in which the space charge is made up of dopant ions. Let the depths of this depletion region at the two sides of the junction be $+x_n$ and $-x_p$ as shown in Fig. A-4b.

a. Built-in potential. As stated before, when equilibrium is reached, the magnitude of the field is such that the tendency of electrons (holes) to *diffuse* from the n-type (p-type) region into the p-type (n-type) region is balanced by the tendency of the electrons (holes) to *drift* in the opposite direction under the influence of the built-in field.

Appendix A

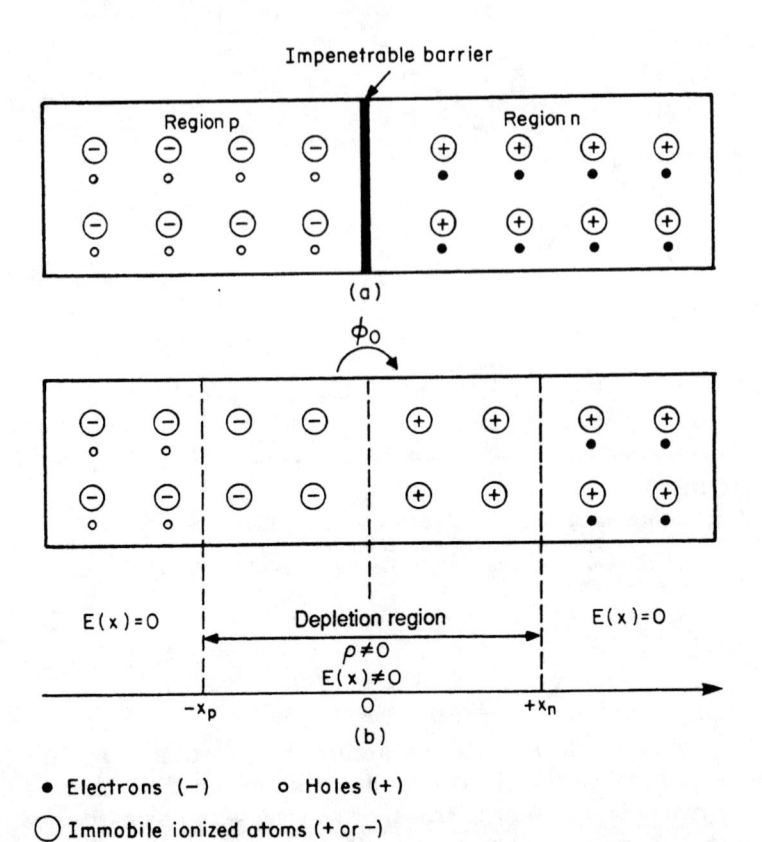

Figure A-4 What happens when a *pn* junction is formed, supposing (*a*) the presence of an imaginary barrier at $x = 0$ and (*b*) the removal of the imaginary barrier (real situation).

This equilibrium condition results in zero electron and hole currents at any point of the structure. Thus, it can be written

$$J_p = J_{p,\text{drift}} + J_{p,\text{diff}} = 0$$
$$J_n = J_{n,\text{drift}} + J_{n,\text{diff}} = 0$$
(A-24)

Taking into account Eqs. (A-19) yields

$$q\left[-D_p \frac{dp(x)}{dx} + \mu_p p(x) E(x)\right] = 0$$
$$q\left[D_n \frac{dn(x)}{dx} + \mu_n n(x) E(x)\right] = 0$$
(A-25)

From Eqs. (A-25),

$$E(x) = \frac{D_p}{\mu_p} \frac{1}{p(x)} \frac{dp(x)}{dx} = -\frac{D_n}{\mu_n} \frac{1}{n(x)} \frac{dn(x)}{dx} \quad \text{(A-26)}$$

Since the electric field is the negative of the gradient of the potential, the built-in potential ϕ_0 is found from Eq. (A-26) to be

$$\begin{aligned}\phi_0 &= -\int_{-x_p}^{x_n} E(x)\, dx = -\frac{D_p}{\mu_p} \int_{-x_p}^{x_n} \frac{1}{p(x)} \frac{dp(x)}{dx} dx \\ &= \frac{D_p}{\mu_p} [\ln p(-x_p) - \ln p(x_n)] \\ \phi_0 &= -\int_{-x_p}^{x_n} E(x)\, dx = \frac{D_n}{\mu_n} \int_{-x_p}^{x_n} \frac{1}{n(x)} \frac{dn(x)}{dx} dx \\ &= \frac{D_n}{\mu_n} [\ln n(x_n) - \ln n(-x_p)]\end{aligned} \quad \text{(A-27)}$$

As the carrier concentrations outside the depletion region are known [see Eqs. (A-22) and (A-23)], the total built-in potential ϕ_0 can be written as

$$\begin{aligned}\phi_0 &= \frac{D_p}{\mu_p} \ln \frac{p_{p0}}{p_{n0}} \simeq \frac{kT}{q} \ln \frac{N_A N_D}{n_i^2} \\ \phi_0 &= \frac{D_n}{\mu_n} \ln \frac{n_{n0}}{n_{p0}} \simeq \frac{kT}{q} \ln \frac{N_A N_D}{n_i^2}\end{aligned} \quad \text{(A-28)}$$

where Einstein relations have been used. The built-in potential lies in the range 0.2 to 1 V for typical pn-junction diodes. Because of the logarithmic dependence, wide variations of the product $N_A N_D$ are needed to obtain appreciable variation of ϕ_0.

Inspection of Eqs. (A-28) shows that ϕ_0 can be resolved into two components:

$$\phi_n = \frac{kT}{q} \ln \frac{N_D}{n_i} \quad \text{(A-29a)}$$

where ϕ_n is the potential at the neutral edge of the depletion region in the n-type semiconductor, and

$$\phi_p = -\frac{kT}{q} \ln \frac{N_A}{n_i} \quad \text{(A-29b)}$$

where $\phi_p < 0$ is the potential at the neutral edge of the depletion region in the p-type semiconductor.

Thus the total potential change ϕ_0 from the neutral p-type region to the neutral n-type region is

$$\phi_0 = \phi_n - \phi_p = \frac{kT}{q}\ln\frac{N_D}{n_i} + \frac{kT}{q}\ln\frac{N_A}{n_i} = \frac{kT}{q}\ln\frac{N_A N_D}{n_i^2} \quad \text{(A-30)}$$

just as obtained in Eqs. (A-28).

The built-in potential ϕ_0 depends on the dopant concentration in each region.

The major portion of the potential change occurs in the region with the lower dopant concentration, and the depletion region is wider in the same region (as shown later).

Besides, it must be noted that the potential at the junction plane ($x = 0$) is not exactly zero unless the junction is symmetrical (that is, $N_A = N_D$).

From Eqs. (A-28) we can also determine the carrier concentrations at the edges of the depletion region (i.e., at $-x_p$ and at x_n). Then

$$n_{p0}(-x_p) = n_{n0}(x_n)\, e^{-q\phi_0/kT}$$
$$p_{n0}(x_n) = p_{p0}(-x_p)\, e^{-q\phi_0/kT} \quad \text{(A-31)}$$

These equations are known as *Boltzmann's relations* for the pn junction.

b. Depletion region*. In the *depletion approximation* it is assumed that the semiconductor may be divided into distinct regions that are either neutral or completely depleted of mobile carriers. These regions join each other at the edges of the depletion region, where the majority-carrier density is assumed to change abruptly from the dopant concentration to zero. The depletion approximation will appreciably simplify the solution of *Poisson's equation*.

Since the carrier concentrations are assumed to be much less than the net ionized dopant density in the depletion region, Poisson's equation can be written as

$$\frac{d^2\phi}{dx^2} = -\frac{\rho}{\varepsilon_s} = -\frac{q}{\varepsilon_s}(p - n + N_D - N_A) \simeq -\frac{q}{\varepsilon_s}(N_D - N_A) \quad \text{(A-32)}$$

In general, N_D and N_A may be functions of position, and thus Eq. (A-32) cannot be solved explicitly; ε_s is the permittivity of the semi-

---

*The material in this section is derived in part from R. S. Muller and T. I. Kamins, *Device Electronics for Integrated Circuits*, copyright © 1977 by John Wiley & Sons, Inc. Reprinted by permission.

conductor ($\varepsilon_s = 1.04 \times 10^{-12}$ F/cm for Si). Equation (A-32), however, can be solved for the *step junction* shown in Fig. A-5, where the dopant concentration changes abruptly from N_A to N_D at $x = 0$. Using the depletion approximation, it is assumed that the region between $-x_p$ and x_n is totally depleted of mobile carriers, as shown in Fig. A-5b, and that the mobile majority-carrier densities abruptly become equal to the respective dopant concentrations at the edges of the depletion region. The charge density is, therefore, zero everywhere except in the depletion region, where it takes the value of the ionized dopant concentration (see Fig. A-5c). Under this assumption, Eq. (A-32) becomes [5]

$$\frac{d^2\phi}{dx^2} = -\frac{dE(x)}{dx} = -\frac{qN_D}{\varepsilon_s} \quad \text{for } 0 < x < x_n \quad \text{(A-33a)}$$

$$\frac{d^2\phi}{dx^2} = -\frac{dE(x)}{dx} = \frac{qN_A}{\varepsilon_s} \quad \text{for } -x_p < x < 0 \quad \text{(A-33b)}$$

which may be integrated from an arbitrary point in the n-type (p-type) depletion region to the edge of the depletion region at x_n ($-x_p$), where the semiconductor becomes neutral and the field vanishes.

Carrying through this integration, we can find the *field* as

$$E(x) = -\frac{qN_D}{\varepsilon_s}(x_n - x) \quad \text{for } 0 < x < x_n \quad \text{(A-34a)}$$

The field is negative throughout the depletion region and varies linearly with x, having its maximum magnitude at $x = 0$ (see Fig. A-5d). The direction of the field toward the left is physically reasonable since the force it exerts must balance the tendency of the negatively charged electrons to diffuse toward the left out of the neutral n-type semiconductor.

The field in the p-type region is similarly found to be

$$E(x) = -\frac{qN_A}{\varepsilon_s}(x + x_p) \quad \text{for } -x_p < x < 0 \quad \text{(A-34b)}$$

The field in the p-type region is also negative in order to oppose the tendency of the positively charged holes to diffuse toward the right.

At $x = 0$, the field must be *continuous*, so that

$$N_A x_p = N_D x_n \quad \text{(A-35)}$$

Thus, the width of the depleted region on each side of the junction varies inversely with the magnitude of the dopant concentration; the *higher* the dopant concentration, the *narrower* the depletion region.

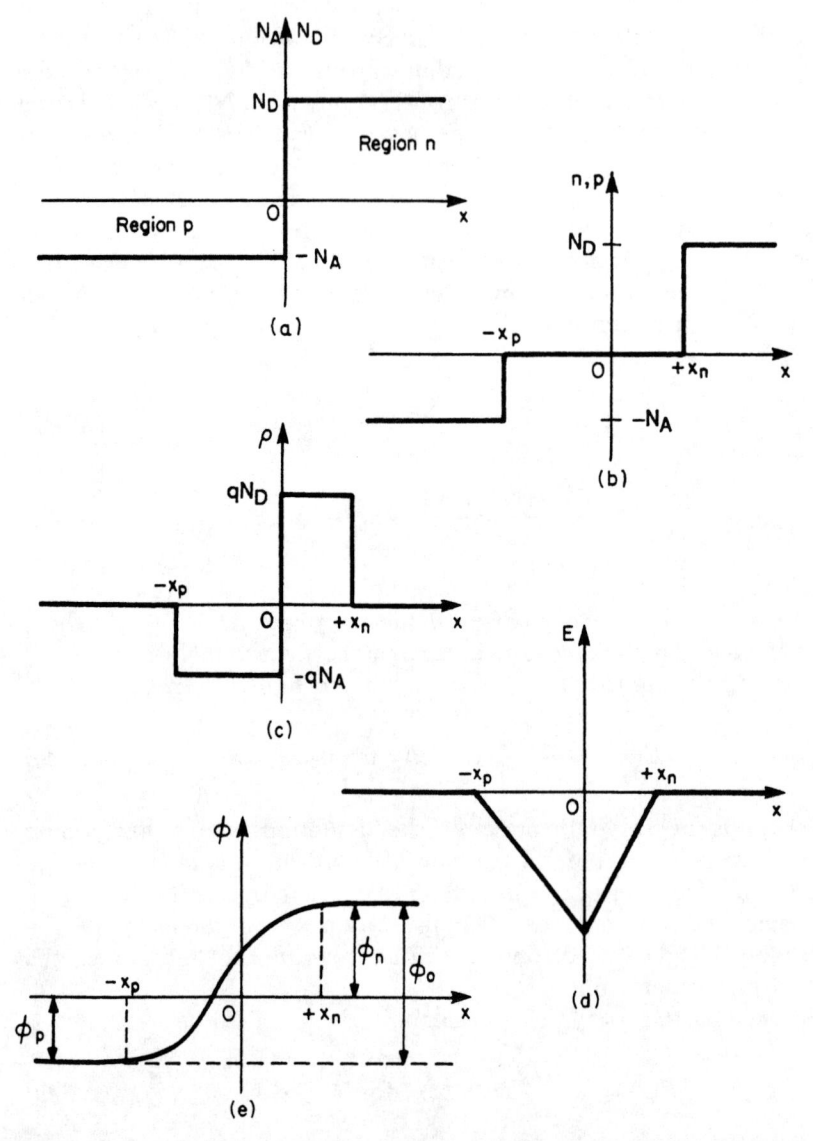

Figure A-5 Properties of a step junction as functions of position using the depletion approximation: (a) net dopant concentration, (b) carrier densities, (c) space charge used in Poisson's equation, (d) electric field found from first integration of Poisson's equation, and (e) potential obtained from second integration. (*From R. S. Muller and T. I. Kamins, Device Electronics for Integrated Circuits, copyright © 1977 by John Wiley & Sons, Inc. Used by permission.*)

In a highly asymmetrical junction where the dopant concentration on one side of the junction is much higher than that on the other side, the depletion region penetrates primarily into the lightly doped material, and the width of the depletion region in the heavily doped material may often be neglected. The expressions for the field may be integrated again to obtain the potential variation across the junction.

In the n-type semiconductor

$$\phi(x) = \phi_n - \frac{qN_D}{2\varepsilon_s}(x_n - x)^2 \quad \text{for } 0 < x < x_n \tag{A-36a}$$

as shown in Fig. A-5e, where ϕ_n is given by Eq. (A-29a).

Similarly, in the p-type semiconductor

$$\phi(x) = \phi_p + \frac{qN_A}{2\varepsilon_s}(x + x_p)^2 \quad \text{for } -x_p < x < 0 \tag{A-36b}$$

where ϕ_p is given by Eq. (A-29b).

Taking into account Eqs. (A-28), (A-29), (A-35), and (A-36), the total *depletion-region width* is found to be

$$W = x_n + x_p = \sqrt{\frac{2\varepsilon_s}{q} \phi_0 \left(\frac{1}{N_A} + \frac{1}{N_D}\right)} \tag{A-37}$$

Thus, the depletion-region width depends most strongly on the semiconductor with the lighter doping and varies approximately as the inverse square root of the smaller dopant concentration.

The widths of the p-type and n-type portions of the depletion region are

$$\begin{aligned} x_p &= W \frac{N_D}{N_A + N_D} \\ x_n &= W \frac{N_A}{N_A + N_D} \end{aligned} \tag{A-38}$$

Often, in practice, we must consider the case of a *one-sided* step junction of the type n^+p (or np^+), i.e., a step junction for which $N_D \gg N_A$ (or $N_A \gg N_D$); these relations imply $x_n \ll x_p$ (or $x_p \ll x_n$) which means that practically all the depletion region extends into the p side (or n side).

In addition to the abrupt junction, there is another doping profile that can be treated exactly and that gives useful results for approximating real pn junctions. It is the *linearly graded* junction where the net dopant concentration varies linearly from the p-type semiconductor to the n-type semiconductor. A description of this type of junction can be found in Refs. [2,3,5].

A.2.2 *PN* junction in nonequilibrium: Effect of a bias voltage

As for equilibrium, to analyze the junction under a bias voltage V_D, we will again make use of the depletion approximation together with the following assumptions about the applied bias.

1. *Ohmic contacts* connect the p and n regions to the external voltage source so that negligible voltage is dropped at the contacts (see Sec. C.1.2).
2. *Small currents* flow through the neutral regions and cause only very low voltage drops.
3. The n-region is *grounded* and voltage V_D is applied to the p region.

With these assumptions, the entire applied voltage appears *across* the junction. Furthermore, under these assumptions, the solutions of Poisson's equation that were found for thermal equilibrium conditions in the previous section apply also to the junction under bias. Only the total potential across the junction changes from the built-in value ϕ_0 to $\phi_0 - V_D \equiv \phi_J$.

If a *positive voltage* V_D is applied to the p region, the junction is said to be *forward-biased*. Under forward bias, appreciable currents can flow even for small values of V_D.

If a *negative voltage* V_D is applied to the p region, the junction is said to be *reverse-biased*. Under reverse bias, there is very little current flow through the junction.

For an abrupt pn junction, we can find the depletion-region width as a function of voltage by replacing the built-in potential ϕ_0 in Eq. (A-37) with $\phi_0 - V_D$ so that

$$W = x_n + x_p = \sqrt{\frac{2\varepsilon_s}{q}(\phi_0 - V_D)\left(\frac{1}{N_A} + \frac{1}{N_D}\right)} \qquad \text{(A-39)}$$

It is important to know the maximum field, $E_{\max}$, at the junction and its relationship to applied voltage. For the step-junction case, taking into account that the field varies linearly with distance [see Eqs. (A-34)], and that the area under the field curve represents the potential, by inspection of Fig. A-5d, it can be written that

$$\tfrac{1}{2} E_{\max} W = \phi_0 - V_D \qquad \text{(A-40)}$$

so that

$$E_{\max} = \frac{2(\phi_0 - V_D)}{W} \qquad \text{(A-41)}$$

A.2.3 Currents in the PN junction

The subject of this section is the current flow across a *pn* junction under both forward and reverse bias.

As a first step toward analyzing current flow, we write the *continuity equation* for free carriers, i.e., an equation that takes account of the various mechanisms affecting the population of carriers inside a semiconductor [2–5].

Through the use of the continuity equation we can be in a position to find expressions for the current in a *pn* junction under bias.

Solutions of the continuity equation in the neutral regions give carrier densities in terms of position and time. Expressions for current are then obtained directly by using Eqs. (A-19), which define carrier flow in terms of carrier densities.

a. Continuity equation. The *continuity equation* can be written in terms of carrier densities as follows.

For *electrons*:

$$\frac{\partial n}{\partial t} = \mu_n n(x) \frac{\partial E(x)}{\partial x} + \mu_n E(x) \frac{\partial n(x)}{\partial x} + D_n \frac{\partial^2 n(x)}{\partial x^2} + (G_n - R_n) \quad \text{(A-42a)}$$

For *holes*:

$$\frac{\partial p}{\partial t} = -\mu_p p(x) \frac{\partial E(x)}{\partial x} - \mu_p E(x) \frac{\partial p(x)}{\partial x} + D_p \frac{\partial^2 p(x)}{\partial x^2} + (G_p - R_p) \quad \text{(A-42b)}$$

where G_n, R_n and G_p, R_p represent the generation and recombination rates ($cm^{-3} s^{-1}$) for electrons and holes, respectively, associated with external disturbances such as light, radiation, or magnetic fields.

In Eqs. (A-42) it has been assumed that mobility μ and diffusion constant D are not functions of x. Although this assumption is not valid in some cases, the major physical effects are included in Eqs. (A-42) and the more exact formulations are seldom considered [5].

If the electric field is zero or negligible in the region under consideration, the first two terms on the right-hand side of Eqs. (A-42) can be neglected and the analysis is greatly simplified. Even if the field is not negligible, some of the terms in Eqs. (A-42) may be unimportant. For example, if the field is constant, the first term on the right-hand side in each equation drops out.

Rarely is it necessary to deal with the full complexity of Eqs. (A-42).

b. Current-voltage characteristics.

The *total current* consists, in general, of the sum of four components: hole and electron drift currents and hole and electron diffusion currents.

We now consider a *pn*-junction diode, connected to a voltage source with the n region grounded and the p region at V_D volts relative to ground. The diode structure has a constant cross-sectional area A_J and it is sketched in Fig. A-6. The junction is not illuminated, and carrier densities within the diode are influenced only by the applied voltage. The applied voltage V_D is dropped partially across the neutral regions and partially across the junction itself. Since the voltage drops across the neutral regions are ohmic, they are very small at low currents [5].

If it is assumed, then, that V_D is sustained entirely at the junction, the total junction voltage will be $\phi_0 - V_D = \phi_J$ under bias, where ϕ_0 is the built-in voltage.

If V_D is *positive*, i.e., for *forward bias*, the applied voltage reduces the barrier to the diffusion flow of majority carriers at the junction. The reduced barrier, in turn, permits a net transfer rate of holes from the p side into the n side and of electrons from the n side into the p side. When these transferred carriers enter the neutral regions, they are minority carriers and are very quickly neutralized by majority carriers that enter the neutral regions from the ohmic contacts at the ends. Once the minority carriers have been injected across the depletion region, they tend to diffuse into the neutral interior region [5].

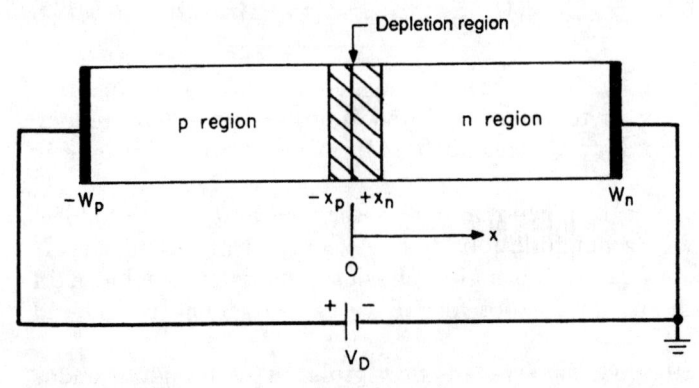

Figure A-6 Sketch of a *pn*-junction diode structure used in the discussion of currents. (*Adapted from Muller and Kamins [5]. Copyright © 1977 by John Wiley & Sons, Inc. Used by permission.*)

Note: The material in Secs. A.2.3b to e is derived primarily from R. S. Muller and T. I. Kamins, *Device Electronics for Integrated Circuits*, copyright © 1977 by John Wiley & Sons, Inc. Reprinted by permission.

If V_D is negative, i.e., for *reverse bias*, the barrier height to diffusing majority carriers is increased. Since equilibrium is disturbed, minority carriers near the depletion region tend to be depleted. The majority-carrier concentration is likewise reduced.

From these few remarks, it can be seen that the minority-carrier densities deserve special attention because they really determine what currents flow in a *pn* junction. The majority carriers act only as suppliers of the injected minority-carrier current or as charge neutralizers in the neutral regions. It is helpful to consider the majority carriers as willing "slaves" of the minority carriers. Accordingly, solutions of the continuity equation for the minority-carrier densities in each of the neutral regions will be sought [5].

c. Boundary values of minority-carrier densities. To write the solutions of the continuity equation in a useful fashion, however, it will be necessary to relate the boundary values of the minority-carrier densities to the applied voltage V_D. The most straightforward way of doing this is to make two additional assumptions: first, that the applied bias leads to low-level injection and, second, that the applied bias is small enough so that the detailed balance between majority and minority concentrations across the junction regions is not appreciably disturbed.

The first assumption implies that the majority-carrier concentrations are negligibly changed at the edge of the neutral regions by the applied bias. The assumption that detailed balance nearly applies allows the use of Eq. (A-31) across the junction where the potential difference is known to be $\phi_0 - V_D$.

Both of these assumptions are certainly valid when V_D is very small ($|V_D| \ll \phi_0$). Their validity at higher biases deserves more careful consideration, as given in Chap. 1.

Hence, the concentrations at the depletion region boundaries are

$$p_n(x_n) = p_p(-x_p) e^{-q(\phi_0 - V_D)/kT}$$
$$n_p(-x_p) = n_n(x_n) e^{-q(\phi_0 - V_D)/kT} \quad \text{(A-43)}$$

Under the condition of low-level injection, i.e., in each of the neutral regions the minority-carrier concentration is much less than the majority-carrier concentration, we will find that $p_p(-x_p) \simeq p_{p0}$ and $n_n(x_n) \simeq n_{n0}$. Therefore, Eqs. (A-43) become

$$p_n(x_n) = p_{p0}(-x_p) e^{-q(\phi_0 - V_D)/kT}$$
$$n_p(-x_p) = n_{n0}(x_n) e^{-q(\phi_0 - V_D)/kT} \quad \text{(A-44)}$$

We can eliminate ϕ_0 from these equations if we take into account Eqs. (A-31). Then Eqs. (A-44) become

$$p_n(x_n) = p_{n0}(x_n) e^{qV_D/kT}$$
$$n_p(-x_p) = n_{p0}(-x_p) e^{qV_D/kT} \tag{A-45}$$

Equations (A-45) are called the *law of the junction*, in that they relate the minority-carrier concentration enhancement or diminution at the edge of the depletion region with the polarity of the applied voltage V_D. Figure A-7 shows the carrier concentrations under for-

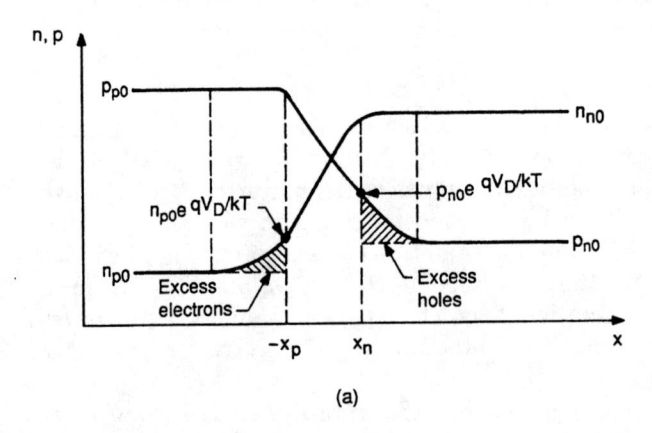

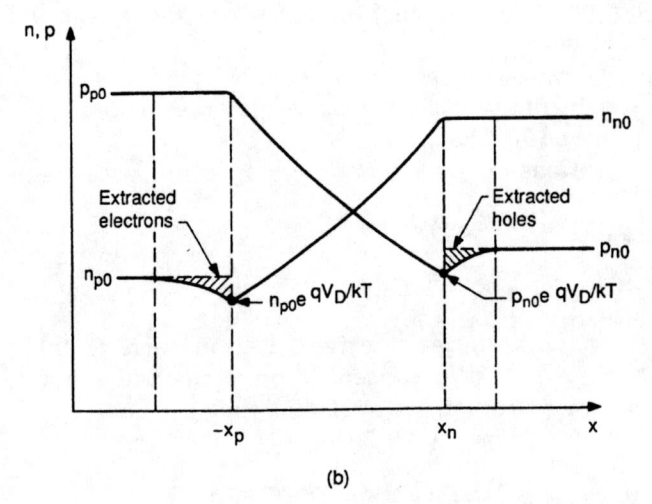

Figure A-7 Carrier concentrations under (*a*) forward bias, and (*b*) reverse bias.

ward and reverse bias. It can be seen, for example, that with forward bias, holes moving to the n region and electrons to the p region become *excess* minority carriers. This because the concentration is now greater than the thermal equilibrium level.

If we define the excess densities by

$$n'_p \equiv n_p - n_{p0}$$
$$p'_n \equiv p_n - p_{n0}$$
(A-46)

then the *excess minority-carrier* concentrations at the boundaries in terms of their thermal equilibrium values are

$$n'_p(-x_p) = n_{p0}(-x_p)(e^{qV_D/kT} - 1)$$
$$p'_n(x_n) = p_{n0}(x_n)(e^{qV_D/kT} - 1)$$
(A-47)

Equations (A-47) are extremely important results that can be used to define specific solutions for the continuity equation for minority carriers in the neutral regions near a pn junction.

d. Long-base diode analysis. Taking into account what was described in the previous sections, we can now consider solutions of the continuity equations [see Eqs. (A-42)] in the neutral region.

First consider excess holes injected into the n region, where bulk recombination through generation-recombination centers is dominant. Thus, the term R_p in Eq. (A-42b) can be expressed through Eqs. (A-21) because of the assumption of low-level injection and assuming $G_p = 0$. Therefore, the continuity equation becomes

$$\frac{\partial p_n}{\partial t} = D_p \frac{\partial^2 p_n}{\partial x^2} - \frac{p_n - p_{n0}}{\tau_p}$$
(A-48)

where the subscript n emphasizes that the holes are in the n region.

Supposing a constant donor density along x, as frequently encountered in practice, and considering the steady state $(\partial p/\partial t = 0)$, Eq. (A-48) can be rewritten as a total differential equation in terms of the excess density p'_n, which was defined in Eq. (A-46). Equation (A-48) becomes

$$0 = D_p \frac{d^2 p'_n}{dx^2} - \frac{p'_n}{\tau_p}$$
(A-49)

and it has the following exponential solution

$$p'_n(x) = A\, e^{-(x-x_n)/\sqrt{D_p \tau_p}} + B\, e^{(x-x_n)/\sqrt{D_p \tau_p}}$$
(A-50)

where A and B are constants determined by the boundary conditions of the problem.

The characteristic length $\sqrt{D_p \tau_p}$ in Eq. (A-50) is called the *diffusion length* and is denoted by L_p. (The diffusion length of an electron in a p-type region is denoted by L_n.) For specific applications of the solution to the continuity equation given in Eq. (A-50), two limiting cases based upon the length W_n of the n region from the junction to the ohmic contact (see Fig. A-6) can be considered: the *long-base diode* and the *short-base diode*.

Now if the neutral n-type region length W_n is very long compared to the diffusion length L_p, essentially all the injected holes recombine before traveling completely across it. This case is known as the *long-base diode*.

For the long-base diode, L_p is the average distance traveled in the neutral region before an injected hole recombines [5]. Since p'_n must decrease with increasing x, the constant B in Eq. (A-50) must be zero. The constant A in the solution is determined by applying Eq. (A-47), which specifies $p'_n(x_n)$ as a function of applied voltage. The complete solution is therefore

$$p'_n(x) = p_{n0}(e^{qV_D/kT} - 1) e^{-(x - x_n)/L_p} \qquad (A\text{-}51)$$

as shown in Fig. A-8. Thus, an expression for hole current can now be obtained.

The hole current flows only by diffusion, since it has been assumed that the field is negligible in the neutral regions; therefore, from Eqs.

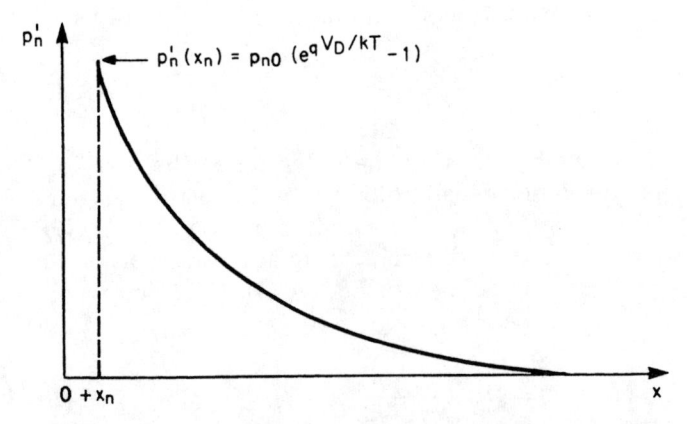

Figure A-8 Spatial variation of holes in the neutral n region of a long-base diode under forward bias V_D. The excess density p'_n has been calculated from Eq. (A-51). (*Adapted from Muller and Kamins* [5]. *Copyright © 1977 by John Wiley & Sons, Inc. Used by permission.*)

(A-19) and (A-51),

$$J_p(x) = -qD_p \frac{dp'_n}{dx} = qD_p \frac{p_{n0}}{L_p}(e^{qV_D/kT} - 1)\, e^{-(x-x_n)/L_p}$$

$$= qD_p \frac{n_i^2}{N_D L_p}(e^{qV_D/kT} - 1)\, e^{-(x-x_n)/L_p} \tag{A-52}$$

The hole current is therefore a maximum at $x = x_n$ and decreases away from the junction because the hole gradient decreases as carriers are lost by recombination.

The minority-carrier electron current that is injected into the p region can be found by a treatment analogous to the analysis used to obtain Eq. (A-52). Under the assumption that the neutral p-type region length $W_p \gg L_n \equiv \sqrt{D_n \tau_n}$, then

$$J_n(x) = qD_n \frac{n_i^2}{N_A L_n}(e^{qV_D/kT} - 1)\, e^{(x+x_p)/L_n} \tag{A-53}$$

Since the origin for x has been chosen at the physical junction (see Fig. A-6), x is a negative number throughout the p region. Hence, J_n decreases away from the junction as did J_p in the n region. To obtain an expression for the total current J_D, it is sufficient to sum the minority-carrier components at $-x_p$ and $+x_n$, respectively, as expressed in Eqs. (A-52) and (A-53):

$$J_D = J_p(x_n) + J_n(-x_p) = qn_i^2 \left(\frac{D_p}{N_D L_p} + \frac{D_n}{N_A L_n}\right)(e^{qV_D/kT} - 1)$$

$$= J_S(e^{qV_D/kT} - 1) \tag{A-54}$$

where J_S is called *reverse saturation current density*.

e. Short-base diode analysis. The second limiting case (that of the *short-base diode*) occurs when the lengths W_n and W_p of the n- and p-type regions are much shorter than the diffusion lengths L_p and L_n, respectively. In this case, there is a little recombination in the bulk of the neutral regions. In the limit, all injected minority carriers recombine at the ohmic contacts at either end of the diode. In this case, we can obtain solutions most easily by approximating the exponentials in Eq. (A-50) by the first two terms of a Taylor series expansion, so that

$$p'_n(x) = C + D\frac{(x-x_n)}{L_p} \tag{A-55}$$

where C and D are constants determined by the boundary conditions. Because of the ohmic contact assumption at $x = W_n$, $p'_n(W_n) = 0$, while the boundary condition at $x = x_n$ is given by Eq. (A-47) as for the long-base diode analysis. The solution for the excess hole concentration in the n region is then

$$p'_n(x) = p_{n0}(e^{qV_D/kT} - 1)\left(1 - \frac{x - x_n}{W'_n}\right) \quad \text{(A-56)}$$

where $W'_n = (W_n - x_n)$ is the neutral n-type region length (Fig. A-6). From Eq. (A-56), the excess hole concentration decreases linearly with distance across the n-type region (see Fig. A-9).

The assumption $W_n \ll L_p$ implies that all the holes diffuse across the n-type region before recombining (no recombination occurs in this region). The assumption of no recombination in the n-type region is also obtained by letting the lifetime approach infinity in Eq. (A-49). The differential equation that results has a linear solution [5]. A linearly varying concentration implies that the hole current remains constant throughout the n-type region. Therefore

$$J_p(x_n) = -qD_p\left.\frac{dp'_n}{dx}\right|_{x=x_n} = qD_p\frac{p_{n0}}{W'_n}(e^{qV_D/kT} - 1)$$

$$= qD_p\frac{n_i^2}{N_D W'_n}(e^{qV_D/kT} - 1) \quad \text{(A-57)}$$

The minority-carrier electron current that is injected into the p region can be found by a treatment analogous to the analysis used

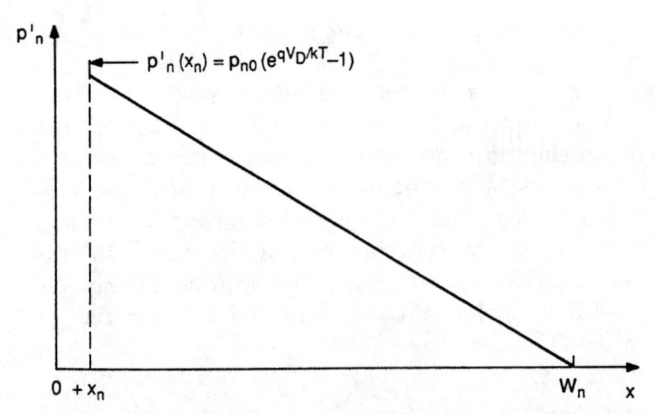

Figure A-9 Spatial variation of holes in the neutral n region of a short-base diode under forward V_D. The excess density p'_n has been calculated from Eq. (A-56). (*Adapted from Muller and Kamins [5]. Copyright © 1977 by John Wiley & Sons, Inc. Used by permission.*)

to obtain Eq. (A-57). If $W_p \ll L_n$, then

$$J_n(-x_p) = qD_n \frac{n_i^2}{N_A W_p'}(e^{qV_D/kT} - 1) \qquad (A-58)$$

As in the long-base diode, the total current in the short-base diode is due to electrons injected into the p region and holes injected into the n region. Hence

$$J_D = J_p(x_n) + J_n(-x_p) = qn_i^2 \left(\frac{D_p}{N_D W_n'} + \frac{D_n}{N_A W_p'}\right)(e^{qV_D/kT} - 1)$$

$$= J_S(e^{qV_D/kT} - 1) \qquad (A-59)$$

where J_S is called the *reverse saturation current density*.

Comparing Eqs. (A-54) and (A-59) for the currents in the long- and short-base diodes, we can see that the solutions obtained are similar.

In the long-base diode, the characteristic length is that of the minority-carrier diffusion; in the short-base diode, it is the length of the neutral region. Of course, a given diode may be approximated by a combination of these two limiting cases; that is, it may be short-base in the p region and long-base in the n region or vice versa [5].

A.2.4 Charge-storage capacitances in a PN junction

Charge storage in the depletion region owing to the dopant concentrations and charge storage owing to the minority-carrier charges injected into the neutral regions give rise to two small-signal capacitances usually called the *junction capacitance* (denoted C_J) and the *diffusion capacitance* (denoted C_d), respectively.

a. Junction capacitance. The *junction capacitance*, often also called the depletion or transition capacitance, is associated with the charge dipole formed by the ionized donors and acceptors in the junction depletion region. Specifically, the junction capacitance relates the changes in the charge at the edges of the depletion region to changes in the junction voltage. Thus, if the voltage applied to the pn junction is changed by a small dV_D, then the depletion-region charge will change by an amount dQ_J', where Q_J' is the charge per unti area. Hence a depletion region or junction capacitance can be defined as

$$C_J' = \frac{dQ_J'}{dV_D} \qquad (A-60)$$

Since $Q'_J = qN_D x_n = qN_A x_p$ in the depletion approximation, then when all the impurities in the depletion region are assumed to be ionized,

$$C'_J = qN_D \frac{dx_n}{dV_D} = qN_A \frac{dx_p}{dV_D} \qquad (A\text{-}61)$$

From Eqs. (A-38) and (A-39), it follows that

$$\frac{dx_n}{dV_D} = \frac{1}{N_D} \sqrt{\frac{\varepsilon_s}{2q(\phi_0 - V_D)(1/N_A + 1/N_D)}} \qquad (A\text{-}62)$$

and

$$C'_J = \sqrt{\frac{q\varepsilon_s}{2(\phi_0 - V_D)(1/N_A + 1/N_D)}} = \frac{C'_J(0)}{\sqrt{1 - V_D/\phi_0}} \qquad (A\text{-}63)$$

In this equation, the first formulation defines the junction capacitance in terms of the physical parameters of the device. For the circuit designer, however, it is more convenient to express this capacitance in terms of electrical parameters (as the second formulation) where $C'_J(0)$ is the capacitance at equilibrium, i.e., at $V_D = 0$ [6].

Figure A-10 shows a plot of the junction capacitance as a function of the bias voltage. For $V_D > \phi_0$, the capacitance of a step pn junction [see Eq. (A-63)] will decrease approximately inversely with the square root of the reverse bias.

Using Eq. (A-39) in Eq. (A-63) yields

$$C'_J = \frac{\varepsilon_s}{W} \qquad (A\text{-}64)$$

which is the general relationship for a parallel-plate capacitor and it is valid for an arbitrarily doped junction.

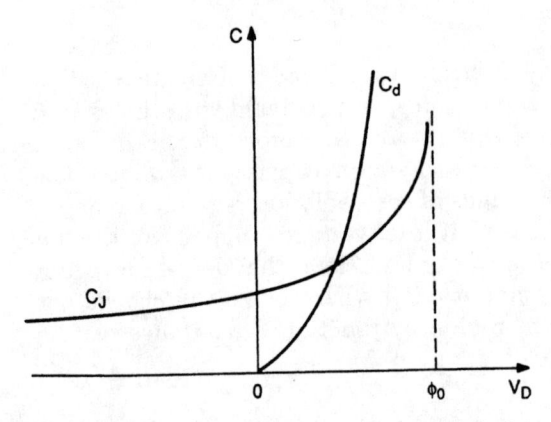

Figure A-10 Junction capacitance C_J and diffusion capacitance C_d as a function of bias voltage V_D.

b. Diffusion capacitance. The *diffusion capacitance*, often also called the storage capacitance, is associated with the excess minority-carrier charge injected into the neutral regions under forward bias. The total injected minority-carrier charge per unit area stored in the neutral n-type region can be found by integrating the excess hole concentration across the neutral region

$$Q'_p = q \int_{x_n}^{W_n} p'_n(x)\, dx \qquad (A\text{-}65)$$

An equation similar to Eq. (A-65) can be written for the excess electron charge per unit area Q'_n in the neutral p-type region.

We first consider the *ideal long-base diode*. Inserting Eq. (A-51) into Eq. (A-65) and integrating from the edge of the depletion region across the neutral region, we find the charge Q'_p due to holes stored in the n-type region to be

$$Q'_p = qL_p p_{n0}(e^{qV_D/kT} - 1) \qquad (A\text{-}66)$$

By using the expression for the hole current [Eq. (A-52)] at $x = x_n$ in Eq. (A-66), we have

$$Q'_p = \frac{L_p^2}{D_p} J_p(x_n) = \tau_p J_p(x_n) \qquad (A\text{-}67)$$

Thus, the amount of stored charge is the product of the current density and the minority-carrier lifetime.

We consider now the *ideal short-base diode*. Inserting Eq. (A-56) into Eq. (A-65) and integrating from the edge of the depletion region across the neutral region, we find the charge Q'_p due to holes stored in the n-type region to be

$$Q'_p = \frac{q(W_n - x_n)}{2} p_{n0}(e^{qV_D/kT} - 1) \qquad (A\text{-}68)$$

where W_n is the length of the n-type region. Again, the stored hole charge can be expressed in terms of the hole current [Eq. (A-57)]:

$$Q'_p = \frac{(W_n - x_n)^2}{2D_p} J_p = \tau_{dp} J_p(x_n) \qquad (A\text{-}69)$$

Even if the term in front of J_p in Eq. (A-69) has the dimensions of time, it is not the lifetime as was found in Eq. (A-67) for the long-base diode. It represents the average time τ_{dp} of a hole moving through the n-type region of the short-base diode.

As for the charge storage in the depletion region, the variation of stored minority-carrier charge in the neutral regions under forward bias can be modeled by a small-signal capacitance usually called the *diffusion capacitance* (denoted C_d). The contribution of the stored holes in the n-type region to the diffusion capacitance per unit area C'_d is

$$C'_{dp} = \frac{dQ'_p}{dV_D} = \frac{q^2}{kT} \frac{D_p}{L_p} \tau_p p_{n0} e^{qV_D/kT} \qquad \text{(A-70)}$$

for the *long-base diode*, and

$$C'_{dp} = \frac{dQ'_p}{dV_D} = \frac{q^2}{kT} \frac{D_p}{W_n - x_n} \tau_{dp} p_{n0} e^{qV_D/kT} \qquad \text{(A-71)}$$

for the *short-base diode*.

It is straightforward to add a C'_{dn} component to C'_d, which represents the electron storage in the p-type region.

Under reverse bias, storage in the neutral regions is negligible, and C'_J dominates. Under forward bias, even if C'_J increases (because the depletion-region width decreases), the exponential dependence of C'_d makes it dominant. This is shown in Fig. A-10.

A.2.5 Transient behavior of the PN junction

So far our study of the electrical and physical properties of the *pn* junction has been concerned mostly with the static, or steady-state, characteristics. Now the switching transients in the *pn* junction (the action of the diode turning on and off) are considered.

The ideal diode equation is a fundamental equation that relates the diode current to the diode voltage. The diode current may also be related to the excess minority-carrier charge. This is another fundamental equation, which introduces the concept of charge control. This concept is a most useful one for analysis of switching transients in both diodes and bipolar junction transistors.

We consider a long-base diode with one side of the junction much more heavily doped than the other. It is assumed that $N_A \gg N_D$; then $p_{n0} \gg n_{p0}$ and $Q'_p \gg Q'_n$. Henceforth, the minority-carrier holes in the neutral n region are considered.

From Eq. (A-66), it can be seen that the excess minority-carrier charge is related to the diode voltage. Furthermore, this charge is positive for forward bias and negative for reverse bias.

Note: The material in Sec. A.2.5 was taken from Hodges and Jackson [6]. Used by permission.

From Eq. (A-54), and eliminating the exponential term by substitution from Eq. (A-66), it follows that [6]

$$J_D = Q'_p \frac{D_p}{L_p^2} = \frac{Q'_p}{\tau_p} \tag{A-72}$$

This equation relates, on a static basis, the diode current density to the excess minority-carrier charge and the excess minority-carrier lifetime τ_p. In the steady state, the current density J_D supplies holes to the neutral n region at the same rate as they are being lost by recombination.

To obtain a dynamic, or time-dependent, relation, we must include the time rate of change of the excess minority-carrier charge. Then an equation for the time-dependent current density can be obtained as [6]

$$J_D(t) = \frac{Q'_p}{\tau_p} + \frac{dQ'_p}{dt} \tag{A-73}$$

This is a fundamental *charge-control equation* for the *pn*-junction diode. It states that current density $J_D(t)$ supplies holes to the neutral n region at the rate at which the stored charge increases plus the rate at which holes are being lost by recombination. Use of Eq. (A-73) can be made in the calculation of the switching times for a *pn*-junction diode.

A.2.6 Approximation considerations

Though Eqs. (A-54) and (A-59) have been obtained on three assumptions—first, that ohmic drops in the neutral regions are small so that V_D is sustained entirely across the depletion region; second, that applied bias does not greatly alter the detailed balance between drift and diffusion tendencies that exist at thermal equilibrium; and third, that hole and electron current components are constant throughout the depletion region (i.e., no carrier generation-recombination processes are present in this region)—it can be shown [5] that these assumptions are reasonable for may practical cases.

With regard to the first assumption, however, it must be noted that if the applied forward voltage is nearly as large as the built-in potential, the barrier to majority-carrier flow is substantially reduced and large currents may flow. In that case a significant portion of the

---

Note: The material in Sec. A.2.6 is derived in part from R. S. Muller and T. I. Kamins, *Device Electronics for Integrated Circuits*, copyright © 1977 by John Wiley & Sons, Inc. Reprinted by permission.

applied voltage is dropped across the neutral regions, and the series resistance of the neutral regions must be considered. The effect of series resistance can be included using circuit analysis (see Chap. 1).

With regard to the second assumption, it proves it is reasonable to treat the case of small and moderate biases by considering only slight deviations from thermal equilibrium. This means that we can relate the carrier concentrations on either side of the depletion region by considering the effective barrier height to be $\phi - V_D$ and by using Eqs. (A-47). These equations depend also on the validity of the low-level injection assumption. The current tendencies at pn junctions that are in detailed balance at thermal equilibrium are so large relative to currents that flow in practice that a more general relationship than Eqs. (A-47) is valid. This relationship, which will not be proved here, is that the carrier densities at either side of a biased pn junction (and also within the depletion region) are dependent on applied voltage according to the equation

$$pn = n_i^2 \, e^{qV_D/kT} \tag{A-74}$$

In the low-level injection case, Eq. (A-74) reduces to the boundary conditions for minority carriers, which were obtained previously. Equation (A-74) has special utility in cases where injected carrier densities approach the thermal-equilibrium densities of majority carriers.

The analysis of the pn junction that led to Eqs. (A-54) and (A-59) was based on events in the neutral regions. The depletion region was treated solely as a barrier to the diffusion of majority carriers, and it played a role only in the establishment of minority-carrier densities at its boundaries [see Eqs. (A-47)]. This is a reasonable first-order description of events, and the equations derived from it are referred to as the *ideal diode* equations [5]. The effects of the generation-recombination on the diode behavior are considered in Chap. 1 to obtain the real diode model.

References

1. P. E. Gray and C. L. Searle, *Electronic Principles: Physics, Models, and Circuits*, Wiley, New York, 1969.
2. A. S. Grove, *Physics and Technology of Semiconductor Devices*, Wiley, New York, 1967.
3. S. M. Sze, *Physics of Semiconductor Devices*, Wiley, New York, 1969.
4. P. E. Gray, D. DeWitt, A. R. Boothroyd, and J. F. Gibbons, *Physical Electronics and Circuit Models of Transistors*, vol. 2, Semiconductor Electronics Education Committee, Wiley, New York, 1964.
5. R. S. Muller and T. I. Kamins, *Device Electronics for Integrated Circuits*, Wiley, New York, 1977.
6. D. A. Hodges and H. G. Jackson, *Analysis and Design of Digital Integrated Circuits*, McGraw-Hill, New York, 1983.

Appendix B

MOS Junction

B.1 MOS Junction

A structure often referred to as a *MOS junction* is here considered. Fabrication of this structure starts with a *p*- or *n*-type semiconductor material, uniformly doped, called the *substrate*, and is assumed to be silicon from now on. An insulating layer is formed on top of the substrate. This layer is often silicon dioxide (SiO_2), simply called *oxide*, and its thickness is usually 0.01 to 0.1 μm. A third layer, referred to as the *gate*, is then formed on top of the insulator. The gate is made either of metal (usually aluminum) or of polycrystalline silicon.

The described structure is often referred to by the acronym *MOS* (*M*etal-*O*xide-*S*emiconductor), independently of whether the gate is made of metal or whether the insulator is silicon dioxide. A more proper acronym is *MIS* (*M*etal-*I*nsulator-*S*emiconductor).

A detailed study of the MOS junction can be found in Refs. [1,2]; here we will only present those results directly relevant to the aim of the book.

A brief introduction to the basic results of the *contact potential* theory will be presented first.

B.1.1 Contact potentials

Consider the junction of two different neutral materials M_1 and M_2; either material can be a semiconductor or a metal. When the two materials are brought together, at first carriers diffuse from one to the other because the concentration of these carriers is, in general,

---

Note: The material in this appendix is taken from the book by Y. P. Tsividis, *Operation and Modeling of the MOS Transistor*, McGraw-Hill, New York, 1987.

different in M_1 and M_2, and no opposing field exists in the initially neutral materials. However, as each charged carrier crosses the junction, it leaves behind a net charge of opposite polarity, and an electric field is thus established in the vicinity of the junction, which tends to inhibit the movement of carriers. Eventually, the field intensity increases to the point that it counteracts the tendency of carriers to diffuse, and a balance is achieved at which there is no more net carrier movement; an electrostatic potential change is then encountered when going from one material, through the junction, to the other material. The total potential drop in going from M_1 to M_2 is called the *contact potential* of material M_1 to material M_2, and will be denoted by $\phi_{M1,M2}$.

If a material is chosen as a reference material, then the contact potential of material M_1 to material M_2 can be expressed by

$$\phi_{M1,M2} = \phi_{M1} - \phi_{M2} \tag{B-1}$$

where ϕ_{M_i} is the contact potential of a generic material M_i to the reference material; ϕ_{M_i} can be indirectly evaluated and tabulated.

If n materials in series are considered, the potential ϕ_{M1,M_n} between the first and the last material is expressed in terms of contact potentials in the loop. Using Eq. (B-1) yields

$$\phi_{M1,M_n} = (\phi_{M1} - \phi_{M2}) + (\phi_{M2} - \phi_{M3}) + \cdots + (\phi_{M_{n-1}} - \phi_{M_n}) \tag{B-2}$$

It is clear that, with the exception of ϕ_{M1} and ϕ_{M_n}, each contact potential appears twice in the sum: once with a plus sign and once with a minus sign. Therefore

$$\phi_{M1,M_n} = \phi_{M1} - \phi_{M_n} \tag{B-3}$$

Thus no matter how many materials are in the loop, the electrostatic potential difference between its two ends depends *only* on the first and the last material.

B.1.2 Flat-band voltage

The *MOS junction* is shown in Fig. B-1a. Here it is assumed that the gate is made of the same material as the substrate (in this case, p-type silicon) with the same doping concentration. It is also assumed that somehow the same material is used to connect the gate to the substrate. No charges are shown in the silicon, since to each positively charged hole of the p-type material there corresponds a negatively charged acceptor atom from which the hole has originated and there is no reason for holes to pile up in any region; therefore the charges cancel and the material is shown as electrically neutral everywhere. Assuming no parasitic charges have been introduced during fabrication, then no field can exist in the insulator, and there is no reason for carriers to be attracted toward the insulator-substrate interface.

We now consider a realistic case, as shown in Fig. B-1b. The gate is made of a certain material, not necessarily the same as the substrate, and a metal is used to contact the gate material and form the *gate terminal G*. The substrate is contacted through a back metal plate; this plate is in turn contacted through some metal, which thus forms the *substrate terminal B*.

a. Effect of contact potentials. Now the gate terminal is shorted to the substrate terminal with a wire. Going from the gate, through the external connection, to the substrate and taking into account that the sum of the encountered *contact potentials* depends only on the first and the last material [see Eq. (B-3)], being independent of any material in between, it can be written [1]

$$\sum_{\text{gate}}^{\text{substrate}} (\text{contact potentials}) = \phi_{\text{gate material}} - \phi_{\text{substrate material}} \quad \text{(B-4)}$$

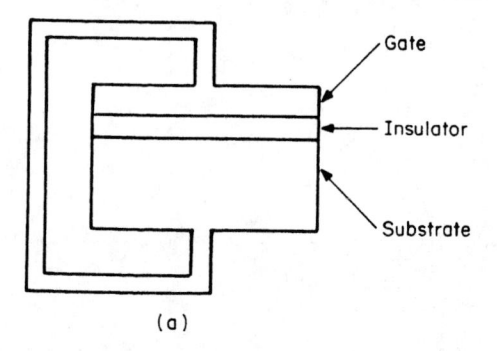

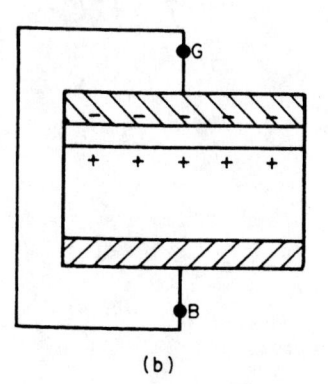

Figure B-1 (a) A MOS junction with gate, substrate, and shorting external connection; (b), (c) effects of contact potentials; and (d), (e) effects of the interface charge [1].

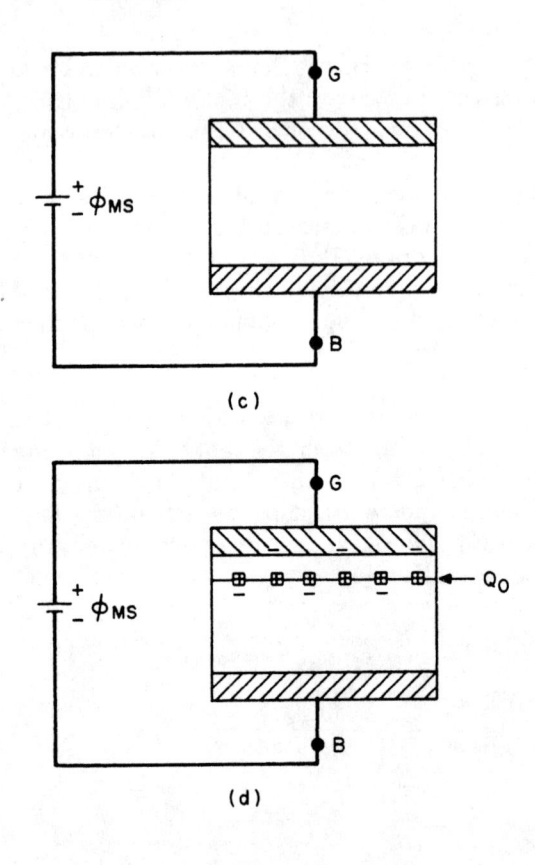

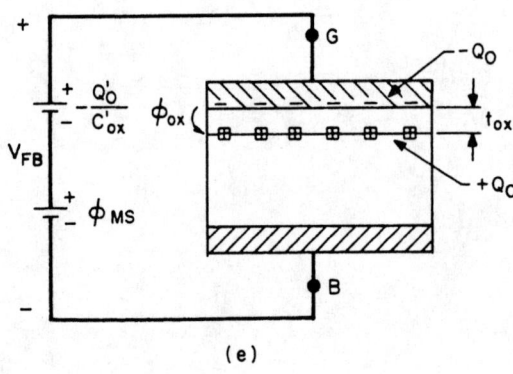

Figure B-1 (*Continued*)

The existence of a nonzero potential between the gate and the substrate causes net charges to appear on both sides of the oxide. If, for example, this potential is negative, the polarity of the charges will be as shown in Fig. B-1b. These net charges disappear if the total potential from the gate, through the external connection, to the substrate equals zero. This is possible if a voltage of an external source is applied to cancel the sum of the contact potentials (see Fig. B-1c).

From Eq. (B-4), the external voltage source must have a value given by

$$V_{GB} = \phi_{MS} \qquad (B\text{-}5)$$

where ϕ_{MS} is a widely used symbol defined as

$$\phi_{MS} = -(\phi_{\text{gate material}} - \phi_{\text{substrate material}}) \qquad (B\text{-}6)$$

b. Effect of parasitic charges. The effect of contact potentials is not the only one that can cause a net concentration of charges in the semiconductor in the absence of external bias. Another cause are the *parasitic charges* that exist within the oxide and at the oxide-semiconductor interface. These charges are of four types [1]:

1. An *oxide fixed charge* exists near the oxide-semiconductor interface due to the mechanisms of oxide formation. This charge is found to be rather independent of oxide thickness, doping type (p or n), and doping concentration.

2. An *oxide trapped charge* can exist throughout the oxide, but usually close to either of its interfaces to the substrate or the gate. This charge can be acquired through radiation, photoemission, or the injection of high-energy carriers from the substrate.

3. A *mobile ionic charge* can exist within the oxide due to contamination by alkali ions (often Na) introduced by the environment. This charge can move within the oxide under the presence of an electric field.

4. An *interface trap charge* (also called *fast surface-state charge*) exists at the oxide-semiconductor interface. It is caused by defects at that interface, which give rise to charge traps; these can exchange mobile carriers with the semiconductor, acting as donors or acceptors.

We will assume that all parasitic charges are located at the oxide-semiconductor interface and that their value, denoted by Q_0 is fixed.

These charges are shown inside little squares in Fig. B-1d. A battery of value ϕ_{MS} is used in that figure to cancel the effect of the contact potentials discussed previously, so that we can study the effect of Q_0 by itself. The charge Q_0 is called the *equivalent interface charge* and is always positive, for both p-type and n-type substrates. The equivalent interface charge Q_0 will cause a total charge $-Q_0$ to appear in the system, as required from charge neutrality. As shown in Fig. B-1d, part of that charge will appear at the gate and part in the semiconductor. If it is desired to eliminate this effect, if all the required balancing charge $-Q_0$ were provided on the gate, no charge would be induced in the semiconductor. To provide a charge $-Q_0$ on the gate, a battery can be connected in series with the external circuit, with the negative terminal toward the gate. Since at the gate and substrate ends of the oxide the charge must be $-Q_0$ and Q_0, respectively, the potential drop across the oxide, ϕ_{ox}, is exactly the voltage that must be provided by the battery, as shown in Fig. B-1e.

Rather than use the total charge Q_0, it is more convenient to use the equivalent interface charge per unit area, $Q'_0 \equiv Q_0/A_J$, where A_J is the cross-sectional area. Similarly, instead of the total capacitance C_{ox}, we use the oxide capacitance per unit area, $C'_{ox} \equiv C_{ox}/A_J$.

From basic physics, it can be written that

$$C'_{ox} = \frac{\varepsilon_{ox}}{t_{ox}} \tag{B-7}$$

where t_{ox} is the *oxide thickness* and ε_{ox} is the *permittivity of the oxide*, given by

$$\varepsilon_{ox} = \varepsilon_r \varepsilon_0 \tag{B-8}$$

in which ε_0 is the *permittivity of free space* ($\varepsilon_0 = 8.85 \times 10^{-14}$ F/cm), and ε_r is the *dielectric constant of the oxide* ($\varepsilon_r = 3.9$ for SiO_2).

The potential drop, ϕ_{ox}, across the oxide can now be written as

$$\phi_{ox} = -\frac{Q'_0}{C'_{ox}} \tag{B-9}$$

Therefore it has been seen that an external voltage is required between the gate and substrate terminals, to keep the semiconductor neutral everywhere by canceling the effects of the contact potentials and Q'_0.

This voltage is called the *flat-band voltage* and is denoted by V_{FB}. From Fig. B-1e, the expression for the flat-band voltage becomes

$$V_{FB} = \phi_{MS} - \frac{Q'_0}{C'_{ox}} \tag{B-10}$$

The flat-band voltage can be considered as the value of the gate-substrate voltage (V_{GB}) at which electron and hole concentrations are equal to their equilibrium values. Of the two terms on the right-hand side of the above equation, the first takes care of the contact potentials and the second takes care of the equivalent interface charge.

B.1.3 Potential balance and charge balance for arbitrary V_{GB}

We will now discuss how the substrate is affected when the externally applied voltage V_{GB} assumes values different from the flat-band voltage V_{FB}. The general case is shown in Fig. B-2. An arbitrary value of V_{GB} in general causes charges to appear in the semiconductor; practically all these charges are contained within a region adjacent to the surface of the semiconductor, called the *space-charge region*, shaded in Fig. B-2a.

Outside the space-charge region, the substrate is neutral, and since that region contains charges, a potential change is expected across it. The total potential drop across the space-charge region, defined from the surface to a point in the bulk outside that region, is called *surface potential* ϕ_s.

Four kinds of potential changes are encountered in the loop, as seen in Fig. B-2 [1]:

1. The voltage of the external source, V_{GB}
2. The potential drop across the oxide, ϕ_{ox}
3. The surface potential, ϕ_s

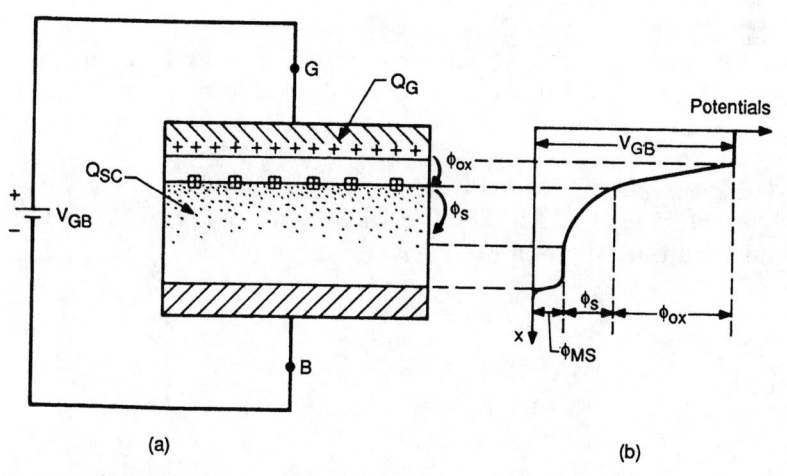

Figure B-2 (a) A MOS junction under general gate-substrate bias and (b) potential distribution [1].

4. Several contact potentials; their sum, when going clockwise, is ϕ_{MS}, which is known from Eqs. (B-4) and (B-6)

Going around the loop, it can be written that

$$V_{GB} = \phi_{ox} + \phi_s + \phi_{MS} \tag{B-11}$$

The potential balance expressed by the above equation is shown in Fig. B-2b. In Eq. (B-11), ϕ_{MS} is a known constant and any changes in V_{GB} must be balanced by changes in ϕ_{ox} and ϕ_s.

Three kinds of charge are encountered in the type of system shown in Fig. B-2:

1. The charge on the gate, Q_G
2. The equivalent interface charge, Q_0
3. The charge in the space-charge region, Q_{SC}

These charges must balance one another for overall charge neutrality in the system.

Using charges per unit area yields

$$Q'_G + Q'_0 + Q'_{SC} = 0 \tag{B-12}$$

If Q'_G is changed, the balance required by Eq. (B-12) will be achieved through a change in Q'_{SC}, since the equivalent interface charge Q'_0 is practically fixed.

B.1.4 Effect of V_{GB} and V_{FB} on surface condition

Depending on whether V_{GB} is equal to, less than, or greater than the flat-band voltage V_{FB}, three cases can be distinguished, as indicated in the following sections.

a. Flat-band condition. The flat-band condition has already been discussed in detail in Sec. B.1.2 and is illustrated in Fig. B-1e. In the flat-band condition, therefore,

$$V_{GB} = V_{FB}$$
$$Q'_{SC} = 0 \tag{B-13}$$
$$\phi_s = 0$$

b. Accumulation. The condition in which $V_{GB} < V_{FB}$ is considered now. The negative change of V_{GB} (relative to the flat band) will cause a negative change in Q'_G which, according to Eq. (B-12), must be balanced by a positive change in Q'_{SC} above the value given by Eq. (B-13). Thus, holes will accumulate at the surface (oxide-semiconductor interface). This condition is called *accumulation* and is illustrated in Fig. B-3a. The negative change in V_{GB} will be shared by negative changes in ϕ_{ox} and ϕ_s, so that Eq. (B-11) will remain valid.

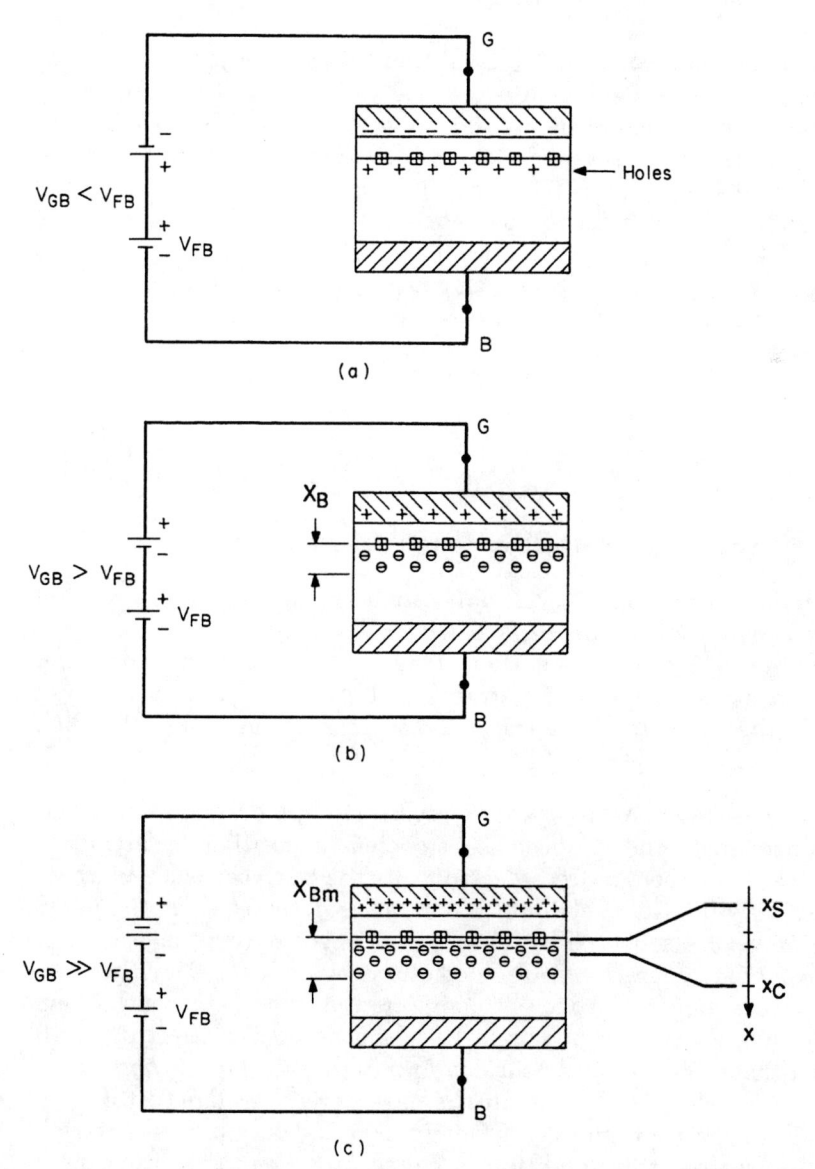

Figure B-3 A MOS junction in (a) accumulation, (b) depletion, and (c) inversion [1].

In *accumulation*, therefore,

$$V_{GB} < V_{FB}$$
$$Q'_{SC} > 0 \quad \text{(B-14)}$$
$$\phi_s < 0$$

c. Depletion and inversion. The condition in which $V_{GB} > V_{FB}$ is considered now. The total charge on the gate Q'_G will become more positive than the value in the flat band (that value is $-Q_0$ in Fig. B-1e). An example is shown in Fig. B-3b, where it is assumed that the resulting Q'_G is positive. The positive change in Q'_G (relative to the flat band) must be balanced by a negative change in Q'_{SC} from the value of Eq. (B-13), so that Eq. (B-12) is still valid. Also, the positive change in V_{GB} will be shared among ϕ_{ox} and ϕ_s, while Eq. (B-11) remains valid. In *depletion* and *inversion*, therefore,

$$V_{GB} > V_{FB}$$
$$Q'_{SC} < 0 \quad \text{(B-15)}$$
$$\phi_s > 0$$

Now the nature of the negative charge Q'_{SC} is considered.

Depletion condition. If V_{GB} is only slightly higher than V_{FB}, the positive potential at the surface with respect to the substrate will drive holes away from the surface, leaving it depleted; this condition is called *depletion* and is illustrated in Fig. B-3b. The space charge Q'_{SC} is then due to the *uncovered* acceptor atoms, each of which contributes a charge $-q$.

Inversion condition. As V_{GB} is increased further, more acceptor atoms are uncovered, and ϕ_s becomes sufficiently positive to attract a significant number of free electrons to the surface; each of these electrons will also contribute a charge $-q$ to Q'_{SC}. In the MOS junction we are considering, these electrons come from the process of electron-hole generation in the depletion region, caused by the thermal vibration of the lattice (assuming no radiation is present). Eventually, with a sufficiently high V_{GB}, the density of electrons will exceed that of holes at the surface forming a thin layer (*channel*) of minority carriers (electrons, in the case here considered); this is a situation opposite from that normally expected in a *p*-type material. For this reason, this condition is called *inversion*. The situation is illustrated in Fig. B-3c.

B.1.5 General analysis of the surface space-charge region

In this section, we will derive the relations between the surface potential ϕ_s and the space charge Q'_{SC} for any value of applied voltage V_{GB}, be it in accumulation, depletion, or inversion. These relations will then be used to derive the capacitance-voltage characteristics of the ideal MOS junction.

Let us refer to Fig. B-2. The charge density at depth x in the semiconductor, is given by Eq. (A-1), repeated here for convenience,

$$\rho(x) = q[p(x) - n(x) + N_D - N_A] \tag{B-16}$$

From the Boltzmann equations [Eqs. (A-10)], the electron and hole concentrations as a function of ϕ and at x are given by

$$n_p(x) = n_{p0}\, e^{q\phi(x)/kT}$$
$$p_p(x) = p_{p0}\, e^{-q\phi(x)/kT} \tag{B-17}$$

where $\phi(x)$ is the potential with respect to the bulk of the semiconductor at x; n_{p0} and p_{p0} are the equilibrium densities of electrons and holes, respectively, in the bulk of the semiconductor (p-type, in the case here considered). At the surface, Eqs. (B-17) reduce to

$$n_s = n_{p0}\, e^{q\phi_s/kT} = n_i\, e^{q(\phi_s - \phi_p)/kT} = N_A\, e^{q(\phi_s - 2\phi_p)/kT}$$
$$p_s = p_{p0}\, e^{-q\phi_s/kT} = n_i\, e^{-q(\phi_s - \phi_p)/kT} \tag{B-18}$$

where ϕ_s is the surface potential (i.e., the vlaue of $\phi(x)$ at the semiconductor surface). The second form of Eqs. (B-18) are written by taking into account Eqs. (A-23) and (A-29b), and the convention on potential used in this chapter.

In the bulk of the semiconductor, far from the surface, charge neutrality must exist; therefore $\rho(x) = 0$ and $\phi(x) = 0$. From these conditions, Eq. (B-16) gives

$$n_{p0} - p_{p0} = N_D - N_A \tag{B-19}$$

Using the above equation and Eqs. (B-17) in Eq. (B-16), and the result in Poisson's equation [Eq. (A-11)], we have

$$\frac{d^2\phi}{dx^2} = -\frac{q}{\varepsilon_s}[p_{p0}(e^{-q\phi(x)/kT} - 1) - n_{p0}(e^{q\phi(x)/kT} - 1)] \tag{B-20}$$

Multiplying both sides of Eq. (B-20) by $2(d\phi/dx)$, the resulting left-hand side can be recognized as $(d/dx)(d\phi/dx)^2$. Replacing x by a dummy variable, integrating from a point deep in the bulk (theoretically at infinity, where $\phi = 0$ and $d\phi/dx = 0$) to a point x, solving for $d\phi/dx$ at a point x, and recalling that $E(x) = -d\phi/dx$, we obtain [3]

$$E(x) = -\frac{d\phi}{dx} = \pm\frac{2kT}{qL_D} F\left(\frac{q\phi}{kT}, \frac{n_{p0}}{p_{p0}}\right) \tag{B-21}$$

where

$$L_D \equiv \sqrt{\frac{2kT\varepsilon_s}{q^2 p_{p0}}} \tag{B-22}$$

is the extrinsic Debye length for holes, and

$$F\left(\frac{q\phi}{kT}, \frac{n_{p0}}{p_{p0}}\right) \equiv \left[\left(e^{-q\phi/kT} - 1 + \frac{q\phi}{kT}\right) + \frac{n_{p0}}{p_{p0}}\left(e^{q\phi/kT} - 1 - \frac{q\phi}{kT}\right)\right]^{1/2} \geq 0 \tag{B-23}$$

Equation (B-21) is considered with positive sign for $\phi > 0$ and negative sign for $\phi < 0$.

The electric field at the surface is determined from Eq. (B-21), by letting $\phi = \phi_s$. Thus

$$E_s = \pm\frac{2kT}{qL_D} F\left(\frac{q\phi_s}{kT}, \frac{n_{p0}}{p_{p0}}\right) \tag{B-24}$$

By Gauss's law, the space charge for unit area required to produce this field is

$$Q'_{SC} = -\varepsilon_s E_s = \mp\frac{2\varepsilon_s kT}{qL_D} F\left(\frac{q\phi_s}{kT}, \frac{n_{p0}}{p_{p0}}\right) \tag{B-25}$$

where the negative sign in front of F must be used with $\phi_s > 0$ (depletion or inversion) and the positive sign with $\phi_s < 0$ (accumulation). This function behaves as shown in Fig. B-4.

Using the above equation, the differential capacitance per unit area of the semiconductor space-charge region is given by

$$C'_{SC} \equiv \frac{dQ'_{SC}}{d\phi_s} = \pm\frac{\varepsilon_s}{L_D} \frac{\left[1 - e^{-q\phi_s/kT} + \frac{n_{p0}}{p_{p0}}(e^{q\phi_s/kT} - 1)\right]}{F\left(\frac{q\phi_s}{kT}, \frac{n_{p0}}{p_{p0}}\right)} \tag{B-26}$$

Equations (B-25) and (B-26) are valid in all conditions (accumulation, depletion, and inversion).

To fully characterize the behavior of the MOS junction, we must consider the charge per unit area above the oxide, Q'_G, that can be related to the potential across the oxide, ϕ_{ox}, and the oxide-capacitance per unit area C'_{ox}, by

$$Q'_G = C'_{ox}\phi_{ox} \tag{B-27}$$

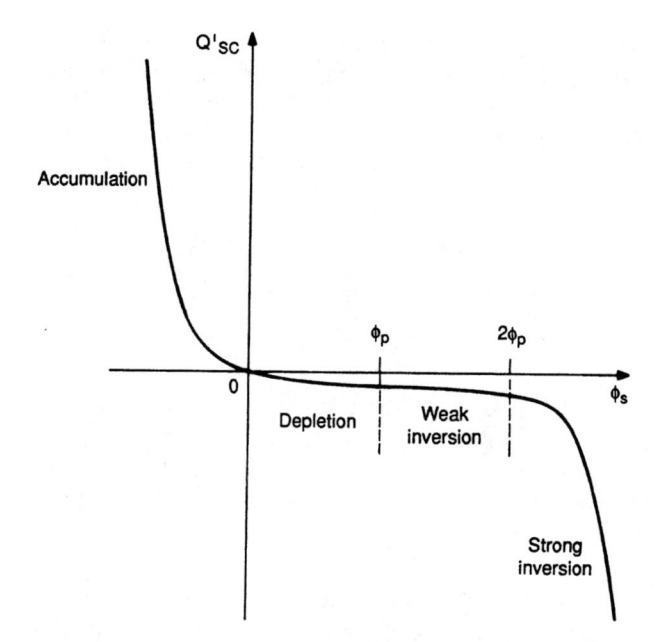

Figure B-4 Variation of the space-charge density Q'_{SC} in the semiconductor of a MOS junction as a function of the surface potential ϕ_s for a p-type silicon.

Thus, four equations characterize the MOS junction. These equations are (B-11), (B-12), (B-25), and (B-27) and contain four unknown variables: ϕ_{ox}, ϕ_s, Q'_{SC}, and Q'_G. This system of equations can be solved numerically to give the values of the above variables for a given MOS junction (i.e., given ϕ_{MS}, $p_{p0} \simeq N_A$ and n_{p0}) and given externally applied voltage V_{GB}. When ϕ_s is obtained in this way, the variables $\phi(x)$, $n_p(x)$, $p_p(x)$, and $\rho(x)$ can be determined from Eqs. (B-21), (B-17), and (B-16), respectively.

The above analysis is complicated, but certain approximations become possible, which simplify the analysis of the MOS junction, if each condition of operation is considered separately. A detailed analysis of each charge contribution at each condition of operation can be found in Ref. [1]. Here, only the considerations useful to the aim of this book are taken into account; in particular we will investigate in the next section the inversion condition which is responsible for current conduction in MOS transistors.

However, before going to the next section, some considerations about the function F [Eq. (B-23)] and Fig. B-4 must be made. From Fig. B-4 we note that for $\phi_s < 0$, Q'_{SC} is positive and corresponds to the accumulation region; the function F is dominated by the first term in Eq. (B-23), i.e., $Q'_{SC} \simeq e^{q|\phi_s|/2kT}$. For $\phi_s = 0$, we have the flat-band condition and $Q'_{SC} = 0$. For $0 < \phi_s < \phi_p$, Q'_{SC} is negative and corresponds to the depletion region; the function F is dominated by the second term, i.e., $Q'_{SC} \simeq \sqrt{\phi_s}$. For $\phi_s \gg \phi_p$, Q'_{SC} is negative and corresponds to the inversion region; the function F is dominated by the fourth term, i.e., $Q'_{SC} \simeq -e^{q\phi_s/2kT}$. It must also be noted that the strong inversion begins at a surface potential

$$\phi_s(\text{inv}) \simeq 2\phi_p \simeq \frac{2kT}{q} \ln\left(\frac{N_A}{n_i}\right) \simeq \frac{kT}{q} \ln\left(\frac{p_{p0}}{n_{p0}}\right) \qquad (\text{B-28})$$

B.1.6 Inversion

The inversion condition is investigated in this section. Taking into account Eqs. (B-15), the discussion focuses on the relative contributions of uncovered acceptor atoms and of free electrons to Q'_{SC}, which are functions of the surface potential ϕ_s. Expressions for this functional dependence will then be derived [1].

a. Charge due to the free electrons Q'_i. The surface electron concentration against the surface potential is shown in Fig. B-5, using Eq. (B-18). At $\phi_s = \phi_p$, n_s becomes equal to the intrinsic concentration;

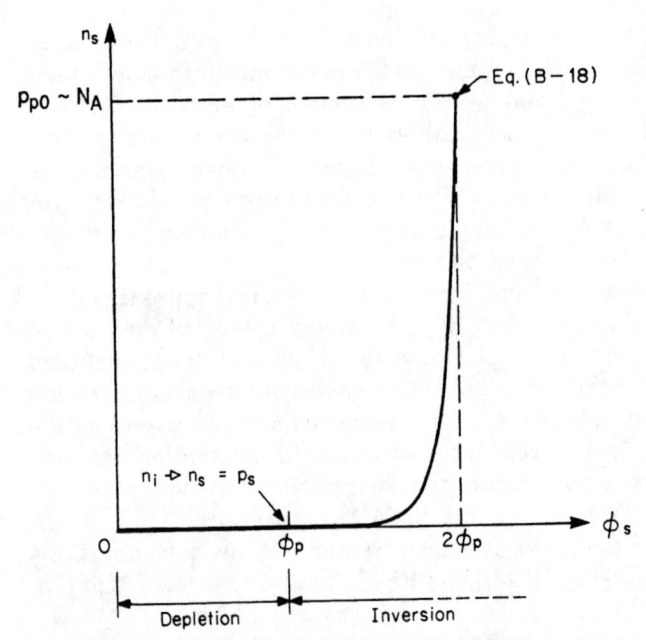

Figure B-5 Electron concentration at the surface vs. suface potential [1].

then $n_s = p_s$. This is defined as the limit point between the depletion and inversion regimes as indicated in the figure.

Equation (B-18) gives the electron concentration at the surface. At any other point x, the electron concentration $n(x)$ will be given by equations like (B-18), when ϕ_s is replaced by $\phi(x)$, the potential at point x with respect to the substrate. Going away from the surface, $\phi(x)$ decreases toward zero and $n(x)$ vanishes rapidly due to its exponential dependence on $\phi(x)$. Hence a point $x = x_c$, below which the electron concentration will be negligible, can be chosen. Practically all the free electrons are then contained in a narrow layer (the inversion layer, or *channel*) between $x = x_s$ and $x = x_c$, x_s being a point at the semiconductor surface (see Fig. B-3c).

Therefore, the inversion-layer charge per unit area, Q'_I, due to the free electrons, is given by

$$Q'_I = -q \int_{x_c}^{x_s} n(x)\, dx \qquad \text{(B-29)}$$

b. Charge due to the uncovered acceptor atoms Q'_B. Evaluating Q'_I in this manner can be done only through numerical integration. Another way is to determine an accurate expression for Q'_B (the charge of the ionized acceptor atoms in the depletion region) and then evaluate Q'_I.

The depletion region is then considered. As in the case of the *pn* junction, we will regard this region as being defined by a sharp boundary at a depth X_B below the surface. The inversion layer is at the top of this region; numerical calculations have shown that most of this layer is concentrated very close to the surface. Since the depth X_B of the depletion region is usually much larger, it is assumed that the inversion layer is a sheet of negligible thickness [1]. This has been called the *charge sheet approximation* and implies that practically all the depletion region is free of minority carriers.

For a negligible thickness, the potential drop across the inversion layer will also be negligible, and it can be assumed that all the surface potential ϕ_s is dropped across the depletion region in the *p*-type substrate. Furthermore, we will assume that the mobile carrier concentrations are negligible in comparison to the acceptor concentration inside the depletion region; this has been called the *depletion approximation* in Sec. A.2. Thus from Eqs. (A-38), with the assumption that $N_D \gg N_A$, it follows that

$$X_B \simeq \sqrt{\frac{2\varepsilon_s}{qN_A} \phi_s} \qquad \text{(B-30)}$$

Let Q_B be the charge due to the uncovered acceptor atoms in the depletion region, and $Q'_B \equiv Q_B/A_J$ the corresponding charge per unit area. As the total charge per unit area due to these uncovered acceptor atoms can be expressed as $-qN_A X_B$, then

$$Q'_B = -\sqrt{2q\varepsilon_s N_A \phi_s} \qquad (B\text{-}31)$$

The total charge per unit area in the space-charge region, Q'_{SC}, is then the sum of the charge of electrons in the inversion layer, Q'_I, and the charge of the ionized acceptor atoms in the depletion region, Q'_B. Thus

$$Q'_{SC} = Q'_I + Q'_B \qquad (B\text{-}32)$$

The evaluation of Q'_{SC} can be made directly (i.e., without regarding Q'_I and Q'_B individually) by using Eqs. (A-11), (B-16), (B-18), and (B-28) with the above assumptions for the inversion condition; the result is

$$Q'_{SC} = -\sqrt{2q\varepsilon_s N_A} \sqrt{\phi_s + \frac{kT}{q} e^{q(\phi_s - 2\phi_p)/kT}} \qquad (B\text{-}33)$$

It should be noted that the above expression is derived by assuming that no holes are present in the space-charge region; thus, it will be valid in inversion as well as in the upper part of the depletion region. However, in accumulation or in the bottom part of the depletion region, holes are present in significant numbers and the above expression will not hold; a more general expression is then needed (see Sec. B.1.5).

The inversion region charge Q'_I can then be evaluated by taking into account Eqs. (B-31) to (B-33). Thus

$$Q'_I = -\sqrt{2q\varepsilon_s N_A} \left(\sqrt{\phi_s + \frac{kT}{q} e^{q(\phi_s - 2\phi_p)/kT}} - \sqrt{\phi_s} \right) \qquad (B\text{-}34)$$

$|Q'_B|$ and $|Q'_I|$ vs. ϕ_s are plotted in Fig. B-6, using Eqs. (B-31) and (B-34); their sum, $|Q'_{SC}|$, is also shown.

It is convenient to divide the inversion region into three subregions: these are marked *weak*, *moderate*, and *strong inversion* in Fig. B-6. ϕ_{MO} and ϕ_{HO} indicate the onset of moderate and strong inversion, respectively. It is thus seen that for surface potentials less than about $2\phi_p$, practically all the surface charge is due to the charge in the depletion region. The corresponding inversion-layer charge is too small to be shown in the scale of the figure, but can nevertheless

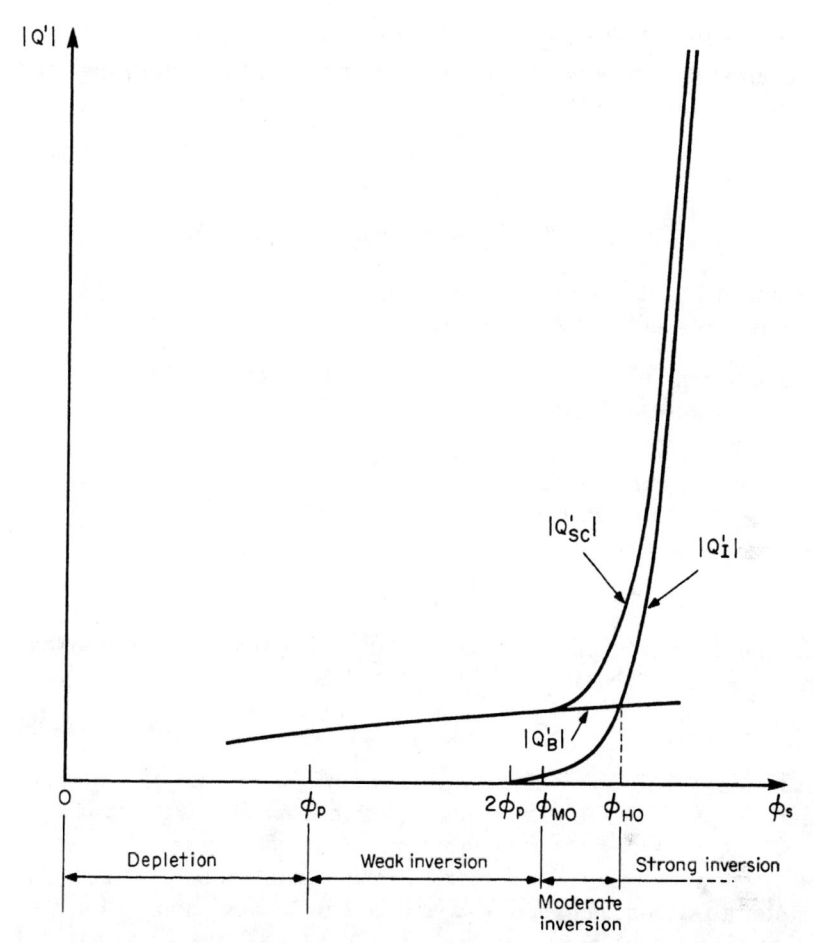

Figure B-6 Magnitude of inversion-layer charge, depletion-region charge, and their sum (all per unit area) vs. surface potential [1].

cause nonnegligible conduction when the MOS junction is part of a transistor. As ϕ_s is raised above $2\phi_p$, $|Q'_I|$ starts becoming significant due to the exponential in Eq. (B-34). For ϕ_s exceeding $2\phi_p$ by a few kT/q, Q'_I becomes a very strong function of ϕ_s. From Fig. B-5 it can be seen that at $\phi_s = 2\phi_p$ the surface electron concentration is already as high as the concentration of acceptor atoms; an increase of ϕ_s above $2\phi_p$ by several kT/q would be enough to provide very large n_s due to the exponential dependence of the latter on ϕ_s [see Eq. (B-18)]. The concentration $n(x)$ at points very close to the surface would also increase drastically, and Q'_I would rise to the point that it would dominate Q'_B.

B.1.7 Relations of charges to their potentials

In the previous sections, it has been found that the equations that characterize MOS systems are

$$V_{GB} = \phi_{ox} + \phi_s + \phi_{MS} \quad \text{(potential balance)} \quad (B-35)$$

and

$$Q'_G + Q'_0 + Q'_I + Q'_B = 0 \quad \text{(charge balance)} \quad (B-36)$$

Here, three equations that relate each charge to the potential that causes its presence are analyzed:

1. The charge on the top of the *oxide capacitor* is related to the potential across the oxide by

$$Q'_G = C'_{ox} \phi_{ox} \quad (B-37)$$

2. The *inversion-layer charge* is related to its cause, the surface potential, by Eq. (B-34), which is of the form

$$Q'_I = Q'_I(\phi_s) \quad (B-38)$$

3. The *depletion-region charge* is related to the potential across that region by Eq. (B-31), which is of the form

$$Q'_B = Q'_B(\phi_s) \quad (B-39)$$

Equations (B-35) to (B-39) contain six variables, three of which are potentials ($V_{GB}, \phi_{ox}, \phi_s$) and three of which are charges per unit area (Q'_G, Q'_I, Q'_B). Among the five equations, four of the variables can be eliminated and a relation between the other two can be obtained. For example, a relation between V_{GB} and ϕ_s can be developed. Eliminating the other variables among Eqs. (B-35) to (B-39), it follows that [1]

$$V_{GB} = V_{FB} + \phi_s + \gamma \sqrt{\phi_s + \frac{kT}{q} e^{q(\phi_s - 2\phi_p)/kT}} \quad (B-40)$$

where

$$\gamma = \frac{\sqrt{2q\varepsilon_s N_A}}{C'_{ox}} \quad (B-41)$$

The parameter γ is called the *body-effect coefficient*, or *substrate-effect coefficient*. Equation (B-40) is plotted in Fig. B-7 as ϕ_s vs. V_{GB}.

The goal in this development is to find a relation of the form

$$Q'_I = Q'_I(V_{GB}) \quad (B-42)$$

The reason is that the inversion-layer charge is the cause of the conduction when the MOS junction is part of a MOS transistor and

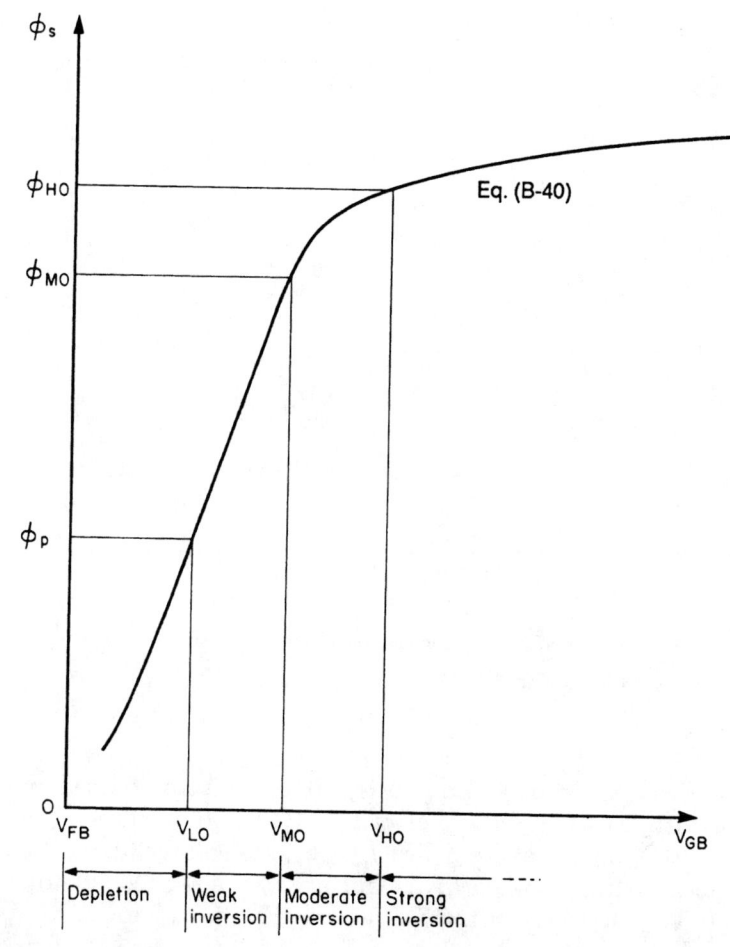

Figure B-7 Surface potential versus gate-substrate bias [1].

the gate-to-substrate voltage is an external bias voltage at the disposal of the circuit designer. Equation (B-42) can be used to derive complete expressions that relate currents to voltages in MOS transistors. Unfortunately, if one attempts to derive a relation in the form of Eq. (B-42) from Eqs. (B-35) to (B-39), an implicit expression results; i.e., Q'_I cannot be expressed in closed form as a function of V_{GB}. The solution of the resulting complicated equation has to be obtained numerically. However, if Q'_G and ϕ_{ox} are eliminated among Eqs. (B-35) to (B-37), and if Eqs. (B-31) and (B-10) are used, then [1]

$$Q'_I = -C'_{ox}(V_{GB} - V_{FB} - \phi_s - \gamma\sqrt{\phi_s}) \qquad (B-43)$$

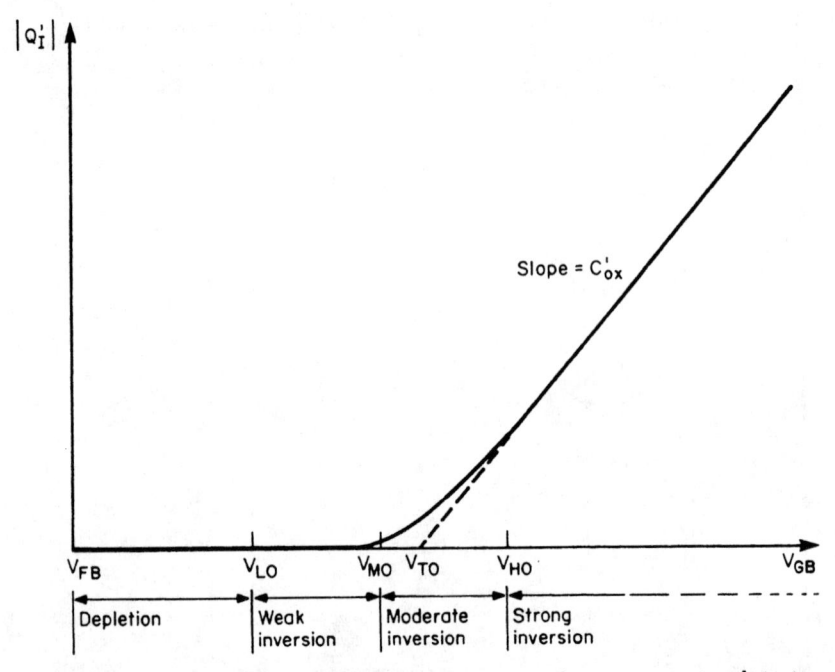

Figure B-8 Magnitude of inversion-layer charge per unit area vs. gate-substrate bias [1].

where γ has been defined in Eq. (B-41). If Eq. (B-43) is used in conjunction with Eq. (B-40), the plot of Fig. B-8 is obtained.

From Eq. (B-43), we can express directly the gate voltage necessary to induce a conducting channel at the surface of the semiconductor. This voltage, known as the *threshold voltage* V_{TO}, is defined as the gate voltage that results in $Q'_I = 0$. From Eq. (B-43), with the assumption of Eq. (B-28), an expression for V_{TO} can be written as

$$V_{TO} = V_{FB} + 2\phi_p + \gamma\sqrt{2\phi_p} \qquad (B\text{-}44)$$

The terms in Eq. (B-44) have the proper qualitative behavior. First, V_{TO} contains V_{FB} because a gate voltage equal to V_{FB} is necessary to bring the semiconductor to a charge-neutral condition. Second, $2\phi_p$ V must be applied to cause the MOS junction to be in the strong inversion condition. Third, the square-root term accounts for the uniform distribution of space charge in the depletion region.

B.1.8 Capacitance-voltage characteristics

If V_{GB} is increased by a small amount dV_{GB} in Fig. B-2a, a positive charge dQ'_G will flow into the gate terminal. For charge neutrality, a charge of equal value must flow out of the substrate terminal, or, equivalently, a charge of value $-dQ'_G$ must flow into the substrate terminal. An incremental (*small-signal*) capacitance per unit area, C'_{GB}, can thus be defined to relate charge changes to voltage changes. Thus

$$C'_{GB} = \frac{dQ'_G}{dV_{GB}} \qquad (B\text{-}45)$$

The charge $-dQ'_G$ flowing into the substrate changes the space charge Q'_{SC} by an amount dQ'_{SC}. Furthermore, the gate-substrate voltage change (dV_{GB}) will be distributed partly across the oxide (as $d\phi_{ox}$) and partly across the semiconductor (as a change $d\phi_s$ in the surface potential). Inverting Eq. (B-45) and taking into account what was stated above, it follows that

$$\frac{1}{C'_{GB}} = \frac{d\phi_{ox}}{dQ'_G} + \frac{d\phi_s}{dQ'_G} \qquad (B\text{-}46)$$

The preceding equation can be written as

$$\frac{1}{C'_{GB}} = \frac{1}{dQ'_G/d\phi_{ox}} + \frac{1}{-dQ'_{SC}/d\phi_s} \qquad (B\text{-}47)$$

To interpret this equation, it must be noted that the denominator of the last fraction in Eq. (B-47) can be interpreted as a capacitance

$$C'_{SC} = -\frac{dQ'_{SC}}{d\phi_s} \qquad (B\text{-}48)$$

corresponding to the semiconductor space-charge region (it relates the changes of the potential across that region to the resulting changes in its charge), while the denominator of the first fraction can be interpreted as a capacitance, C'_{ox}, following Eq. (B-37). Thus Eq. (B-47) becomes

$$\frac{1}{C'_{GB}} = \frac{1}{C'_{ox}} + \frac{1}{C'_{SC}} \qquad (B\text{-}49)$$

Therefore the small-signal capacitance C'_{GB} is the same as that exhibited by two capacitors of values C'_{ox} and C'_{SC}, connected in series as in Fig. B-9a. C'_{SC} can be evaluated from Eqs. (B-48) and (B-26).

It is interesting to consider the individual contributions of the depletion-region and inversion-layer charges to C'_{SC}. Thus, from Eqs.

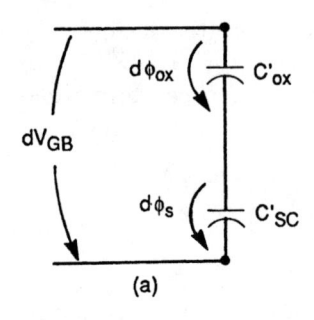

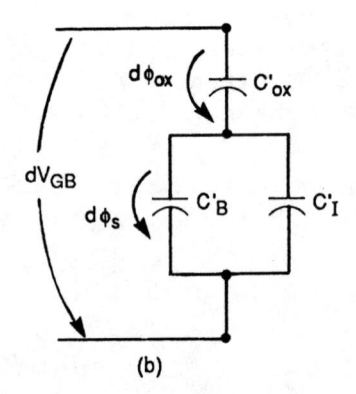

Figure B-9 Small-signal capacitances for a MOS junction [1].

(B-32) and (B-48), it can be written that

$$C'_{SC} = -\frac{dQ'_B}{d\phi_s} + \left(-\frac{dQ'_I}{d\phi_s}\right) \tag{B-50}$$

The total space-charge capacitance C'_{SC} is then divided into two components: one due to the depletion-region charge and one due to the inversion-layer charge.

Analogously to Eq. (B-48), a depletion-region small-signal capacitance per unit area can be defined as

$$C'_B = -\frac{dQ'_B}{d\phi_s} \tag{B-51}$$

This capacitance relates changes of the potential across the depletion region to the resulting changes of the charge in it.

Finally, a capacitance per unit area associated with the inversion layer can be defined. This capacitance should relate changes in the charge of that layer to the potential changes that cause them. In analogy with Eq. (B-51) it can be written that

$$C'_I = -\frac{dQ'_I}{d\phi_s} \tag{B-52}$$

Using Eqs. (B-51) and (B-52) in Eq. (B-50), it follows that

$$C'_{SC} = C'_B + C'_I \tag{B-53}$$

and Eq. (B-49) becomes

$$\frac{1}{C'_{GB}} = \frac{1}{C'_{ox}} + \frac{1}{C'_B + C'_I} \tag{B-54}$$

which can be represented by the circuit of Fig. B-9b. The plot of C'_{GB} components vs. ϕ_s is shown in Fig. B-10.

Equation (B-54) has been obtained by assuming that the equivalent interface charge Q'_0 is fixed and independent of voltage. This may not be accurate in some devices (especially those fabricated with the use of older techniques) in which a significant density of interface traps may exist [1]. These traps, located at the oxide-silicon interface, can exchange carriers with the substrate. The charge trapped in them depends on the value of the surface potential ϕ_s. If Q'_{it} represents the fraction of Q'_0 associated with interface traps, we can define, in analogy with Eqs. (B-51) and (B-52), the interface trap capacitance [1]

$$C'_{it} = -\frac{dQ'_{it}}{d\phi_s} \tag{B-55}$$

C'_{it} will appear in parallel with C'_B and C'_I in Fig. B-9b.

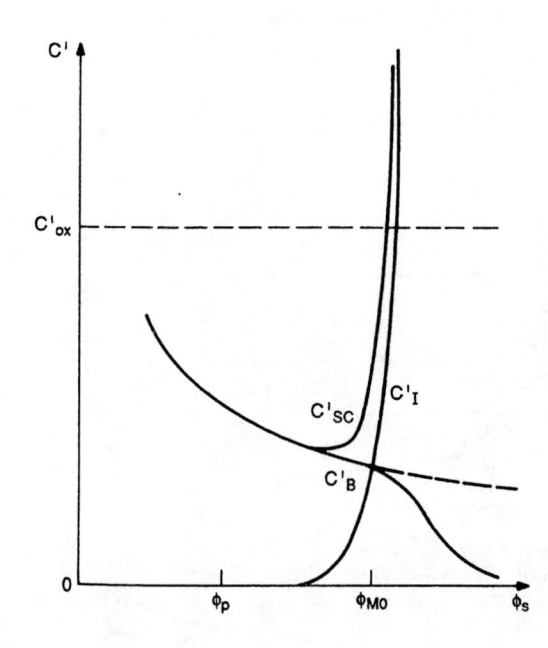

Figure B-10 Small-signal capacitances per unit area vs. surface potential: C'_I is the inversion-layer capacitance; C'_B is the depletion-region capacitance (solid line is exact; broken line is as predicted by the charge sheet model); C'_{SC} is the semiconductor charge region capacitance; C'_{ox} is the oxide capacitance [1].

Measurements performed on MOS transistors have shown that, with modern fabrication processes, C'_{it} is often much smaller than C'_B and may be neglected.

The above considerations lead us to obtain an important result: the *C-V plot* [1–5], i.e., the plot of the total capacitance seen externally (C'_{GB}) versus the total external applied bias (V_{GB}). This can be done as follows [1]. For a given ϕ_s, C'_{SC} is determined; then C'_{GB} is found from Eq. (B-54). The result is shown by the solid curve in Fig. B-11. We see that deep in accumulation, C'_{GB} approaches C'_{ox}, as we intuitively predicted. For V_{GB} in the weak-inversion region (except for points close to the upper limit of the region), the inversion-layer capacitance is negligible (see Fig. B-10). From Eq. (B-54), then, C'_{GB} is given by the series combination of C'_{ox} and C'_B. As V_{GB} is increased, C'_B becomes smaller (see Fig. B-10), and therefore, the series combination of C'_{ox} and C'_B also decreases, as shown in Fig. B-11. Above weak inversion, C'_I becomes significant and drastically increases if V_{GB} is raised further. This capacitance is in parallel with C'_B. The last fraction in Eq. (B-54) decreases drastically, and C'_{GB} approaches C'_{ox}.

The above discussion has been in terms of *static* changes, i.e., it is assumed that after V_{GB} is changed by a small amount dV_{GB}, it

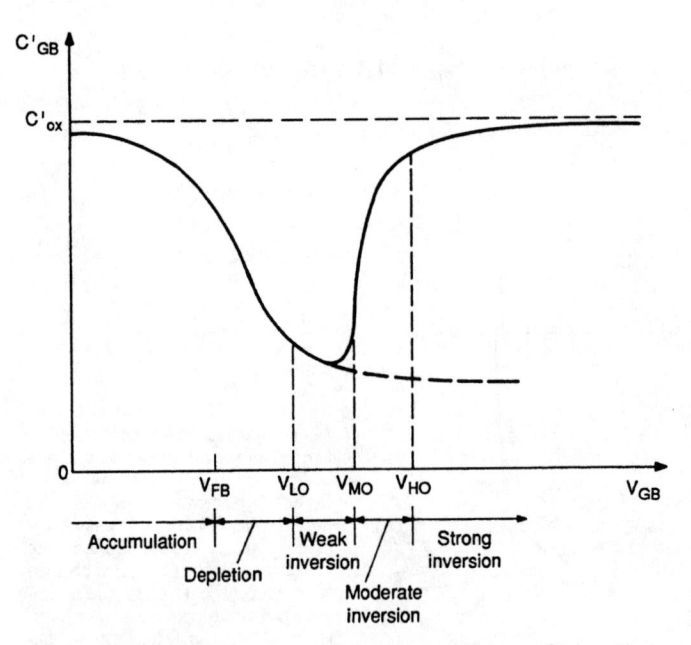

Figure B-11 Gate-substrate capacitance per unit area C'_{GB} vs. gate-substrate voltage V_{GB}. Solid line: *static* behavior; broken line: *high-frequency* behavior [1].

remains fixed at its new value for long enough before a new equilibrium is reached. If dV_{GB} is a sinusoidal voltage, the steady-state charge changes will also be sinusoidal. They will correspond to equilibrium values if the frequency is low enough (e.g., 1 Hz). However, if the frequency is high (e.g., 10 kHz), things will be different, as shown by the broken line in Fig. B-11. The inversion-layer charge now cannot keep up with the fast-changing dV_{GB}, and the required charge changes must be provided by covering or uncovering acceptor atoms at the bottom of the depletion region, just as in the case of depletion operation. The reason the inversion-layer charge cannot follow fast enough is that it is in a sense isolated from the outside world by the oxide on top and the depletion region below. Therefore, the electron concentration there can only be changed by the mechanisms of thermal generation and recombination, which in this case are very slow (no external irradiation is assumed) [1]. If, instead, communication with the outside world were possible, in the sense that inversion-layer charge could be provided or removed externally, then the behavior exhibited by the solid curve in strong inversion would persist for much higher frequencies. This communication with the outside world will be provided by the source and drain regions in a MOS transistor.

Finally, a general remark must be made: all the results have been obtained by considering a p-type substrate. If the substrate is made of n-type semiconductor, the inversion layer (channel) now consists of holes, which will be attracted to the surface if V_{GB} is sufficiently negative. The immobile charge in the depletion region will consist of positive charge ionized donor atoms. With V_{GB} sufficiently positive, electrons will pile up at the surface of the semiconductor and we will have accumulation. Since the case of n-type substrate is complementary to that of p-type substrate, its analysis will follow the same approach discussed in this chapter.

References

1. Y. P. Tsividis, *Operation and Modeling of the MOS Transistor*, McGraw-Hill, New York, 1987.
2. E. H. Nicollian and J. R. Brews, *MOS Physics and Technology*, Wiley, New York, 1982.
3. S. M. Sze, *Physics of Semiconductor Devices*, Wiley, New York, 1969.
4. A. S. Grove, *Physics and Technology of Semiconductor Devices*, Wiley, New York, 1967.
5. DeWitt, G. Ong, *Modern MOS Technology: Processes, Devices, and Design*, McGraw-Hill, New York, 1984.

Appendix C

MS Junction

C.1 MS Junction

A structure made by forming a contact between a metal and a semiconductor is defined as a *Metal-Semiconductor (MS) junction* (see Fig. C-1a). A detailed understanding of the MS junction involves energy-band theory which is beyond the aim of the book. A comprehension of the MS junction behavior, however, can be developed through considerations of the electron energy and electrostatic potential distribution as shown in Fig. C-1b. This figure can be compared with the corresponding potential distribution for a *pn*-junction as shown in Fig. A-5e.

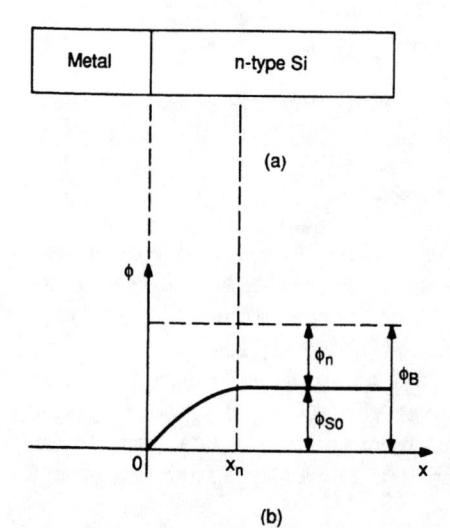

Figure C-1 (a) MS junction structure, (b) electrostatic potential.

Each MS junction is characterized by a potential barrier ϕ_B (*Schottky barrier*) which depends only on the two materials and temperature and it is not a function of the semiconductor doping. The Schottky barrier blocks the flow of electrons from the metal toward the semiconductor.

C.1.1 MS junction as a diode

We consider the structure of Fig. C-1a, and we suppose the n-type semiconductor is lightly doped ($N_D \leq 10^{16}$ cm^{-3}). In the neutral semiconductor, far from the junction, and under the condition of no externally applied bias, the electron potential is given by Eq. (A-29a), while the potential difference

$$\phi_B - \phi_n = \phi_{S0} \tag{C-1}$$

represents the metal-semiconductor electron potential difference in equilibrium. In Eq. (C-1), ϕ_{S0} is analogous to ϕ_0 for the *pn*-junction. A reverse bias V_D applied to the MS structure adds to ϕ_{S0} and widens the depletion-region width x_n, which extends entirely into the semiconductor. On the other hand, a forward bias V_D applied to the MS structure subtracts from ϕ_{S0} and narrows the depletion-region width x_n, allowing the electrons to flow from the n-type semiconductor into the metal. These mechanisms are identical to those encountered in the analysis of the *pn* junction.

Various theories of current transport in the MS junction have been proposed [1], but are beyond the scope of this text. Suffice it to say that the current I_D flowing through the MS junction is given by

$$I_D = I_S(e^{qV_D/kT} - 1) \tag{C-2}$$

where V_D is the externally bias applied voltage, T is the absolute temperature, and I_S is given by

$$I_S = AT^2 e^{-q\phi_B/kT} \tag{C-3}$$

In Eq. (C-3), A is a constant which depends on the choice of theory [1], and ϕ_B is the barrier voltage between the metal and the semiconductor. Some typical approximate values for the barrier voltage ϕ_B of a metal-n silicon junction are given in Table C-1 [2–4].

Equation (C-2) is similar to Shockley's equations [Eqs. (A-54) and (A-59)] for the *pn* junction. However, Eq. (C-3), representing the saturation current, is quite different from the analogous equation of the *pn*-junction (see Sec. A.2.3), and it is markedly a function of the barrier voltage ϕ_B.

TABLE C-1   Metal-n Silicon Barrier Voltage ϕ_B

Metal	Barrier voltage ϕ_B (V)
Zr	0.55
Mo	0.58
Ta*	0.59
Ti*	0.60
Ni*	0.66
W	0.66
Al	0.69
Pd*	0.74
Au	0.79
Pt*	0.85

*Reacts with silicon to form a thin layer of silicide.

The general conclusion of most theories [1] is that the current transport through the MS junction is due to the emission of majority carriers. As a consequence, carrier storage effects in the structure, due to storage of minority carriers, are negligible.

At sufficiently high reverse voltages, breakdown occurs, similar to that observed in the *pn* junction (see Chap. 1).

C.1.2   MS junction as an ohmic contact

Not all MS junctions have diode characteristics. When the semiconductor is heavily doped ($N_D \gg 10^{17}$ cm^{-3}), the depletion region becomes so narrow that carriers can travel in either direction through the potential barrier by a quantum-mechanical carrier transport mechanism known as *tunneling*. Under these conditions, current flows in either direction, resulting in what is usually described as an *ohmic contact* [4].

Although the preceding discussion is based on the *n*-type semiconductor, similar considerations apply for *p*-type material, therefore, suffice it to consider the appropriate signs for the involved quantities.

References

1. S. M. Sze, *Physics of Semiconductor Devices*, Wiley, New York, 1969.
2. R. S. Muller and T. I. Kamins, *Device Electronics for Integrated Circuits*, Wiley, New York, 1977.
3. D. L. Pulfrey and N. G. Tarr, *Introduction to Microelectronics Devices*, Prentice-Hall, Englewood Cliffs, N.J., 1989.
4. D. A. Hodges and H. G. Jackson, *Analysis and Design of Digital Integrated Circuits*, McGraw-Hill, New York, 1983.

Index

Abrupt junction, 421
Acceptors, 413
Accumulation, 170, 229, 452
Avalanche breakdown, 9

Base:
  charge in, 84
  current-induced, 111
  pushout factor of, 115
Base resistance, 58, 89
Base transit time, 64
Base-width modulation, 58
Baum model [see Metal-oxide-
    semiconductor transistor (MOST)]
Bipolar junction transistor (BJT), 45–130
  action of, 45
  area dependence of model parameters,
    108, 124–125, 128, 129
  base resistance, 58, 89
  base-width modulation, 58
  collector resistance, 57
  current-crowding effects in, 89
  current gain in, 49, 50, 75
  current-induced base of, 111
  depletion approximation, 82
  diffusion capacitance, 62, 64
  distortion, 321
  distributed base-collector capacitance,
    96
  Early effect, 58, 61, 86
  Ebers-Moll models:
    large-signal, 62–70, 118, 121, 128
    small-signal, 70–74, 118, 121, 128
    static, 48–62, 118, 119–121, 128
  emitter resistance, 57

  excess phase, 100
  Gummel-Poon models:
    large-signal, 96–101, 118, 121, 128
    small-signal, 101–105, 118, 121, 128
    static, 74–95, 118, 119–121, 128
  junction capacitance, 65
  measurements of, 251–266
  noise in, 304, 309
  power model, 109–118
  pushout factor of, 115
  quasi-saturation model, 113–118,
    125–127
  quasi-saturation region, 111
  regions of operation, 47, 109
  saturation current, 49, 79
  substrate capacitance, 67
  symbols and conventions for, 46
  temperature dependence of model
    parameters, 105–107, 121–124, 127,
    129
  transit time, 64, 97
Body effect, 172, 462
Boltzmann relations, 415, 426
Breakdown, 9, 15, 33, 41
  avalanche, 9
  voltage, 10
  Zener, 11
BSIM1, BSIM2 models [see
    Metal-oxide-semiconductor
    transistor (MOST)]

Capacitance:
  diffusion, 20, 22, 34, 62, 64, 441
  distributed base-collector, 96
  gate-drain, 147, 195, 210

Capacitance (*Cont.*):
  gate-source, 147, 195, 210
  gate-substrate, 195, 210
  junction, 20, 22, 34, 65, 200, 439
  junction periphery, 34
  metal, 35
  oxide, 168, 450
  poly, 35
  substrate, 67
Carriers:
  excess, 420, 435
  immobile, 411, 412
  intrinsic, 412
  lifetime of, 421
  majority, 413
  minority, 413
  mobile, 411, 412
Channel-length modulation, 139, 143, 175, 187, 209, 240
  narrow, 193, 207
  pinch-off, 137
  short, 188, 204
Collector resistance, 57
Contact potentials, 445–446
Continuity equation, 431
Convergence, 328, 354
Current crowding, 89
Current density:
  electron, 417, 418
  hole, 417, 418
Curtice model [*se* Metal-semiconductor field-effect transistor (MESFET)]

Dang model [*see* Metal-oxide-semiconductor transistor (MOST)]
Depletion, 170, 454
Depletion approximation, 82, 426, 459
Depletion-region width, 429
Diffusion coefficient, 418, 419
Diffusion current, 418
Diffusion length, 436
Diode, 1–43
  area dependence of model parameters, 31, 37–39, 42
  breakdown characteristics of, 9, 15, 33, 41
  diffusion capacitance, 20, 22, 34
  distortion, 320
  forward characteristic, 11, 13, 32, 40
  Fowler-Nordheim model, 39–40
  generation-recombination in depletion region, 4, 13
  high-level injection, 6, 14
  ideal behavior of, 1
  junction capacitance, 20, 22, 34
  junction periphery capacitance, 34
  large-signal model, 19–23, 31, 34–35, 41
  long-base, 3
  low-level injection, 6
  metal capacitance, 35
  noise in, 303, 307, 309
  poly capacitance, 35
  resistance of, 9, 14
  reverse characteristic, 12, 16, 33, 41
  reverse saturation current, 2, 3
  Schottky, 26–28
  short-base, 3
  small-signal model, 23–26, 31, 35, 41
  static model, 1–19, 31, 32–34, 40–41
  temperature dependence of model parameters, 28–30, 35–37, 42
  transit time, 20, 22, 28
  voltage drop in neutral regions of, 8
  Zener, 11, 17–19
Distortion, 309–324
  BJT, 321
  diode, 320
  harmonic coefficients of:
    at high frequencies, 311
    at low frequencies, 309
    Taylor series, 310, 313, 315, 318
    Volterra series, 313, 318, 319
Donors, 413
Drain resistance, 140, 212
Drift current, 416
Drift velocity, 416

Early effect, 58, 61, 86
Ebers-Moll models [*see* Bipolar junction transistor (BJT)]
Einstein relations, 419
Electrons, 412
Emitter resistance, 57
Excess carriers, 420, 435
  lifetime of, 421
Excess phase, 100

Flat-band voltage, 168, 450
Fourier analysis, 323
Fowler-Nordheim model [*see* Diode]

Generation-recombination, 4, 13, 420
Gradual-channel approximation, 134

Index

Gummel-Poon models [see Bipolar junction transistor (BJT)]

High-level injection, 6, 14, 420
Holes, 411

Inversion, 171, 454, 458–461
  moderate, 186, 460
  strong, 457, 460
  weak, 183, 457, 460
Ion-sensitive field-effect transistor (ISFET), 390–399
  EIS structure model, 391–394
  Gouy-Chapman-Stern model, 393
  large-signal model, 395, 398
  site-binding model, 392
  slow responses, 396–397, 398
  static model, 394–395, 397
  temperature dependence of model parameters, 395–396, 398

Junction field-effect transistor (JFET), 131–159
  abrupt gate-channel junction, 134
  area dependence of model parameters, 151, 156, 158
  breakdown of, 141
  channel-length modulation, 139, 143
  depletion-region width, 135, 139
  drain resistance, 140
  gate resistance, 156
  gradual-channel approximation, 134
  large-signal model, 146–148, 152, 153–154, 157
  noise in, 305, 308, 309
  pinch-off voltage, 137
  small-signal model, 148–150, 152, 154, 157
  source resistance, 140
  square-law characteristic of, 137
  static model, 132–145, 152, 153, 156–157
  temperature dependence of model parameters, 150–151, 155–156, 158
  threshold voltage, 137
  turn-off voltage, 139

Lifetime of excess carriers, 421
Long-base diode, 435–437
Low-level injection, 420

Measurements:
  of BJT parameters, 251–265
  of MOST parameters, 267–298
Metal-oxide-semiconductor transistor (MOST), 161–249
  accumulation in, 170, 229
  Baum-Beneking model, 192
  body effect in, 172
  BSIM1 model, 216–234
  BSIM2 model, 234–240
  channel-length modulation, 175, 187, 209, 240
  Dang model, 202
  depletion, 170
  diode model, 242
  drain resistance, 212
  fabrication of, 162
  fixed charge effects in, 168
  flat-band voltage, 168
  gate capacitances, 195, 210
  gate resistance, 246
  impact ionization, 243
  inversion in, 171
  junction capacitance, 200
  large-signal model, 195–202, 210–211, 228–234, 238–239, 241, 247
  measurements of model parameters, 267–298
  Meyer model, 181, 195, 197
  mobility, 182, 191, 203, 207, 235
  moderate inversion region of, 186
  narrow-channel effect, 193, 207
  noise in, 306, 308, 309
  operating regions of, 166
  oxide capacitance, 168
  saturation voltage, 173
  Shichman-Hodges model, 171
  short-channel effect in, 188, 204
  small-signal model, 213–214, 234, 241, 247
  source resistance, 212
  space-charge region of, 168–170
  static model, 168–195, 202–210, 221–228, 235–238, 241, 246
  subthreshold current, 222, 229, 237
  surface potential of, 171
  Swanson-Meindl model, 183
  temperature dependence of model parameters, 215, 244, 247
  threshold voltage, 171, 175, 176, 195, 204, 221
  types of, 168

478 Index

Metal-oxide-semiconductor transistor
    (MOST) (*Cont.*):
  Ward-Dutton model, 198
  weak inversion region of, 183
  Yau model, 189
Metal-semiconductor field-effect
    transistor (MESFET), 375–390
  Curtice model, 378–381
  large-signal model, 380, 382, 387–388,
    389–390
  noise in, 309, 388, 390
  small-signal model, 380, 383, 388, 390
  static model, 379, 381, 385–387, 388–389
  Statz model, 381–384
Meyer model [*see* Metal-oxide-
    semiconductor transistor (MOST)]
Mobility, 182, 191, 203, 207, 235, 416
Moderate inversion, 186, 460
MOS junction, 445–469
  accumulation condition of, 452
  body effect in, 462
  capacitance-voltage characteristics,
    464–469
  charge balance of, 452
  contact potential effects in, 445, 447
  depletion condition of, 454
  equivalent interface charge of, 450
  fixed charge effects in, 449–450
  flat band condition of, 452
  flat-band voltage, 450
  inversion condition of, 454, 458–461
  inversion regions:
    moderate, 460–461
    strong, 460–461
    weak, 460–461
  oxide capacitance, 450
  parasitic charge effects in, 449
  potential balance of, 452
  space-charge region of, 451, 455–458
  surface potential, 451
  threshold voltage, 464
MS junction, 471–473
  diode, 472
  ohmic contact, 473

Narrow-channel effect, 193, 207
Noise, 299–324
  BJT model, 304, 309
  burst, 302
  diode model, 303, 307, 309
  flicker, 301

JFET model, 305, 308, 309
MESFET model, 309, 388, 390
MOST model, 306, 308, 309
shot, 300
spectral density of, 299
thermal, 300

Pinch-off, 137
PN junction, 411–444
  abrupt or step, 421
  bias of, 430, 432–433
  built-in potential of, 423–426
  continuity equation, 431
  current-voltage characteristics of,
    432–433
  depletion approximation, 426
  depletion-region width of, 426–429
  diffusion capacitance, 441
  in equilibrium, 422–429
  long-base diode analysis of, 435–437
  maximum field at, 430
  minority-carrier boundary values of,
    433–435
  in nonequilibrium, 430–439
  junction capacitance, 439
  saturation current, 437, 439
  short-base diode analysis of, 437–439
  transient behavior of, 442
Poisson's equation, 415
Potential:
  built-in, 423–426
  contact, 445–446
  surface, 451

Quasi-saturation model, 113–118, 125–127
Quasi-saturation region, 111

Resistance:
  base, 58, 89
  collector, 57
  in diode, 9, 14
  drain, 140, 212
  emitter, 57
  gate, 156, 246
  source, 140, 212

Saturation current, 2, 3, 29, 49, 79, 105,
    437, 439
Schottky diode, 26–28

Semiconductor:
  intrinsic, 413
  n-type, 413
  p-type, 413
Shichman-Hodges model [see Metal-oxide-semiconductor transistor (MOST)]
Short-base diode, 437–439
Short-channel effect, 188, 204
Source resistance, 140, 212
Statz model [see Metal-semiconductor field-effect transistor (MESFET)]
Swanson-Meindl model [see Metal-oxide-semiconductor transistor (MOST)]
SPICE, 325–373
  analysis, types of, 326
  capabilities of, 325
  circuit description of, 329
  convergence in, 328, 354
  flags in, 368
  input format for, 328
  linked-list specification in, 361
  structure of, 333
Step junction, 421

Strong inversion, 460
Surface potential, 451

Threshold voltage, 137, 171, 175, 176, 195, 204, 207, 221, 464
Thyristor, 400–409
  models, 405–408
  triggering, 404–405
  two-BJT analogy, 401–403
Transit time, 20, 22, 28, 64, 97
Turn-off voltage, 139

Ward-Dutton model [see Metal-oxide-semiconductor transistor (MOST)]
Weak inversion, 183, 457, 460

Yau model [see Metal-oxide-semiconductor transistor (MOST)]

Zener breakdown, 11
Zener diode, 17–19

About the Authors

Giuseppe Massobrio is a research associate in the department of electronic engineering (DIBE) at the University of Genova. He has done research in the areas of semiconductor device modeling as well as circuit simulation and optimization. He is the author of several papers on semiconductor device models for CAD applications. He is currently researching the development of semiconductor-based biosensor models for biomedical technologies.

Paolo Antognetti is a professor and chairman of the department of electronic engineering (DIBE) at the University of Genova. His extensive background in semiconductor modeling includes teaching and research positions at Stanford University and MIT. He has contributed to more than fifty papers in the field of electronics and VSLI design and simulation. He has edited several books, including *Power Integrated Circuits: Physics, Design, and Applications*, also published by McGraw-Hill.